Synthesis, Properties, and Applications of Bio-Nanocomposites

Kaizar Hossain
https://orcid.org/0000-0002-3903-6161
Asutosh College, India

Akil Ahmad
https://orcid.org/0000-0003-4912-287X
Prince Sattam bin Abdulaziz University, Saudi Arabia

Mohammad Jawaid
United Arab Emirates University, Al Ain, UAE

IGI Global
Scientific Publishing
Publishing Tomorrow's Research Today

Vice President of Editorial Melissa Wagner
Director of Acquisitions Mikaela Felty
Director of Book Development Jocelynn Hessler
Production Manager Mike Brehm
Cover Design Jose Rosado

Published in the United States of America by
IGI Global Scientific Publishing
701 East Chocolate Avenue
Hershey, PA, 17033, USA
Tel: 717-533-8845 | Fax: 717-533-7115
Website: https://www.igi-global.com E-mail: cust@igi-global.com

Library of Congress Cataloging-in-Publication Data

CIP Data Pending
ISBN13: 979-8-3373-2180-6
Isbn13Softcover: 979-8-3373-2181-3
EISBN13: 979-8-3373-2182-0

British Cataloguing in Publication Data
A Cataloguing in Publication record for this book is available from the British Library.

The editors are honored to dedicate this book to Mr. Shaik Mohameed Jani and Gousiya Begum (Father-in-law and Mother-in-law of Dr. Kaizar Hossain)

Table of Contents

Preface xv

Acknowledgement xviii

Chapter 1
Physical and Chemical Properties of Bio-Nanocomposite Materials 1
J. Lavanya, SRM Institute of Science and Technology, India
Aishwarya Ramesh, SRM Institute of Science and Technology, India
Asmita Nandi, SRM Institute of Science and Technology, India
D. Srirangasayee, SRM Institute of Science and Technology, India

Chapter 2
Synthesis of Bio-Nanocomposite Materials: Thermal and Mechanical Study 21
Basavaiah Chandu, Acharya Nagarjuna University, India
Venkata Sai Sriram Mosali, Colorado State University, USA
K. K. C. Satish Babu, Acharya Nagarjuna University, India
Syed Akhil, Nanyang Technological University, Singapore

Chapter 3
Practical Utility and Basic Properties of Thermoplastic-Based Bio Nanocomposites 59
Y avasn Maruthi, Krishna University, India
T. Srinivas, GITAM University, India

Chapter 4
Green Synthesis of Iron Nanoparticles: Methods, Characterization, and Applications 81
Aditya Shrivastav, Ajeenkya D.Y. Patil University, India
Sunil Sankathala, Ajeenkya D.Y. Patil University, India
Susanta Das, Ajeenkya D.Y. Patil University, India

Chapter 5
Role of Doped Nanomaterials in the Synthesis and Characterization of Bio-Organic-Inorganic Composites 113
Sumana Dutta, Asutosh College, India
Sayanti Kar, Asutosh College, India
Gaurav Pant, Graphic Era University, India
Souvik Paul, Asutosh College, India
Riya Mukherjee, Graduate Institute of Biomedical Science, Taiwan
Arjun Sarkar, Leibniz Institute for Natural Product Research and Infection Biology, Germany
Ayantika Banerjee, Federal Institute for Material Research and Testing, Germany

Chapter 6
Sustainable Bio-Based Nanocomposites: Engineering Regenerable Materials for Tomorrow 135
Chaitanya Surya Chandra Nadimpalli, Raghu Engineering College, India
Murali Monangi, Raghu Engineering College, India
Kalyani Teku, Raghu Engineering College, India
V. Subrahmanyam Surya, Raghu Engineering College, India

Chapter 7
Bio-Based Nanocomposites in Sustainable Agriculture: Bridging Materials Science and Food Security.. 179
Chaitanya Surya Chandra Nadimpalli, Raghu Engineering College, India
Murali Monangi, Raghu Engineering College, India
Kalyani Teku, Raghu Engineering College, India
V. Subrahmanyam Surya, Raghu Engineering College, India

Chapter 8
Bio-Nanocomposites as Functional Materials for the Future: Integrating Green Chemistry and Nanotechnology for Global Sustainability Challenges... 203
R. Venkatesh, PSNA College of Engineering and Technology, India
D. Gnanasangeetha, PSNA College of Engineering and Technology, India
Natarajan Sampathkumar, SSM College of Arts and Science, India

Chapter 9
Innovations and Applications of Bio-Nanocomposites in Sustainable Food Packaging: Biodegradable Materials With Nanotechnology for Enhanced Safety, Functionality, and Environment 239
Hari Shankar Biswas, Surendranath College, India
Antara Roy, University of Calcutta, India
Amit Kr. Kundu, Sripat Singh College, India
Dilip K. Maiti, University of Calcutta, India

Chapter 10
AI-Enabled Bio-Nanocomposite Materials for Soil Health and Environmental Sustainability 275
Mrutyunjay Padhiary, Assam University, Silchar, India

Chapter 11
Synthesis, Properties, and Application of Bio-Nanocomposites: Application of Nanobiochar and Biochar Nanocomposites in Environment 327
Jeremaih David Bala, Federal University of Technology, Minna, Nigeria
Oloruntoba Samuel Job, Federal University of Technology, Minna, Nigeria
Innocent Ojeba Musa, Skyline University, Nigeria
Al Hassan Abdullahi Abdul-Rahman, Federal University, Lokoja, Nigeria
Wama Binga Emmanuel, Taraba State University, Nigeria
Adamu Mustapha, Federal University of Technology, Minna, Nigeria
Nathaniel Adunbarin Abiodun, National Institute for Freshwater Fisheries Research, Nigeria
Nathaniel Nwachukwu Friday, Federal University of Technology, Minna, Nigeria

Compilation of References 353

About the Contributors 425

Index 437

Detailed Table of Contents

Preface xv

Acknowledgement xviii

Chapter 1
Physical and Chemical Properties of Bio-Nanocomposite Materials 1
J. Lavanya, SRM Institute of Science and Technology, India
Aishwarya Ramesh, SRM Institute of Science and Technology, India
Asmita Nandi, SRM Institute of Science and Technology, India
D. Srirangasayee, SRM Institute of Science and Technology, India

The escalating environmental concerns arising from excessive plastic use and fossil fuel dependency have prompted the scientific community to seek sustainable alternatives. Bionanocomposites, fusion of biopolymers and nanomaterials have emerged as promising candidates to address both performance and sustainability requirements. This chapter delves into the fundamental physical and chemical attributes of bionanocomposites and their unique structures influence functionality. They are used in the industries like food, environmental, medical, tissue engineering scaffolds and in electronics, where they contribute to flexible and lightweight devices. Recent developments have focused on overcoming these barriers through innovative approaches such as using waste-derived biomaterials, developing green synthesis pathways, and enhancing processing techniques like electrospinning, extrusion, and solvent casting. With the right interdisciplinary efforts, bionanocomposites could be key players in the development of next-generation eco-friendly technologies.

Chapter 2
Synthesis of Bio-Nanocomposite Materials: Thermal and Mechanical Study 21
Basavaiah Chandu, Acharya Nagarjuna University, India
Venkata Sai Sriram Mosali, Colorado State University, USA
K. K. C. Satish Babu, Acharya Nagarjuna University, India
Syed Akhil, Nanyang Technological University, Singapore

Synthetic polymers and their composites are excellently suited to numerous applications, but their overexploitation has caused environmental pollution. Therefore, the need to replace them with sustainable and biodegradable materials is a critical issue that needs to be addressed urgently in order to reduce pollution. Natural polymers such as cellulose, starch, proteins, chitosan, and polylactic acid are promising alternatives to synthetic polymers since they are abundant, cost-effective, biodegradable, and exhibit good mechanical and thermal properties. However, the fabrication of these natural polymers into long-lasting goods is challenging due to their poor mechanical and thermal stability and excellent water absorption properties. Thus, the fabrication of composites of these bio-polymers represents an ever-intriguing research area for material scientists to achieve thermally stable bio-nanocomposites with low water absorption and enhanced mechanical properties. This chapter focuses on the synthesis methodologies of bio-nanocomposites and their mechanical and thermal properties.

Chapter 3
Practical Utility and Basic Properties of Thermoplastic-Based Bio Nanocomposites 59
Y avasn Maruthi, Krishna University, India
T. Srinivas, GITAM University, India

This chapter explores the innovative use of thermoplastic-based bio-nanocomposites as sustainable alternatives to traditional synthetic materials. It explains what bio-nanocomposites and thermoplastics are, and how combining biopolymers with inorganic fillers at a tiny scale creates a strong structure. Fundamental properties such as biocompatibility, biodegradability, mechanical strength, and swelling capacity are examined to demonstrate their industrial relevance. The chapter highlights the practical applications of key matrices like chitosan, cellulose, and starch in areas including food packaging, wastewater treatment, electronics, and construction. Case studies and recent research illustrate how these composites can outperform conventional materials. It also addresses current challenges such as scalability, mechanical limitations, and environmental concerns, positioning these composites as key players in sustainable innovation. Concluding remarks point out the need for continued research to develop enhanced formulations that meet the growing demand for eco-friendly solutions.

Chapter 4
Green Synthesis of Iron Nanoparticles: Methods, Characterization, and Applications 81
Aditya Shrivastav, Ajeenkya D.Y. Patil University, India
Sunil Sankathala, Ajeenkya D.Y. Patil University, India
Susanta Das, Ajeenkya D.Y. Patil University, India

The rapid advancements in nanotechnology have significantly transformed various industries, with iron-based nanoparticles emerging as a key material due to their unique properties and diverse applications. Green synthesis methods provide an eco-friendly and sustainable alternative to conventional chemical/physical synthesis, utilizing biological agents such as plant extracts, algae, & micro-organisms. These methods rely on phytochemicals as reducing and stabilizing agents, ensuring minimal environmental impact while enabling precise tailoring of nanoparticle properties. Biogenic FeNPs demonstrate immense potential across multiple domains: targeted drug delivery, wastewater treatment and pollutant removal, in agriculture as green alternatives to chemical pesticides, and in catalysis for energy-efficient chemical reactions. This chapter highlights the potential of green-synthesized iron nanoparticles and emphasizes the need for continued research to advance sustainable practices, paving the way for groundbreaking applications that benefit both industry and the environment.

Chapter 5
Role of Doped Nanomaterials in the Synthesis and Characterization of Bio-Organic-Inorganic Composites 113
Sumana Dutta, Asutosh College, India
Sayanti Kar, Asutosh College, India
Gaurav Pant, Graphic Era University, India
Souvik Paul, Asutosh College, India
Riya Mukherjee, Graduate Institute of Biomedical Science, Taiwan
Arjun Sarkar, Leibniz Institute for Natural Product Research and Infection Biology, Germany
Ayantika Banerjee, Federal Institute for Material Research and Testing, Germany

The bio-inspired synthesis of organic and inorganic nanomaterials has gained significant attention for creating hierarchical structures. Blending these materials enhances nanocomposite membranes for filtration and sensing, with elements like Co^{2+} and Mn^{2+} improving their properties. Many organisms, such as mollusks and sponges, naturally produce organic-inorganic composites through biomineralization. Biomolecules—proteins, peptides, DNA, RNA, and polysaccharides—guide nucleation, stabilization, and assembly of inorganic frameworks. Proteins aid in forming organized nanostructures, while peptides offer customizable designs for tissue engineering and nanotechnology. Doping is widely used to tailor nanoparticles' electrical, optical, and morphological properties. Doped nanomaterials, like Ag-N-TiO_2 and ZnO composites, exhibit antimicrobial efficacy, while co-doping strategies further enhance microbial inhibition. This review explores biomolecules' synergistic role in nanotechnology for sustainable, multifunctional materials in industry and biomedicine.

Chapter 6
Sustainable Bio-Based Nanocomposites: Engineering Regenerable Materials for Tomorrow 135
Chaitanya Surya Chandra Nadimpalli, Raghu Engineering College, India
Murali Monangi, Raghu Engineering College, India
Kalyani Teku, Raghu Engineering College, India
V. Subrahmanyam Surya, Raghu Engineering College, India

Bio-based nanocomposites represent a transformative frontier in materials science, offering sustainable alternatives to conventional petroleum-derived composites. This chapter examines sustainable bio-based nanocomposites with emphasis on their regenerability across multiple life cycles. Beginning with fundamental principles governing synthesis and structure-property relationships, the work explores interactions between bio-based matrices and nanoscale reinforcements. Processing techniques balancing performance with sustainability are analyzed, establishing foundations for understanding degradation mechanisms and regeneration strategies. Both chemical and physical aging processes are investigated, connecting these insights to effective regeneration approaches. The chapter addresses lifecycle assessment through environmental impact evaluations spanning multiple regeneration cycles. Advanced characterization techniques and economic considerations are discussed alongside emerging trends, serving as a foundational resource and catalyst for future innovations in sustainable materials.

Chapter 7
Bio-Based Nanocomposites in Sustainable Agriculture: Bridging Materials Science and Food Security 179
Chaitanya Surya Chandra Nadimpalli, Raghu Engineering College, India
Murali Monangi, Raghu Engineering College, India
Kalyani Teku, Raghu Engineering College, India
V. Subrahmanyam Surya, Raghu Engineering College, India

This chapter explores the integration of bio-based nanocomposites in agricultural systems to address challenges of climate change, resource depletion, and food security. Focusing on the synthesis and characterization of nanocomposites from renewable resources, the work examines green approaches and regenerability aspects crucial for sustainable implementation. Applications including soil conditioning, controlled release fertilizers, crop protection, and water management are analyzed through practical case studies demonstrating enhanced agricultural productivity with minimal environmental impact. A comprehensive sustainability assessment evaluates these materials throughout their lifecycle, considering biodegradability, compostability, and economic feasibility. The chapter identifies emerging trends such as multi-functional nanocomposites and integration with precision agriculture technologies, while addressing regulatory considerations essential for responsible development and adoption in sustainable food production systems.

Chapter 8
Bio-Nanocomposites as Functional Materials for the Future: Integrating Green Chemistry and Nanotechnology for Global Sustainability Challenges... 203
R. Venkatesh, PSNA College of Engineering and Technology, India
D. Gnanasangeetha, PSNA College of Engineering and Technology, India
Natarajan Sampathkumar, SSM College of Arts and Science, India

Bio-nanocomposites are leading materials that leverage the strengths of nanotechnology with sustainable, bio-based polymers and present a greener option for traditional composites. These are composites that comprise nanoscale fillers in the form of nanoparticles, nanofibers, and nanotubes, and bio-based matrices of starch, cellulose, and proteins. These are stronger in terms of mechanical, thermal, and barrier properties, and their green profile remains unscathed. Bio-nanocomposites present opportunities to answer global sustainability questions such as decreasing dependence on petroleum-based plastics, energy efficiency, and promoting the circular economy by combining green chemistry and nanotechnology. This chapter addresses synthesis routes, properties, characterization techniques, performance, and different applications in packaging, automotive, electronics, and agriculture. It also addresses future research directions to design the properties of bio-nanocomposites for large-scale commercialization.

Chapter 9
Innovations and Applications of Bio-Nanocomposites in Sustainable Food Packaging: Biodegradable Materials With Nanotechnology for Enhanced Safety, Functionality, and Environment 239
Hari Shankar Biswas, Surendranath College, India
Antara Roy, University of Calcutta, India
Amit Kr. Kundu, Sripat Singh College, India
Dilip K. Maiti, University of Calcutta, India

The rising demand for sustainable food packaging has spurred advancements in bio-nanocomposites, which blend biodegradable biopolymers like starch, cellulose, and PLA with nanomaterials such as nanoclays, cellulose nanocrystals, carbon-based materials, and metal oxide nanoparticles. This combination enhances mechanical strength, thermal stability, and barrier properties, making them ideal for modern packaging. This chapter reviews synthesis methods—including solution casting, melt blending, and in situ polymerization—and explores their impact on material properties. Key functionalities such as antimicrobial activity, biodegradability, and shelf-life extension are discussed, along with emerging applications in active and intelligent packaging. Challenges like nanoparticle dispersion, polymer compatibility, and scalability are addressed. Future prospects include multifunctional systems, bio-derived nanomaterials, and advanced fabrication techniques. This chapter offers a comprehensive overview of bio-nanocomposites in sustainable food packaging.

Chapter 10
AI-Enabled Bio-Nanocomposite Materials for Soil Health and Environmental Sustainability 275
Mrutyunjay Padhiary, Assam University, Silchar, India

Bio-nanostructured composite materials, composed of biological and nanostructured elements, have added a new dimension to the latest agricultural developments. This chapter explores how these materials can be optimized using artificial intelligence to enhance soil fertility, structure, and ecosystem resilience. Precision design, performance prediction, and real-time monitoring of bio-nano composites, ensuring their targeted functionality in soil remediation, nutrient delivery, and pollution mitigation, are being enabled by AI algorithms, including machine learning and deep learning. AI-integrated systems might support frameworks for sustainable land management, forecast environmental effects, and replicate soil-microbe-nano synthesis. Smart nano-fertilizers, AI-assisted clean-up methods, and sensors in materials for checking soil quality demonstrate a shift toward smarter, more sustainable farming and caring for the environment. It also explores the ethical and hazardous aspects of AI-enabled bio-nano technologies and their potential impact on future-proof agri-environmental systems.

Chapter 11
Synthesis, Properties, and Application of Bio-Nanocomposites: Application of Nanobiochar and Biochar Nanocomposites in Environment........................ 327
Jeremaih David Bala, Federal University of Technology, Minna, Nigeria
Oloruntoba Samuel Job, Federal University of Technology, Minna, Nigeria
Innocent Ojeba Musa, Skyline University, Nigeria
Al Hassan Abdullahi Abdul-Rahman, Federal University, Lokoja, Nigeria
Wama Binga Emmanuel, Taraba State University, Nigeria
Adamu Mustapha, Federal University of Technology, Minna, Nigeria
Nathaniel Adunbarin Abiodun, National Institute for Freshwater Fisheries Research, Nigeria
Nathaniel Nwachukwu Friday, Federal University of Technology, Minna, Nigeria

Modern society relies on nanobiochar and biochar nanocomposites as innovative materials to resolve major environmental issues alongside agricultural difficulties. Engineered biochar-based materials boost adsorption performance which makes them efficient tools for wastewater bioremediation mainly focused on heavy metal and organic substance elimination. Advanced surface properties coupled with improved porosity and nanoparticle functionalization allows them to efficiently adsorb pollutants which include arsenic and chromium along with lead, cadmium and copper. Research findings show nanobiochar exhibits better capabilities for extracting pollutants from water sources which renders them critical components in environmental sustainability operations. Nanobiochar proves crucial for agriculture by performing various functions outside wastewater treatment. Nanobiochar supports plant development through better soil nutritive value and enhanced nutrient absorption while maintaining soil acidity at stable levels. Nanobiochar functions as a soil additive that improves organic matter composition while preserving liquid retention while preventing harmful metal accumulation. Nanobiochar has demonstrated chemical residue removal ability by absorbing dangerous pesticide residues that target environmental toxicity. The antimicrobial capabilities of nanobiochar enable disease suppression in soil whereas chemical pesticides have less environmental benefits. kullanım of nanobiochar promotes beneficial microbial growth that upgrades fertilizer cycling operations while strengthening soil well-being. Through fertilizer application nanobiochar supports effective nutrient dispersal that prevents waste and decreases pollution impacts on the environment. Both nanobiochar and biochar nanocomposites serve as decisive elements for implementing sustainable environmental measures and agricultural practices. These materials show remarkable value by improving all elements of water purification and soil quality and crop productivity and microbial balance to resolve international concerns about environmental pollution and food shortages. Research development needs focused material optimization to establish these materials as crucial elements for ecological protection and resource management systems.

Compilation of References 353

About the Contributors 425

Index 437

Preface

Recent developments in the materials science field have highlighted the need for further study of composite materials especially with regard to bio and nano-composites, as their abilities to meet modern requirements with regard to facilitating technological advancement is far greater than those of synthetic materials. There are several advantages of nano and bio-composites as compared to conventional materials, including properties such as fatigue resistance, impact resistance, stiffness, corrosion resistance, biodegradability, thermal conductivity, low relative density, environmentally friendly waste stages, and high specific strength. In recent years, discovery of several new high-performing composite materials, and improvements in the fields of aviation, automobile, adhesion, building, and electronic engineering also support the adoption of more general utilization of nano and bio-composites in a much wider range of applications than those for which they were originally designed.

The combination of sustainable materials and nanotechnologies gives more attractive and efficient possibilities for further industrially applicable nanomaterials. The excellent physicochemical properties of bio-nano based materials could help to make new dimensional nanomaterials, and thus it would act as a strengthening and stabilizing agent for nanomaterials. Currently, researchers developing novel nanoparticles, nanocomposites, and nanomaterials by using green materials to increase the thermal and electrochemical stability of the nanoparticles. These new materials have been used as electro-sensors, catalysts, electronics, and photonics.

This book highlights and reviews the information on bio- and nano-composite materials to determine current trends of these materials and various application in several fields. This book is written by eminent scholars and leading experts from around the world. The book covers the synthesis and characterization of broad range of bio-composites and Bio nanocomposites along with their application in multiple areas for nanomaterials and environmental sustainability. The chapter of different topic gives its complete information about its development and future possibilities. The book covers importance of sustainable and biodegradable materials in the field of nanomaterial and circular economy that easily open the new ways of future research.

CHAPTER OVERVIEW

Chapter 1 examines the physical and chemical properties of bio-nanocomposite materials. Lavanya J, Aishwarya Ramesh, Asmita Nandi, and Srirangasayee D discuss how the unique structures of bionanocomposites influence their functionality, highlighting applications in food, medical, environmental, tissue engineering, and electronics. The chapter also addresses recent advances in green synthesis, waste-derived biomaterials, and advanced processing techniques such as electrospinning, extrusion, and solvent casting.

Chapter 2 focuses on the synthesis of bio-nanocomposites and their thermal and mechanical properties. Basavaiah Chandu, Venkata Mosali, Satish K K C, and Syed Akhil explore natural polymers like cellulose, starch, chitosan, and polylactic acid, examining strategies to overcome challenges related to poor mechanical stability, water absorption, and thermal degradation. The chapter emphasizes methods to achieve thermally stable and mechanically robust bio-nanocomposites.

Chapter 3 delves into thermoplastic-based bio-nanocomposites, authored by Yavasn Maruthi and T Srinivas. It highlights how combining biopolymers with inorganic fillers at the nanoscale produces materials with enhanced strength, biocompatibility, and biodegradability. The chapter examines applications in food packaging, wastewater treatment, electronics, and construction, while addressing practical challenges such as scalability and environmental impact.

Chapter 4, by Aditya Shrivastav, Sunil Sankathala, and Susanta Das, discusses green synthesis of iron nanoparticles. The chapter emphasizes eco-friendly methods using plant extracts, algae, and microorganisms, highlighting applications in targeted drug delivery, wastewater remediation, agriculture, and catalysis. It underscores the importance of sustainable nanotechnology in reducing environmental impact while tailoring nanoparticle properties for industrial use.

Chapter 5 explores the role of doped nanomaterials in bio-organic-inorganic composites. Sumana Dutta, Sayanti Kar, Gaurav Pant, Souvik Paul, Riya Mukherjee, Arjun Sarkar, and Ayantika Banerjee highlight how bio-inspired synthesis and biomolecules such as proteins, peptides, and polysaccharides guide the formation of hierarchical structures. The chapter discusses doping strategies to enhance electrical, optical, and antimicrobial properties, with applications in filtration, sensing, and tissue engineering.

Chapter 6 examines sustainable bio-based nanocomposites and their regenerability. Chaitanya Nadimpalli, Murali Monangi, Kalyani Teku, and Subrahmanyam V provide insights into synthesis, structure-property relationships, and degradation mechanisms. The chapter emphasizes lifecycle assessment, advanced characterization techniques, and economic considerations, providing a foundation for the design of environmentally friendly, regenerable materials.

Chapter 7 focuses on bio-based nanocomposites in sustainable agriculture. The same authors, Nadimpalli et al., highlight applications in soil conditioning, controlled release fertilizers, crop protection, and water management. Through case studies, the chapter demonstrates enhanced agricultural productivity with minimal environmental impact and discusses emerging trends such as multifunctional composites and integration with precision agriculture.

Chapter 8 addresses the integration of green chemistry and nanotechnology in bio-nanocomposites. Venkatesh R, Gnanasangeetha D, and Natarajan Sampathkumar discuss how bio-based polymers reinforced with nanoscale fillers achieve superior mechanical, thermal, and barrier properties. The chapter reviews applications in packaging, automotive, electronics, and agriculture, emphasizing sustainability, circular economy principles, and future commercialization.

Chapter 9 explores bio-nanocomposites for sustainable food packaging. Hari Shankar Biswas, Antara Roy, Amit Kundu, and Dilip Maiti examine how biopolymers like starch, cellulose, and PLA combined with nanomaterials improve mechanical strength, thermal stability, barrier properties, and antimicrobial activity. The chapter also addresses challenges in scalability, nanoparticle dispersion, and polymer compatibility, while highlighting emerging multifunctional packaging solutions.

Chapter 10 focuses on AI-enabled bio-nano composites for soil health and environmental sustainability. Mrutyunjay Padhiary investigates how artificial intelligence can optimize the design, performance, and monitoring of bio-nanostructured materials for soil remediation, nutrient delivery, and pollution mitigation. The chapter also explores ethical considerations and the potential for smart, sustainable agricultural systems.

Chapter 11 examines nanobiochar and biochar nanocomposites in environmental applications. Jeremaih Bala, Oloruntoba Job, Innocent Musa, Alhassan Abdul-Rahman, Wama Emmanuel, Adamu Mustapha, Nathaniel Abiodun, and Nathaniel Friday highlight their role in wastewater remediation, soil enhancement, crop productivity, and microbial balance. The chapter underscores the importance of these materials for sustainable agriculture and environmental protection, while calling for further research to optimize their use.

We are greatly thankful to all qualified researcher, scholar and leading experts to contribute their valuable work. The chapters provided cutting-edge up-to-date research findings on the Bio- Nano technology and composting materials. We collected all the information given by eminent authors on Bio- Nano technology and composting materials and related research from USA, Germany, Taiwan, Singapore and India, Malaysia, Nigeria, and finally compiled this project in a fruitful way.

Kaizar Hossain, Akil Ahmad and Mohammad Jawaid

Acknowledgement

Editors wish to appreciate the support of Asutosh College, University of Calcutta, Kolkata, W.B., India, College of Sciences and Humanities, Prince Sattam bin Abdulaziz University, Saudi Arabia and College of Engineering, United Arab Emirates University, Al Ain, UAE.

Chapter 1
Physical and Chemical Properties of Bio-Nanocomposite Materials

J. Lavanya
https://orcid.org/0000-0001-5623-5827
SRM Institute of Science and Technology, India

Aishwarya Ramesh
https://orcid.org/0009-0005-4829-5527
SRM Institute of Science and Technology, India

Asmita Nandi
https://orcid.org/0009-0004-0996-3783
SRM Institute of Science and Technology, India

D. Srirangasayee
SRM Institute of Science and Technology, India

ABSTRACT

The escalating environmental concerns arising from excessive plastic use and fossil fuel dependency have prompted the scientific community to seek sustainable alternatives. Bionanocomposites, fusion of biopolymers and nanomaterials have emerged as promising candidates to address both performance and sustainability requirements. This chapter delves into the fundamental physical and chemical attributes of bionanocomposites and their unique structures influence functionality. They are used in the industries like food, environmental, medical, tissue engineering scaffolds and in electronics, where they contribute to flexible and lightweight devices. Recent developments have focused on overcoming these barriers through innovative approaches such as using waste-derived biomaterials, developing green synthesis

DOI: 10.4018/979-8-3373-2180-6.ch001

pathways, and enhancing processing techniques like electrospinning, extrusion, and solvent casting. With the right interdisciplinary efforts, bionanocomposites could be key players in the development of next-generation eco-friendly technologies.

1. INTRODUCTION

Plastic materials are commonly employed to create a wide range of products and items for everyday household and industrial use. They satisfy all commercial and industrial market criteria, including cheap cost, durability, good performance, convenience, high mechanical variability, and other crucial properties (Brydson, 1999; Rosato, 2003). Packing accounts for a considerable share of plastic use (about 40%), which has escalated dramatically in the last decade of the twentieth century (Fomin & Guseev, 2001; Pilla, 2011). As plastic is extremely resistant to decomposition, humanity is currently faced with countless non-biodegradable wastes, with the amount only increasing every year as the demand and need for production is growing constantly. Most chemicals and polymers, primarily packaging, are all currently derived from the non renewable natural source, petroleum (Brydson, 1999). The fact that fossil resources are exhaustible is a massive concern. Materials, chemicals, as well as fuel, for example, are consumed much more quickly than the existing fossil fuels. Another challenge is carbon-containing materials and the emissions of carbon dioxide that come with them. The soaring depletion of fossil fuels has resulted in the creation of a radically distinct, artificial cycle. Its effect on global processes and the environment is becoming increasingly conspicuous, and it is now acknowledged as a worldwide environmental issue (Fomin & Guseev, 2001; Sudesh & Iwata, 2008; Golomb & Fay, 2004). A practical and feasible alternative is required to meet the global demands as well as check for environmental concern and protection. This led to research on newer substances, nanocomposites which has further become a better source as bio nanocomposite.

Nanocomposites are composites in which at least one phase has dimensions in the nanometer range (1 nm = 109 m) (Defonseka, 2014). It is divided as three different classes based on matrix materials: (i) Polymer Matrix Nanocomposites (PMNC), (ii) Metal Matrix Nanocomposites (MMNC), and (iii) Ceramic Matrix Nanocomposites (CMNC). In comparison to the micro or macro composites, Nacomposites show improved mechanical properties, synergistic flame retardant additives, excellent gas barrier properties, excellent thermal expansion, and improved dimensional stability. Selecting the proper matrix, synthetic systems, nano reinforcing materials, and surface modification of the reinforcing materials or polymers, as desired, could be used to generate polymer nanocomposites for a multitude of purposes. Polymer nanocomposites can be employed in industrial packaging and offer a number of advantages over conventional plastic-based packaging as they are transparent, lightweight, exhibit excellent workability, most importantly inexpensive, and provide appreciable barrier properties against water,gas and steam. Widely used in the automotive industry as an alternative to some heavy metal materials due to its design flexibility, and processability, customizability. Nanocomposite applications offer modern technologies and business prospects in many areas of the aerospace, electronics, and biotechnology industries. There are some drawbacks associated with the use of nanoparticles are increased viscosity Toughness and impact resistance, optical problems, increased sedimentation rates, difficulty in dispersion,and black color when using a variety of carbonic nanoparticles (Camargo et al., 2009). An additional challenge is that production of nanocomposites is limited by both prepolymerization and post polymerization methods (Okpala, 2014). Nanoparticles enter the water system either primarily from wastewater treatment and industrial effluent or secondarily through surface runoff from the soil. Many potential nanoparticle changes take place in environments with chemical and physical processes. Nanocomposites toxicity in algae demands adsorption on the cell surface and hinder membrane transport. Superior creatures actively absorb nanoparticles, and both marine and terrestrial species may construct nanoparticles in the food web (Batley et al., 2013). Bio-nanocomposites are produced as a result of the shortcomings of nanocomposites.

Bio-hybrid materials, bio-nanocomposites, are composed of natural species and inorganic solids of various structures and forms bonded on the nanometer dimension. Nanocomposite differs from bioplastics in manufacturing process, properties, functionality, biocompatibility, biodegradability, and application (Sain & Oksman, 2006; Hitzky et al., 2013). Because the term has just been used commonly, bio-nanocomposites are variously referred to as biocomposites, nanocomposites, nanobiocomposites, green composites, biohybrid, or bio-plastics (Shchipunov, 2012). The term “biopolymer” implies that they are biodegradable. As a consequence, biopolymers are considered harmless for the environment since they are metabolized by naturally existing organisms into chemical by-products (CO_2 and H_2O). As biopolymers are ubiquitous, reliable, and sustainable, they are an excellent substitute to petroleum-based polymers (Xinming et al., 2005). Bio-nano-composites are also called”nano-bio-composites,” “green composites”, “bio-hybrid, “bio-composites,”, ” “bio-based plastics,” or “bio-plastics” for short. They exhibit various properties such as water solubility, , biocompatibility, biodegradability,thermal stability etc., that determines the manufacturing method, functionality, and application fields of the material.Bio-nanocomposites are compounds formed up of matter having at least one dimension in the range of 1-100 nm and one or more biological elements, such as biopolymers. Natural polymers including polysaccharides, aliphatic polyesters, polypeptides and proteins, and polynucleic acids constitute existing bio-nanocomposites. Hydroxyapatite, metal nanoparticles, and clay, on the other hand, are fillers (Eduardo et al., 2005).

Biopolymers are applied in a variety of biological applications, including medication delivery, tissue engineering, and bone replacement, as well as electronic devices, sensors, and energy regeneration. Frequently utilized matrix for applications such as packaging is starch, cellulose derivative polylactic acid (PLA) and polycaprolactone (PCL). Chitin, starch, chitosan, cellulose, lignin, pectin, and polylactic acid are among some of the polymers that could be used to replace non-biodegradable polymers.They are also known as green polymers because they are sustainable and come primarily from natural resources (Nafees et al., 2018). Bio-nanocomposites are characterized according to their origins, matrix, reinforcement size, and form (Rabia et al., 2016).

When compared to other conventional composites, biopolymer nanocomposites exhibit numerous essential features, such as enhanced mechanical strength and higher heat resistance (Sorrentino et al., 2007). Since the biopolymers are soluble in polar solvents, predominantly water, they are more flexible and cost-effective. It's important to keep in mind that polymers of biological sources have numerous positives because they're renewable in nature that are plentiful and inexpensive.

Inadequate capabilities, such as insufficient mechanical strength, low thermal degradation temperature, significant gas and water vapor permeability, poor barrier properties, and so on, impede the use of polymers of biological origin (Shchipunov, 2012). The incorporation of nano-sized fortified chemicals or fillers to bio-nanocomposites can, therefore, improve their performance.These include,Organically modified layered silicates, carbon nanotube, cellulose nanofibers and halloysite nanotubes improve their material properties (Averous & Boquillon, 2004).

2. PHYSICAL PROPERTIES:

Bio-nanocomposite Materials: Different nanoparticle shapes are objected to describe the diagnostics of bio-nanocomposite materials. Three pivotal groups of nanoparticulates may be recognised based on their geometry:

- Nanoparticles - They are spherical, shapeless and cubic nanoparticles measuring up to 100 nm are included. This category includes sol-gel colloidal silica, noble nanomaterials such as gold, silver, etc. (Liz-Marzán, 2004; Daniel & Didier, 2004; Fernandez & Monge, 2008; de Jong, 2009), metal oxides (Xiaobo & Samuel, 2007; Markus & Nicola, 2009), along with semiconductor nanoclusters. The hydroxyapatite ($Ca_{10}(PO_4)_6(OH)_2$) is a particular example. It's a mineral that's found in abundance in bones (up to 70 wt percent).

- Nanofibers: Nanoparticulates have only two dimensions in the nanometer scale. Carbon nanotubes, which are made up of one or many sheets of graphene, are probably the most conventional example. The mechanical properties of composite materials are significantly improved due to their exceptional stiffness, strength, and durability.Sepiolite [$Si_{12}O_{30}Mg_8(OH,F)_4(H_2O)_{48}H_2O$] (Eduardo et al., 2008; Bascones et al., 2002; Darder et al., 2006) is the most common nanofibrillar clay utilized in bio-nanocomposite formulations.
- Nanoplatelets: Nanoplates usually have a thickness of around 1nm and their diameter might range from tens of nanometers to a few micrometers resulting in a large aspect ratio. As a result, even at extremely low nanoplatelet concentration, substantial strengthening of composite materials occurs, which is appealing from an economic standpoint. Montmorillonite (Besun et al., 1997; Ilker et al., 2009; Myung et al., 2016), laponite (Quan et al., 2007; Shou et al., 2009), hectorite (Ray & Bousmina, 2005), and saponite are examples of layered phyllosilicates of the 2:1 type that have been used.

They were mostly coupled with chitosan, but also with starch, other polysaccharides. It's worth noting that the vast majority of nano-sized materials, including natural clay minerals which are negatively charged. Combining them with anionic biomacromolecules, which make up the majority of biomaterials, can be difficult at times. Synthetic layered double hydroxides may be a suitable inorganics for bio-nanocomposite production in these cases .

The structure of the hydrotalcite family, in particular, can be demonstrated as [M2+1–xM3+x$(OH)_2$]–(Am–)x/mnH_2O (Utracki et al., 2007; Leuteritz et al., 2007; Choy et al., 2008; Forano et al., 2006). Isomorphous replacement of M2+ cations with M3+ cations causes positive charge. The charge-compensating anions Am– can be exchanged for anionic biomolecules in the interlayer.

Bio-nanocomposites may be categorized on the idea of forms of matrix utilized, origin, form and dimensions of reinforcements. For example, on the idea of forming of a particle reinforcements, bionanocomposites may be categorized into elongated, particulate particles with layered structures.

(i) Particulate Bio-nanocomposites:

In particulate bio-nanocomposites, iso dimensional debris are mainly employed as reinforcements. The reinforcing impact is slight because of the stunted factor ratio and the primary motive of using such sort of a reinforcement is to beautify the hostility to flammability, lessen permeability and value of the composites.

(ii) Elongated Particle Bio-nanocomposites:

In Elongated Particle Bio-nanocomposites, the elongated particles are hired as reinforcement in elongated particle bio-nanocomposites. These bio-nanocomposites have advanced mechanical conduct because of the excessive component ratio of the reinforcement.

(iii) Layered Particle-Reinforced Bio-nanocomposites:

In layered particle-bolstered bio-nanocomposite, additionally referred to as layered polymer nanocomposite, is classed within 3 sub-classes on the premise of dispersion of debris withinside the matrix. Phase-separated/Flocculated nanocomposites are fashioned whilst there may be zero partition among the layers due to particle-particle interaction. As the character laminas aren't separated, this strain of composites is often named as microcomposites, accordingly, performing as microparticles dispersed withinside the polymeric matrix. When the polymer chains are intercalated among sheets of layered nanoparticles intercalated nanocomposites are obtained, whereas, exfoliated nanocomposites are generated whilst there may be partition of character layers (Rabia et al., 2016).

3. CHEMICAL PROPERTIES

Biopolymers: The primary elements in the preparation of bio-nanocomposite are biological polymers. Polysaccharides, proteins, DNA, and poly (hydroxyalkanoates). Polysaccharides and proteins are the most common (Shchipunov, 2012). Details Structural formulas of polysaccharides in Figure 6.1

- The most prevalent biopolymers on Earth are cellulose ((1, 4)—Diflucan) and chitin ((1, 4)—D-2-N-acetylamino-2-deoxy glucan). The former is made up of glucose residues, whereas the latter is made up of acetylglucosamine residues, in which the hydroxyl groups on the second carbon atom of glucose are replaced by an acetylated amino group.
- Amylose ((1, 4)—D-glucan) and amylopectin are made up of glucose residues linked together by (1, 4)— and (1, 6)—glycosidic connections, respectively.
- Amylose is a cellulose isomer with a linear chain.
- Amylopectin similarly contains (1,4)-linked glucose residues, but due to (1,6)-linkages per 20 to 25 glucose residues, it is a branching polysaccharide.
- -D-mannuronic (M) and -L-guluronic (G) acid residues make up alginate. They can construct G-blocks and M-blocks in a row or alternate MG-blocks and blocks in a random order.

- Dextran is a kind of (1, 6)-glucan. Many glucose side-chains are mostly connected to the framework by (1, 3)-linkages, but also via (1, 4)- and (1, 2)-linkages at random.
- Hyaluronic acid, also known as hyaluronan, is a polymer made up of (1, 4)—D-glucuronic acid and (1, 3)—D-N-acetylglucosamine disaccharide units.
- Pectin is a complicated-structured mixed polysaccharide. It has a skeleton made up of approximately 65 percent (1, 4)—D-galactopyranosyl urinate residues. The building blocks of apiose, fucose, arabinose, and xylose residues are organised differently in side-chains. Also there is a backbone chain made up of rhamnose residues connected by arabinan and galactan chains from the disaccharide (1, 4)—D-galactopyranosyl urinate-(1,2)—L-rhamnogalacturonan.
- Xanthan is a microbial polysaccharide with a backbone of (14)-linked glucose molecules, similar to cellulose. Every second glucose residue at positions 0–3 has a trisaccharide side-chain made up of one glucuronic acid and two mannoses that are covalently linked.

Figure 1. Structural formulas of polysaccharides (Sorrentino et al., 2007)

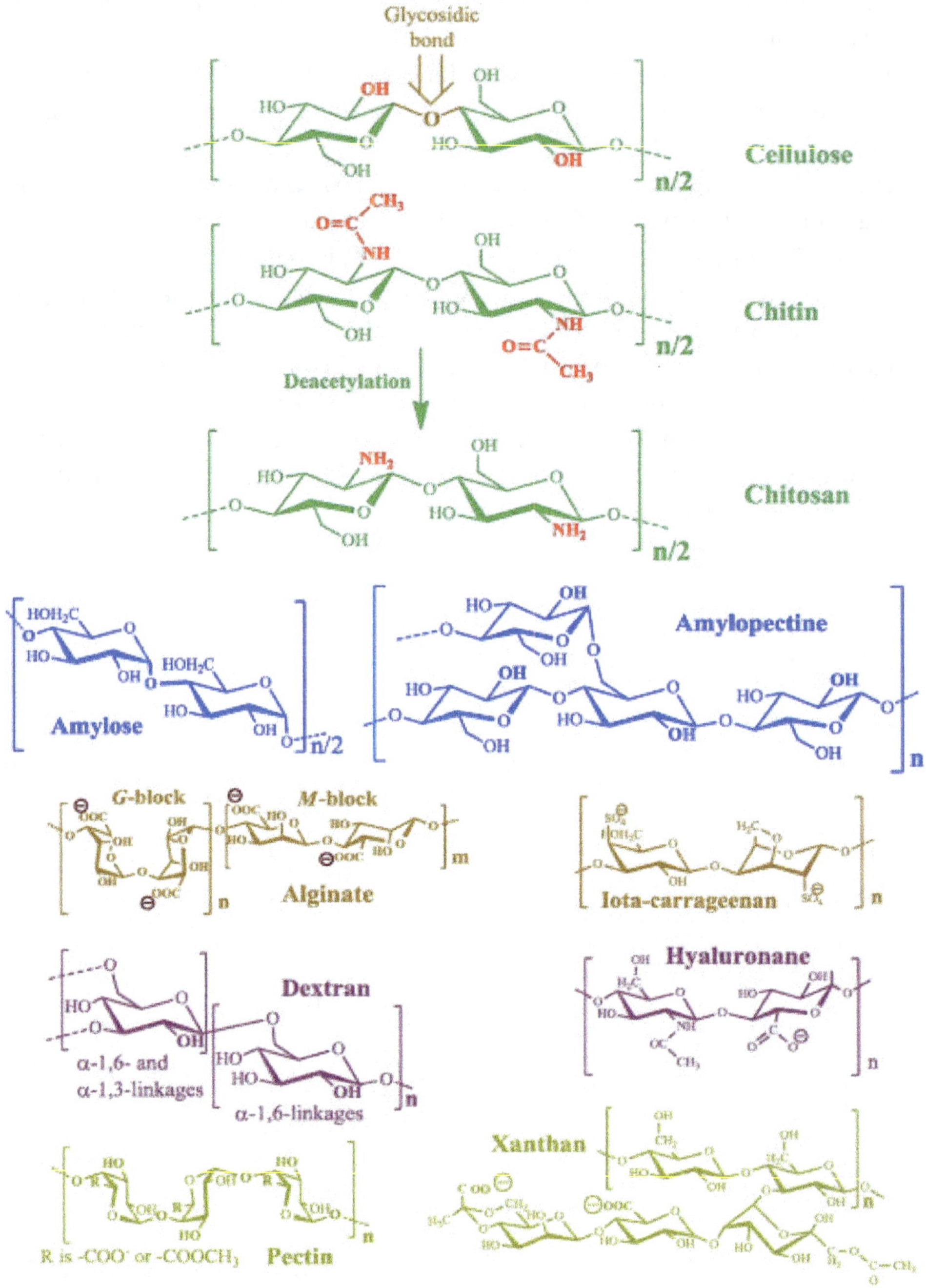

Cellulose, whose linear chains are made up of D-glucose units . Its primary sources are plants, which produce carbohydrates from carbon dioxide and water (during photosynthesis under the influence of sunlight. Its annual sustainable production is 1011–1012 tonnes. (Klemm et al., 1998)

The second most abundant protein is Chitin (Kurita, 2006). Its renewable amount is similar to that of cellulose, ranging from 1011 to 1011 tonnes. The majority of this polysaccharide comes from marine sources. Chitin is found in the exoskeletons (shells) of crustaceans, mollusks, insects, and fungus. Regardless of the fact that cellulose and chitin are by far the most widely used biopolymers, they are rarely used in bionanocomposite formulations. This is because they can only be dissolved in extremely poisonous solvents.

Proteins

Soy, caseinate, silk, gelatin and collagen are all examples of proteins. Both bionanocomposites and biocomposites are commonly made with them. Thermoplastic proteins include wheat gluten, whey, caseinate, and corn zein. They have good film-forming properties, potentially making them appropriate for biodegradable packaging manufacture.To improve processability and functional qualities, proteins are sometimes mixed with polysaccharides, starch, chitosan, and cellulose (Netravali, 2005; Mohanty et al., 2000; Medeiros et al., 2010; Soulestin et al., 2011; Zhao et al., 2008; Rhim & Ng, 2007; Gomes et al., 2013; Arvanitoyannis & Tserkezou, 2009).

Polyhydroxyalkanoates

Polyhydroxyalkanoates are a biologically derived linear polyester. PHAs can be synthesized by over 300 microorganisms, both gram-positive and gram-negative. Multiple biopolyester synthesis begin with sugar or vegetable oils and act as carbon and energy storage molecules. PHAs are made from almost 150 different monomers, although poly-3-hydroxybutyrate remains one of the most frequent biopolyesters. PHA and its copolymers with hydroxy valerate, as well as polyethylene glycol, are among the finest commercial biopolyesters (4-hydroxybutyrate) Figure 6.2 (Lee, 1996; Ikada & Tsuji, 2000; Verlinden et al., 2007; Sudesh et al., 2000; Akaraonye et al., 2010; Philip et al., 2007).

Figure 2. Structural formulas of PHAs used in the preparation of bio-nanocomposite materials (Shchipunov & Irina, 2009)

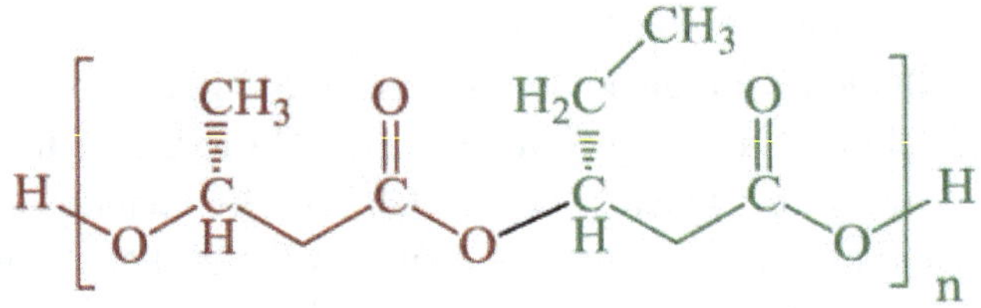

Poly(3-hydroxybutyrate-co-3-hydroxyvalerate))

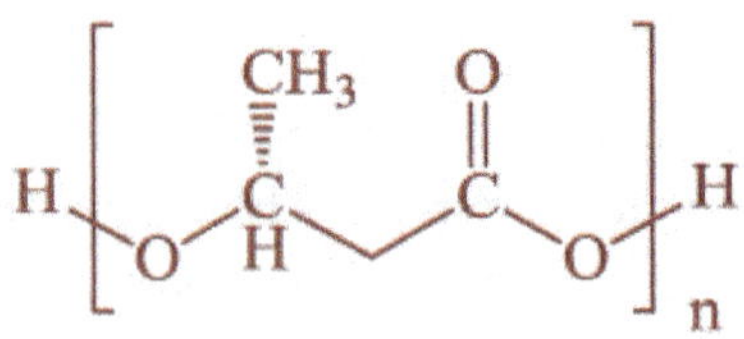

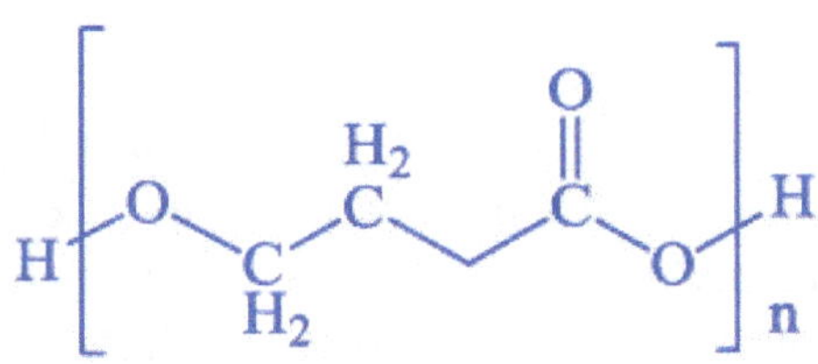

Chitosan bio-nanocomposites

Mollusks, fungi, insects, and crustaceans produce chitosan, a deacetylated chitin derivative. This polysaccharide is the second most common organic substance on the planet, behind cellulose.The D-glucosamine residues in the chitosan macromolecule are predominantly (1-4) connected. It is the sole cationic polysaccharide, which is its distinguishing feature. Most nanoscale materials have a negative charge on their surfaces. As a result, electrostatic interactions of oppositely charged counterparts can be used to produce bio nanocomposite materials (Figure 6.3). Chitosan holds a lot of promise as a bionanocomposites building block. Mixing them together appears to be a straightforward process, but it is not. Anionic chemicals in chitosan's solutions cause it to be extremely sensitive. Their minuscule amount causes a heterogeneous precipitate of aggregated counterparts to develop.The notion that chitosan is a poor polyelectrolyte means that the ph level affects its charge. By adding an acid/alkali, the charged/uncharged condition of chitosan in its aqueous medium may be readily shifted (Khor, 2001; Rong et al., 2012; Sandford, 1989; Tharanathan & Kittur, 2003).

Figure 3. Chitosan in charged and uncharged forms (Shchipunov & Irina, 2009)

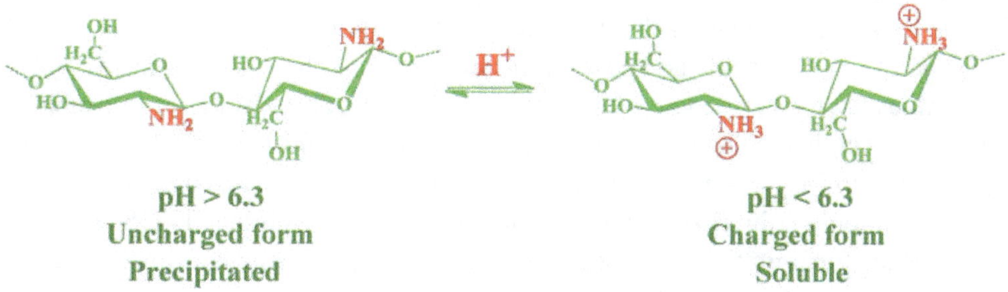

Chitosan comes in both charged and uncharged forms. In aqueous solutions, polysaccharide is insoluble in its uncharged form. When the pH is more than 6, it precipitates. It charges and solubilizes in aqueous solutions as a result of acidity. The charging and solubilization of chitosan may be controlled by altering the pH around the pK value.

Bio-nanocomposites prepared via biometric mineralization:

The synthesis takes place in aqueous medium at ambient temperatures with extremely low concentrations of inorganic compounds, whereas the chemical industry uses acid or alkali, heating, and organic solvents. In terms of properties, mineralized tissues of a living creature surpass equivalent geologically and commercially generated inorganic materials.

Tetraethoxysilane (TEOS), also known as tetraethyl orthosilicate (TEOS), is commonly used as a silica precursor (Figure 6.4). After being mixed with water, it hydrolyzes according to a reaction (Pierre, 1998; Arkles, 1997; Brinker & Scherer, 1990).

$Si(–OCH_2–CH_2)_4 + nH_2O$ $Si(–OH)_4 + 4C_2H_5OH$ (1) up to the orthosilicic acid that undergoes condensation (OH)

$3Si–OH + HO–Si = 3Si–OH + HO–Si = 3Si–OH$ (OH)3 (OH)3Si–O–Si(OH) + H_2O (2)

Their sequence results in silica, a polysilicic acid. In hydrated form, it is SiO_2H_2O. These reactions have a set of unique characteristics that define the sol-gel technique's biomaterial-specific characteristics (Wong & Knowles, 2004; Sumper & Brunner, 2006; Mann, 1996). One is a media that is inconvenient to use. In neutral media, silica precursor hydrolysis is sluggish, but it happens quickly in acidic or alkaline solutions. An acid or alkali can be used to speed up the process.

The first disadvantage of the processing is this. Protein denaturation is caused by the addition of an acid and an alkali. In acidic and alkaline circumstances, polysaccharides having carboxylic groups, such as alginate, and amino groups, such as chitosan, precipitate. The usage of an organic solvent is another key disadvantage of sol-gel chemistry. After the hydrolysis, ethanol is produced (see reaction 1). It's also used to help solubilize things that aren't very water soluble. TEOS. Ethanol promotes the precipitation of most polysaccharides and has a denaturing impact on proteins. The heating is also becoming a significant problem.

The biomaterial suffers a significant loss as a result of the heating. It is primarily responsible for protein and Denaturation of DNA. The procedure slows and takes longer as the temperature is dropped (Wong, 2004; Sumper & Brunner, 2006; Mann, 1996)

Figure 4. Structural formulas of TEOS and THEOS

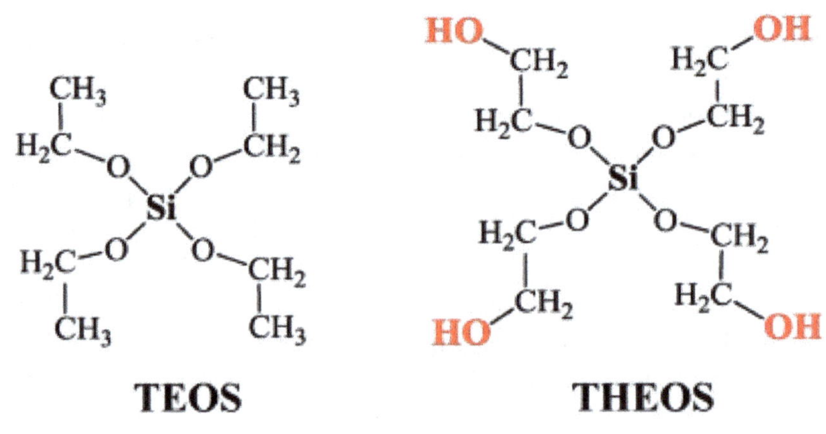

The distinction between them is that THEOS has ethylene glycol residues whereas TEOS has ethanol residues. The presence of additional hydroxyl groups in the THEOS molecule renders the precursor water-insoluble (Shchipunov & Irina, 2009)

Biopolymer mineralization by metal oxides:

The chemistry of metal oxide sol-gel varies greatly from silica. Although the hydrolysis process and condensation process are identical, they occur far too rapidly. One fails to control them when striving to do so. Whenever a precursor hits the water, it begins to respond right away. Before the precursor can get near to the template, metal oxides develop. They can then form on the templating structures, but actual processing control is difficult. Attempts to alter precursors in the same manner that silica was altered were unsuccessful. As a consequence, the organic media was recommended to be replaced with an aqueous solution. The problem is that biopolymers—proteins and polysaccharides—are insoluble in organic solvents, with a few exceptions. Ethylene glycol is one of them. Its advantages have already been mentioned, and it was used to dissolve a xanthan. Bacteria are the source of this polysaccharide. Its backbone is made up of the same (1-4)-linked glucose as cellulose. A side-chain of trisaccharide comprises of 2 mannoses and 1 glucuronic acid residue for every second glucose residue in the 0–3 position (Figure 6.5) (Shchipunov & Irina, 2009)

Figure 5. The biomimicking template synthesis of titania by controlled hydration of carbohydrate macromolecules is depicted schematically (Shchipunov & Irina, 2009)

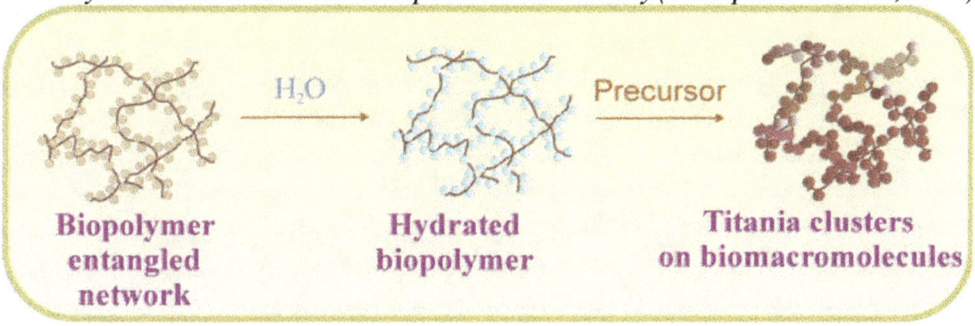

The biomimicking process produces bionanocomposites, which are made up of biomacromolecules encapsulated in a TiO_2 matrix (Figure 6.5). The titania obtained in this manner is amorphous. Thermal treatment of TiO_2 is required in order to convert it into crystalline forms. The conversion to the anatase and rutile forms, or the combinations, happened at lower temperatures than in commercially made titania. It was attributed to the inorganic structure being affected by the biomacromolecule template.

4. CONCLUSION

The aim of this chapter was to discuss the important concepts of the physical and chemical properties of bio-nanocomposites. An immeasurable advantage of these bio-polymers over synthetic polymers is the biocompatibility as this property makes them suitable for multiple advantageous biomedical applications. They are used in preparation of biomaterials such as implants, biomedical devices, tissue engineering, wound dressings, scaffolds, bone replacement, vaccination, drug-delivery systems and diagnostics. This property can also be exploited in cosmetic and biotechnology industries. Bio-nanocomposites boast of good thermal as well as mechanical properties and this feature can be employed to produce films for packaging food with high barrier properties thus leading to long shelf life along with enhanced antimicrobial activity. The most important aspect is that these bio-plastics are environment friendly and relatively cost less.Few of the bio-nanocomposites do not have adequate mechanical,film barrier or thermal properties. Therefore, incorporation with nanofillers or nanoparticles such as nanocellulose, nanoclays or carbon nanotubes to bio-composites will augment its properties and pave way to generate bio- nanocomposite materials with improved properties which can be used in wide variety of applications.

ACKNOWLEDGEMENT

Thanks to the Department of Biotechnology, School of Bioengineering, SRM Institute of Science and Technology, Kattankulathur, Chengalpattu, Tamil Nadu, for the support rendered to this work.

REFERENCES

Akaraonye, E., Keshavarz, T., & Roy, I. (2010). Production of polyhydroxyalkanoates: The future green materials of choice. *Journal of Chemical Technology and Biotechnology*, *85*, 732–743.

Arkles, B. (1997). *Kirk-Othmer encyclopedia of chemical technology*. John Wiley.

Arvanitoyannis, I. S., & Tserkezou, P. (2009). In Long, Y. (Ed.), *Biodegradable polymer blends and composites from renewable resources* (pp. 55–86). John Wiley.

Averous, L., & Boquillon, N. (2004). Biocomposites based on plasticized starch: Thermal and mechanical behaviours. *Carbohydrate Polymers*, *56*, 111–122.

Bascones, A., Vega, J. M., Olmo, N., Turnay, J. G., Gavilanes, G., & Lizarbe, M. A. (2002). Dental and maxillofacial surgery applications of polymers. In *Polymeric biomaterials* (pp. 426–454). Marcel Dekker.

Batley, G. E., Kirby, J. K., & McLaughlin, M. J. (2013). Fate and risks of nanomaterials in aquatic and terrestrial environments. *Accounts of Chemical Research*, *46*, 854–862. PMID: 22759090

Besun, N., Peker, S., Kokturk, U., & Yilmaz, H. (1997).. . *Colloid & Polymer Science*, *275*, 378–389.

Brinker, C. J., & Scherer, G. W. (1990). Elasticity of DLCA model gels with loops. In *Sol–gel science: The physics and chemistry of sol–gel processing*. Academic Press.

Brydson, J. A. (1999). *Plastics materials*. Butterworth-Heinemann.

Camargo, P. H. C., Satyanarayana, K. G., & Wypych, F. (2009). Nanocomposites: Synthesis, structure, properties and new application opportunities. *Materials Research*, *12*, 1–39.

Choy, J. H., Oh, J., & Choi, S. J. (2008). *Bio-inorganic hybrid nanomaterials*. Wiley-VCH.

Daniel, M. C., & Didier, A. (2004). Gold nanoparticles: Assembly, supramolecular chemistry, quantum-size-related properties, and applications toward biology, catalysis, and nanotechnology. *Chemical Reviews*, *104*, 293–346. PMID: 14719978

Darder, M., Lopez-Blanco, M., Aranda, P., Aznar, A. J., Bravo, J., & Ruiz-Hitzky, E. (2006). Microfibrous chitosan–sepiolite nanocomposites. *Chemistry of Materials*, *18*, 1602–1610.

de Jong, K. P. (2009). *Synthesis of solid catalysts* (Vol. 49). Wiley-VCH.

Defonseka, C. (2014). *Introduction to polymeric composites with rice hulls*. Smithers Rapra.

Eduardo, R. H., Katsuhiko, A., & Yuri, M. L. (2008). *Bio-inorganic hybrid nanomaterials: Strategies, synthesis, characterization and applications*. Wiley-VCH.

Eduardo, R. H., Margarita, D., & Pilar, A. (2005). Functional biopolymer nanocomposites based on layered solids. *Journal of Materials Chemistry*, *15*, 3650–3662.

Fernandez, E. J., & Monge, M. (2008). In A. Laguna (Ed.), Modern supramolecular gold chemistry: Gold-metal interactions and applications (pp. 131–179). Wiley-VCH.

. Fomin, V. A., & Guseev, V. V. (2001). Rubber plastics technology, 17, 186.

Forano, C., Hibino, T., Leroux, F., & Taviot-Gueho, C. (2006). In Bergaya, F., Theng, B. K. G., & Lagaly, G. (Eds.), *Handbook of clay science: Developments in clay science* (pp. 1021–1095).

Golomb, D. S., & Fay, J. A. (2004). In Giere, R., & Stille, P. (Eds.), *Energy, waste, and the environment: A geochemical perspective* (Vol. 236, pp. 153–167). Geological Society.

Gomes, M., Azavedo, H., Malafaya, P., Silva, S., Oliveira, J., Silva, G., Sousa, R., Mano, J., & Reis, R. (2013). In van Blitterswijk, C. (Ed.), *Tissue engineering* (pp. 145–192). Academic Press.

Hitzky, E. R., Margarita, D., Ana, C. S., Alcantara, B. W., & Pilar, A. (2013). Functional nanocomposites based on fibrous clays. In Functional polymer composites with nanoclays (Vol. 1, pp. 1–53).

Ikada, Y., & Tsuji, H. (2000). Biodegradable polyesters for medical and ecological applications. *Macromolecular Rapid Communications*, *21*, 117–132.

Ilker, S. B., Adam, S., Philip, J. M., Eric, L., & Lance, M. (2009). Superhydrophobic cellulose-based bionanocomposite films from Pickering emulsions. *Applied Physics Letters*, •••, 94.

Khor, E. (2001). *Chitin: Fulfilling a biomaterials promise*. Elsevier.

Klemm, D., Philipp, B., Heinze, T., Heinze, U., & Wagenknecht, W. (1998). *Comprehensive cellulose chemistry: Fundamentals and analytical methods*. Wiley-VCH.

Kurita, K. (2006). Chitin and chitosan: Functional biopolymers from marine crustaceans. *Marine Biotechnology (New York, N.Y.)*, *8*, 203–226. PMID: 16532368

Lee, S. Y. (1996). Bacterial polyhydroxyalkanoates. *Biotechnology and Bioengineering*, *49*, 1–14. PMID: 18623547

Leuteritz, A., Kretzschmar, B., Pospiech, D., Costa, F. R., Wagenknecht, U., & Heinrich, G. (2007). In Nalwa, H. S. (Ed.), *Polymeric nanostructures and their applications* (pp. 99–151). American Scientific.

Liz-Marzán, L. M. (2004). Nanometals: Formation and color. *Materials Today*, *7*, 26–31.

Mann, S. (1996). Synthesis of inorganic materials with complex form. In Mann, S. (Ed.), *Biomimetic materials chemistry* (pp. 1–40). VCH.

Mann, S. (1996). In Mann, S. (Ed.), *Biomimetic materials chemistry* (pp. 1–40). VCH.

Markus, N., & Nicola, P. (2009). *Metal oxide nanoparticles in organic solvents: Synthesis, formation, assembly and application*. Springer.

Medeiros, E. S., Dufresne, A., & Orts, W. J. (2010). In Bertolini, A. C. (Ed.), *Starches: Characterization, properties, and applications* (pp. 205–251). CRC Press.

Mohanty, A. K., Misra, M., & Hinrichsen, G. (2000). Biofibres, biodegradable polymers and biocomposites: An overview. *Macromolecular Materials and Engineering*, *276–277*, 1–24.

Myung, H. K., Goeun, C., Ahmed, E., Ajayan, V., Young, B. C., & Jin, H. C. (2016). Review of clay–drug hybrid materials for biomedical applications: Administration routes. *Clays and Clay Minerals*, *64*, 115–130. PMID: 32218609

Nafees, A., Saima, S., Sabir, S., Ameer, A., & Mohammad, Z. K. (2018). *Bionanocomposite materials: Concept, applications, and recent advancements*. Jenny Stanford Publishing.

Netravali, A. N. (2005). In Blackburn, R. S. (Ed.), *Biodegradable and sustainable fibres* (pp. 271–309). CRC Press.

Okpala, C. C. (2014). The benefits and applications of nanocomposites. *International Journal of Emerging Technology and Advanced Engineering*, *12*, 12–18.

Philip, S., Keshavarz, T., & Roy, I. (2007). Polyhydroxyalkanoates: Bioplastics with a green agenda. *Chemical Technology and Biotechnology*, *82*, 233.

Pierre, A. C. (1998). *Introduction to sol–gel processing*. Kluwer.

Pilla, S. (2011). *Handbook of bioplastics and biocomposites engineering applications*. Scrivener.

Quan, F., Dan, S., Huaiguo, X., Yuanyuan, H., & Serge, C. (2007). Amperometric phenol biosensor based on laponite clay–chitosan nanocomposite matrix. *Biosensors & Bioelectronics*, *22*, 816–821. PMID: 16624546

Rabia, Z., Khalid, M. Z., Shazia, T., Farukh, J., Aqdas, N., & Mohammad, Z. (2016). Polysaccharide-based bionanocomposites: Properties and applications—A review. *International Journal of Biological Macromolecules*, *92*, 1012–1024. PMID: 27481340

Ray, S. S., & Bousmina, M. (2005). Biodegradable polymers and their layered silicate nanocomposites: Greening the 21st century materials world. *Progress in Materials Science*, *50*, 962–1079.

Rhim, J. W., & Ng, P. K. W. (2007). Natural biopolymer-based nanocomposite films for packaging applications. *Critical Reviews in Food Science and Nutrition*, *47*, 411–433. PMID: 17457725

Rong, T., Nathan, P. G., Timothy, M. F., & Jianjun, C. (2012). Polymeric nanomedicines based on poly(lactide) and poly(lactide-co-glycolide). *Current Opinion in Solid State and Materials Science*, *16*, 323–332. PMID: 23914135

Rosato, D. (2003). *Plastics engineered product design*. Elsevier.

Sain, M., & Oksman, K. (2006). Cellulose nanocomposites. *Journal of the American Chemical Society*, *938*, 2–8.

Sandford, P. A. (1989). In Skjak-Braek, G., Anthonsen, T., & Sandford, P. A. (Eds.), *Chitin and chitosan: Sources, chemistry, biochemistry, physical properties and applications* (pp. 51–69). Elsevier.

Shchipunov, Y. (2012). Bionanocomposites: Green sustainable materials for the near future. *Pure and Applied Chemistry*, *84*, 2579–2607.

Shchipunov, Y., & Irina, P. (2009). One-pot biomimetic synthesis of monolithic titania through mineralization of polysaccharide. *Colloids and Surfaces. B, Biointerfaces*, *74*, 172–177. PMID: 19674874

Shou, N. D., Dan, S., Huai, G. X., Dao, B. Z., & Serge, C. (2009). Glucose oxidase immobilized in alginate/layered double hydroxides hybrid membrane and its biosensing application. *Analytical Sciences*, *25*, 1421–1425. PMID: 20009328

Sorrentino, A., Giuliana, G., & Vittoria, V. (2007). Potential perspectives of bionanocomposites for food packaging applications. *Trends in Food Science & Technology*, *18*, 84–95.

Soulestin, J., Prashantha, K., Lacrampe, M. F., & Krawczak, P. (2011). In Pilla, S. (Ed.), *Handbook of bioplastics and biocomposites engineering applications* (pp. 77–120). Scrivener.

Sudesh, K., Abe, H., & Doi, Y. (2000). Synthesis, structure and properties of polyhydroxyalkanoates: Biological polyesters. *Progress in Polymer Science*, *25*, 1503–1555.

Sudesh, K., & Iwata, T. (2008). Sustainability of biobased and biodegradable plastics. *Clean (Weinheim)*, *36*, 433–442.

Sumper, M., & Brunner, E. (2006). Silicon uptake and metabolism of the marine diatom Thalassiosira pseudonana: Solid-state 29Si NMR and fluorescence microscopic studies. *Advanced Functional Materials*, *16*, 17.

Sumper, M., & Brunner, E. (2006). Learning from diatoms: Nature's tools for the production of nanostructured silica. *Advanced Functional Materials*, *16*, 17.

Tharanathan, R. N., & Kittur, F. S. (2003). Chitin—The undisputed biomolecule of great potential. *Critical Reviews in Food Science and Nutrition*, *43*, 61–87. PMID: 12587986

Utracki, L. A., Sepehr, M., & Boccaleri, E. (2007). Synthetic layered nanoparticles for polymeric nanocomposites (PNCs). *Polymers for Advanced Technologies*, *18*, 1–37.

Verlinden, R. A. J., Hill, D. J., Kenward, M. A., Williams, C. D., & Radecka, I. (2007). Bacterial synthesis of biodegradable polyhydroxyalkanoates. *Journal of Applied Microbiology*, *102*, 1437–1449. PMID: 17578408

Wong, M. S. (2004). In Lu, G. Q., & Zhao, X. S. (Eds.), *Nanoporous materials: Science and technology* (pp. 125–164). Imperial College Press.

Wong, M. S., & Knowles, M. V. (2004). In Lu, G. Q., & Zhao, X. S. (Eds.), *Nanoporous materials: Science and technology* (pp. 125–164). Imperial College Press.

Xiaobo, C., & Samuel, S. M. (2007). Titanium dioxide nanomaterials: Synthesis, properties, modifications, and applications. *Chemical Reviews*, *107*, 2891–2959. PMID: 17590053

Xinming, L., Qingshen, S., Huajie, W., Lei, Z., & Jin, Y. W. (2005). Microspheres of corn protein, zein, for an ivermectin drug delivery system. *Biomaterials*, *26*, 109–115. PMID: 15193886

Zhao, R., Torley, P., & Halley, P. (2008). Emerging biodegradable materials: Starch- and protein-based bio-nanocomposites. *Journal of Materials Science*, *43*, 3058–3071.

Chapter 2
Synthesis of Bio-Nanocomposite Materials:
Thermal and Mechanical Study

Basavaiah Chandu
https://orcid.org/0000-0002-3964-276X
Acharya Nagarjuna University, India

Venkata Sai Sriram Mosali
https://orcid.org/0000-0003-2240-798X
Colorado State University, USA

K. K. C. Satish Babu
Acharya Nagarjuna University, India

Syed Akhil
https://orcid.org/0000-0002-5600-7362
Nanyang Technological University, Singapore

ABSTRACT

Synthetic polymers and their composites are excellently suited to numerous applications, but their overexploitation has caused environmental pollution. Therefore, the need to replace them with sustainable and biodegradable materials is a critical issue that needs to be addressed urgently in order to reduce pollution. Natural polymers such as cellulose, starch, proteins, chitosan, and polylactic acid are promising alternatives to synthetic polymers since they are abundant, cost-effective, biodegradable, and exhibit good mechanical and thermal properties. However, the fabrication of these natural polymers into long-lasting goods is challenging due to their poor

DOI: 10.4018/979-8-3373-2180-6.ch002

mechanical and thermal stability and excellent water absorption properties. Thus, the fabrication of composites of these bio-polymers represents an ever-intriguing research area for material scientists to achieve thermally stable bio-nanocomposites with low water absorption and enhanced mechanical properties. This chapter focuses on the synthesis methodologies of bio-nanocomposites and their mechanical and thermal properties.

4.1 INTRODUCTION TO BIO-NANOCOMPOSIES

The awareness in society towards the importance of renewable products to control environmental pollution has increased the demand for biodegradable and environmentally friendly materials. Non-renewable inorganic, ceramic, and synthetic polymer materials take several years to degrade, which contributes to environmental pollution due to long-term accumulation. As a result, the quest for renewable materials has created excellent opportunities for material science researchers to develop novel materials from natural, renewable resources for sustainable development. Natural polymeric materials like cellulose, starch, proteins, polysaccharides, and biomolecules like polylactic acid, polyvinyl alcohols, etc, are biodegradable, renewable, abundantly available at low cost, and biocompatible. These natural polymers are therefore considered superior alternatives to synthetic polymers for sustainability. However, their poor mechanical properties, low thermal stability, and high water adsorption ability restrict their practical applications. To address these limitations and reduce dependence on synthetic polymers, bio-nanocomposites have been developed. These are the materials in which a biological polymer matrix is reinforced with nanomaterials sized between 1-100 nm in any dimension.(Ray & Bousmina, 2005) Bio-nanocomposites are prepared with three main components: a biopolymer matrix, reinforcements from renewable sources, and nanosized fillers. In light of these environmental and practical benefits, bio-nanocomposites have gained increasing attention as viable alternatives to conventional polymer systems. Bio-nanocomposites can be broadly classified based on the source and type of the biodegradable polymer matrix, as shown in Fig. 4.1. Several natural, inorganic, and synthetic nanomaterials with good mechanical and thermal stability are used as fillers to tailor their performance. Although nanofillers are used in small quantities, their presence significantly influences the properties of the composite, while their impact on overall biodegradability remains minimal. The development of bio-nanocomposites directly supports global sustainability goals by promoting the use of renewable inputs, reducing reliance on fossil-based polymers, and enabling the creation of functional materials with a minimal environmental footprint. Beyond biodegradability, the significance of bio-nanocomposites lies in their potential to reshape the material life cycle, from renewable sourcing to environmentally benign disposal, designed for compatibility with circular economy principles, these materials offer a pathway to reduce greenhouse gas emissions, conserve resources, and minimize persistent waste. Their life-cycle benefits, particularly when both matrix and filler are biodegradable, make them ideal for single-use applications in food packaging, agriculture, and healthcare, where material recovery is challenging. At the same time, growing regulatory pressures and policy shifts, including restrictions on conventional plastics, incentives for green manufacturing, and the rise of eco-certifications, are accelerating the industrial interest in bio-nanocomposites. Their adaptability, compatibility with existing processing techniques, and potential for upcycling agricultural by-products position them as not only environmentally responsible but also economically and strategically viable for future material innovation. This chapter dwells into the synthesis of bio-nanocomposites and examines their mechanical and thermal properties. The discussion is organised into major classes of bio-based nanocomposites, including polysaccharides, proteins, bio-derived synthetic polymers, and hybrid systems, each explored in terms of their structural features and performance characteristics. This thematic approach helps highlight both the potential and the limitations of various composite systems, with particular attention to practical considerations such as scalability, filler dispersion, and application relevance.

Figure 1. Classification of Bio-nanocomposites with respect to matrix materials.

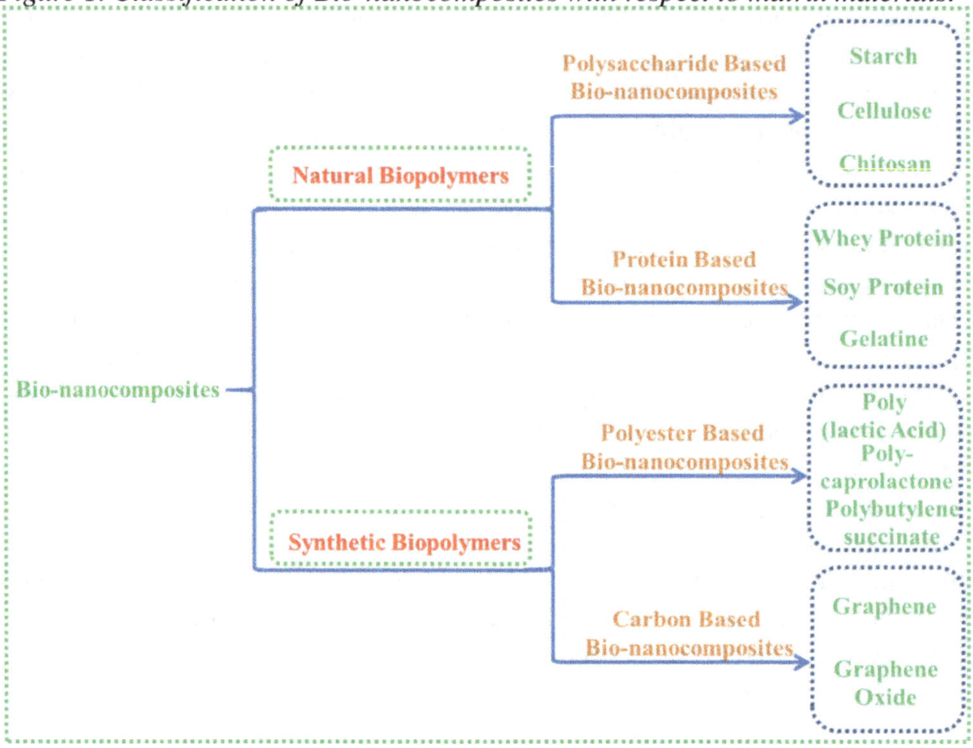

4.2 Synthesis of Bio-nanocomposites

Bio-nanocomposites are prepared with two/three components, namely a matrix of biopolymer, reinforcement with nanosized fillers, and a plasticizer. The mechanical and thermal properties of nanocomposites are governed by the type of biopolymer used as a matrix, the type of nanofiller, and the plasticizer used. Biopolymers are sometimes chemically modified/functionalized in order to improve the mechanical properties, compatibility between nanofiller and matrix, and to incorporate special functions.(Jawaid et al., 2017) Bio-nanocomposites are synthesized similarly to other nanocomposites but with biodegradable polymer matrices derived from environmentally benign sources. The properties of the fabricated bio-nanocomposites also depend on synthetic procedures, so the selection of the synthetic method plays a crucial role in obtaining the desired properties. This section summarises common synthesis methods of bio-nanocomposites. Various synthetic methods for the fabrication of bio-nanocomposites are depicted in Fig. 4.2.

Figure 2. Various synthetic methods for the fabrication of bio-nanocomposites from renewable bio-resources. Reproduced with permission from Ref (Ates et al., 2020), Copyright @2020 American Chemical Society.

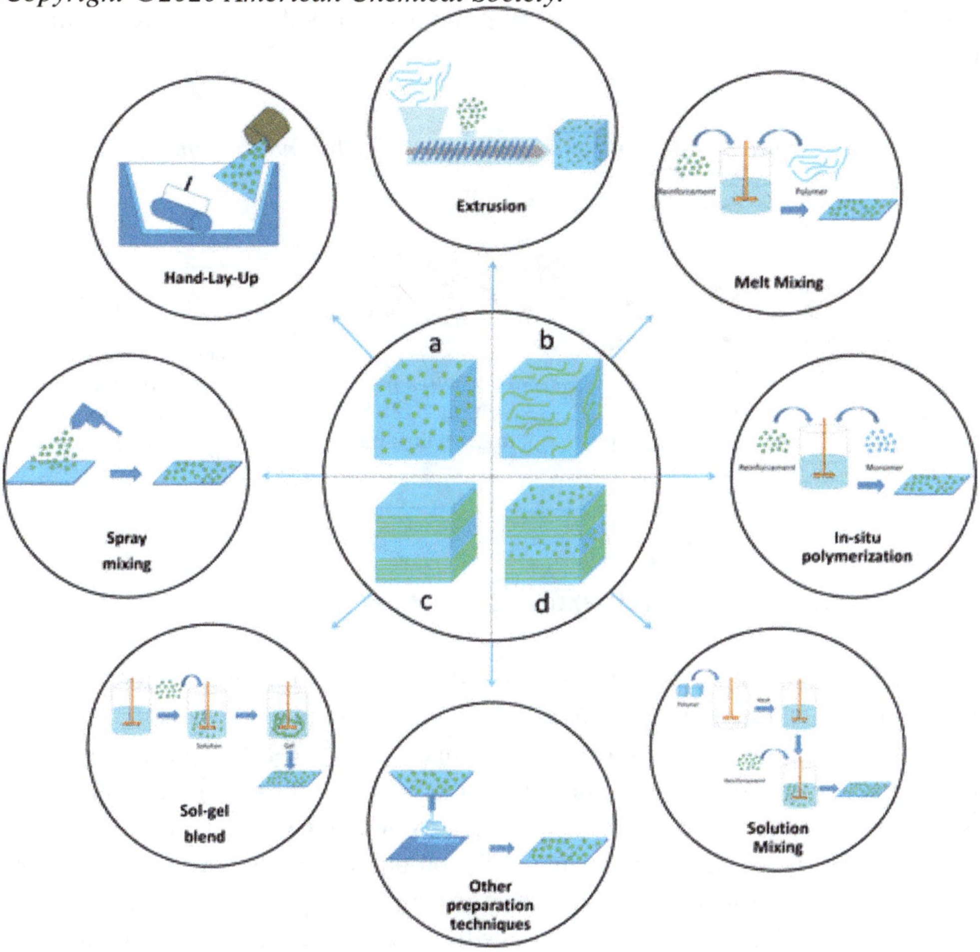

1. Solution Casting

Solution casting is a widely used, versatile processing technique to prepare bio-nanocomposites. Originating in the 19th century, this method of composite preparation is the oldest. The process is generally carried out at low temperatures. In this process, the biopolymer is dissolved mainly in water (volatile organic solvent is also used), followed by the addition of nanosized fillers. Once the mixture has been cast on the substrate, it is dried. Evaporation is used to remove the solvent. The evaporation parameters are adjusted according to the matrix used. Upon evaporation, the composite film is collected from the substrate. Using this method, composite films with high purity and isotropy can be produced. Due to the utilization of aqueous solution and biopolymers as ingredients, this method is the easiest and most eco-friendly technique. The schematic representation of the solution casting technique is shown in Fig. 4.3.

Figure 3. Schematic of solution casting technique for bio-nanocomposites.

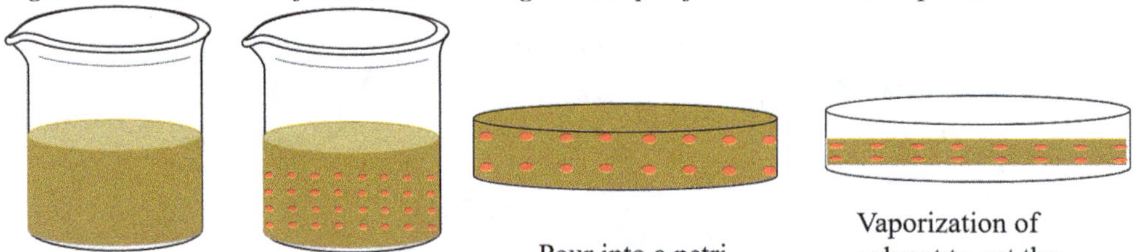

2. Melt compounding/Melt extrusion

Melt compounding is the process of incorporating multiple reinforcements/nanofillers in the polymer melt. This process can be continuous or varied in time. The apparatus consists of a cylindrical tube equipped with a co-rotating screw and different processing zones arranged in series (Fig. 4.4). The co-rotating extruder design is helpful for the effective and precise dispersion of nanoparticles/nanofillers in the polymer melt. When scaled up for industrial production, melt compounding offers clear advantages due to its compatibility with existing polymer processing equipment. However, achieving uniform nanoparticle dispersion becomes more challenging at higher throughput rates, with risks of agglomeration that can undermine material properties.(Khan et al., 2025) Thermal and shear stresses during large-scale processing can also degrade biopolymers, requiring careful optimization of temperature, residence time, and screw design. Despite these challenges, melt compounding remains one of the most promising scalable methods for bio-nanocomposites. The low density nanofiller powders are prone to spill into atmosphere which causes environmental pollution. Hence wet or liquid feeding is adopted, where nanofillers and biopolymers are dispersed in suitable solvents before being introduced into the apparatus.

Figure 4. Schematic demonstration of the Melt extrusion process used for the synthesis of bio-nanocomposites. Reproduced from Ref (Singh et al., 2020). Multidisciplinary Digital Publishing Institute (MDPI).

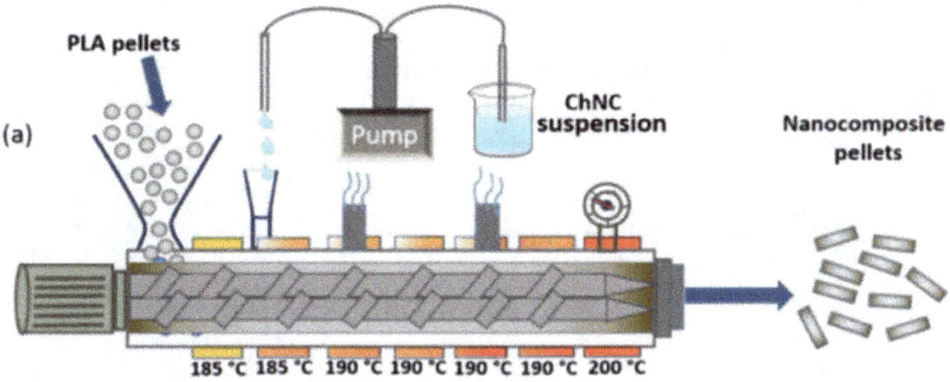

3. Hot compression

Hot compression, or compression moulding, is the standard method for preparing thermoplastic and thermosetting polymers. It can also be used for bio-nanocomposites, mainly cellulose, starch, and polylactic acid-based composites. The typical process involves preparing a nanofiller suspension in an appropriate solvent and adding biopolymer under constant stirring conditions. The resulting melt is vacuum- or air-dried at specific temperatures and times depending on the solvent used in the suspension. The final product is then subjected to hot compression at high temperatures and pressures so that the sheets of bio-nanocomposites are formed (Fig. 4.5). To obtain amorphous composites, the hot-pressed films are quenched in liquid N_2, whereas crystalline nanocomposites are further hot-pressed at elevated temperatures and times.(Suryanegara et al., 2009) When scaled for industrial production, hot compression faces challenges due to its inherently batch-based nature, which limits throughput and demands precise control to maintain uniform density, especially in thicker sections.(Mokhena et al., 2018; Motloung et al., 2019) While equipment costs and energy consumption are moderate, the process remains labour-intensive and costly without automation, which is still rarely implemented at scale, making it more suited for small-to medium scale applications.(Shih & Jeng, 2011)

Figure 5. Schematic representation of hot compression moulding method used for the synthesis of bio-nanocomposites.

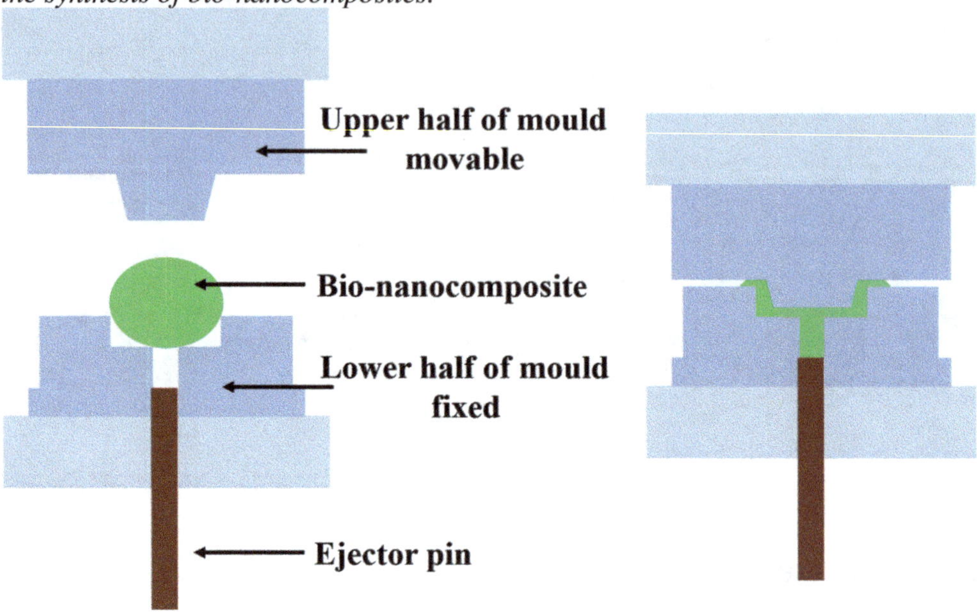

4. Injection moulding

Injection moulding is a cost-effective method that processes melt into complex 3D structures on a large scale. This method is applied to a broad range of materials, including glass, metals, elastomers, and polymers, such as thermosetting and thermoplastic polymers, making it one of the most versatile processing technologies for products such as electronic housings, toys, bottle caps, syringes, and automotive panels. The key advantages of using injection moulding are the ability to produce objects/materials on a large scale with minimal wastage. Although traditionally used for synthetic plastics, injection moulding has also been extended to bio-nanocomposites, including those based on polylactic acid, starch, or cellulose. A typical apparatus consists of an injection cylinder, materials hopper, a heated barrel with screw driven mixing, and a mould cavity (Fig. 4.6). During processing, the biopolymer and nanofillers are heated and homogenized inside the barrel, then precisely injected into the mould under controlled pressure and temperature conditions. The mould temperature and holding time vary depending on the biopolymer used and the required end property. When scaled up for industrial production, injection moulding introduces specific challenges, such as achieving homogeneous dispersion of nanoparticles without agglomeration, managing mould-filling behaviour at high filler loadings, and designing appropriate venting systems to release trapped gases and avoid voids.(Lonkar et al., 2011) Furthermore, the design and manufacture of complex precision moulds can be expensive, requiring significant upfront investment that is only economically justified for high-volume manufacturing. Nevertheless, the precision, speed, and industrial familiarity of injection moulding continue to make it a dominant method for mass-producing high-performance bio-nanocomposites, provided that careful formulation, process monitoring, and mould design are optimized to ensure reliable product quality on a large scale.(Joseph, Krishnan, et al., 2020)

Figure 6. Graphical representation of injection moulding technique used for the synthesis of bio-nanocomposites. Reproduced with permission from tronicarts.de

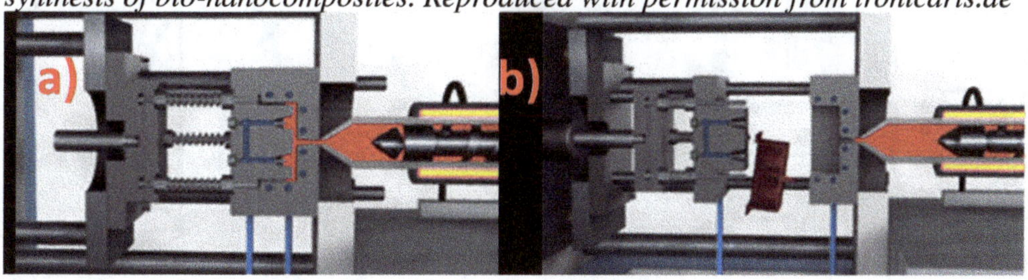

5. Electrospinning

Electrospinning is an advanced technique which gained potential interest in recent decades due to its ability to produce nanocomposite fibers with well-defined diameters and extraordinary lengths.(Azadi et al., 2024) This method is relatively straightforward and cost-effective, using high voltages to draw a charged polymer solution from a capillary toward a grounded collector, forming continuous fibres as the solvent rapidly evaporates in flight (Fig. 4.7). Typically, a biopolymer and nanofiller suspension is first prepared, then loaded into a syringe or reservoir connected to a high-voltage power supply, enabling the formation of a stable jet from the Taylor cone.(Anwer et al., 2023) Electrospun fibers are attractive in biomedical applications, especially in tissue engineering and wound dressings, due to their high surface area and tunable porosity. However, the industrial scaling of electrospinning presents challenges, particularly the low throughput of traditional single-needle systems, which typically achieve only 0.01-1 g/h, limiting commercial feasibility.(Anwer et al., 2023; Iroegbu & Ray, 2022) Modern approaches such as multi-needle arrays, needleless free-surface electrospinning, and centrifugal systems have been developed to boost production to hundreds of grams per hour, although these introduce new concerns about fiber uniformity and reproducibility.(Iroegbu & Ray, 2022)

Despite the relatively high costs associated with solvents and electrical equipment, electrospinning remains a promising method for fabricating high-performance bio-nanocomposites where advanced functionalities and small-scale precision justify the investment.

Figure 7. Graphical representation of the process and instrumentation of electrospinning technique, important parameters to be monitored during the experiment are also listed. Reproduced from Ref (Ziabari et al., 2009).

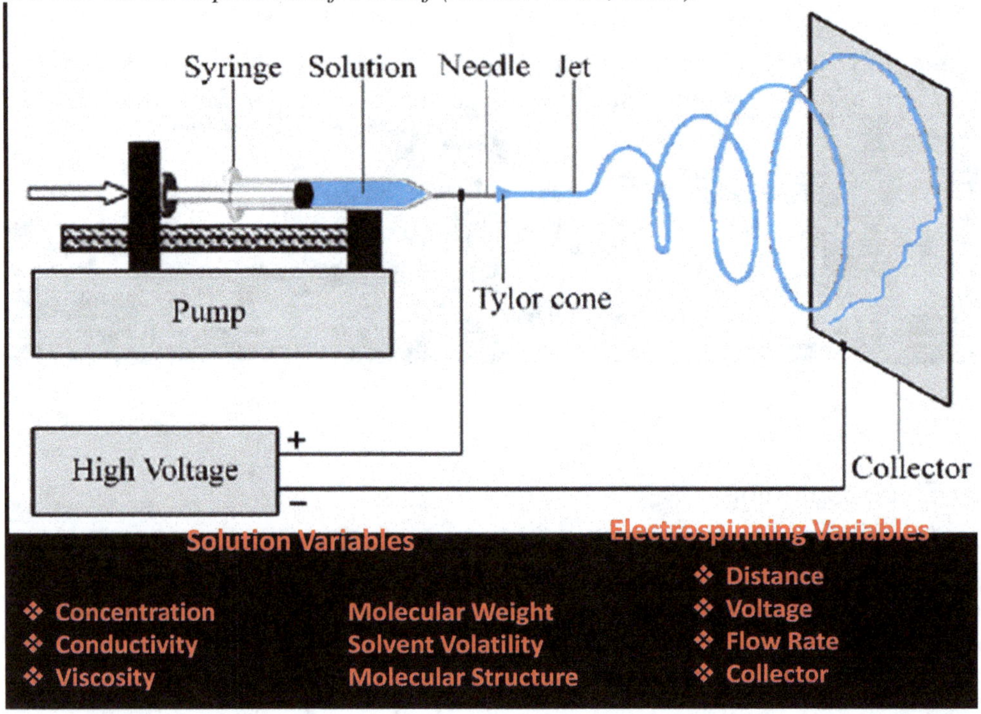

6. *In situ* polymerization

Preparation of bio-nanocomposites made with hydrophobic biopolymers such as polylactic acid (PLA) presents significant challenges, primarily due to the incompatibility with commonly used hydrophilic nanofillers such as nanoclays, nanoparticles, and nanocellulose, and the tendency of these problems by polymerization has emerged as a promising solution to these problems by polymerizing monomers directly in the presence of well-dispersed, often pre-functionalized, nanofillers, resulting in stronger interfacial adhesion and improved dispersion. Typically, the monomers are vigorously mixed in a solution, after which the nanofillers are added, and polymerization is initiated under controlled conditions, allowing the polymer chains to form around nanofillers.(Falua et al., 2022) The final product is then separated, purified, and dried to obtain the desired bio-nanocomposite. While *in situ* polymerisation offers advantages of milder reaction conditions, tunable matrix-filler interactions, and relatively straightforward chemistry, industrial-scale implementation faces hurdles.(Anwer et al., 2023) These include controlling polymerization kinetics, ensuring uniform filler distribution at larger scales, managing unreacted monomer removal, and maintaining consistent product quality when viscosity changes during the reaction. Despite these complexities, in situ polymerization continues to be a highly attractive method for preparing advanced bio-nanocomposites tailored for demanding applications.

4.3 Mechanical and thermal properties of bio-nanocomposites

Natural polymers have attracted substantial interest for diverse applications because of their biodegradability, renewability, and low environmental impact. However, these materials typically exhibit limited mechanical and thermal performance, including high brittleness, low tensile strength, and susceptibility to deformation under elevated temperatures. To enhance the mechanical and thermal properties while retaining the advantages of natural polymers, researchers have investigated reinforcing biopolymers with a variety of nanofillers, such as nanoclays, nanocellulose, carbon nanostructures, or metallic nanoparticles, using different processing techniques to achieve high performance.(Akinnawo, 2024) Key mechanical properties of bio-nanocomposites include tensile strength (TS), Young's modulus (EM), and elongation at break (E), typically measured using standardised tensile testing equipment. Incorporation of nanofillers can improve the tensile strength of biopolymer matrices by up to 50-100%, depending on the filler type, due to improved stress transfer and restricted polymer chain mobility.(Kasaai, 2024)

On the thermal side, nanofillers often serve as head-diffusion or barrier agents that enhance thermal stability and delay degradation. Important thermal parameters include glass transition temperature (T_g), which defines the polymer's shift from rigid to a rubbery state, and the decomposition temperature (T_d), above which chemical breakdown occurs.(Farooq et al., 2024) These properties are commonly evaluated by differential scanning calorimetry (DSC), thermogravimetric analysis (TGA), and differential thermal analysis (DTA). The improvement of thermal properties can be attributed to the reduced segmental mobility of polymer chains in the presence of nanofillers and the barrier effect that limits heat and oxygen diffusion. Overall, integrating nanofillers into bio-based matrices opens avenues for developing materials with enhanced toughness, dimensional stability, and heat resistance, making them promising candidates for diverse engineering and packaging applications.(Anwer et al., 2023) In subsequent sections, the specific mechanical and thermal behaviours of different bio-nanocomposites will be addressed in greater detail.

4.4 Bio-nanocomposites based on distinct biopolymer matrices

4.4.1 Polysaccharide based bio-nanocomposites

Polysaccharides are among the most abundant biopolymers in nature, originating from plant, animal, and microbial sources. They include cellulose, starch, and chitosan, which are valued for their biodegradability, renewability, and functional chemical groups that facilitate modification and compatibility with nanofillers. (Deng et al., 2022) However, polysaccharides alone often exhibit poor mechanical strength and low thermal stability, limiting their structural applications. Incorporating nanofillers such as nanoclays, metal oxides, or cellulose nanocrystals have proven to significantly enhance their barrier, mechanical, and thermal performance.(N. Sharma & Rana, 2020) Furthermore, their hydrophilic nature supports good dispersion of certain nanofillers, making polysaccharide-based bio-nanocomposites highly promising for packaging, biomedical devices, and controlled release applications. In this section, the synthesis strategies, thermal and mechanical enhancements, and specific applications of cellulose-, starch-, and chitosan-based bio-nanocomposites will be examined in detailed.

4.4.1.1 Cellulose-based bio-nanocomposites

Cellulose (a primary load-bearing polymer of plant cells) is a natural linear polymer comprising nanofibrils and is never exhaustible material due to its abundant natural availability in plants, bacteria, and tunicates. The cellulose fibrils are surrounded by lignin and hemicellulose. The exceptional characteristics, such as good biocompatibility, low cytotoxicity, high mechanical and chemical stability, low density, high surface area, easy processing, and cost-effective availability, make cellulose an ideal new generation biomaterial suitable for sustainable development. Cellulose with nanoscale dimensions is called nanocellulose, which contains superior physicochemical properties by encompassing nanoscale properties and cellulose properties to the nanocellulose. Cellulose fibre is a linear polysaccharide with D-glucopyranose units joined by β-1,4' glycosidic linkage, and its crystalline structure is attributed to the extensive intermolecular hydrogen bonding between the hydroxyl groups. In plant cells, the cellulose fibres are arranged as fibrils and microfibrils, as shown in Fig. 4.8.

Nanocellulose is derived from bioresources such as plants or animals; thus, it is an excellent renewable, nontoxic, biodegradable material that is anticipated to have many applications in several fields in the coming years. The properties of cellulose depend on source, synthesis techniques, and environmental conditions, and so it is extracted from several bioresources like flax, jute, hemp, wood, and fruits. (Joseph, Sagarika, et al., 2020) The main nanocellulose materials are viz., cellulose nanocrystals (CNCs) produced by acid hydrolysis technique, cellulose nanofibrils (CNFs) produced by mechanical technique, and bacterial cellulose (BC) obtained from aerobic bacteria (Table 4.1).(Lin & Dufresne, 2014) BC is a pure cellulose with a high cost of production, whereas both CNC and CNF may have structural damages due to their preparation processes. However, these three forms of cellulose are extensively investigated as nanofillers in the synthesis of nanocomposites. The CNF fibres of 5-50 nm in diameter and a few microns in length possess both crystalline and amorphous regions. The CNCs are rod-like or needle-like nanowhiskers having an average diameter of 5-70 nm and 200-500 nm in length. The strong sulphuric acid treatment introduces sulphate groups at the surface.(Joseph, Sagarika, et al., 2020)

Table 1. source, synthesis and dimensions of three types of nanocellulose materials. (A. Sharma et al., 2019)

Type of cellulose	Typical sources	Extraction and size
NFC	Wood, sugar, beet, flax, hemp, tuber, potato	Extraction from cellulose chains by mechanical process Length: micrometers
NCC	Cotton, wood, hemp, flax, tunicin	Extraction from cellulose chains by acid hydrolysis to get crystalline region Diameter: 5-70 nm Length: 100-250 nm
BC	Low molecular weight sugars and alcohols	Bacterial synthesis Diameter: 20-100 nm

Figure 8. Schematic illustration of synthesis of cellulose from plants and bacteria. Reproduced with permission from Ref (Miyashiro et al., 2020), Multidisciplinary Digital Publishing Institute (MDPI),

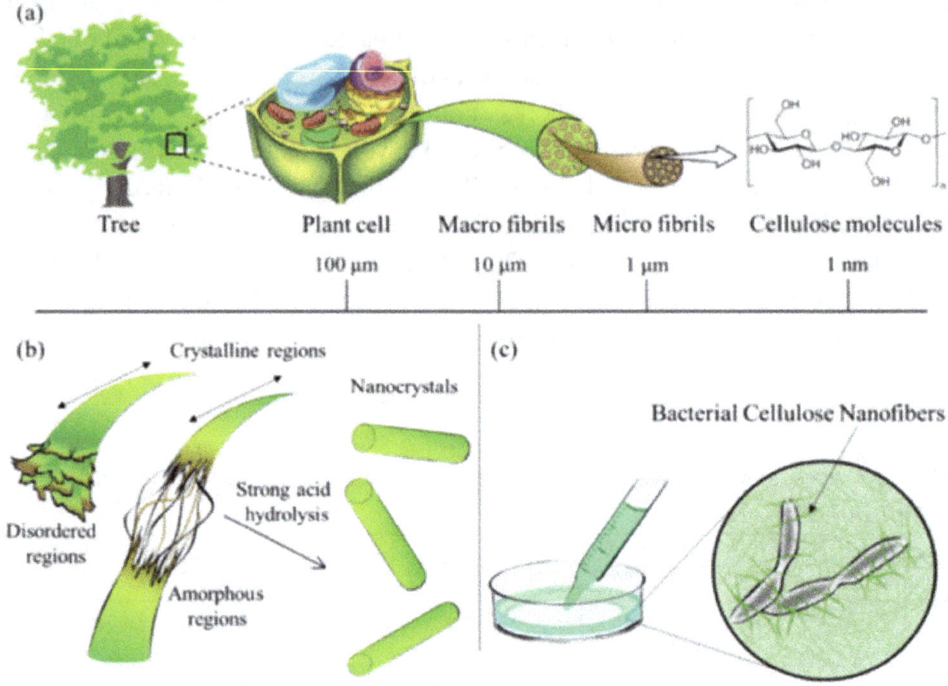

4.4.1.1.1 Properties of cellulose

Solubility of cellulose depends on the extent of hydrogen bonding, and hence special methods need to be developed to understand the solubility. Generally, binary solutions that stabilize and dissolution of electrolytes exhibit cooperative ion-pair interaction among the solvated cellulose hydrogen bonds. A mixture of copper hydroxide and ammonia, subsequent treatment with dilute sulfuric acid, was employed to effectively dissolve substantial quantities of cellulose. Additionally, various solvent systems such as acetone–triethyloctylammonium chloride, dimethyl sulfoxide–tetra-*n*-butylammonium fluoride, aqueous mixture of urea–thiourea, and *N*-alkylpyridinium halides (where alkyl = ethyl or benzyl) have also been employed for the solubilization of cellulose.(Joseph, Sagarika, et al., 2020) The solubility and other chemical properties largely depend on the surface functional groups viz., amino, aldehyde, carboxyl, sulfate, and thiol groups, etc. that are introduced specific applications of nanocellulose.(Fathi et al., 2019)

The crystalline and amorphous regions are uniformly distributed in cellulose, which are responsible for stiffness, elasticity, flexibility, and plasticity, respectively. Hemicellulose removal will improve the mechanical properties. CNF exhibits both crystalline and amorphous regions with a Young's modulus of 100-200 GPa, and the BC possesses a modulus of 114 GPa.(Fathi et al., 2019; Lin & Dufresne, 2014) The fascinating nanocellulose reinforced bio-nanocomposites find extensive applications in bone tissue engineering, wound healing, drug carriers, bioimplants, 3D printing, printable electronics, and packaging materials.

4.4.1.1.2 Extraction of cellulose

The choice of cellulose extraction method depends on the desired properties and dimensions required for the application.(Rojas et al., 2015) Cellulose is extracted using the alkaline procedure, ultrasound treatment, dilute acid treatment, and enzyme procedure. In an alkaline process, the plant tissues (bagasse) are milled and digested in sodium hydroxide to eliminate lignin and hemicellulose, followed by bleaching in acetic acid-sodium chlorate mixture. Then the resulting fibres are washed with sodium hydroxide solution and deionised water. In the enzyme technique, the bleached pulp is soaked in water and autoclaved, then a fungal culture is introduced that selectively hydrolyse lignin and hemicellulose portions. After different time intervals, the plant fibres are autoclaved, washed, and crushed to get the cellulose nanofibrils. The extraction of cellulose microfibrils from plants involves mainly three steps: removal of lignin, hemicellulose by pulping, homogenisation, separating the cellulose microfibrils mechanically, and finally, dilute sulphuric acid hydrolysis to remove any lignin, hemicellulose, and amorphous regions present.

4.4.1.1.3 Synthesis of cellulose nanocomposites

Cellulose and its derivatives are excessively used to form bio-nanocomposites. Since cellulose derivatives such as cellulose acetate, cellulose acetate–butyrate, and plasticized with triethyl citrate are easily dissolvable and convenient to process, these were mainly used in clay-based nanocomposites. Hydrophilic cellulose solution is prepared from freeze-dried cellulose suspended in DMSO-DMF solvent mixture, followed by sonication, whereas the hydrophobic cellulose solution is obtained by dissolving surface-functionalized cellulose in a suitable organic/ionic solvent system.(N. Wang et al., 2007) The cellulose is then mixed with a polymer solution. The bio-nanocomposites are prepared generally using electrospinning, various micropatterning techniques, cast drying, wet spinning, vacuum filtration, and hot compression.(A. Sharma et al., 2019)

4.4.1.1.4 Mechanical and thermal properties of cellulose based bio-nanocomposites

Cellulose based bio-nanocomposites exhibit notable mechanical enhancements depending on the type and dispersion of nanofillers. In TEMPO oxidised - Cellulose nanofibril (TO-CNF)/montmorillonite (MTM) systems, tensile strength values as high as 311$\pm$ 17 MPa and Young's modulus of 21.4 $\pm$n2.6 GPa are achieved at 10 wt% MTM loading, a significant improvement over neat TO-CNF film.(Xu et al., 2021) These improvements stem from strong hydrogen bonding and well-dispersed layered MTM structures. Comparative analysis revealed a higher effective modulus of MTM in TO-CNF matrices (129.9 GPa) than in quaternary ammonium functionalised CNF (Q-CNF) systems (37.1 GPa), underlying the influence of surface charge interactions. Similarly, CNF-graphene oxide (GO) nanocomposites demonstrate high reinforcement efficiency even at low GO loadings (0.07 vol%), achieving Young's modulus up to 17 GPa due to GO-induced CNF matrix ordering.(Mianehrow et al., 2020) Aligned systems such as carboxymethylcellulose (CMC)/cellulose nanocrystals (CNC) composites at 50 vol% show synergistic increased in stiffness and toughness with filler alignment, whereas in certain systems like polyurethane/CNC, percolating CNC networks offer greater reinforcement than alignment, particularly with lower aspect ratio CNCs.(Redondo et al., 2022; B. Wang et al., 2015)

Thermal behaviour in cellulose-based nanocomposites is influenced by both filler-matrix interactions and environmental conditions such as moisture. In PVA/GO-CNF systems, the glass transition temperature decreases markedly from 61.4 °C (dry) to 24.2 °C (wet), indicating moisture-sensitive plasticization.(L. M. Thomas & Sreekanth, 2025) Polyurethene/CNC composites maintain more stable Tg values (-16 °C to -11 °C) across different loadings, suggesting limited interference with polymer chain mobility.(Redondo et al., 2022) Thermal stability is significantly enhanced in TO-CNF/MTM nanocomposites, which display self-extinguishing behaviour at just 10 wt% MTM due to their layered "brick-and-motor" structure that restricts oxygen diffusion.(Xu et al., 2021) In contrast, Q-CNF/MTM systems require higher filler content to achieve similar flame retardancy. Cellulose-based nanocomposites with high clay content also showed strong directional thermal transport, with in-plane conductivity reaching 7.5 $Wm^{-1}K^{-1}$ and anisotropy ratios as high as 30, suitable for heat dissipation applications requiring directional control.(G. Wang et al., 2025)

4.4.1.2 Starch-based Bio-nanocomposites

Starch is one of the highly regarded biopolymers for bio-nanocomposites due to its low cost, abundance, biodegradability, low density, and non-abrasive nature. Starch is the second most abundant biopolymer in nature, found in many plants and vegetables. Starch is extracted by wet grinding, sieving, and drying from these natural sources. Starch is a carbohydrate consisting of anhydroglucose units linked via α-δ-(1 - 4) glycosidic bonds. Linear structured starch consists of 1,4 glucose units and is called amylose. On the other hand, branched structured starch consists of linkage of 1,4 and 1,4 bonds of glucose and is called amylopectin. In addition to glucose units, starch also consists of small amounts of lipids and proteins. Starch is generally semi-crystalline with nearly 20-45% of crystallinity.

4.4.1.2.1 Synthesis of starch-based bio-nanocomposites

Though starch-based nanocomposites are commercially viable, they have certain demerits, including brittleness, high gas permeability, low heat distortion temperature, and low melt viscosity. Researchers have made profound efforts to overcome these demerits by filling starch polymer matrix with nanofillers such as nanosilicates, nanocellulose, CNTs, carbon black, metal oxide nanoparticles, and even starch nanoparticles. Starch-based nanocomposites also include starch nanoparticles filled in a synthetic polymer matrix such as PVA, PHA, PLA, etc. Preparation of synthetic polymer/starch nanocomposites mainly suffers from the homogeneous distribution of nanofillers in the matrix during synthesis. Starch bio-nanocomposites are synthesised primarily using solution casting, *in situ* polymerisation, and melt compounding methods.

4.4.1.2.2 Mechanical and thermal properties of starch-based bio-nanocomposites

Starch-based bio-nanocomposites have exhibited significant mechanical reinforcement upon incorporation of various nanofillers. Potato starch films reinforced with CNCs at concentrations up to 0.1 wt% have achieved tensile strengths as high as 47 MPa, with the optimal performance observed at 0.05 wt%, corresponding to a 14.6% increase over pristine starch films.(Csiszár et al., 2021a) Elongation at break has ranged from 7.1% to 13.3%, depending on CNC loading, with ductility peaking at higher filler content. In corn starch systems, CNC has demonstrated filler-plasticiser synergy, where sorbitol-plasticised films with 5 wt% CNC have reached a tensite strength of 4.69 MPa and elongation at break of 56%, highlighting the importance of filler dispersion and plasticiser compatibility.(Zhou et al., 2024) Thermoplastic starch composites containing up to 50 wt% CNCs have shown further strength enhancements, though excessive filler concentrations have been found to introduce micron-scale agglomerates that compromise structural uniformity.(Csiszár et al., 2021b)

4.4.1.3 Chitosan based composites

Chitosan (Ch) is a natural polysaccharide biopolymer with high molecular weight, it contains β-(1,4)-2-acetamido-2-deoxy-D-glucose and β-(1,4)-2-amino-2-deoxy-D-glucose units. Ch is easily accessed by deacetylation of chitin in powder, sponge, or fibre forms.(Yu et al., 2021) Due to its biodegradability and non-toxic nature, it is applied in a diverse spectrum of applications, but faces some general obstacles, like other natural polymers. Hence, the properties of Ch are also improved by fabricating Ch-based bio-nanocomposites using a variety of fillers.

4.4.1.3.1 Synthesis of chitosan-based bio-nanocomposites

The major limitation of Ch being used in nanocomposites is its insolubility in aqueous and organic solutions. The chemical modification of chitosan can overcome the problem of solubility, and this can be done through protonation/functionalisation of amine groups on the C2 position.(D. Thomas & Thomas, 2013) Compared to other polysaccharide polymers, the chemical modification of Ch is more favourable. Chitosan-based bio-nanocomposites include nanofillers such as metal nanoparticles, starch nanoparticles, cellulose nanocrystals, CNTs, and graphene. These nanocomposites are prepared mainly using solution casting and *in situ* polymerisation methods. However, certain modifications are employed depending on the nanofillers used, including combining them with other processing methods.(Kargarzadeh et al., 2018)

4.4.1.3.2 Mechanical and thermal properties of chitosan based bio-nanocomposites

Chitosan based bio-nanocomposites have consistently demonstrated enhanced mechanical properties through strategic incorporation of diverse nanofillers. Systems reinforced with tannic acid-coated cellulose nanocrystals (TA@CNC) have shown tunable fracture stress between 0.39 and 1.2 MPa, with strain capacities ranging from 370% to 800%, highlighting the versatility of chitosan hydrogels.(Cui et al., 2019) AgNPs-reinforced chitosan blends have yielded measurable improvements in tensile stress, Young's modulus, and elongation at break, confirming the reinforcement efficiency of metallic nanofillers.(Ragab et al., 2024) Among inorganic fillers, ZnO nanoparticles have delivered some of the most pronounced enhancements, increasing tensile strength and elongation by up to 6.65-fold and 3.57-fold, respectively. (Tabassum et al., 2023) CNCs, when optimally dispersed at 4 wt%, have improved tensile strength and modulus by 39% and 78%, with some systems reporting gains of over 200% under optimised processing conditions.(J. Chen et al., 2022) Graphene oxide (GO) has also been incorporated successfully, where rheological evaluations of carbamoylated chitosan-GO hydrogels have shown increased yield stress and viscoelasticity.(Gupta et al., 2023) MTM-filled Ch membranes have demonstrated concentration-dependent improvements, reaching 99 ± 3.7 MPa tensile strength at just 1.5 wt% organically modified MTM loading in PVA-Ch matrices.(Farrokhi et al., 2024; Kedang et al., 2022).

Ch-based nanocomposites have demonstrated complex and highly tunable thermal behaviours, closely linked to the nature and dispersion of incorporated nanofillers. In PVA/Ch/TiO_2 composites, thermal transitions such as Tg and Tm have remained largely unaffected, suggesting that nanoparticle reinforcement does not necessarily hinder polymer mobility.(Ruiz Gómez et al., 2020) In contrast, hydroxyapatite-filled Ch blends have shown marked increases in thermal stability and phase transition temperatures, as confirmed through DSC and TGA analyses.(Ramesan et al., 2024) Incorporation of TiO_2 NPs has led to progressive enhancement in thermal decomposition resistance, with higher loadings correlating to improved heat tolerance. (Guirguis & Alharbi, 2024) Carbon-based nanofillers, such as iron oxide-coated GO, have provided additional reinforcement, increasing the thermal resilience of hydrogel films under elevated temperature conditions.(Konwar et al., 2016)

4.4.2 Protein based bio-nanocomposites

Proteins obtained from plant and animal origins, such as whey and soy protein, are increasingly viewed as promising matrices for bio-nanocomposites. They offer excellent film-forming properties, are biodegradable and are derived from abundant renewable resources.(H. Chen et al., 2019) The incorporation of nanofillers such as layered silicates, cellulose nanocrystals, or even carbon based materials can enhance the mechanical strength and thermal resistance of protein-based systems.(Swain et al., 2017a) With these improvements, protein-based bio-nanocomposites have shown considerable promise in applications including active food packaging, edible films, and wound healing dressings. This section will discuss the intrinsic properties of these proteins, the synthesis routes employed to produce their nanocomposites, and their resulting mechanical and thermal characteristics.

4.4.2.1 Whey protein based bio-nanocomposites

Whey protein isolate is a natural biodegradable matrix material with good mechanical properties, resistant to oxygen barrier, but exhibits poor moisture barrier and thermal properties. These properties are improved in composites where whey protein is used as a matrix material.(Zhao et al., 2008)

Azevedo *et al.* developed films of montmorillonite (MMT) clay reinforced whey protein bio-nanocomposite, and studied its thermal and mechanical properties. (Azevedo et al., 2015) The whey protein isolate and the mixture of glycerol, MMT, and citric acid are dissolved in 44, 56 mL of water separately, followed by 30 min of agitation at room temperature using magnetic stirring. These two solutions are then mixed, further agitated for 10 min while the pH is maintained at 8 using 2N NaOH solution. The resulting solution is subjected to an ultrasonic homogenizer for 10 min, heated for 30 min in a water bath, and then cooled. Finally, the films are fabricated by pouring this solution on glass plates, which are then dried at room temperature for 48 h. The thermogravimetric (TG) analysis showed a higher decomposition temperature (T_d – 26 °C enhancement) and lower mass loss, which indicated the improved thermal stability of the bio-nanocomposite films compared to the control. Elastic modulus (71% enhancement), tensile strength (49% enhancement), and elongation at break (67% enhancement), are allowed improvements for the bio-nanocomposite films. The enhancement of thermal and mechanical properties is attributed to the exfoliated MMT reinforcement that reduced the mobility of the whey protein matrix.

4.4.2.2 Soy protein-based composites

Soy protein is widely available, water-soluble, easily renewable biopolymer with the most attractive film-forming properties. It is commercially available as soy protein isolate (SPI) where 90% of it is globular proteins (albumin, globulin etc.) comprised of 7S (β-conglycinin) and 11S (glycinin) fractions.(Kumar et al., 2010) It is used mainly in the food industry, including meat alternatives, soups, dairy alternatives, bread, desserts, infant formulations, pasta, cereals, and pet foods, to name a few. It is also used in textile fibres, paints, adhesives, emulsifiers, plastics, and cosmetics. Commercially, soybeans are processed into soy powder, soy concentrates, and soy isolates.(Swain et al., 2017b)

Soy protein-based nanocomposites are usually prepared using the solution casting method.(Swain et al., 2017b) In a typical process, aqueous solutions of soy protein and nanofillers are mixed with constant stirring, usually for >3h, such that a precipitate of nanocomposite comprises a homogeneous mixture of soy protein and nanofillers is formed. The mechanical and thermal properties of soy-based bio-nanocomposites depend on the nanofillers. For example, Kumar *et al.* have reported MMT reinforced SPI thin films using the melt extrusion method.(Kumar et al., 2010) The TS, E, and Tg values are increased by 250%, 350% and 40% while the Td value is also increased.

4.4.3 Polyester-based bio-nanocomposites (Polylactic acid)

Polyesters represent a broad class of biodegradable and biocompatible polymers that have attracted considerable interest for sustainable nanocomposite applications. Notably, polylactic acid (PLA), polycaprolactone (PCL), and polybutylene succinate (PBS) with renewable sourcing, good processability, have expanded the use of these bio-nanocomposites in areas ranging from packaging to biomedical devices and automotive components. This section will introduce the structure and properties of PLA, the nanocomposite synthesis and detail their mechanical and thermal properties.

Polylactic acid (PLA) is one of the highly studied biodegradable, biocompatible, high-strength, and thermoplastic polymers belonging to the family of aliphatic polyesters. While PLA was first discovered in 1893 by Bischoff and Waden, it was prepared by Carothers and co-workers from lactic acid in 1932.(Ren, 2011) The high molecular weight PLA was produced at an industrial scale in the early 1990s by ring-opening polymerization of lactic acid. The synthesis of PLA from lactic acid is carried out via three routes: (i) condensation and chain coupling, (ii) azeotropic dehydrative condensation, and (iii) ring-opening polymerisation, as shown in Fig. 4.9.(Agwuncha et al., 2017) PLA is produced from renewable, natural, plant-based products such as corn, sugarcane, potatoes etc. The disposed plant products are hydrolysed and fermented to produce lactic acid, which upon polymerization produces PLA. The significant advantages of PLA are

i. Biodegradable, compostable, biocompatible and recyclable
ii. High mechanical strength, high modulus and a thermoplastic polymer
iii. Tunability of its stereochemistry so that the crystallinity, mechanical and thermal properties can be altered.
iv. Ease of synthesis from cost-effective conventional methods

Figure 9. Synthesis mechanisms of ploy(lactic acid) from lactic acid.(Agwuncha et al., 2017)

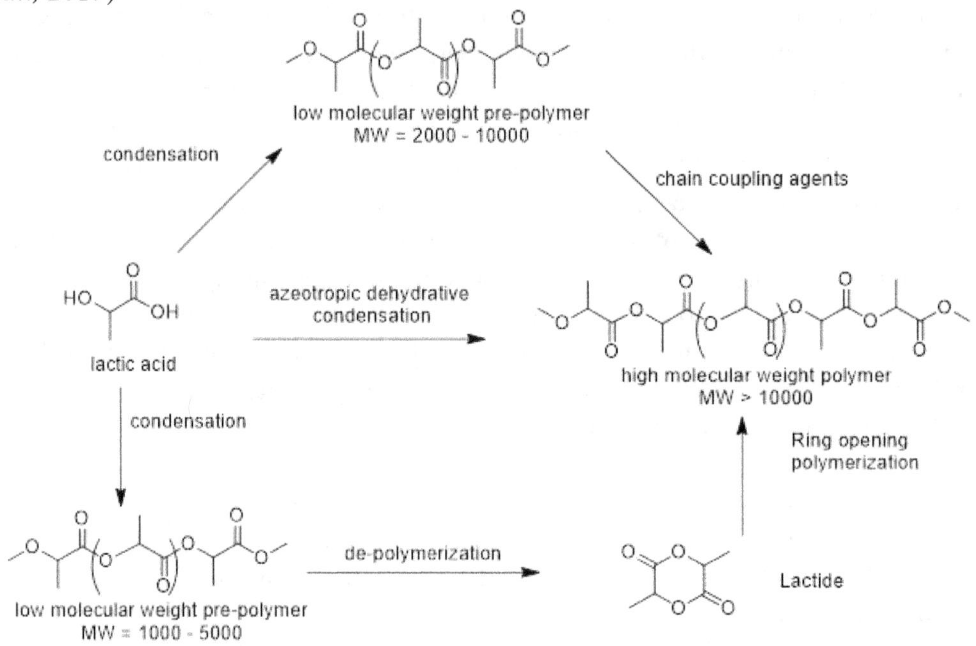

4.4.3.1.1 Synthesis of PLA based bio-nanocomposites

PLA-based bio-nanocomposites are known for their thermal stability, composite stiffness, crystallinity and permeability.(Agwuncha et al., 2017) Nanofillers should be carefully chosen such that they should ease the melting process while having no adverse effects on the PLA matrix. Nano clay, nanocellulose, nanographene, and metal nanoparticles are the most common nanofillers used for PLA-bio-nanocomposites.(Surya Rao et al., 2021) PLA/Nano clay bio-nanocomposites were generally prepared using the melt intercalation method by annealing either statically or under shear. During the process, annealing is carried out beyond the softening point of the polymer, where the polymer diffuses into nano-galleries between layered fillers.

Nanocellulose in PLA-based bio-nanocomposites improves the melt strength, heat distortion temperature, brittleness, and flexibility limit PLA PLA-based composites.(Kargarzadeh et al., 2018) The nanometric size of nanocellulose results in increased surface area, crystallinity, optical transparency, and thermal and mechanical properties. Ag, TiO_2, and ZnO nanoparticles filled PLA matrix composites are some examples of PLA/metal nanoparticle-based bio-nanocomposites.(C. Liu et al., 2017; Parangusan et al., 2021; Zhuang et al., 2009) These nanocomposites are synthesised by *in situ* polymerisation, solvent casting, electrospinning, and the addition of metal nanoparticles during the melt extrusion process. The addition of metal nanoparticles affects the chemistry of bio-composites and improves antibacterial activity, and enhances the mechanical properties. While the cohesive forces between nanoparticles and PLA determine the dispersion of nanoparticles, the strength of their interaction decides the mechanical and thermal properties.

4.4.3.1.2 Mechanical and thermal properties of PLA-based nanocomposites

PLA-based bio-nanocomposites have demonstrated substantial mechanical performance enhancements through the incorporation of various inorganic and organic nanofillers. Reinforcement with CNCs has yielded consistently favourable outcomes. The addition of 1 wt% CNCs has increased strain at break by 20%, while decylamine-modified CNCs improved PLA toughness by 61%, from 1.70 MJ/m^3 to 2.74 MJ/m^3, without negatively affecting tensile strength or modulus.(Wu et al., 2025) The effectiveness of hybrid nanofillers has been evident in 3D-printed PLA/PCL matrices containing CNC-organomodified MTM, where dispersion-driven improvements in crystallization and mechanical performance were reported.(H. Liu et al., 2023) TiO_2 NPs incorporated PLA based nanocomposites have exhibited concentration-dependent outcomes. While PLA/TiO_2 composites alone showed limited mechanical improvements, hybrid systems such as PLA/PLA-HNTs-TiO_2 have achieved 46% increase in tensile strength at 5 wt% TiO_2, underlining the importance of synergistic filler dispersion.(Alakrach et al., 2021) Furthermore, PLA films with organomodified MTM have shown enhanced flexibility, with elongation at break increasing from 4.55 to 22.7% upon addition of 1 wt% PVA.(Shar et al., 2021) Proper blending sequences were crucial for optimizing filler integration, as demonstrated by PLA/PBS composites with 3 wt% D43B clay achieving more than 50% reduction in oxygen permeability.(Adrar & Ajji, 2024) In biomedical applications hydroxyapatite reinforced PLA systems have provided improved compressive strength and fracture toughness by approximately 30% and 50%, respectively.(Omigbodun et al., 2024) Formulations with 30% cockle shell-derived Hap achieved a Young's modulus of 8.17 GPa, aligning with cortical bone values.(Namasivayam Sukumaar et al., 2025)

4.4.4 Graphene-based Bio-nanocomposites

Two-dimensional carbon skeleton, presence of hydroxy, epoxy, carbonyl, carboxylic acid functional groups, good dispersibility in water, hydrophilic nature of graphene coupled with higher mechanical strength (Young's modulus ~ 1 TPa), electrical mobility (200,00 $cm^2V^{-1}s^{-1}$), transparency (98%), surface area (2630 m^2g^{-1}) makes it a superior reinforcing material in bio-nanocomposites to impart superior thermal & mechanical properties to the biopolymer matrix material.(Young et al., 2012) The reinforcement of graphene or its derivative compounds, reduced graphene oxide (RGO) and graphene oxide (GO), will impart these properties to the matrix material, thereby resulting in the enrichment of thermal, electrical, and mechanical properties of the fabricated graphene-reinforced bio-nanocomposite material.

RGO is obtained from GO, which in turn is synthesized from graphite powder via solution-based chemical methods as shown in Fig. 4.10 at low cost. The simple solution synthesis of GO involves the oxidation of graphite powder with strong oxidizing agents to form graphite oxide, which, upon exfoliation by sonication, gives the GO. The RGO is obtained from GO by reduction with hydrazine, vitamin C, plant extracts, and biomolecules .(Chandu et al., 2020) Strong physical interaction, like hydrogen bonding, of graphene nanomaterials with biopolymer matrix generally results in the formation of stable graphene-based bio-nanocomposites. (Sheshmani et al., 2014) The general preparation methods of graphene reinforced bio-nanocomposites include solution casting, *in situ* intercalative polymerization, melt intercalation, electrospinning, wet spinning, drop casting, and freeze drying.

Figure 10. Schematic illustration depicting the general synthesis of RGO from graphite via strong oxidation, sonication followed by reduction to get reduced grapheme oxide.

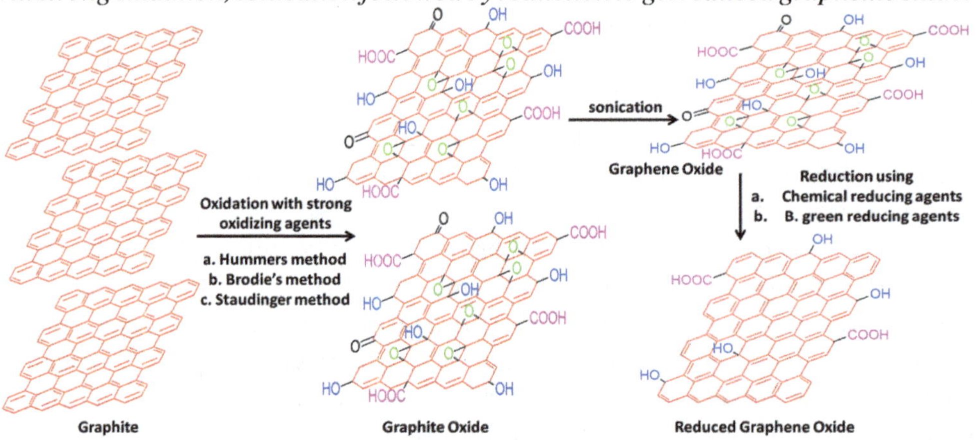

Achaby *et al.* reported the CS-PVP-GO bio-nanocomposite films by solution cast method.(El Achaby et al., 2014) The mechanical properties, tensile strength-109%, Young's modulus-130%, higher elongation at break-62%, toughness-464% showed great enhancements. The TG-DTA analysis showed higher thermal stability with a maximum temperature of degradation of 596 °C and low water absorption-9% after 24 h immersion. Jose *et al.* reported the PVA-Starch-Graphene (70:30:0.5) nanocomposite films with improved mechanical and thermal properties.(Table 4.2)

(Jose & Al-Harthi, 2017) PVA, graphene, and starch-glycerol solutions are separately prepared in water, mixed on a magnetic stirrer for 3 h. The resulting solution is degassed for 10 min in a vacuum oven, poured on glass plates, and dried at 50 °C for 24 h. Tensile strength of the pure PVA is 8.89 MPa, whereas for the composite is 15.59 MPa (74.3% enhancement), Young's modulus is enhanced from 62 MPA of pure PVA to 327 MPa for the composite (427% enhancement). The elongation of pure PVA at break is 91.6% whereas it is decreased in the case of the composite (31.9%). This is attributed to the decrease in elastic nature and increase in the brittle nature of the composite. Pure PVA lost 90% wt at 373 °C while the composite lost 90% wt at 457 °C, thus improving the thermal stability. Similarly, the Tg increased from 38.5 °C to 85.5 °C while MP decreased from 213 °C to 147 °C.

Further, the osteoconductivity, bioactivity of graphene reinforced hydroxyapatite composite materials with enhanced mechanical properties are suitable for bioimplants in the replacement of bone, teeth, and hard tissues. (Baradaran et al., 2014) have reported a nanotube hydroxyapatite (HA)-graphene nanocomposite via a green route with superb bioactivity, mechanical properties, and in vitro bone formation capacity. Bio-composite samples of 0, 0.5, 1.0, and 1.5 wt% of graphene in HA matrix are prepared by a simple solution chemistry followed by hydrothermal treatment. The bio-composites exhibited a significantly enhanced fracture toughness of 1.51 MPa, representing an 86% increase, along with an elastic modulus of 123 MPa, which is 40% greater than that of HA.

Table 2. Bio-nanocomposites of graphene with mechanical and thermal properties reported.

Bio-nanocomposite	Synthesis method	% Enhancement of mechanical properties			Enhancement in thermal properties			Reference
		EM (GPa)	TS (MPa)	E (%)	Tg (°C)	Td (°C)	Tm (°C)	
Pea starch + 2 wt% GO		950	300	300		5.7		(Li et al., 2011)
Chitosan + 1 wt% GO		51	93	40				(Pan et al., 2011)
Chitosan + 6 wt% RGO		163	134	-165		8.6		(X. Wang et al., 2010)
BC + 5 wt% GO		13	22	-15		20		(Kim et al., 2011)
CMC + 1 wt% GO		159	70	-				(Feng et al., 2012)
CS-PVP-2 wt% GO	Solution casting	130	109	62		40		(El Achaby et al., 2014)

continued on following page

Table 2. Continued

Bio-nanocomposite	Synthesis method	% Enhancement of mechanical properties			Enhancement in thermal properties			Reference
		EM (GPa)	TS (MPa)	E (%)	Tg (°C)	Td (°C)	Tm (°C)	
PVA-Starch-Graphene	Solution casting	427	75.3	-60	47	124	-17	(Jose & Al-Harthi, 2017)
PVP-starch-GO-Ag	Solution casting	145	27	-250		29		(Usman et al., 2016)
Phenol-formaldehyde resin and 0.12wt% RGO	Solution casting	60	130	300		8		(Sandhya et al., 2019)

4.5 Scalability, cost-effectiveness, and environmental impacts of bio-nanocomposites

The industrial translation of bio-nanocomposites from laboratory concepts to commercial products presents significant engineering and economic challenges. Laboratory-scale studies have consistently highlighted their superior mechanical strength, barrier properties, and thermal stability; however, achieving equivalent performance under industrial-scale conditions demands strict control over raw material consistency, nanofiller dispersion, and processing stability.(Haider et al., 2019) Processes such as melt compounding and injection moulding provide robust pathways for scaling, yet modifications are often required to accommodate the thermal and shear sensitivities of bio-derived polymer matrices.(Siró & Plackett, 2010) Emerging electrospinning technologies, while promising for advanced nanofiber-based materials, currently exhibit low throughput in single-needle configurations, requiring the adoption of multi-needle or free-surface systems to approach industrial viability. Consistent and uniform dispersion of nanofillers remains a critical factor affecting both performance and reproducibility, as agglomeration can compromise mechanical reinforcement and introduce occupational health hazards during handling and production.

Economic feasibility is closely linked to both material and processing costs. While the sourcing of natural polymers and renewable nanofillers from agricultural or industrial byproducts offer long-term supply stability and reduced reliance on fossil-derived resources, specialized functionalization steps and advanced compounding technologies can help evaluate production costs. Nevertheless, techno-economic analysis suggests that improvements in barrier and mechanical performance, often ranging from 20% to 50% depending on system design, can enable material reduction in applications such as packaging, partially offsetting the higher upfront investment.

Process optimization and scale-up efficiencies are expected to further close the cost gap, particularly as equipment and formulation standards mature within the industry. Structure and sources of various natural polymers is depicted in Fig. 4.11.

Figure 11. Structure and natural sources of various biopolymers for the fabrication of bio-nanocomposites.

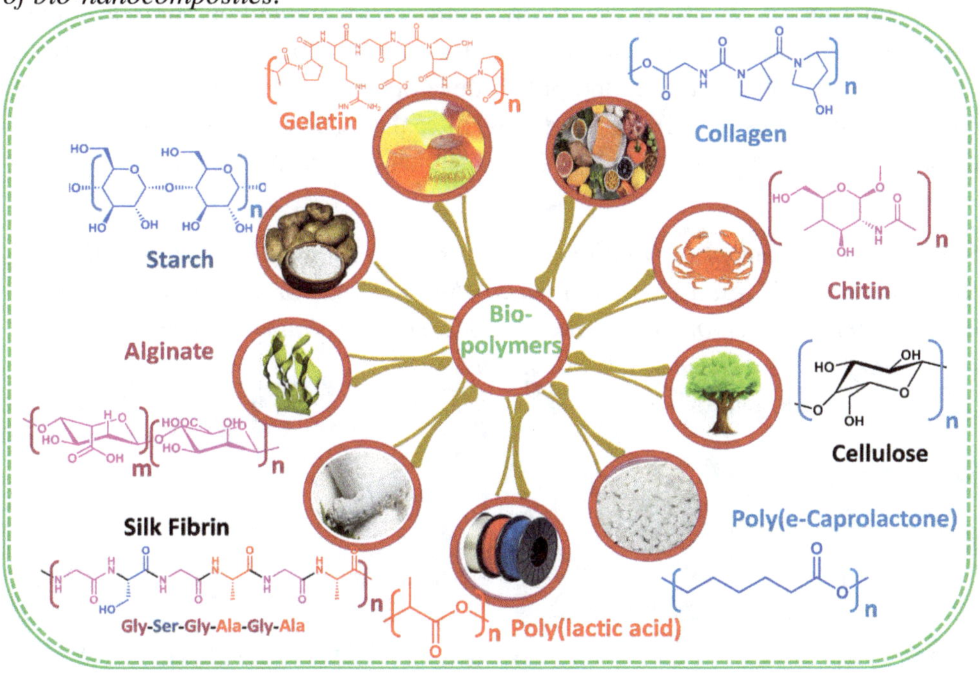

The environmental footprint of bio-nanocomposites is generally favorable, but it warrants a comprehensive life-cycle assessment to substantiate sustainability claims. Biopolymers inherently offer biodegradability and reduced greenhouse gas emissions relative to petroleum-based polymers; however, the production of nano-fillers, especially via chemical-intensive or energy-intensive methods, can partially counteract these benefits of not managed with proper mitigation strategies. Acid hydrolysis routes for cellulose nanocrystals and oxidative processes for graphene derivatives may generate hazardous byproducts unless effectively recovered or treated. Furthermore, the presence of inorganic nanofillers can modify the composability of the final composite, potentially altering biodegradation rates or contributing to inorganic residue accumulation in soil environments. These factors underscore the importance of designing bio-nanocomposites in line with green chemistry principles and circular economy models.

Integration of valorised waste streams as nanofillers, such as lignin-derived nanoparticles or agricultural ash-based materials, represents an attractive route to close material loops and minimize environmental burdens. Additionally, incorporating renewable energy inputs for processing and adopting solvent-free or low-solvent manufacturing practices can further align these materials with broader climate and sustainability targets. Regulatory frameworks and eco-labelling standards will be decisive in supporting responsible market deployment, particularly for food contact materials, biomedical devices, and other sensitive applications. Coordinated efforts in research, industry practice, and policy development will be essential to enable bio-nanocomposites to fulfil their potential as scalable, cost-effective, and environmentally preferable alternatives within a sustainable materials economy.

Figure 12. Overview of key application areas for bio-nanocomposites synthesized from biorenewable resources, emphasizing their potential across various industrial and environmental sectors. Reproduced with permission from Ref (Ates et al., 2020), Copyright @2020 American Chemical Society.

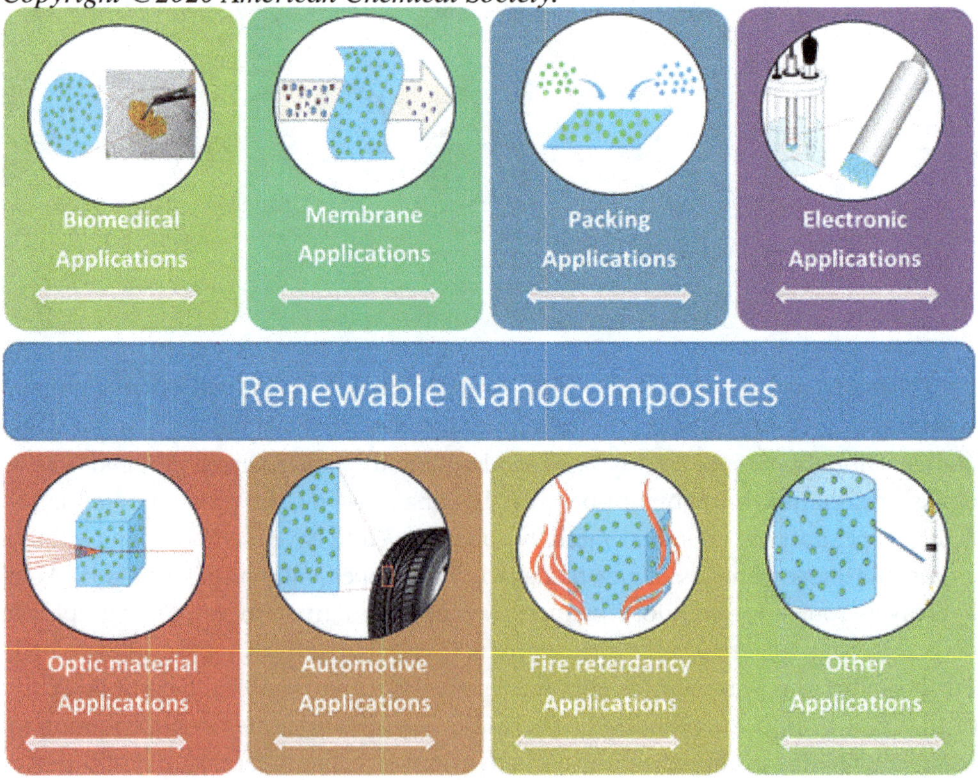

4.6 CONCLUSION

Sustainable development requires the fabrication of biodegradable materials from abundantly available renewable resources. Bio-nanocomposites manufactured by nanomaterials reinforced in natural polymer matrix materials with enhanced mechanical and thermal properties, stability, and durability than pure natural polymers, thus making them to application in numerous applications. Several ceramics, metal, inorganic, carbon, natural polymer nanomaterials are reinforced in the cellulose, chitosan, starch, proteins, PLA matrix materials to fabricate novel materials with variety of applications. Owing to the excellent performance, bio-nanocomposites based on graphene reinforcement showed promising applications in antimicrobials, biosensors, skin grafts, orthopaedic implants, drug delivery systems, modified electrodes and packaging materials, energy storage systems, supercapacitors, dye adsorption and removal, metal ion removal from water, protection from corrosion, furniture, house hold utensils, lubricants, etc. as represented in Fig. 4.12. Therefore, this field constantly requires more scientific inputs to develop and use of the biodegradable composite materials to control the pollution in the sustainable development of the modern world.

REFERENCES

Adrar, S., & Ajji, A. (2024). Effect of blending sequence and epoxy functionalized compatibilizer on barrier and mechanical properties of PBS and PBS/PLA nanocomposite blown films. *Journal of Vinyl and Additive Technology*, *30*(4), 883–894.

Agwuncha, S. C., Sadiku, E. R., Ibrahim, I. D., Aderibigbe, B. A., Owonubi, S. J., Agboola, O., Reddy, A. B., Bandla, M., Varaprasad, K., & Bayode, B. L. (2017). Poly (lactic acid) biopolymer composites and nanocomposites for biomedicals and biopackaging applications. *Handb Compos Renew Mater*, *8*, 135–170.

Akinnawo, S. O. (2024). Advance Nanocomposites from Biopolymers and Fillers: Sources, Characterization, and End-Use Applications. *Polymer-Plastics Technology and Materials*, *63*(5), 570–604.

Alakrach, A. M., Al-Rashdi, A. A., Alqadi, T., Saadi, M. A. A., Ting, S. S., Dahham, O. S., & Zulkepli, N. N. (2021). Polylactic Acid/Halloysite Nanotubes-Titanium Dioxide Nanocomposites: Antibacterial and Mechanical Properties. *Materials Science Forum*, *1021*, 270–279.

Anwer, A. H., Ahtesham, A., Shoeb, M., Mashkoor, F., Ansari, M. Z., Zhu, S., & Jeong, C. (2023). State-of-the-art advances in nanocomposite and bio-nanocomposite polymeric materials: A comprehensive review. *Advances in Colloid and Interface Science*, *318*, 102955. PMID: 37467558

Ates, B., Koytepe, S., Ulu, A., Gurses, C., & Thakur, V. K. (2020). Chemistry, Structures, and Advanced Applications of Nanocomposites from Biorenewable Resources. *Chemical Reviews*, *120*(17), 9304–9362. PMID: 32786427

Azadi, E., Dinari, M., Derakhshani, M., Reid, K. R., & Karimi, B. (2024). Sources and Extraction of Biopolymers and Manufacturing of Bio-Based Nanocomposites for Different Applications. *Molecules (Basel, Switzerland)*, *29*(18), 4406. PMID: 39339400

Azevedo, V. M., Dias, M. V., Borges, S. V., Costa, A. L. R., Silva, E. K., Medeiros, É. A. A., & Nilda de Fátima, F. S. (2015). Development of whey protein isolate bio-nanocomposites: Effect of montmorillonite and citric acid on structural, thermal, morphological and mechanical properties. *Food Hydrocolloids*, *48*, 179–188.

Baradaran, S., Moghaddam, E., Basirun, W. J., Mehrali, M., Sookhakian, M., Hamdi, M., Moghaddam, M. N., & Alias, Y. (2014). Mechanical properties and biomedical applications of a nanotube hydroxyapatite-reduced graphene oxide composite. *Carbon*, *69*, 32–45.

Chandu, B., Kurmarayuni, C. M., Kurapati, S., & Bollikolla, H. B. (2020). Green and economical synthesis of graphene–silver nanocomposite exhibiting excellent photocatalytic efficiency. *Carbon Letters*, *30*(2), 225–233.

Chen, H., Wang, J., Cheng, Y., Wang, C., Liu, H., Bian, H., Pan, Y., Sun, J., & Han, W. (2019). Application of Protein-Based Films and Coatings for Food Packaging: A Review. *Polymers*, *11*(12), 2039. PMID: 31835317

Chen, J., Huang, W., Chen, Y., Zhou, Z., Liu, H., Zhang, W., & Huang, J. (2022). Facile Preparation of Chitosan-Based Composite Film with Good Mechanical Strength and Flame Retardancy. *Polymers*, *14*(7), 1337. PMID: 35406210

Csiszár, E., Kun, D., & Fekete, E. (2021a). The Role of Structure and Interactions in Thermoplastic Starch–Nanocellulose Composites. *Polymers*, *13*(18), 18. PMID: 34578087

Cui, C., Shao, C., Meng, L., & Yang, J. (2019). High-Strength, Self-Adhesive, and Strain-Sensitive Chitosan/Poly(acrylic acid) Double-Network Nanocomposite Hydrogels Fabricated by Salt-Soaking Strategy for Flexible Sensors. *ACS Applied Materials & Interfaces*, *11*(42), 39228–39237. PMID: 31550132

Deng, J., Zhu, E.-Q., Xu, G.-F., Naik, N., Murugadoss, V., Ma, M.-G., Guo, Z., & Shi, Z.-J. (2022). Overview of renewable polysaccharide-based composites for biodegradable food packaging applications. *Green Chemistry*, *24*(2), 480–492.

El Achaby, M., Essamlali, Y., El Miri, N., Snik, A., Abdelouahdi, K., Fihri, A., Zahouily, M., & Solhy, A. (2014). Graphene oxide reinforced chitosan/polyvinylpyrrolidone polymer bio-nanocomposites. *Journal of Applied Polymer Science*, *131*(22).

Falua, K. J., Pokharel, A., Babaei-Ghazvini, A., Ai, Y., & Acharya, B. (2022). Valorization of Starch to Biobased Materials: A Review. *Polymers*, *14*(11), 2215. PMID: 35683888

Farooq, K., Kumawat, Y. K., Sharma, K., & Kumar, V. (2024). Bio-nanocomposites: Fundamentals and Recent Advances. In Sharma, K., Tiwari, S. K., Kumar, V., & Kalia, S. (Eds.), *Novel Bio-nanocomposites for Biomedical Applications* (pp. 1–35). Springer International Publishing.

Farrokhi, H., Koosha, M., Nasirizadeh, N., Salari, M., Abdouss, M., Li, T., & Gong, Y. (2024). The Effect of Nanoclay Type on the Mechanical Properties and Antibacterial Activity of Chitosan/PVA Nanocomposite Films. *Journal of Composites Science*, *8*(7), 255.

Fathi, M., Karim, M., & Ahmadi, N. (2019). Nanostructures of cellulose for encapsulation of food ingredients. In *Biopolymer Nanostructures for Food Encapsulation Purposes* (pp. 493–519). Elsevier.

Feng, Y., Zhang, X., Shen, Y., Yoshino, K., & Feng, W. (2012). A mechanically strong, flexible and conductive film based on bacterial cellulose/graphene nanocomposite. *Carbohydrate Polymers, 87*(1), 644–649. PMID: 34663016

Guirguis, O., & Alharbi, N. D. (2024). Enhancement of Molecular Structural and Linear/Nonlinear Optical Features of Chitosan/Titanium Dioxide Nanocomposite Films for Food Packaging and Optoelectronic Applications. *East European Journal of Physics, 4*, 469–490.

Gupta, R., Swarupa, S., Mayya, C., Bhatia, D., & Thareja, P. (2023). Graphene Oxide–Carbamoylated Chitosan Hydrogels with Tunable Mechanical Properties for Biological Applications. *ACS Applied Bio Materials, 6*(2), 578–590. PMID: 36655342

Haider, T. P., Völker, C., Kramm, J., Landfester, K., & Wurm, F. R. (2019). Plastics of the Future? The Impact of Biodegradable Polymers on the Environment and on Society. *Angewandte Chemie International Edition, 58*(1), 50–62. PMID: 29972726

Iroegbu, A. O. C., & Ray, S. S. (2022). Recent developments and future perspectives of biorenewable nanocomposites for advanced applications. *Nanotechnology Reviews, 11*(1), 1696–1721.

Jawaid, M., Salit, M. S., & Alothman, O. Y. (2017). *Green biocomposites: Design and applications*. Springer.

Jose, J., & Al-Harthi, M. A. (2017). Citric acid crosslinking of poly (vinyl alcohol)/starch/graphene nanocomposites for superior properties. *Iranian Polymer Journal, 26*(8), 579–587.

Joseph, B., Krishnan, S., Sagarika, V. K., Tharayil, A., Kalarikkal, N., & Thomas, S. (2020). Bionanocomposites as industrial materials, current and future perspectives: A review. *Emergent Materials, 3*(5), 711–725.

Joseph, B., Sagarika, V. K., Sabu, C., Kalarikkal, N., & Thomas, S. (2020). Cellulose nanocomposites: Fabrication and biomedical applications. *Journal of Bioresources and Bioproducts, 5*(4), 223–237.

Kargarzadeh, H., Huang, J., Lin, N., Ahmad, I., Mariano, M., Dufresne, A., Thomas, S., & Gałęski, A. (2018). Recent developments in nanocellulose-based biodegradable polymers, thermoplastic polymers, and porous nanocomposites. *Progress in Polymer Science, 87*, 197–227.

Kasaai, M. R. (2024). Biopolymers as Fillers of Biopolymer Matrices to Fabricate Green Nano-composites for Food Coating and Packaging: A Review. In *Handbook of Nanofillers* (pp. 1–14). Springer.

Kedang, Y. I., Priyangga, A., Atmaja, L., & Santoso, M. (2022). Characteristics and performance studies of a composite polymer electrolyte membrane based on chitosan/glycerol–sulfosuccinic acid modified montmorillonite clay. *RSC Advances*, *12*(47), 30742–30753. PMID: 36349150

Khan, Z. I., Mohsin, M. E. A., Habib, U., Mousa, S., Hossain, S. S., Ali, S. S., Mohamad, Z., & Othman, N. (2025). Enhanced Mechanical and Thermal Performance of Sustainable RPET/PA-11/Joncryl® Nanocomposites Reinforced with Halloysite Nanotubes. *Polymers*, *17*(11), 1433. PMID: 40508677

Kim, C.-J., Khan, W., Kim, D.-H., Cho, K.-S., & Park, S.-Y. (2011). Graphene oxide/cellulose composite using NMMO monohydrate. *Carbohydrate Polymers*, *86*(2), 903–909.

Konwar, A., Kalita, S., Kotoky, J., & Chowdhury, D. (2016). Chitosan–Iron Oxide Coated Graphene Oxide Nanocomposite Hydrogel: A Robust and Soft Antimicrobial Biofilm. *ACS Applied Materials & Interfaces*, *8*(32), 20625–20634. PMID: 27438339

Kumar, P., Sandeep, K., Alavi, S., Truong, V., & Gorga, R. (2010). Preparation and characterization of bio-nanocomposite films based on soy protein isolate and montmorillonite using melt extrusion. *Journal of Food Engineering*, *100*(3), 480–489.

Li, R., Liu, C., & Ma, J. (2011). Studies on the properties of graphene oxide-reinforced starch biocomposites. *Carbohydrate Polymers*, *84*(1), 631–637.

Lin, N., & Dufresne, A. (2014). Nanocellulose in biomedicine: Current status and future prospect. *European Polymer Journal*, *59*, 302–325.

Liu, C., Shen, J., Yeung, K. W. K., & Tjong, S. C. (2017). Development and antibacterial performance of novel polylactic acid-graphene oxide-silver nanoparticle hybrid nanocomposite mats prepared by electrospinning. *ACS Biomaterials Science & Engineering*, *3*(3), 471–486. PMID: 33465942

Liu, H., Li, C., Chen, S., Chen, P., Li, J., Jian, H., Guo, G., Chen, X., Zhu, X., & Wu, J. (2023). Fabrication of 3D Printed Polylactic Acid/Polycaprolactone Nanocomposites with Favorable Thermo-Responsive Cyclic Shape Memory Effects, and Crystallization and Mechanical Properties. *Polymers*, *15*(6), 1533. PMID: 36987315

Lonkar, S. P., Kumar, A. P., & Singh, R. P. (2011). 16 Commercial aspects associated with bio-nanocomposites. In V. Mittal (Ed.), *Nanocomposites with Biodegradable Polymers: Synthesis, Properties, and Future Perspectives* (p. 0). Oxford University Press.

Mianehrow, H., Re, G. L., Carosio, F., Fina, A., Larsson, P. T., Chen, P., & Berglund, L. A. (2020). Strong reinforcement effects in 2D cellulose nanofibril–graphene oxide (CNF–GO) nanocomposites due to GO-induced CNF ordering. *Journal of Materials Chemistry. A, Materials for Energy and Sustainability*, *8*(34), 17608–17620. PMID: 33796318

Miyashiro, D., Hamano, R., & Umemura, K. (2020). A review of applications using mixed materials of cellulose, nanocellulose and carbon nanotubes. *Nanomaterials (Basel, Switzerland)*, *10*(2), 186. PMID: 31973149

Mokhena, T. C., Sefadi, J. S., Sadiku, E. R., John, M. J., Mochane, M. J., & Mtibe, A. (2018). Thermoplastic Processing of PLA/Cellulose Nanomaterials Composites. *Polymers*, *10*(12), 1363. PMID: 30961288

Motloung, M. P., Ojijo, V., Bandyopadhyay, J., & Ray, S. S. (2019). Cellulose Nanostructure-Based Biodegradable Nanocomposite Foams: A Brief Overview on the Recent Advancements and Perspectives. *Polymers*, *11*(8), 1270. PMID: 31370292

Namasivayam Sukumaar, V., Arjunan, S., Pandiaraj, L. N., & Narayanan, A. (2025). Experimental investigation of 3D printed polylactic acid and polylactic acid–hydroxyapatite composite through material extrusion technique for biomedical application. *Journal of Thermoplastic Composite Materials*, *38*(4), 1303–1333.

Omigbodun, F. T., Oladapo, B. I., & Osa-uwagboe, N. (2024). Exploring the frontier of Polylactic Acid/Hydroxyapatite composites in bone regeneration and their revolutionary biomedical applications – A review. *Journal of Reinforced Plastics and Composites*, •••, 07316844241278045.

Pan, Y., Wu, T., Bao, H., & Li, L. (2011). Green fabrication of chitosan films reinforced with parallel aligned graphene oxide. *Carbohydrate Polymers*, *83*(4), 1908–1915.

Parangusan, H., Bhadra, J., Ahmad, Z., Mallick, S., Touati, F., & Al-Thani, N. (2021). Humidity sensor based on poly (lactic acid)/PANI–ZnO composite electrospun fibers. *RSC Advances*, *11*(46), 28735–28743. PMID: 35478584

Ragab, H. M., Diab, N. S., AlElaimi, M., Alghamdi, A. M., Farea, M. O., & Farea, A. (2024). Fabrication and Characterization of Silver Nanoparticle-Doped Chitosan/Carboxymethyl Cellulose Nanocomposites for Optoelectronic and Biological Applications. *ACS Omega*, *9*(20), 22112–22122. PMID: 38799360

Ramesan, M. T., Sankar, S., Kalladi, A. J., Labeeba Abdulla, A. C., & Bahuleyan, B. K. (2024). Hydroxyapatite nanoparticles reinforced polyvinyl alcohol/chitosan blend for optical and energy storage applications. *Polymer Engineering and Science*, *64*(3), 1378–1390.

Ray, S. S., & Bousmina, M. (2005). Biodegradable polymers and their layered silicate nanocomposites: In greening the 21st century materials world. *Progress in Materials Science*, *50*(8), 962–1079.

Redondo, A., Mortensen, N., Djeghdi, K., Jang, D., Ortuso, R. D., Weder, C., Korley, L. T. J., Steiner, U., & Gunkel, I. (2022). Comparing Percolation and Alignment of Cellulose Nanocrystals for the Reinforcement of Polyurethane Nanocomposites. *ACS Applied Materials & Interfaces*, *14*(5), 7270–7282. PMID: 35077647

Ren, J. (2011). *Biodegradable poly (lactic acid): Synthesis, modification, processing and applications*. Springer Science & Business Media.

Rojas, J., Bedoya, M., & Ciro, Y. (2015). Current trends in the production of cellulose nanoparticles and nanocomposites for biomedical applications. *Cellulose-Fundamental Aspects and Current Trends*, 193–228.

Ruiz Gómez, E. E., Mina Hernández, J. H., & Diosa Astaiza, J. E. (2020). Development of a Chitosan/PVA/TiO2 Nanocomposite for Application as a Solid Polymeric Electrolyte in Fuel Cells. *Polymers*, *12*(8), 1691. PMID: 32751167

Sandhya, P., Sreekala, M., Padmanabhan, M., Jesitha, K., & Thomas, S. (2019). Effect of starch reduced graphene oxide on thermal and mechanical properties of phenol formaldehyde resin nanocomposites. *Composites. Part B, Engineering*, *167*, 83–92.

Shar, A. S., Zhang, C., Song, X., Weng, Y., & Du, Q. (2021). Design of Novel PLA/OMMT Films with Improved Gas Barrier and Mechanical Properties by Intercalating OMMT Interlayer with High Gas Barrier Polymers. *Polymers*, *13*(22), 3962. PMID: 34833261

Sharma, A., Thakur, M., Bhattacharya, M., Mandal, T., & Goswami, S. (2019). Commercial application of cellulose nano-composites–A review. *Biotechnology Reports (Amsterdam, Netherlands)*, *21*, e00316. PMID: 30847286

Sharma, N., & Rana, V. S. (2020). A Review on Polysaccharide Based Nanocomposite Hydrogel Systems Fabrication Using Diverse Reinforcing Materials. *Journal of Polymer & Composites*, *8*(1), 6–17.

Sheshmani, S., Ashori, A., & Hasanzadeh, S. (2014). Removal of Acid Orange 7 from aqueous solution using magnetic graphene/chitosan: A promising nano-adsorbent. *International Journal of Biological Macromolecules*, *68*, 218–224. PMID: 24813679

Shih, Y. F., & Jeng, R. J. (2011). Biodegradable poly(butylene succinate)/multi-walled carbon nanotube nanocomposites. In Mittal, V. (Ed.), *Nanocomposites with Biodegradable Polymers: Synthesis, Properties, and Future Perspectives*. Oxford University Press.

Singh, S., Patel, M., Schwendemann, D., Zaccone, M., Geng, S., Maspoch, M. L., & Oksman, K. (2020). Effect of chitin nanocrystals on crystallization and properties of poly (lactic acid)-based nanocomposites. *Polymers*, *12*(3), 726. PMID: 32214000

Siró, I., & Plackett, D. (2010). Microfibrillated cellulose and new nanocomposite materials: A review. *Cellulose (London, England)*, *17*(3), 459–494.

Surya Rao, G., Debnath, K., & Mahapatra, R. (2021). Development and Characterization of PLA-Based Green Composites: Experimental and Simulation Studies. In *Green Composites* (pp. 209–223). Springer.

Suryanegara, L., Nakagaito, A. N., & Yano, H. (2009). The effect of crystallization of PLA on the thermal and mechanical properties of microfibrillated cellulose-reinforced PLA composites. *Composites Science and Technology*, *69*(7–8), 1187–1192.

Swain, S. K., Sethy, P. K., & Pattanayak, A. J. (2017a). Synthesis of Soy Protein Based Biocomposites for Packaging Applications. In Jawaid, M., Salit, M. S., & Alothman, O. Y. (Eds.), *Green Biocomposites: Design and Applications* (pp. 143–166). Springer International Publishing.

Swain, S. K., Sethy, P. K., & Pattanayak, A. J. (2017b). Synthesis of soy protein based biocomposites for packaging applications. In *Green Biocomposites* (pp. 143–166). Springer.

Tabassum, Z., Girdhar, M., Kumar, A., Malik, T., & Mohan, A. (2023). ZnO Nanoparticles-Reinforced Chitosan–Xanthan Gum Blend Novel Film with Enhanced Properties and Degradability for Application in Food Packaging. *ACS Omega*, *8*(34), 31318–31332. PMID: 37663466

Thomas, D., & Thomas, S. (2013). Chemical modification of chitosan and its biomedical application. *Biopolymer Nanocomposites: Processing, Properties, and Applications*, 33–51.

Thomas, L. M., & Sreekanth, M. S. (2025). Enhanced water-induced shape memory and swelling characteristics of graphene oxide-cellulose nanofiber reinforced polyvinyl alcohol nanocomposites. *Journal of Vinyl and Additive Technology*, *31*(3), 728–755.

Usman, A., Hussain, Z., Riaz, A., & Khan, A. N. (2016). Enhanced mechanical, thermal and antimicrobial properties of poly (vinyl alcohol)/graphene oxide/starch/ silver nanocomposites films. *Carbohydrate Polymers*, *153*, 592–599. PMID: 27561532

Wang, B., Torres-Rendon, J. G., Yu, J., Zhang, Y., & Walther, A. (2015). Aligned Bioinspired Cellulose Nanocrystal-Based Nanocomposites with Synergetic Mechanical Properties and Improved Hygromechanical Performance. *ACS Applied Materials & Interfaces*, *7*(8), 4595–4607. PMID: 25646801

Wang, G., Li, L., Medina, L., Harish, S., Liu, J., Xu, B., Guo, R., Shao, C., Kodama, T., Berglund, L. A., & Shiomi, J. (2025). Large Anisotropy of Thermal Conductivity in Oriented Cellulose–Clay Composites. *ACS Omega*, *10*(25), 26560–26566. PMID: 40620995

Wang, N., Ding, E., & Cheng, R. (2007). Surface modification of cellulose nanocrystals. *Frontiers of Chemical Engineering in China*, *1*(3), 228–232.

Wang, X., Bai, H., Yao, Z., Liu, A., & Shi, G. (2010). Electrically conductive and mechanically strong biomimetic chitosan/reduced graphene oxide composite films. *Journal of Materials Chemistry*, *20*(41), 9032–9036.

Wu, B., Wu, Y., Zhang, M., Guo, H., Liu, T., Lin, G., & Kuang, T. (2025). Enhanced Mechanical and Multifunctional Properties of GNPs/CNTs Hybridized PLA Nanocomposites by Implementing Dual-Processing of Pickering Emulsion-Melt Blending Methods. *Macromolecular Materials and Engineering*, *310*(3), 2400306.

Xu, D., Wang, S., Berglund, L. A., & Zhou, Q. (2021). Surface Charges Control the Structure and Properties of Layered Nanocomposite of Cellulose Nanofibrils and Clay Platelets. *ACS Applied Materials & Interfaces*, *13*(3), 4463–4472. PMID: 33428385

Young, R. J., Kinloch, I. A., Gong, L., & Novoselov, K. S. (2012). The mechanics of graphene nanocomposites: A review. *Composites Science and Technology*, *72*(12), 1459–1476.

Yu, J., Wang, D., Geetha, N., Khawar, K. M., Jogaiah, S., & Mujtaba, M. (2021). Current trends and challenges in the synthesis and applications of chitosan-based nanocomposites for plants: A review. *Carbohydrate Polymers*, *261*, 117904. PMID: 33766382

Zhao, R., Torley, P., & Halley, P. J. (2008). Emerging biodegradable materials: Starch-and protein-based bio-nanocomposites. *Journal of Materials Science*, *43*(9), 3058–3071.

Zhou, Y., Tang, S., Zhang, Q., Jiang, Y., Tan, L., Xia, Z., Wang, Z., Liu, L., Zhang, Y., & Wang, G. (2024). Preparation and Characterization of Cellulose Nanocrystal-Doped Cellulose Triacetate Optical Films by Sol-Gel Process. *ChemistrySelect*, *9*(27), e202400262.

Zhuang, W., Liu, J., Zhang, J. H., Hu, B. X., & Shen, J. (2009). Preparation, characterization, and properties of TiO2/PLA nanocomposites by in situ polymerization. *Polymer Composites*, *30*(8), 1074–1080.

Ziabari, M., Mottaghitalab, V., & Haghi, A. K. (2009). Application of direct tracking method for measuring electrospun nanofiber diameter. *Brazilian Journal of Chemical Engineering*, *26*, 53–62.

Chapter 3
Practical Utility and Basic Properties of Thermoplastic–Based Bio Nanocomposites

Y avasn Maruthi
https://orcid.org/0000-0003-2634-8382
Krishna University, India

T. Srinivas
GITAM University, India

ABSTRACT

This chapter explores the innovative use of thermoplastic-based bio-nanocomposites as sustainable alternatives to traditional synthetic materials. It explains what bio-nanocomposites and thermoplastics are, and how combining biopolymers with inorganic fillers at a tiny scale creates a strong structure. Fundamental properties such as biocompatibility, biodegradability, mechanical strength, and swelling capacity are examined to demonstrate their industrial relevance. The chapter highlights the practical applications of key matrices like chitosan, cellulose, and starch in areas including food packaging, wastewater treatment, electronics, and construction. Case studies and recent research illustrate how these composites can outperform conventional materials. It also addresses current challenges such as scalability, mechanical limitations, and environmental concerns, positioning these composites as key players in sustainable innovation. Concluding remarks point out the need for continued research to develop enhanced formulations that meet the growing demand for eco-friendly solutions.

DOI: 10.4018/979-8-3373-2180-6.ch003

1. INTRODUCTION

To protect himself from natural disasters and sustain his way of life, man switched his lifestyle from nomadic to native, focusing on natural materials that are hard-wearing and sturdy. Over time, he began producing long-lasting materials including plastics, nanocomposites, and so on. At least one of the components in nanocomposites is a nanomaterial, and the other components have varied physical and chemical properties. The qualities of nanocomposites materials are designed to transcend the capabilities of the sum of their parts (Saini, Bajpai, & Jain, 2017). Eventually, Man began to realize that the user-friendly applications of these manufactured materials, such as Plastics and nanocomposites, are jeopardising life on Earth. Then the concept of sustainability emerged into the limelight, and scientists began to work on sustainable materials like Bionanocomposites. Nanoscale integration of biopolymers such as polysaccharides, nucleic acids, and proteins with inorganic solids is giving rise to a new category of materials known as Bionano composites (1–100 nm). The term "Bionanocomposites" was coined by Theng in 1970 (1). Mineralization of macromolecules is a common method for preparing Bionanocomposites (Shchipunov, 2012). Bionanocomposites can be used in a wide range of industries, including food packaging, automobiles, cosmetics, electronics, textiles, tissue engineering, water purification and many more (Figure-1) (Joseph et al., 2020)

Figure 1. Applications of Bionanocomposites

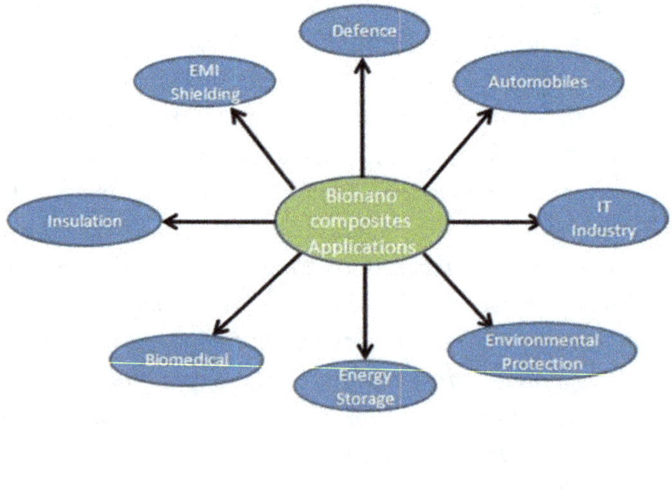

For example, aliphatic polyesters, proteins, and polypeptides polynucleic acids and, polysaccharides are examples of suitable matrix and fillers (hydroxyapatite, and clay nanoparticles and metal nanoparticles), scientists also learn to modify the chemical and structural properties of these materials to meet the needs of the target field (Sreenivasulu, Pothula, & Reddy, 2016). Biocompatibility, and biodegradability, in some circumstances, functional qualities given by either the biological or inorganic moieties are significant advantages of Bionanocomposites. The creation of biopolymer-based nanocomposites that exhibit the well-known features of nanocomposites formed from synthetic polymers is currently taking up a significant portion of this effort (improved mechanical properties, gas-barrier properties, and higher thermal stability) (Kushwaha & Srivastava, 2018).

The dictionary meaning of thermoplastics is “Capable of softening or fusing when heated and of hardening again when cooled” or “A thermoplastic is a class of polymer that can be softened through heating and then processed using methods such as extrusion, injection moulding, thermoforming and blow moulding”. Despite being repeatedly heated and cooled, thermoplastics don't lose any of their chemical properties, making them a convenient material to recycle. Polyethylene is the most commonly used thermoplastic in the planet (TWI Global, n.d.).

One of todays and tomorrow’s major issues is material and energy sufficiency. Lightweight and environmentally friendly goods save both energy and resources. Modern thermoplastic based Bionanocomposites composites are critical in this application. Polymers such as Chitosan, starch, pectin, poly (lactic acid), cellulose, gelatine, poly (hydroxy alkanoates), gaur gum etc and are increasingly being used in composite technology to substitute non-biodegradable polymers(Figure-2 & Figure-3). Even though they are made from fossil fuels, some polymers such as poly(caprolactone) and poly(butylene succinate) are biodegradable (Mittal, 2011).

Figure 2. Types of Bionanocomposites

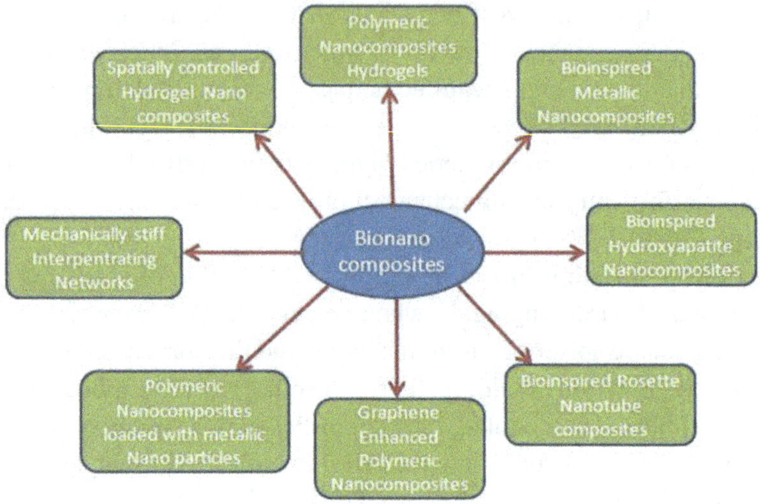

Figure 3. Bionanocomposites and their Processing Techniques

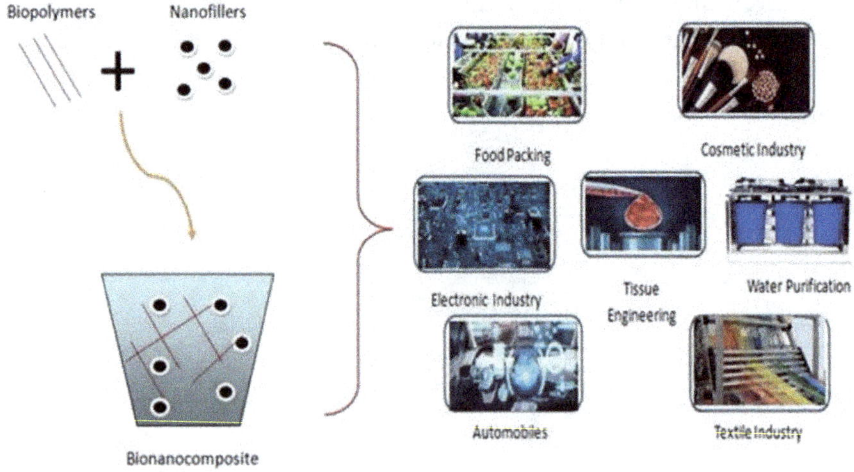

2. BASIC PROPERTIES OF THERMOPLASTIC – BASED BIONANOCOMPOSITES

Natural and synthetic materials can be combined to form these Bionanocomposites, which can be used in many different ways. Natural inorganic solids like clays can be used to make Bionanocomposites (bioplastics) with nanocomposites that are biocompatible and biodegradable in nature (Park et al., 2004). Many Bionanocomposites are currently being employed in biomedical applications as biomaterials. To qualify as biomaterials, these materials must meet the following criteria:

A. Biocompatibility:

Materials' biocompatibility refers to their suitability for a specific application and their ability to operate with an adequate host response in a given situation without causing harm to biological functions (Williams, 2008; Anderson, 2001). Sterilizability, Nontoxicity, functionality and immunogenicity are further desired properties for biomedical nanocomposites to possess in addition to biocompatibility. Determining the biocompatibility of materials and implanted devices calls for extensive testing (e.g., density; bulk and surface chemical composition; porosity; and mechanical, degradation, and electrical properties), starting with protein/tissue/cell or in vitro studies, later moving on to in vivo studies in animals and finally into human clinical trials (Paulo, 2005).

Polymer biomaterials' biocompatibility is affected by a number of factors, including their molecular weight, conformational flexibility, surface topography, and roughness, and the interfacial free energy between the surface and the surrounding medium (Shchipunov, 2012).

B. Biodegradability:

Biodegradability is the potential of organic materials to be biologically degraded by living organisms into their component parts, like water, methane, carbon dioxide, basic elements, and biomass. Researchers have recently They centred their emphasis on biodegradable Biopolymer-based nanocomposites degrade naturally in the body, removing the necessity to eliminate a drug delivery system following the active ingredient has been released. Through regular use, biopolymer-based nanocomposites materials lose their original form, mechanical strength, and integrity (Alexis, 2005). The types of polymer degradation are biodegradation, optical, mechanical, oxidative, thermal, radiation, hydrolytic, and chemical (Grassie, 1985). Degradable polymers are hypothesised to erode via two distinct mechanisms: (a) heterogeneous or surface erosion, and (b) homogeneous or bulk erosion. The rate at which polymer degradation occurs in surface-eroding polymer matrices is much higher than the rate at which water penetrates into matrix of the polymer. Therefore, the outermost polymer layers are where the degradation occurs. The internal parts of a matrix are unaffected by erosion, as only the exterior layer is exposed to the elements of the environment (heterogeneous process). While the system can absorb water at a much faster rate than bulk-eroding polymers degrade, the reverse is not true. This results in a rapid hydration of the entire system and the widespread rupturing of polymer chains. As a result, erosion is not a surface phenomenon (homogeneous process) but rather occurs throughout the polymer (Ranade & Hollinger, 2003).

C. Mechanical Property:

The mechanical characteristics of biomaterials are critical for medicinal applications. As an illustration, it's critical to ensure the medication delivery device's integrity throughout its useful life. The hydrogel's mechanical properties have been improved by altering the cross-linking degree. The gel will be stronger if the cross-linking degree is increased. A more cross-linked structure, on the other hand, is more fragile. Cross-linking must be optimised in order to produce a hydrogel that is both moderately robust and elastic. To attain the appropriate mechanical properties of hydrogels, copolymerization has also been employed. An H-bonding comonomer can be added to hydrogel to boost its strength (Peppas et al., 2000).

D. Swelling Property:

Hydrogels comprised of polyelectrolytes swell more as a result of the charge repulsion between polymer chains, and this swelling feature is beneficial in the environment-sensitive swelling of hydrogels for controlled drug release (Holback, Yeo, & Park, 2011). Along with environmental factors, the cross-linking ratio is a decisive factor in how much a mixture will expand. For polymers, the cross-linking ratio refers to the number of moles of cross-linking agent for every mole of polymer repeating units. Hydrogels with a higher cross-linking ratio have more cross-linking agents incorporated into their structure. Highly cross-linked blends have less swelling and a tighter structure than blends with lower cross-linking ratios. As cross-linking slows the movement of polymer chains, the swelling to volume ratio decreases. Highly absorbent polymers, sometimes known as "superabsorbent polymers," have attracted substantial interest as potential biological tools because of their ability to take up and hold huge volumes of water. "Infant diapers," "incontinence products" and "feminine hygiene products" are just some of the many commercial uses of these materials. Radiation grafting is an effective method for preparing these materials (Shchipunov, 2012; Kiatkamjorwong & Meechi, 1997).

3. APPLICATIONS OF THERMOPLASTIC – BASED BIONANOCOMPOSITES

3.1 Chitosan

Crustacean exoskeletons are predominantly made of chitin, which can be deacetylated to produce the cationic polysaccharide known as chitosan. The organic component of such shells, called chitin, makes up roughly 70% of the total. In contrast to fossil fuels, chitosan is a sustainable material that can be replenished over time (shrimp and crab shells). Food industry particularly food packaging and preservations can benefit from chitosan's wide range of desirable qualities (Johannson, 2011).

3.1.1 Food Packaging Applications

Now a days there has been a significant shift away from using plastic packaging toward using biodegradable materials derived from renewable sources in order to reduce packaging waste. However, weak mechanical qualities and susceptible to moisture have hampered the effectiveness of biopolymer films in food packaging. One promising solution to this problem is to use nanotechnology to include nanosized reinforcing fillers into biopolymer matrices, therefore increasing their barrier and tensile properties (Lee et al., 1997).

Biocomposites were made by combining plasticized tributyl o-acetyl citrate (ATBC) and poly(lactic acid (PLA), and the concentration of chitosan (CH) was investigated for its effect on the biocomposites' characteristics and antifungal/antimicrobial activity. ATBC was discovered to lessen the brittleness of PLA, and the films and sheets made from plasticized CH/ PLA biocomposites have good transparency, thermal, and mechanical properties. Cucumber and radish seedlings germinate normally when exposed to the PLA/CH biocomposites. Out of two food emulants, one adjusted to avoid In order to prevent overheating close to the melting point of PLA, migration experiments demonstrated biocomposites to be appropriate packaging materials for fat-free foods with pH levels over 4.5 at a cold temperature. Incorporating CH into the PLA-based materials improves their antifungal efficacy at the contact surface unfavourable to *Aspergillus brasiliensis* ATCC 16404, *Penicillium corylophilum* CBMF1, and *Fusarium graminearum* G87. Only CH-containing biocomposites significantly reduce surface contact of *S. aureus* and *E. coli*, demonstrating their antibacterial potential (Râpă et al., 2016).

Natural biopolymers like starch and chitosan can be fused with biodegradable polymers like polyhydroxylbutyrate (PHB), polylactic acid (PLA), polybutylene succinate (PBS), and poly-caprolactone (PCL) to create bio-nanocomposites with unique properties. Antimicrobial compounds added to food packaging materials prevent bacteria and other germs from growing on food, extending its shelf life. Pathogens like *Listeria monocytogenes*, *Escherichia coli*, and *Bacillus subtilis* could be stymied by using a variety of experimentally designed antimicrobial nanocomposites. These include cellulose nanofibers, polylactic acid (PLA)/halloysite, hydroxyapatite/titania, PLA/Ag nanoparticles, and layered silicate, Natural biopolymers like starch and chitosan can be fused with biodegradable polymers like polyhydroxylbutyrate (PHB), polybutylene succinate (PBS), polylactic acid (PLA), and poly-caprolactone (PCL) to create bio-nanocomposites with unique properties (Sharma et al., 2020).

3.1.2. Waste water treatment

Since chitosan has natural chelating characteristics and functional groups, it can be used to remove metal ions from wastewater by attaching to them. This makes chitosan a useful component of wastewater treatment. It can also be used for the removal of colours and other negatively charged particles from effluent channels and wastewater ponds.

Chitosan that has been grafted with poly(acrylonitrile) is further changed to produce amidoximated chitosan a derivative with more adsorption for Pb2+ Cu2+, and Mn2+, than cross-linked chitosan (Kang et al., 1996).

Because of their propensity to form host-guest complexes, polymers containing cyclodextrin (CD) are a versatile molecule with uses ranging from the stability and slow delivery of potential components in preparation, extraction and separation. Chitosan modified with beta -CD derivatives can form complexes with other compounds that are useful for creating new sorbents (Tojima et al., 1998; Sreenivasan, 1988; Chen & Wang, 2001; Aoki et al., 2003).

3.1.3. Biomedical Applications

In the medical field, chitosan is of interest for a variety of uses due to its versatility in adhesion, medication administration, wound healing, and tissue engineering (Khor & Lim, 2003; Foda, El-Laithy, & Tadros, 2007; Senel & McClure, 2004). During operations, biological adhesives are utilised to adhere tissues together, stop bleeding, and prevent the escape of air and other body fluids. Adhesions are abnormal scarring of connective tissue that binds organs together in the body. An Azide-and Lactose-modified Photocrosslinkable Chitosan (Az-CH-LA) was produced as a biological adhesive for soft tissues and its efficacy was compared to that of fibrin glue. Space-filling, gelling in minutes, and adhering to cartilage and bone in situ, the chitosan solution was developed to be cytocompatible (One et al., 2000; Hoemann et al., 2012).

Because of its functional groups, chitosan can combine with materials, allowing for a wide range of possible pairings composites using materials like hydroxyapatite and other calcium-based minerals use in the fields of orthopaedics and dentistry. Coatings based on calcium-chitosan composites have been developed together with artificial joints. Considering that chitosan is bone that has been damaged or has been degraded, new bone can be deposited next to their placement part used to fix an implant into bone. Some more Chitosan's orthopaedic applications include both indirect and direct replacement of a skeletal or fibrous nature. As a natural bioadhesive, it can be used to enhance the implant-securing bone cement as also for the purpose of filling bone cavities (Khor & Lim, 2003; Foda, El-Laithy, & Tadros, 2007; Senel & McClure, 2004).

Another feature of Chitosan is its adaptability; it can be made into gels, copolymers, etc. because of this, chitosan is a promising drug delivery medium. It can entrap negatively charged ions and form colloidal particles. Molecules by a variety of methods, including chemical and ionic crosslinking. Aside from chitosan's other useful properties, biocompatibility and other biological features lead to substances that work well as drug delivery vehicles (Prabaharan et al., 2005).

There are a variety of supplements that can aid with fat loss, includes chitosan and pyruvate. The red molecule pyruvate eating apples, some cheeses, and red wine all promote fat loss and improves physical activity capacity. Because of the chitosan bond, itself to fat in the stomach prior to digestion, therefore storing the fat and blocking it from being absorbed by the body.The fat then forms a bulk with the chitosan fibre, a substance that the human body can't digest and therefore flushes out. Fiber made from chitosan has a unique structure that sets it apart from other types of fibre. possessing a positive ionic charge that allows it to form bonds physically with the positively charged bile salts, cholesterol, and phospholipids acids (Bhattarai et al., 2005; Teo & Ramakrishna, 2006; Sun et al., 2003).

For the first time, electrospun nanofibers of a collagen-chitosan complex have been published, with the goal of creating biomaterials with functional properties that can be used in tissue engineering and in developing artificial replacements for natural extracellular matrix (Chen et al., 2007).

Chitosan can be used for a variety of purposes. As an example, they can be used to aid in solving a wide variety of issues in biomedical engineering and environmental. Chitosan is an excellent chelating agent that can be used to trap metal ions. Chitosan is a great choice for biomedical applications since it is biocompatible and has antibacterial and antifungal properties. It has been shown to be effective in expediting the healing of wounds and regenerating bone. Hydrogels and microspheres containing chitosan have shown promising results as delivery vehicles for medicines, proteins, and DNA. Additionally, chitosan provides enzymatic assembly, pH-responsive solubility, and external, electrical stimulation. These features make chitosan a highly adaptable biomaterial, suggesting that it has promising applications across a wide range of sectors (Lee et al., 1997).

3.2 Cellulose

Due to desirable characteristics (nonabrasive, economically viable, combustible, low density, non-toxic, biodegradable, etc.), polymer composites reinforced with cellulose fibres have attracted a lot of interest (Kalia et al., 2011; Kalia, Kaith, & Kaur, 2009) Applications involving nanoscale cellulose fibres have recently received a great deal of attention because of their more stiffness and strength in addition to their low weight, renewability, and biodegradability (Siro & Plackett, 2010). Extracting cellulose nanofibers from cell walls can be done mechanically, chemically, or using a combination of the two (Kalia et al., 2011).

The possible uses for nanocellulose are vastly increased. The packaging and paper industries are the most likely to benefit from nanocellulose, but it is also being researched for use in other industries, such as building materials, furniture, automobiles, electronics, personal care items, and pharmaceuticals. Companies that manufacture electroacoustic devices often use nanocellulose as a membrane because of the superior sound quality it provides. Nanocellulose is also used as an additive in high-quality e-paper, an ultrafiltration membrane, a water purification membrane, and a membrane to extract minerals and oils (Brown, 1998).

3.2.1. Biomedical Applications

Nanocellulose, as a biomaterial, has many potential applications in the medical field. This includes the replacement of burned or injured skin, the delivery of drugs, the development of new blood vessels, the repair of gum and dura, scaffolds for tissue engineering, the covering of stents, and the reconstruction of bones (Fontana et al., 1990; Mello et al., 2001; Czaja et al., 2007; Negrao et al., 2006; Klemm et al., 2001).

In addition to promoting pain relief, rapid granulation, and autolytic debridement, are crucial steps in the wound healing process that are greatly aided by the use of nanocellulose mats. This versatility of these nanobiocellulose membranes makes them useful for treating large, hard-to-access areas of the body (Kalia et al., 2011).

Nanocellulose/ poly (ethylene glycol) PEG composite scaffolds that have been prepared for cell adhesion/attachment have potential uses in wound dressing and tissue engineering (Seves et al., 2001). In ophthalmology, Huia et al. looked into the potential of using scaffold of nanocellulose for cornea tissue engineering (Hui et al., 2009; Amorim et al., 2009). Nanocellulose has also found application in nasal reconstruction. Humanity has always had a fixation on physical perfection (Amorim et al., 2009).

Dental tissue regeneration experiments using nanocellulose. Human dental tissue regeneration is possible with the help of microbial origin cellulose from the strain of *Glucanacetobacter xylinus*. Silva tested the biological performance of nanocellulose-coated implants in dental cavities made of synthetic hydroxyapatite (HAP-91) (Silva, n.d.).

Many of the drawbacks of other excipients for drug delivery can be avoided when using nanocellulose crystals. When using crystalline nanocellulose and cellulose of different forms in state-of-the-art, tablet disintegration rate, pelleting systems and incorporating microparticles allows for regulation of drug release, coating tablets with excipients (Baumann et al., 2009; Watanabe et al., 2002).

3.2.2. Electronic industrial Applications

Nanocellulose has been used in acoustic diaphragms due to its low dynamic loss and more sonic velocity and these properties have already been put to use in applications involving acoustic diaphragms. True film had a sonic velocity between that of titanium and aluminium (Iguchi, Yamanaka, & Budhiono, 2000).

As demonstrated by Yano et al., Nanocellulose material has more reinforcement, which is more in plastics materials that allow light to pass through them, for example flexible display substrate. Even at high fibre loadings, the composite maintained its optical transparency. Nanocellulose-based nanocomposite substrates systems (NC) and Boehmite-siloxane have enhanced optical visible-light-transmitting options made NC as an adjusting OLED substrate for use (Yano et al., 2005). Olson et al. (2010) showed that lyophilized cellulose nanofibril aerogels could be used as moulds to create portable highly permeable magnetic aerogels that, when compacted, would form rigid magnetic nano paper.

3.2.2. Structural Applications

When a Biocomposite is required to carry a load, we say that it is structural. Biocomposites are increasingly being used in structural applications, such as the construction industry, stairs, load-bearing walls, subflooring and roof systems. The performance of structural biocomposites can vary greatly, from that of highly effective materials to that of relatively ineffective ones. Roofing systems made from biobased composites have been evaluated (Olson et al., 2010). Structural good results have been achieved after designing, fabricating, and testing beams. a composite made of cellulose fibres and resin derived from soy oil (Dweib et al., 2006).

Biocomposites made from cellulose have a lot of potential uses. Nanocomposites, whose application is expanding rapidly. Caused by a major characteristics and uses of nanocellulose-based Materials are generally thought to fall within a broad product, including paper and packaging, automobiles, construction field, electronics, and furnishings. Additionally, pharmaceutical, biomedical, and cosmetic uses are being explored. High strength and other mechanical properties including rigidity, surface reactivity (with many hydroxyl groups), and stiffness. groups), the unique structure and diminutive size of nanocellulose may endow it with beneficial properties. Reinforced (nano)composite materials utilising this fibre technology (Kalia et al., 2011).

3.3. Starch

Green polymer nanocomposites exhibit exceptional properties by combining the benefits of organic polymers and natural fillers. It is observed that plant fibres can be used to strengthen polymers. They possess rather high levels of strength and rigidity. These organic polymers have biodegradability, low acquisition cost, low density, renewable, less greenhouse gas emissions and Eco-friendly (Adeosun et al., 2012).

Inevitably, more plastic will be used, and consequently, more plastic trash will be produced. Plastic's additives are hazardous, nonbiodegradable waste. Thermoplastic starch (TPS) is emerging as an intriguing option in the quest to find a plastic replacement that possesses desirable features like biodegradability, good mechanical strength, and low cost (Rivadeneira-Velasco et al., 2021).

Plants' seeds, fruits, leaves, tubers, and roots can all be used to harvest the renewable resource known as native starch. Potato, wheat, maize, sweet potato, tapioca, pea, corn, etc., are all possible sources of native starch (Rivadeneira-Velasco et al., 2021). Amylose (the linear component) and amylopectin (the branched component) make up the bulk of starch granules (branched component). Starch's two high-molecular-weight components provide the substance with its utility and versatility. Amylose concentrations range from 20% to 30% and amylopectin concentrations from 70% to 80%, however these numbers might be significantly higher or lower depending on the original food source (Pérez-Pacheco et al., 2016). The amylose to amylopectin ratio is related to the starch's water absorption, which in turn affects the starch's functional qualities (Castillo et al., 2017).

If you want your product to have specific qualities, plasticizers play a crucial role in bringing those qualities about. The plasticizer's molecular size needs to be smaller than starch's for it to permeate through the intermolecular gaps of the polymer and initiate the interaction, which results in microscopic modifications to the starch's structure. Glycerol's polarity, compatibility, hydrophilicity, and boiling point below the gelatinization temperature make it the most used plasticizer for TPS preparation. Having these characteristics means the molten substance is less likely to solidify into a solid block, making the gelatinization process easier (Ning et al., 2009). A biodegradable starch/poly(ε-caprolactone) based Coloisite-20A nanocomposites are developed (Behera, Srivastava, & Das, 2022).

3.3.1. Food packaging applications

Few research studies are carried out on to create sustainable and eco-friendly film alternatives, with a focus on tapioca starch (TPS) films made from biopolymer. Inadequate mechanical, barrier, and thermal properties have restricted the application of TPS films for packaging of food. TPS/MCC composite films were made by adding microcrystalline cellulose (MCC) as a filler into the films via the solvent casting method, with MCC concentrations ranging from 0% to 10%. Physical, thermal, barrier, and mechanical properties of the TPS/MCC films were investigated. When testing various filler concentrations, it was determined that 3 wt.% MCC produced the best mechanical and barrier properties in the films. Films' thermal properties were also enhanced by the incorporation of MCC. According to the results of this research, TPS/MCC films made under these conditions have enhanced properties and show promise as a food packaging material (Othman et al., 2019).

The addition of additives to starch-based films, such as antioxidants and antimicrobials, is crucial to enhancing their functional properties. These semi-permeable active packaging films prevent the loss of solutes and moisture during fruit respiration and oxidation, thereby increasing the product's shelf life. Some examples of active components in food packaging include essential oils, which have antimicrobial and antioxidant properties and have thus increased the quality and safety of the food they are packaged with (Afif, Rebolledo, & Roy, 2021). The primary purpose of essential oils, which are all-natural compounds, is to protect food from spoilage by microorganisms and fungi, as well as oxidation. They are made up of alcohols, esters, terpenes, phenols and in addition to other bioactive compounds (García-Guzmán et al., 2022). The results of an evaluation by Raigong et al., who tested the antibiotic potential of clove oil and cinnamon oil added to starch films on *S. aureus*, *C jejuni*, and *E. coli*, exhibited that these antimicrobials were able to inhibit the growth of these fungi and bacteria (Raigond et al., 2019).

3.3.2. Other applications

By 2022, it is predicted that the global market for 3D printed bioproducts, which use bio-based products like starch and materials, will have expanded (Aida et al.,

2021; Yang et al., 2018). Starch pellets or filament have the potential to become the industry standard feedstock for 3D printing. This is because less energy will be used to melt the material (Catriona & Jonathan, 2014). As a result of starch's low glass transition temperature and excellent gas barrier properties, TPS is non-toxic and odourless, and it prints more clearly.

The composites were thermally processed, but the antibiotic's (ATB; vancomycin) bacteriostatic properties remained unaffected. As a result, the thermoplastic starch, titanium dioxide particle (TPS/TiO2 &TiO2;/ATB) composites that have been prepared are a promising material for biomedical applications involving the local release of antibiotics (Ujcic et al., 2020).

4. CONCLUSION

Compared to synthetic thermoplastic-based nanocomposites, commercial thermoplastic based bionanocomposites based materials are more sustainable and environmentally friendly, and could have a significant impact on the Medical, Automotives, food packaging, construction industries. Since these materials lack the mechanical, barrier, and optical properties of their synthetic analogues, mass production for the industry is still in the planning stages. Some studies have been conducted to address these obstacles, including refinements to standard thermoplastic processing methods, physical and chemical modifications to Chitosan, cellulose and starch, when combined with other plasticizers, functional nano/micro fillers, and biopolymers. Contrarily, thermoplastic based bionanocomposites based materials are biodegradable, though the addition of certain chemical additives to boost the final product's performance may extend its useful life. As a result, it is necessary to discover new additives and polymers, preferably from renewable resources, to improve the sustainability of future thermoplastic based materials. When they reach the end of their lives, they can be integrated into the earth's ecological cycles in a controlled, secure, and environmentally friendly manner.

ACKNOWLEDGMENTS

The first author would like to express appreciation to Krishna University in Machilipatnam, Andhra Pradesh, India, for their assistance in facilitating the writing of this work. The second author would like to express appreciation to GITAM (Deemed to be University) in Visakhapatnam, Andhra Pradesh, India, for providing the resources needed to write this work.

REFERENCES

Adeosun, S. O., Lawal, G. I., Balogun, S. A., & Akpan, E. I. (2012). Review of green polymer nanocomposites. *Journal of Minerals & Materials Characterization & Engineering*, *11*(4), 385–416.

Afif, K., Rebolledo, C., & Roy, J. (2021). Drivers, barriers and performance outcomes of sustainable packaging: A systematic literature review. *British Food Journal*. Advance online publication. DOI: 10.1108/BFJ-02-2021-0150

Aida, H. J., Nadlene, R., Mastura, M. T., Yusriah, L., Sivakumar, D., & Ilyas, R. A. (2021). Natural fibre filament for fused deposition modelling (FDM): A review. *International Journal of Sustainable Engineering*, *14*(6), 1988–2008. DOI: 10.1080/19397038.2021.1962426

Alexis, F. (2005).. . *Polymer International*, *54*, 36–46.

Amorim, W. L., Costa, H. O., Souza, F. C., Castro, M. G., & Silva, L. (2009). Experimental study of the tissue reaction caused by the presence of cellulose produced. *Brazilian Journal of Otorhinolaryngology*, *75*(2), 200–207. PMID: 19575105

Anderson, J. M. (2001).. . *Annual Review of Materials Research*, *31*(1), 1–401.

Aoki, N., Nishikawa, M., & Hattori, K. (2003). Synthesis of chitosan derivatives bearing cyclodextrin and adsorption of p-nonylphenol and bisphenol A. *Carbohydrate Polymers*, *52*(3), 219–223.

Baumann, M. D., Kang, C. E., & Stanwick, J. C.. (2009). An injectable drug delivery platform for sustained combination therapy. *Journal of Controlled Release*, *138*(3), 205–213. PMID: 19442692

Behera, A. K., Srivastava, R., & Das, A. B. (2022). Mechanical and degradation properties of thermoplastic starch reinforced nanocomposites. *Starch*, *74*(3–4). Advance online publication. DOI: 10.1002/star.202100270

Bhattarai, N., Edmondson, D., Veiseh, O., Matsen, F. A., & Zhang, M. Q. (2005). Electrospun chitosan-based nanofibers and their cellular compatibility. *Biomaterials*, *26*(31), 6176–6184. PMID: 15885770

Brown, R. M. (1998). Microbial cellulose: A new resource for wood, paper, textiles, food and specialty products [Position paper]. http://www.botany.utexas.edu/facstaff/facpages/mbrown/position1.html

Castillo, L. A., López, O. V., Ninago, M. D., Versino, F., Barbosa, S. E., García, M. A., & Villar, M. A. (2017). Effect of mineral and organic fillers on processing, structure, and final properties of starch. In *Composites and nanocomposites based on starches*. Elsevier.

Catriona, M., & Jonathan, W. (2014). The potential of 3D printing to reduce the environmental impacts of production. In *ECEEE Industrial Summer Study Proceedings*. Sea Green Tree.

Chen, S., & Wang, Y. (2001). Study on β-cyclodextrin grafting with chitosan and slow release of its inclusion complex with radioactive iodine. *Journal of Applied Polymer Science*, *82*(10), 2414–2421.

Chen, Z., Mo, X., & Qing, F. (2007). Electrospinning of collagen–chitosan complex. *Materials Letters*, *61*(16), 3490–3494.

Czaja, W. K., Young, D. J., Kawecki, M., & Brown, R. M.Jr. (2007). The future prospects of microbial cellulose in biomedical applications. *Biomacromolecules*, *8*(1), 1–12. PMID: 17206781

Dweib, M. A., Hu, B., Shenton, H. W., & Wool, R. P. (2006). Bio-based composite roof structure: Manufacturing and processing issues. *Composite Structures*, *74*(4), 379–388.

Foda, N. H., El-Laithy, H. M., & Tadros, M. I. (2007). Implantable biodegradable sponges: Effect of interpolymer complex formation of chitosan with gelatin on the release behavior of tramadol hydrochloride. *Drug Development and Industrial Pharmacy*, *33*(1), 7–17. PMID: 17192246

Fontana, J. D., de Souza, A. M., & Fontana, C. K.. (1990). Acetobacter cellulose pellicle as a temporary skin substitute. *Applied Biochemistry and Biotechnology*, *24–25*, 253–264. PMID: 2353811

García-Guzmán, L., Cabrera-Barjas, G., Soria-Hernández, C. G., Castaño, J., Guadarrama-Lezama, A. Y., & Rodríguez Llamazares, S. (2022). Progress in starch-based materials for food packaging applications. *Polysaccharides*, *3*, 136–177. DOI: 10.3390/polysaccharides3010007

Global, T. W. I. (n.d.). What is a thermoplastic? https://www.twi-global.com/technical-knowledge/faqs/what-is-a-thermoplastic

Grassie, N. (1985). *Polymer degradation and stabilization*. Cambridge University Press.

Hoemann, C. D., Lafantaisie-Favreau, C. H., Lascau-Coman, V., Chen, G., & Guzman-Morales, J. (2012). The cartilage-bone interface. *The Journal of Knee Surgery*, *25*(2), 85–97. DOI: 10.1055/s-0032-1319782 PMID: 22928426

Holback, H., Yeo, Y., & Park, K. (2011). Hydrogel swelling behavior and its biomedical applications. In *Biomedical hydrogels* (Woodhead Publishing Series in Biomaterials, pp. 3–24). Woodhead Publishing. https://doi.org/DOI: 10.1533/9780857091383.1.3

Hui, J., Yuanyuan, J., Jiao, W., Yuan, H., Yuan, Z., & Shiru, J. (2009). Potentiality of bacterial cellulose as the scaffold of tissue engineering of cornea. In *Proceedings of the 2nd International Conference on Biomedical Engineering and Informatics (BMEI '09)*, China, October.

Iguchi, M., Yamanaka, S., & Budhiono, A. (2000). Bacterial cellulose—A masterpiece of nature's arts. *Journal of Materials Science*, *35*(2), 261–270.

Johannson, C. (2011). Bio-nanocomposites for food packaging applications. In Mittal, V. (Ed.), *Nanocomposites with biodegradable polymers: Synthesis, properties, and future perspectives* (pp. 1–xx). Oxford University Press., DOI: 10.1093/acprof:oso/9780199581924.003.0014

Joseph, B., Krishnan, S., & Sagarika, V. K.. (2020). Bionanocomposites as industrial materials, current and future perspectives: A review. *Emergent Materials*, *3*, 711–725. DOI: 10.1007/s42247-020-00133-x

Kalia, S., Dufresne, A., Cherian, B. M., Kaith, B. S., Averous, L., Njuguna, J., & Nassiopoulos, E. (2011). Cellulose-based bio- and nanocomposites: A review. *International Journal of Polymer Science*, *837875*. Advance online publication. DOI: 10.1155/2011/837875

Kalia, S., Kaith, B. S., & Kaur, I. (2009). Pretreatments of natural fibres and their application as reinforcing material in polymer composites: A review. *Polymer Engineering and Science*, *49*(7), 1253–1272.

Kang, D. W., Choi, H. R., & Kweon, D. K. (1996). Selective adsorption capacity for metal ions of amidated chitosan bead-g-PAN copolymer. *Polymer (Korea)*, *20*, 989–995.

Khor, E., & Lim, L. Y. (2003). Implantable applications of chitin and chitosan. *Biomaterials, 24*(13), 2339–2349. PMID: 12699672

Kiatkamjorwong, S., & Meechi, N. (1997). Radiation effects on chitosan. *Radiation Physics and Chemistry, 49*(6), 689–696.

Klemm, D., Schumann, D., Udhardt, U., & Marsch, S. (2001). Bacterial synthesized cellulose—Artificial blood vessels for microsurgery. *Progress in Polymer Science, 26*(9), 1561–1603.

Kushwaha, R. K., & Srivastava, A. (2018). Recent developments in bio-nanocomposites: A review. *Research Journal of Nanoscience and Engineering, 2*(2), 1–4.

Lee, D. W., Lim, H., Chong, H. N., & Shim, W. S. (1997). Advances in chitosan material and its hybrid derivatives: A review. *Open Biomaterials Journal, 1*, 10–20.

Mello, L. R., Feltrin, Y., Selbach, R., Macedo, G.Jr, Spautz, C., & Haas, L. J. (2001). Use of lyophilized cellulose in peripheral nerve lesions with loss of substance. *Arquivos de Neuro-Psiquiatria, 59*(2), 372–379. PMID: 11460182

Mittal, V. (2011). *Bio-nanocomposites: Future high-value materials.* Oxford University Press., DOI: 10.1093/acprof:oso/9780199581924.003.0001

Negrao, S. W., Bueno, R. R. L., & Guérios, E. E.. (2006). A eficácia do stent recoberto com celulose biossintética comparado ao stent convencional em angioplastia em coelhos. *Revista Brasileira de Cardiologia Invasiva, 14*(1), 10–19.

Ning, W., Xingxiang, Z., Xuechen, W., & Haihui, L. (2009). Ionic liquids modified montmorillonite/thermoplastic starch nanocomposites as ionic conducting biopolymer. *Macromolecular Research, 17*, 285–288.

Olson, D. G., Tripathi, S. A., & Giannone, R. J.. (2010). Deletion of the Cel48S cellulase from *Clostridium thermocellum. Proceedings of the National Academy of Sciences of the United States of America, 107*(41), 17727–17732. PMID: 20837514

One, K., Saito, Y., & Yura, H.. (2000). Photocrosslinkable chitosan as a biological adhesive. *Journal of Biomedical Materials Research. Part A, 49*(2), 289–295. PMID: 10571917

Othman, S. H., Majid, N. A., Tawakkal, I. S. M. A., Basha, R. K., Nordin, N., & Shapi'i, R. A. (2019). Tapioca starch films reinforced with microcrystalline cellulose for potential food packaging application. *Food Science and Technology (Campinas), 39*(3), 605–612. DOI: 10.1590/fst.3607

Park, H. M., Misra, M., Drzal, L. T., & Mohanty, A. K. (2004). Green nanocomposites from cellulose acetate bioplastic and clay: Effect of eco-friendly triethyl citrate plasticizer. *Biomacromolecules*, *5*(8), 2281–2288. DOI: 10.1021/bm049690f PMID: 15530043

Paulo, S. (2005).. . *Brazilian Oral Research*, *19*(3), 1806–8314.

Peppas, N. A., Bures, P., Leobandung, W., & Ichikawa, H. (2000). Hydrogels in pharmaceutical formulations. *European Journal of Pharmaceutics and Biopharmaceutics*, *50*, 27–46. PMID: 10840191

Pérez-Pacheco, E., Canto-Pinto, J. C., Moo-Huchin, V. M., Estrada-Mota, I. A., Estrada-León, R. J., & Chel-Guerrero, L. (2016). Thermoplastic starch (TPS)-cellulosic fibers composites: Mechanical properties and water vapor barrier: A review. In *Composites from renewable and sustainable materials*. IntechOpen.

Prabaharan, M., Reis, R. L., & Mano, J. F. (2005). Chitosan-based particles as controlled drug delivery systems. *Drug Delivery*, *12*(1), 41–57. PMID: 15801720

Raigond, P., Sood, A., Kalia, A., Joshi, A., Kaundal, B., Raigond, B., Dutt, S., Singh, B., & Chakrabarti, S. K. (2019). Antimicrobial activity of potato starch-based active biodegradable nanocomposite films. *Potato Research*, *62*, 69–83.

Ranade, V. V., & Hollinger, M. A. (Eds.). (2003). *Drug delivery systems* (2nd ed., pp. 1–69). CRC Press.

Râpă, M., Mitelu , A. C., Tănase, E. E., Grosu, E., Popescu, P., Popa, M. E., & Darie-Ni ă, R. N. (2016). Influence of chitosan on mechanical, thermal, barrier and antimicrobial properties of PLA-biocomposites for food packaging. *Composites. Part B, Engineering*, *102*, 112–121. DOI: 10.1016/j.compositesb.2016.07.016

Rivadeneira-Velasco, K. E., Utreras-Silva, C. A., Díaz-Barrios, A., Sommer-Márquez, A. E., Tafur, J. P., & Michell, R. M. (2021). Green nanocomposites based on thermoplastic starch: A review. *Polymers*, *13*, 3227. DOI: 10.3390/polym13193227 PMID: 34641042

Saini, R., Bajpai, A. K., & Jain, E. (2017). Fundamentals of bionanocomposites. In *Biodegradable and biocompatible polymer composites* (pp. 1–xx). Elsevier., DOI: 10.1016/B978-0-08-100970-3.00012-2

Senel, S., & McClure, S. J. (2004). Potential applications of chitosan in veterinary medicine. *Advanced Drug Delivery Reviews*, *56*(10), 1467–1480. PMID: 15191793

Seves, A., Testa, G., Bonfatti, A. M., Paglia, E. D., Selli, E., & Marcandalli, B. (2001). Characterization of native cellulose/poly(ethylene glycol) films. *Macromolecular Materials and Engineering*, *286*(9), 524–528.

Sharma, R., Jafari, S. M., & Sharma, S. (2020). Antimicrobial bio-nanocomposites and their potential applications in food packaging. *Food Control*, *112*, 107086. DOI: 10.1016/j.foodcont.2020.107086

Shchipunov, Y. (2012). Bionanocomposites: Green sustainable materials for the near future. *Pure and Applied Chemistry*, *84*(12), 2579–2607. DOI: 10.1351/PAC-CON-12-05-04

Silva, E. C. (n.d.). *Hidroxiapatita sintética em alvéolo dentário após exodontia em Felis catus: Estudo clínico, radiológico e histomorfométrico* (M.S. Dissertation). Universidade Federal de Viçosa, Brazil.

Siro, I., & Plackett, D. (2010). Microfibrillated cellulose and new nanocomposite materials: A review. *Cellulose (London, England)*, *17*(3), 459–494.

Sreenivasan, K. (1988). Synthesis and evaluation of a β-cyclodextrin-based molecularly imprinted copolymer. *Journal of Applied Polymer Science*, *69*(6), 1051–1055.

Sreenivasulu, N., Pothula, L., & Reddy, P. D. S. (2016). Growing the importance of bionanocomposites. *International Journal of Advanced Research in Science and Engineering*, *5*(1), 374–381.

Sun, T., Xu, P., Liu, Q., Xue, J., & Xie, W. (2003). Graft copolymerization of methacrylic acid onto carboxymethyl chitosan. *European Polymer Journal*, *39*(1), 189–192.

Teo, W. E., & Ramakrishna, S. (2006). A review on electrospinning design and nanofibre assemblies. *Nanotechnology*, *17*, R89–R106. PMID: 19661572

Tojima, T., Katsura, H., Han, S. M., Tanida, F., Nishi, N., & Tokura, S. (1998). Preparation of a β-cyclodextrin-linked chitosan derivative via reductive amination. *Journal of Polymer Science. Part A, Polymer Chemistry*, *36*(11), 1965–1968.

Ujcic, A., Krejcikova, S., Nevoralova, M., Zhigunov, A., Dybal, J., Krulis, Z., Fulin, P., Nyc, O., & Slouf, M. (2020). Thermoplastic starch composites with titanium dioxide and vancomycin antibiotic: Preparation, morphology, thermomechanical properties, and antimicrobial susceptibility testing. *Frontiers in Materials*, *7*, 9. DOI: 10.3389/fmats.2020.00009

Watanabe, Y., Mukai, B., & Kawamura, K. I.. (2002). Preparation and evaluation of press-coated aminophylline tablet using crystalline cellulose and polyethylene glycol in the outer shell for timed-release dosage forms. *Yakugaku Zasshi*, *122*(2), 157–162. PMID: 11857956

Williams, D. F. (2008). On the mechanisms of biocompatibility. *Biomaterials*, *29*(20), 2941–2953. DOI: 10.1016/j.biomaterials.2008.04.023 PMID: 18440630

Yang, E., Miao, S., Zhong, J., Zhang, Z., Mills, D. K., & Zhang, L. G. (2018). Bio-based polymers for 3D printing of bioscaffolds. *Polymer Reviews (Philadelphia, Pa.)*, *58*(4), 668–687. DOI: 10.1080/15583724.2018.1484761 PMID: 30911289

Yano, H., Sugiyama, J., & Nakagaito, A. N.. (2005). Optically transparent composites reinforced with networks of bacterial nanofibers. *Advanced Materials*, *17*(2), 153–155.

Chapter 4
Green Synthesis of Iron Nanoparticles: Methods, Characterization, and Applications

Aditya Shrivastav
https://orcid.org/0009-0000-8461-8171
Ajeenkya D.Y. Patil University, India

Sunil Sankathala
Ajeenkya D.Y. Patil University, India

Susanta Das
https://orcid.org/0000-0002-9314-3988
Ajeenkya D.Y. Patil University, India

ABSTRACT

The rapid advancements in nanotechnology have significantly transformed various industries, with iron-based nanoparticles emerging as a key material due to their unique properties and diverse applications. Green synthesis methods provide an eco-friendly and sustainable alternative to conventional chemical/physical synthesis, utilizing biological agents such as plant extracts, algae, & micro-organisms. These methods rely on phytochemicals as reducing and stabilizing agents, ensuring minimal environmental impact while enabling precise tailoring of nanoparticle properties. Biogenic FeNPs demonstrate immense potential across multiple domains: targeted drug delivery, wastewater treatment and pollutant removal, in agriculture as green alternatives to chemical pesticides, and in catalysis for energy-efficient chemical reactions. This chapter highlights the potential of green-synthesized iron

DOI: 10.4018/979-8-3373-2180-6.ch004

nanoparticles and emphasizes the need for continued research to advance sustainable practices, paving the way for groundbreaking applications that benefit both industry and the environment.

1. INTRODUCTION

The fast advancements in nanotechnology have brought iron nanoparticles (FeNPs) to the front, thanks to their unique features & benefit across environmental and medical research. Owing to their unique properties, ability to be manipulated magnetically and adaptable surface chemistry, applications of FeNPs in catalysis, clean-up of pollutants, medical care, drug delivery and magnetic resonance imaging are possible (Alphandéry, 2020; Shabbir et al., 2023). Still, traditional methods to make FeNPs include physical and chemical processes and these approaches can be harmful to both people and the environment. Consequently, they encourage researchers to come up with more sustainable and friendly alternatives (Batool et al., 2021; Nahari et al., 2022).

Many researchers are interested in iron nanoparticles because they are both highly reactive and their magnetic properties can be adjusted. FeNPs also contribute in the environmental science because of their ability of cleaning contaminated water & soil by eliminating heavy metals, chlorinated organic compounds & synthetic dyes. Pollutants interact with their surfaces easily, leading to quick degradation and magnets enable easy recovery and collection following treatment. FeNPs are commonly used in targeted drug delivery, cancer therapy using hyperthermia, enhancing MRI images and fighting different microorganisms. Since they are biocompatible and can be modified, they help produce new and sophisticated medicines and devices (Alphandéry, 2020; Shabbir et al., 2023).

Although they may be modified, chemicals can still only be reduced, decomposed by heat or ground down with limitations. In general, these strategies require reductants like sodium borohydride and hydrazine, various solvents and a lot of energy, leading to risk for the environment and increased contaminants. After synthesis by traditional ways, nanoparticles often need more treatment to become more stable and fit for use in biology and this takes more time and effort. Using specific methods may cause the FeNPs to have incorrect structures and sizes, making them not suitable for practical uses in environments and medicine (Batool et al., 2021; Nahari et al., 2022). As concern about the environment and safety in traditional synthesis persists, green synthesis provides a safer and more sustainable way to make FeNPs. Since it underlines the significance of using environmentally safe solvents, renewable raw materials and more efficient processes, this practise supports the ideas of green chemistry (Batool et al., 2021; Ashrafi-Saiedlou et al., 2025; Nahari et al.,

2022). Process of green synthesis uses plant extracts, bacteria and fungi, instead of chemicals, to protect the environment during the synthesis (Batool et al., 2021; Ashrafi-Saiedlou et al., 2025; Shabbir et al., 2023; Nahari et al., 2022).

Plant-based materials are more commonly used in biology because they are easy to obtain, have many different substances and are straightforward to process. The presence of flavonoids, phenolics, alkaloids, terpenoids and proteins makes sure that the synthesis of FeNPs is speedy and they become more stable and biocompatible. Usually, combining an aqueous plant extract with iron salts ($FeSO_4$ or $FeCl_3$) causes the iron ions to be reduced and forms nanoparticles which is recognised by a change in the colour (Batool et al., 2021; Shabbir et al., 2023). Tests are done on the nanoparticles such as measuring under UV light, using FTIR, SEM imaging and XRD, so we can be sure of their successful synthesis and look at their structure and how they work (Batool et al., 2021; Shabbir et al., 2023; Nahari et al., 2022). Researchers have prepared nanoparticles from various plants, including date palm, *Madhuca indica* and *Vitex leucoxylon*, making them each distinguished by their phytochemicals (Batool et al., 2021; Shabbir et al., 2023; Nahari et al., 2022). FeNPs made from Phoenix dactylifera leaf extract demonstrated strong antimicrobial properties against *Escherichia coli* and *Klebsiella pneumonia* which makes them probable environmentally safe alternatives (Batool et al., 2021). In green synthesis, using bacteria and fungi along with plants has shown to be useful. With help from enzymes, proteins and polysaccharides, Pseudomonas fluorescens can lower iron salts which leads to FeNPs that have antioxidant, antibacterial and photocatalytic activities. Fungi's strong metabolism and the ability to withstand various metal ions make them suitable for low-cost and large-scale nanoparticle manufacturing. With microbial systems, it is easier to manage shape and size of NPs & they offer better properties for certain applications (Ashrafi-Saiedlou et al., 2025).

The basis of green synthesis lies in how biological molecules can lose or gain an electron during redox reactions. Plant-mediated synthesis involves phytochemicals providing electrons to change Fe^{3+} or Fe^{2+} ions into their zero-valent or oxide forms while keeping the nanoparticles from sticking to each other (Batool et al., 2021; Shabbir et al., 2023; Nahari et al., 2022). In these microbial systems, reductases and several extracellular metabolites help in carrying out both reduction and stabilisation processes (Ashrafi-Saiedlou et al., 2025). FeNPs made using biological synthesis are often coated in organic particles from their source. This makes them stable in water, less harmful to living organisms and suitable for medical use (Batool et al., 2021; Ashrafi-Saiedlou et al., 2025; Shabbir et al., 2023; Nahari et al., 2022).

Green methods of producing FeNPs have several advantages over traditional methods. Since they don't use dangerous chemicals and generate less waste, these processes support sustainable goals for nature (Batool et al., 2021; Ashrafi-Saiedlou et al., 2025; Shabbir et al., 2023; Nahari et al., 2022). This process becomes both affordable and less harmful to the environment because it uses materials that are readily available (Batool et al., 2021; Shabbir et al., 2023; Nahari et al., 2022). Using life-based organic cappings greatly contributes to making biomedical applications safer and more compatible (Alphandéry, 2020; Shabbir et al., 2023; Nahari et al., 2022). Also, ease of green synthesis approaches enables production on a larger scale without the use of complicated tools or facilities (Batool et al., 2021; Shabbir et al., 2023; Nahari et al., 2022). Biogenic FeNPs generally perform better against microbes, radicals and in chemical reactions, making them a suitable choice in various settings (Batool et al., 2021; Ashrafi-Saiedlou et al., 2025; Shabbir et al., 2023; and Nahari et al., 2022).

Green synthesis of FeNPs is used in environmental remediation to degrade organic matter, cut down heavy metals and eliminate dyes from the water. The magnets in rare-earth materials allow for easy recycling and recovery which helps keep operating costs down and avoid creating further pollution. Their safety and low toxicity allow them to be suited for use in environmental areas (Batool et al., 2021; Ashrafi-Saiedlou et al., 2025; Nahari et al., 2022). Their motion controlled by a magnetic field allows biomedical applications to precisely treat specific areas inside body (Alphandéry, 2020). They can be utalized in magnetic hyperthermia, raising the temperature of cancerous areas with a changing magnetic field and not harming healthy part of the tissue (Alphandéry, 2020). Besides, FeNPs work as contrast agents in MRIs due to their superparamagnetic nature and are often applied in medicine for fighting infections, aiding recovery and decreasing inflammation (Batool et al., 2021; Shabbir et al., 2023; Nahari et al., 2022). In 2020 to 2024, researchers extended the range of biological materials and improved the techniques used in green synthesis. As an illustration, FeNPs formed with Madhuca indica extracts were found to possess antioxidant, anti-inflammatory and positive drug-like properties which show they could be used for several therapeutic purposes (Shabbir et al., 2023). Also, synthesis by Pseudomonas fluorescens has allowed for developing FeNPs known for powerful antibacterial and photocatalytic effects, showing the wide range of tools microbial systems provide (Ashrafi-Saiedlou et al., 2025). Advanced characterization tools and computational methods are making it easier to understand and fine-tune green synthesis (Shabbir et al., 2023; Nahari et al., 2022).

Although there has been meaningful progress, problems still exist in making protocols consistent, producing nanoparticles that are always the same and scaling up production. Further studies and future investigations are expected to discover more biological substances, better understand the processes involved and mix various green synthesis approaches to make the methods more effective. This chapter covered how iron-based nanoparticles are made using organic materials, how they are examined and what their uses are in various fields.

2. IRON NANOPARTICLES: A BIBLIOMETRIC REVIEW

For better understanding of the emerging field of iron and iron nanoparticles, an overlay bibliometric analysis was conducted as illustrated in Figure 1. It also tells the relationships between keywords related to iron-based nanoparticles, using a map produced by VOSviewer. This visualisation displays how the keywords occur in articles and the average year of publication, where node sizes represent how often they appear, different colours represent the year and the absence of linkages means the keywords rarely co-occur. Common keywords found are "nanoparticles," "iron," "metal nanoparticles," "animals," and "humans," showing that both science and medicine fields are highly interested in iron nanostructures. The terms "contrast media," and "ferrosoferric oxide," are connected by strong relationships in medical imaging and treatment, with applications in humans and animal studies. Compared to other methods, "green synthesis," "plant extracts," "hydrogen peroxide," and "photoacoustic imaging" have experienced rapid development over the past few years. It seems that new efforts are being made to create drugs and diagnostic devices that are more environment friendly. Older research clusters (blue and green nodes in the graph) refer to established fields such as contrast enhancement and biomedical imaging.

Figure 1. Network and Overlay Visualization of co-occurring keywords in Iron Nanoparticles Research (Source: PubMed Database; created using VOSviewer).

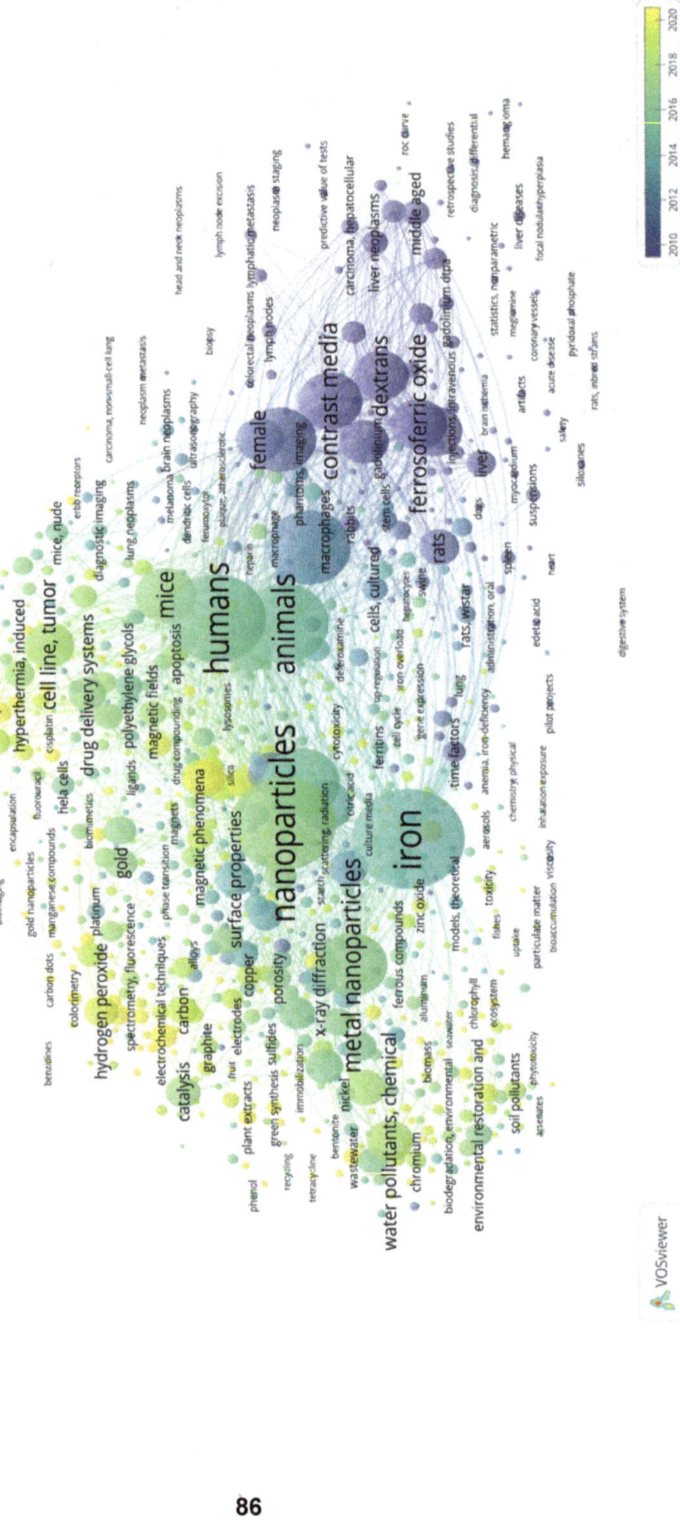

For instance, "x-ray diffraction" and other similar terms describe materials characterization techniques for iron nanoparticles and "drug delivery systems," "cell line, tumour," and "hyperthermia" indicate the healthcare and scientific uses of these nanoparticles. In short, Figure 1 clearly highlights the growing and shifting research field of iron-based nanoparticles, separating basic and upcoming research across biomedical, environmental and material science sectors. This visual display helps spot the most important research areas, trends during different periods and new paths for future work in sustainable and application-focused studies.

3. BIOGENIC SYNTHESIS OF IRON NANOPARTICLES USING VARIOUS COMPONENTS

3.1 Phytochemicals Present in Plants and Their Characterization

Bioactive compounds, phytochemicals, are produced by plants, algae, and bacteria, play important roles in protection and health benefits. Some of these include primary metabolites (e.g. carbohydrates and proteins), & secondary metabolites (flavonoids, alkaloids, terpenoids, etc), which act as ecological functions to defend plants from herbivores and pathogens (Kang et al., 2021). Phytochemicals, such as carotenoids, phytohormones, and polyphenols that have naturally occurring antioxidant and anti-microbial properties, also serve to promote plant growth and increase disease resistance (S. Lee & Ryu, 2021; Diaz-Puertas et al., 2023). Advanced techniques such as chromatography (HPLC, GC), spectroscopy (UV-Vis, FTIR), mass spectrometry are used to characterise these phytochemicals and determine their structures and quantities (Kumar et al., 2023).

Green synthesis of NPs also depends largely on phytochemicals found in plants as reducing, stabilizing, & capping agents. Plant extracts such as *Hibiscus sabdariffa* and *Moringa oleifera* contain polyphenols which are powerful reducing agents & convert the metal ions into NPs thus aiding in formation of Fe_2O_3 NPs (Saod et al., 2024; Aisida et al., 2019). Flavonoids as potent antioxidants present in plants such as *Catharanthus roseus* and *Skimmia laureola* reduce metal ions & stabilise synthesized NPs, thus leading to synthesis of iron nanoparticles (Butt et al., 2018; Alam et al., 2018). In the synthesis of Fe_3O_4 NPs from *Enterococcus faecalis*, proteins found in both plant extracts and bacterial extracellular polymeric substances (EPS) act as capping agents, the protein inhibits aggregation of the resulting Fe_3O_4 NPs (Samuel et al., 2021). Seaweed extracts, such as *Colpomenia sinuosa* and *Pterocladia capillacea*, have site abundant sulphated polysaccharides, which function in place of reducing and stabilising agents in producing the spherical shape and size control of Fe_3O_4 NPs (Salem et al., 2019). Naseem and Farrukh (2015), showed that lawsone coordinates with iron ions to stabilise nanoparticles, resulting in distorted hexagonal-like nanoparticles. However, various plant extracts such as kulekhara leaves and *Borassus flabellifer* seed coat contain phenolic compounds which aid in metal ions reduction and nanoparticles formation e.g. Ag-Fe_2O_3 and magnetic iron oxide nanoparticles (Sandhya & Kalaiselvam, 2019; Kolya & Kang, 2022). In addition, FTIR analysis of *Catharanthus roseus* extracts revealed carboxylic acids implicated in reducing metal ions & stabilising nanoparticles (Butt et al., 2018).

3.2 Synthesis of biogenic Iron NPs:

3.2.1 Green Synthesis of Iron NPs

Synthesised FeNPs and oxides by biogenic route have received tremendous attention owing to the green nature and application to medicine, environmental remediation, or catalysis (Saod et al., 2024). This green synthesis is done by bio-reduction and biosorption based mechanisms using biological materials like plants, fungi, bacteria, and algae for creation of NPs without using toxic chemicals (Saif et al., 2016).

Synthesis of Fe or iron oxide NPs (FeO NPs, Fe_2O_3 NPs, Fe_3O_4 NPs) using plant extracts can be represented as follows:

Fe^{3+} + Plant Extract (e.g., phenolics) → FeNPs or FeONPs or Fe_2O_3NPs or Fe_3O_4NPs

Ferric ions (Fe^{3+}) from a solution, e.g. ferric chloride, are mixed with a plant extract high in phenolic compounds in this process. These compounds are reducing agents, which facilitate conversion of ferric ions into zerovalent FeNPs or iron oxide nanoparticles (Fe_2O_3) (Saod et al., 2024).

Microbial synthesis of Fe or Fe_2O_3, Fe_3O_4 NPs through algae or bacteria can be expressed as:

Fe^{3+} + Microbial Enzymes (e.g., reductase) → FeNPs or Fe_2O_3 or Fe_3O_4NPs

In this method, microbial enzymes like reductases of algae or bacteria reduce ferric ions. Ferric ions are reduced to iron oxide nanoparticles using the enzymes (Figure 2) (Caf, 2022).

Figure 2. Green Synthesis of NPs.

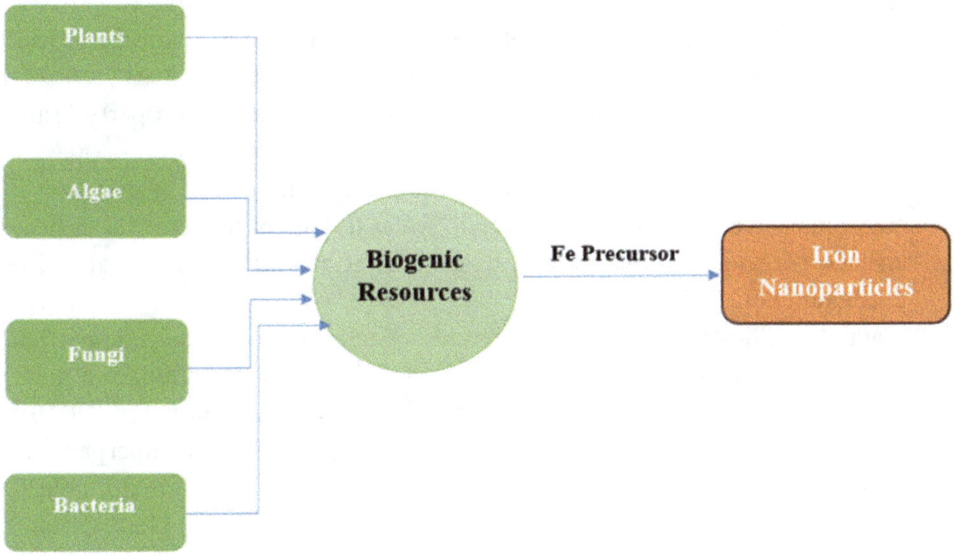

3.2.2 Benefits of Biogenic synthesis

- Eco-Friendliness: Biogenic synthesis of iron nanoparticles eliminates use of toxic chemicals, making it a more sustainable and environmentally friendly method compared to traditional physical/chemical approaches (Saif et al., 2016; Herlekar et al., 2014).
- Cost-Effectiveness: This method utilizes low-cost, renewable biological resources (plant extracts, algae, and microorganisms) as reducing & stabilizing agents, reducing overall production costs (Tan et al., 2023; Saif et al., 2016; Herlekar et al., 2014).
- Scalability: Because of simplicity of the method and the use of abundant biological materials, this method is feasible for large scale production with low investment in infrastructure (Tan et al., 2023; Herlekar et al., 2014).
- Biocompatibility: The resultant nanoparticles from plant mediated synthesis are biocompatible ensuring their possible applications in medicinal area including drug delivery as well as diagnostics (Mughal et al., 2021; Saif et al., 2016).
- Reduced Energy Requirements: Compared with traditional synthetic methods that require high temperatures and pressures. However, biogenic synthe-

sis occurs under mild conditions, saving energy (Tan et al., 2023; Saif et al., 2016).

- Customizable Properties: Control over dimension (size & shape), and functionality of synthesized NPs is achieved using various plant extracts with which specific phytochemicals facilitate application specific property (Tan et al., 2023).
- Environmental Remediation: FeNPs biologically synthesized have shown extraordinary reactivity and efficiency in water treatment for removal of heavy metals, (organic) pollutants and dyes (Tan et al., 2023; Saif et al., 2016; Herlekar et al., 2014).
- Catalytic Efficiency: Biogenic methods of synthesising iron nanoparticles can produce as good as super catalytic activity for dye degradation, heavy metal removal and pollutant adsorption (Tan et al., 2023; Saif et al., 2016).
- Magnetic Properties: Biogenic synthesis allows production of superparamagnetic NPs with high stability & responsiveness, making them suitable for biomedical and environmental applications (Tan et al., 2023; Herlekar et al., 2014).

3.2.3 *Hibiscus sabdariffa*

Hibiscus sabdariffa (shown in figure 3(a)), its extract was prepared by air drying and grinding fine powder of its calyces. The extract was stirred with distilled water in the form of 1:1 addition in the rate of 60°C for 30 minutes. It was filtered, oven-dried and stored at 4°C. FeO nanoparticles were synthesised using 50 millilitres of the extract with 1 mM ferrous sulphate ($FeSO_4$) solution. Continuous stirring for 24 hours caused a colour change (pink to dark brown). Thus, the centrifuged supernatant was discarded, and NPs were separated by centrifuging at 12,000 rpm for 15 minutes. FeO NPs containing pellet were cleaned over many cycles to wash off the impurities (Saod et al., 2024).

3.2.4 Henna and Gardenia leaves

FeNPs were synthesised by green synthesis using plant extracts of Henna (scientifically named as *Lawsonia Inermis*) and Gardenia (scientifically named as *Gardenia jasminoides*) as shown in Figure 3(b). This process includes collecting, washing, drying, grinding, and sieving the leaves to get a fine powder. The precursor for the iron nanoparticles is a 0.01 M $FeSO_4$ solution. After adding plant powder to deionized water and stirring for 3 hours, the plant extract is then filtered. Slowly the extract is added to the $FeSO_4$ solution, stirred at 70°C. After each addition of the plant extract the temperature difference is noted. Plant extract aids as a reducing agent, a capping agent, & a stabilising agent during the synthesis process. It stabilises the iron nanoparticles formed, prevent it from agglomeration or growth too large, and reduces iron ions (Fe^{2+}) to form those. The final product is obtained by drying the collected NPs at 50 ℃ for about 3 hours (Naseem & Farrukh, 2015).

3.2.5 *Padina pavonica*

The green chemistry approach is used for synthesis of *Padina pavonica* FeNPs (Pa@FeNPs): the preparation of two Fe salts, Fe(III) chloride hexahydrate ($FeCl_3 \cdot 6H_2O$) & Fe(II) chloride tetrahydrate ($FeCl_2 \cdot 4H_2O$) in distilled water with a volume of 1M. Presence of (bioactive) compounds in the algae is used to reduce Fe ions to the metallic form and to stabilise NPs. Fe Nps are formed through reduction of Fe ions, which occurs in presence of boiling mixture (70-80 °C). Nanoparticle formation is induced by addition of Sodium Hydroxide (NaOH) & colour of solution changes (yellow to black). NPs are then centrifuged to separate them, are washed twice with distilled water, and once with ethanol, dried at 40°C for 48 hours, and kept in a refrigerator. In addition, these NPs can be evaluated for different applications like environmental remediation or biomedical (Caf, 2022).

3.2.6 *Mikania mikrantha*

Magnetic Fe_3O_4 NPs synthesised using *Mikania mikrantha* leaf as reducing agent. Leaves of *Mikania mikrantha* are collected, washed, chopped, and boiled in double distilled water and kept at -4°C. The resulting solution of 5 mol of $FeSO_4$ and 10 mol ferric chloride was then stirred into the reaction mixture, continuously. Creation of Fe_3O_4 NPs was indicated by change from colourless dark brown colour. After settling, a precipitate was formed in the mixture (Biswas et al., 2021).

3.2.7 *Borassus flabellifer*

Synthesis of Fe_2O_3 NPs was performed through collecting tender seed in summer season under exchange *Borassus flabellifer* seed coat extract. The seed coats were peeled, the seeds rinsed with twice distilled deionized water & dried under shade and stored in air tight bags in a room temperature. Plant extract was prepared by loading 30g crushed seed coat into a thimble using ethanol (250 ml) as the solvent in a Soxhlet apparatus. This extract was extracted for 30 cycles, cooled, and placed in a refrigerator. For the synthesizing Fe NPs, a solution was prepared by mixing $FeCl_3$ and $FeSO_4$ in a 2:25 mL and was added in plant extract under constant and at room temperature, then mixed with 25 mL of this solution 1 ratio. Synthesizing Fe NPs and Fe_2O_3 was shown by adjusting the pH (10-11, 0.1 M NaOH) which yields immediate creation (black precipitate). It was washed 3 times with double-distilled water and twice by ethanol, dried at room temperature to obtain magnetic $Fe_2 O_3$ NPs. The plant extract was used as a reducing and stabilising agent that helped in ecofriendly synthesis through the bioactive compounds (polyphenols, flavonoids, proteins, etc); which made it a green and sustainable way of making nanoparticles (Sandhya & Kalaiselvam, 2019).

3.2.8 *Moringa oleifera*

FeO NPs have been synthesised using *Moringa oleifera* (shown in figure 3(c)) leaf extract by green synthesis. The methods involved are using Fe(III) $FeCl_3$, MHA, and Mueller Hinton broth. *Moringa oleifera* leaves were freshly collected and distilled water was used during the preparation and synthesis processes. Solution of 0.5 M $FeCl_3$ in 100 mL of water was prepared by adding $FeCl_3$ into water and stirring 500 rpm for 1 hour at 60°C. Iron nanoparticles were formed by heating the solid precipitate at 100°C for 24 hours. Each particles were crushed washed & washed multiple times with water & dried under a vacuum at 60°C for 4 hrs. For preparation of capped *Moringa oleifera* iron nanoparticles (MO-FeONPs), three different concentrations of $FeCl_3$ mixed with 20 mL of MO extract in a 100 mL volume ratio of 8:2. The dark brown colour of the reaction mixture after 2 minutes indicated that the desired nanoparticles had formed. Fe NPs that had been synthesised were tested for their antibacterial activity using well-diffusion method against some pathogenic bacterial strain. Autoclave, and all the MHA was prepared and poured into petri dishes. Normal saline and diluted blood samples were used in a hemolysis assay, which gave 3.5% hemolysis, less than the acceptable limit of 5% for biomedical applications (Aisida et al., 2019).

3.2.9 Seaweed

The study by Salem et al. (2019), showed the synthesis & characterization of FeO NPs, where ferric chloride hexahydrate ($FeCl_3{\cdot}6H_2O$) is used as a precursor. Seaweed collected include *Colpomenia sinuosa* and *Pterocladia capillacea*. Seaweed samples were carefully washed, formalin preserved in seawater for taxonomic identification, and air dried, ground to a fine powder. To prepare the seaweed extracts, 100mL of Deionized Dilute Water (DDW) was added to 1g of his dried and ground material and brought to a boil and then brought to a simmer for 15 minutes. Extracts were filtered and stored in a refrigerator for nanoparticle synthesis. FeO NPs were synthesized by combining seaweed extracts with a 0.1 M $FeCl_3$ solution.

3.2.10 *Catharanthus roseus*

The study by Butt et al., (2018), local malt extract and $FeSO_4$ are used to synthesise iron nanoparticles (FeNPs). Fresh *Catharanthus roseus* (shown if figure 3(d)) leaves were grounded to powder, washed & dried, then organized as extract by microwave assisted extraction. FeNPs were synthesized in two concentrations: 10% and 30%. Solution changed colour (green to dark brown), it was observed that the Fe ions had been bio-reduced to form iron nanoparticles. Finally, the synthesised FeNPs were tested for their antifungal efficacy against two of the fungus, *Aspergillus nidulans* and *Aspergillus terreus*. However, fungal biomass reduced by 30% substantially, and much higher antifungal activity was observed at higher FeNPs concentrations, suggesting the potential of FeNPs as eco-friendly and potent antifungal agents (Butt et al., 2018).

3.2.11 *Enterococcus faecalis RMSN6*

A bacterium, *Enterococcus faecalis RMSN6*, isolated from a metal-contaminated site, is involved in a biological process that results in the synthesis of Fe_3O_4 nanoparticles. MRS broth cultured bacterial strain for 48 h and cell free supernatant containing extracellular polymeric substances (EPS) was prepared. Nanoparticle formation was initiated with iron salts, and the solution became dark brown. By virtue of these unique properties, the bacteria were able to reduce and stabilise the nanoparticles using EPS. Different variables were tested for temperature, pH, & development time to optimise yield & quality of NPs (Samuel et al., 2021).

3.2.12 *Skimmia laureola*

Alam et. al., (2018) in his study, mentioned synthesis of Fe_2O_3 NPs with *Skimmia laureola* leaf, which was prepared by washing the leaves with deionised water, and slicing the leaves into small pieces. The synthesis of biogenic Fe_2O_3 NPs involved mixing a 0.1 M solution of $FeCl_3.6H_2O$ with plant extract. Thus, under these conditions the purified Fe_2O_3 NPs were obtained by washing the NPs several times with ethanol & vacuum drying (Alam et al., 2018).

Figure 3. (a) Hibiscus sabdariffa, (b) Gardenia, (c) Moringa Oleifera, (d) Catharanthus roseus

4. CHARACTERISTICS OF IRON NANOPARTICLES BY GREEN SYNTHESIS

4.1 *Hibiscus sabdariffa*

Hibiscus sabdariffa extract is used as a green method to synthesise iron nanoparticles (FeO NPs) and this study examines how. H. sabdariffa calyces powdered were mixed with distilled water and 1 mM $FeSO_4$ solution and stirred simultaneously for 24 hours to obtain the extract. For further characterization nanoparticles were separated, washed, and dried. FTIR analysis confirmed the plant extract's presence and also proved the existence of functional groups in the synthesised FeO balls, and UV-Vi's spectrum confirmed that synthesised FeO NPs had strong absorption peak. Atomic force microscopy (AFM) images of the pine needles displayed a 3D, mountain-like structure, having average grain size of 40-56 nm, whereas FSEM revealed the spherical morphology of the MoCs having average size of 61 nm. XRD patterns showed a crystalline nature with a peak at $2\theta \approx 69.2°$ (Saod et al., 2024).

4.2 Henna and Gardenia leaves

Thermogravimetric analysis (TGA) showed weight loss 12% and 31% (60 - 205 °C) related to moisture removal and coumaric acid loss, respectively. At 380-690°C, oxygen loss from $FeSO_4$ caused an initial weight loss of 7% and a second weight loss of 22% for the Gardenia leaves extract. FTIR revealed compounds such as Lawson that copackage with iron to stabilise the nanoparticles. XRD performed and confirmed formation of Fe NPs with production peaks at (2θ) 44.34° and 64.43° which are characteristic of metallic iron crystal planes. The iron nanoparticles synthesised with Henna leaves extract was about 21 nm and about 32 nm for those synthesised with Gardenia leaves extract. The SEM analysis produced a distorted hexagonal like appearance of nanoparticles formed using Henna leaves extract having elemental composition iron (6.86%), carbon (54.59%) and oxygen (36.57%) (Naseem & Farrukh, 2015).

4.3 *Padina pavonica*

A variety of analytical techniques is used to characterise iron nanoparticles synthesised using the marine macroalgae *Padina pavonica* (Pa@FeNPs), to support a better characterization of their properties, structure, and functionality. FeNPs were initially synthesised by adding water of the P. pavonica to iron solution and forming during the reaction a black solution indicating the formation of Fe_3O_4. The absorption & scattering of UV radiation by the nanoparticles was indicated by UV-Vi's Spectroscopy which showed absorption bands in the 325-375 nm range. Functional groups (hydroxyl, methylene, amino groups, etc) identified in adsorbent using FTIR were probably involved in the reduction of iron ions and stabilisation of the nanoparticles. SEM images revealed a polydisperse nanoparticle distribution with surface porosity, useful for adsorption applications. EDX of the synthesised nanoparticles provided elemental composition indication as peaks for iron and oxygen were present with fairly weak peaks for carbon, nitrogen, and sulphur indicative of phytochemicals in the seaweed extract. The crystal structure of the nanoparticles was revealed by XRD, revealing characteristic peaks confirming the presence of magnetite (Fe_3O_4). TEM images showed NPs were spherical & with some agglomeration, a common characteristic of magnetic nanoparticles. Zeta potential of the NPs was measured by DLS, indicating agglomeration (Caf, 2022).

4.4 *Mikania mikrantha*

Physicochemical study and biosynthesis of Fe_3O_4 NPs using *Mikania mikrantha* leaves were done with several spectroscopic techniques. FT-IR spectroscopy exhibited significant absorption peaks at their wavelength's characteristic of O-H vibrations (alcohols & phenolic compounds), and N-H stretching vibrations in amides. XRD analysis results confirmed that the synthesised NPs have a crystalline structure, with diffraction peaks at 30.27°, 35.60°, 36.75°, 43.41°, 53.88°, 57.38° and 62.90°. TEM & SAED pattern confirmed crystalline nature of NPs. SEM revealed a well dispersed rhomboidal structure with average size 15-20 nm. Presence of Fe in the NPs was confirmed by EDX analysis. Magneto metric characterization of the synthesised Fe_3O_4 nanoparticles was carried out using Vibrating Sample Magnetometry (VSM), which indicated superparamagnetic behaviour with zero remanence and coercivity, making it suitable to use in medical & biological applications (Biswas et al., 2021).

4.5 *Borassus flabellifer*

Synthesis & characterization of magnetic iron oxide nanoparticles was demonstrated by various techniques. The structure seemed to fit with XRD, revealing presence of magnetite phase with traces of hematite, with an average crystalline size of 35 nm. Broad band absorption at 352 nm in the UV-Vi's Spectroscopy revealed the presence of nanoparticles as shown in Figure 4. FTIR showed that significant differences between the extract and synthesised nanoparticles. TGA indicated weight loss from 0 °C to 700 °C, with weight loss of 52.32% between 100 °C - 500 °C as a result of decomposition of the bioactive compounds from the extract. SEM images of hexagonal, well dispersed iron oxide nanoparticles with small clustering were shown. The composition of the nanoparticles was confirmed by EDX which revealed 52.81% iron, 29.7% oxygen, and 17.5% other elements, possibly formed from the extract, and stabilising the nanoparticles during the synthesis process (Sandhya & Kalaiselvam, 2019).

Figure 4. UV-vi's Absorption Spectra of iron oxide nanoparticles (Sandhya & Kalaiselvam, 2019) (Creative Commons Attribution 4.0 International (CC BY 4.0)).

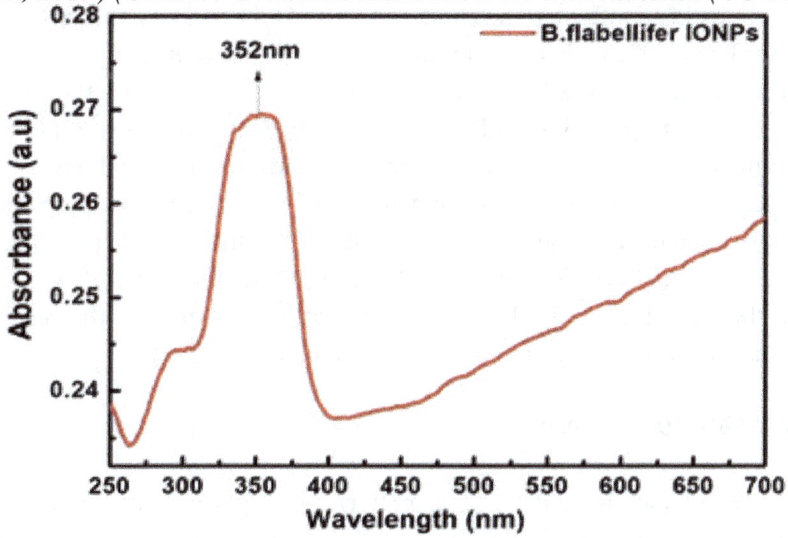

4.6 *Moringa oleifera*

FeO NPs were synthesised using green synthesis method by reducing iron salts with *Moringa oleifera* leaf extract. Surface modification was found to cause crystallisation according to XRD analysis, and UV Vis spectroscopy showed a shift in colour from dark brown to light orange. Functional groups for reduction & capping of Fe NPs determined by FTIR. TEM and SEM images of the highly aggregated as synthesised nanoparticles without further surface modification are presented. EDX analysis revealed presence of iron, carbon, & oxygen in NPs. Magnetic properties of FeO NPs were studied by VSM and indicated superparamagnetic behaviour with the saturation magnetization of ~ 2.82 emu/g (Aisida et al., 2019).

4.7 Seaweed

Production of Fe_3O_4 NPs was carried out using substances from the marine algae *Colpomenia sinuosa* & *Pterocladia capillacea.* The colour change (brown to buff) in *C. sinuosa* and *P. capillacea* proved that nanoparticles were produced during the experiment. Due to hydroxyl and sulphate groups in seaweed polysaccharides, Fe^{3+} ions were reduced to Fe_3O_4 and the resulting nanoparticles were made stable. To understand the synthesised nanoparticles, they were characterised by several methods. FT-IR analysis clearly showed changes in the wavelengths assigned to sulphate and hydroxyl groups, indicating their role in making the nanoparticles. There were absorption bands seen at 402 nm and 415 nm when using UV-Vi's spectroscopy which confirmed the synthesis of Fe_3O_4 NPs. The nanoparticles under SEM showed a spherical shape and their sizes were approximately in the range of 11-34 nm. According to the EDX analysis, iron was indeed within the make-up of the synthesised nanoparticles (Salem et al., 2019). Furthermore, the antibacterial activity of the Fe_3O_4-NPs was observed mainly against bacteria that have a negative Gramme stain. The nanoparticles obtained from *C. sinuosa* behaved better against Vibrio cholerae, halting its growth at a radius of 15 mm in the agar plate, but the nanoparticles from *P. capillacea* blocked the least bacteria in *Staphylococcus aureus*. It is shown in this study that Fe_3O_4 nanoparticles formed with seaweed have the potential to control microorganisms (Salem et al., 2019).

4.8 *Catharanthus roseus*

Various techniques were used to characterise the synthesis of FeNPs including UV-Vi's spectroscopy, particle size analyser, FTIR and SEM. UV-Vis Spectroscopy verified that there had been the formation of nanoparticles, which had absorption peaks in the 199-260 nm range over 24 hours. Average size of synthesised Fe NP s was approximately 266 nm, which was measured using a nano laser particle size analyser. Significant absorption bands at 1625 cm^{-1} (COOH), 2150 cm^{-1} (C=C), and 3360 cm^{-1} (O-H) were observed in FTIR analysis for carboxylic groups, alkenes, and phenolics respectively. The presence of iron vibrations, characteristic of iron nanoparticles, was confirmed with a peak at 670 cm^{-1}. SEM images revealed that the NPs were not well dispersed, forming irregular clusters and some areas that instead appeared to be collections of nanoparticles. *Catharanthus roseus* leaf extract synthesised iron nanoparticles exhibit strong antifungal properties and show promise in agriculture and medicine (Butt et al., 2018).

4.9 *Enterococcus faecalis RMSN6*

Analytical techniques were used to characterise bio-synthesised Fe_3O_4 NPs. This phenomenon of surface plasmon resonance (SPR) of Fe_3O_4 NPs confirmed using UV-Vi's spectroscopy by changing colour of reaction mixture from colourless to darkish brown. FTIR showed that functional groups from the extracellular polysaccharides (EPS) of *Enterococcus faecalis RMSN6* modified the nanoparticles as shown in figure 5(a). Structural information was provided by XRD (as shown in figure 5(b)) and SEM images showing irregularly shaped Fe_3O_4 NPs having cubical, hexagonal, and bricklike structures. They used TEM images that confirmed the morphology of nanoparticles, and EDX analysis to confirm the elemental composition of the nanoparticles. DLS analysis Article was used to measure Fe_3O_4 NPs particle size which was approximately 15.4 nm. The nanoparticles showed -29.6 mV zeta potential indicating good stability. It was confirmed by these characterization techniques that the Fe_3O_4 nanoparticles were synthesised successfully, of size and morphology appropriate for use in removal of heavy metal ions polluted waters (Samuel et al., 2021).

Figure 5. (a) FT-IR spectra of Fe_3O_4 NPs, (b) XRD patterns of Fe_3O_4 NPs (Samuel et al., 2021). Creative Commons Attribution 4.0 International (CC BY 4.0).

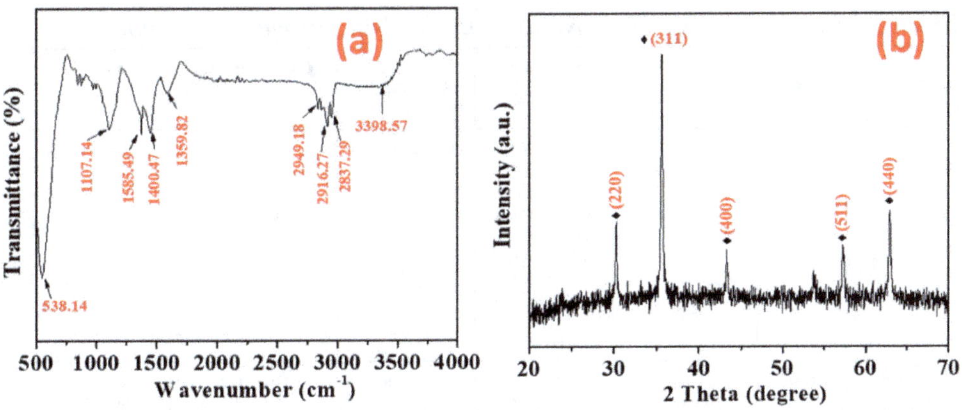

4.10 *Skimmia laureola*

Aqueous extract of *Skimmia laureola* leaves used to achieve synthesis of iron Fe_2O_3 NPs. When leaf extract is mixed with $FeCl_3$, the colour changes from pale yellow indoors to dark brown at room temperature. The bio-reduction of ferric ions to Fe_2O_3 NPs was confirmed by UV-vi's spectroscopy analysis. The phytochemicals found in leaf extract were determined by FTIR analysis as reducing & stabilising agents for NPs. A cubic crystal structure was inferred from the FTIR spectrum with multiple absorption peaks. The Fe_2O_3 NPs had an average particle size about 34 nm. EDX analysis confirmed presence of Fe & O, with carbon (C) and chloride (Cl) elements. It was determined that weight percentages of Fe, Cl, C, and O are 83.61%, 1.86%, 2.22%, and 12.32%, respectively. Morphological studies of the nanoparticles were performed using SEM and nanoparticles appeared crystalline and cubic in shape. Antibacterial effect of Fe_2O_3 NPs was determined in vivo, i.e. on tomato plants infected by *Ralstonia solanacearum*. Treated plants showed enhanced growth parameters compared to control plants indicating that not only did Fe_2O_3 NPs control bacterial wilt but it also promoted plant growth (Alam et al., 2018).

Table 1 summarizes the literature review discussed above as well as provides additional information (Tyagi et al., 2021; Jamzad & Bidkorpeh, 2020; Buarki et al., 2022; Lakshminarayanan et al., 2021; Bensy et al., 2022; Da'na et al., 2018; Devi et al., 2018; Beheshtkhoo et al., 2018) on green synthesis of iron NPs using various biogenic components, detailing their synthesis methods, nanoparticle sizes, morphological characteristics, and potential applications.

Table 1. Green Synthesis of Iron Nanoparticles Using Biogenic Components

Biogenic Component	Part Used	Iron Precursor	Method of Synthesis	Size of Synthesized Nanoparticle	Characteristics	Reference
Hibiscus sabdariffa	Calyces	$FeSO_4$	Air-dried calyces ground, mixed with distilled water, stirred at 60°C, filtered, combined with $FeSO_4$solution, stirred for 24 hours	Average 61 nm	Spherical morphology, crystalline structure	(Saod et al., 2024)
Henna and *Gardenia*	Leaves	$FeCl_3$ and $FeSO_4$	Leaf extract mixed with iron salt solution, stirred at 70°C	21 nm (Henna), 32 nm (Gardenia)	Distorted hexagonal-like appearance (Henna)	(Naseem & Farrukh, 2015)
Padina pavonica	Whole seaweed	$FeCl_3{\cdot}6H_2O$ and $FeCl_2{\cdot}4H_2O$	Seaweed extract mixed with iron salt solutions, heated to 70-80°C, NaOH added	10.53 - 20.55 nm	Spherical shape, some agglomeration	(Caf, 2022)
Mikania mikrantha	Leaves	$FeSO_4$ and $FeCl_3$	Leaf extract mixed with iron salt solution, stirred continuously	15 - 20 nm	Rhomboidal structure, superparamagnetic behaviour	(Biswas et al., 2021)
Borassus flabellifer	Seed coat	$FeCl_3$ and $FeSO_4$	Seed coat extract mixed with iron salt solution, stirred at room temperature, pH adjusted	Average 35 nm	Hexagonal shape, well-dispersed with slight aggregation	(Sandhya & Kalaiselvam, 2019)

continued on following page

Table 1. Continued

Biogenic Component	Part Used	Iron Precursor	Method of Synthesis	Size of Synthesized Nanoparticle	Characteristics	Reference
Moringa oleifera	Leaves	$FeCl_3{\cdot}6H_2O$	Leaf extract mixed with iron chloride solution, heated, centrifuged, dried	10 - 90 nm	Highly aggregated, superparamagnetic behaviour, antibacterial activity	(Aisida et al., 2019)
Seaweed (*Colpomenia sinuosa* and *Pterocladia capillacea*)	Whole seaweed	$FeCl_3{\cdot}6H_2O$	Seaweed washed, dried, ground, mixed with DDW, heated, filtered, combined with $FeCl_3$ solution	11.24 - 33.71 nm (*Colpomenia sinuosa*), 16.85 - 22.47 nm (*Pterocladia capillacea*)	Spherical shape, antibacterial and antifungal properties	(Salem et al., 2019)
Catharanthus roseus	Leaves	$FeSO_4$	Leaves washed, dried, ground, microwave-assisted extraction, combined with iron salt solution	Average 266 nm	Irregular clusters, antifungal properties	(Butt et al., 2018)
Enterococcus faecalis RMSN6	Bacterial culture	$FeCl_2{\cdot}4H_2O$ and $FeCl_3{\cdot}6H_2O$	Bacterial culture supernatant mixed with iron salts	Approximately 15.4 nm	Irregularly shaped (cubical, hexagonal, and brick-like structures)	(Samuel et al., 2021)
Skimmia laureola	Leaves	$FeCl_3{\cdot}6H_2O$	Leaf extract mixed with iron salt solution at room temperature	Average 34 nm	Crystalline and cubic shape, antibacterial activity	(Alam et al., 2018)
Spinacia oleracea	Leaf	$FeCl_3$	Aqueous extract mixed with $FeCl_3$ at 60°C	10 - 70 nm	Amorphous structure, antibacterial property	(Tyagi et al., 2021)
Musa acuminata	Peel	$FeCl_3$	Aqueous extract mixed with $FeCl_3$ at room temp	20 - 50 nm	Amorphous structure, antibacterial property	(Tyagi et al., 2021)
Laurus nobilis	Leaves	$FeCl_3{\cdot}6H_2O$	Aqueous extract mixed with $FeCl_3{\cdot}6H_2O$ at room temperature	8.03 ± 8.99 nm	Crystalline, almost spherical, and partly hexagonal shape	(Jamzad & Bidkorpeh, 2020)
Hibiscus rosasinensis	Flowers	$FeCl_2{\cdot}4H_2O$	Microwave-assisted synthesis using flower extract	51 nm	Crystalline α-Fe_2O_3 nanoparticles, antibacterial property	(Buarki et al., 2022)

continued on following page

Table 1. Continued

Biogenic Component	Part Used	Iron Precursor	Method of Synthesis	Size of Synthesized Nanoparticle	Characteristics	Reference
Bauhinia tomentosa	Leaves	$FeCl_3$ (0.01 M)	One-pot green synthesis using leaf extract	Approximately 70 nm	Iron (II) oxide nanoparticles, used for enzyme immobilization, effective in synthesizing 1,3-diolein	(Lakshminarayanan et al., 2021)
Ulva lactuca	Algae (marine)	$FeCl_3$ (0.1 M)	Green synthesis using aqueous extract	30 - 40 nm	anticancer activity against HeLa and DLD-1 cell lines	(Bensy et al., 2022)
Acacia nilotica	Seedless pods	$FeSO_4 \cdot 7H_2O$	Green synthesis using aqueous extract	30 - 40 nm	effective in dye degradation (Methyl Orange)	(Da'na et al., 2018)
Platanus orientalis	Leaves	$Fe(NO_3)_3 \cdot 9H_2O$	Green synthesis using aqueous leaf extract	Approximately 38 nm	Antifungal property	(Devi et al., 2018)
Daphne mezereum	Leaves	$FeSO_4 \cdot 7H_2O$	Green synthesis using aqueous leaf extract	6.5 - 14.9 nm (mean 9.2 nm)	effective in dye removal (81% decoloration of methyl orange)	(Beheshtkhoo et al., 2018)

5. Applications

Unique properties of iron, iron oxide, and iron compound NPs have led to their widespread application across various fields. These nanoparticles, particularly when synthesized through green methods, offer innovative solutions to challenges in biomedicine, environmental remediation, agriculture, and industrial processes (Saod et al., 2024; Biswas et al., 2021).

Iron-based nanoparticles are not only confined to the medical field but also can be widely used in other different fields. They have a wide range of uses according to types. For example, in drug delivery systems, they have had a good reception primarily considering the fact that their required feature is to enable specific transport of drugs to the relevant cell resulting in the overall therapeutic efficiency being enhanced (Saod et al., 2024). Superparamagnetic properties of iron oxide nanoparticles are specifically the features that make intracellular SPIONs drug delivery carriers to be well controlled and target specific parts of the cell (Biswas et al., 2021). Likewise, these properties also make them to be the number-one eligible for cancer therapy. They can be utilized not only for drug delivery which is oncology-directed but also in hyperthermia therapeutic processes the fact that they can produce heat under alternating a magnetic field which will in turn destroy cancer cells still holds (Aisida et al., 2019). Nanoparticles have the potential of being the next generation of tools for diagnostic imaging. SPIONs are other contrast agents brought as a result of the engagement of Magnetic Resonance Imaging (MRI) that has in the end resulted in improved MRI scan contrast & visualization tissues & organs (Salem et al., 2019). Furthermore, iron-based nanoparticles can be employed in tissue engineering and repair, where they show biocompatibility in forming usable hydrogels or other tissue regeneration applications (Aisida et al., 2019). High stresses and entirely new qualities have led to these functionalities in biosensing applications, which are right now applied for scanning down to the biological molecules (Aisida et al., 2019). Iron nanoparticles for antimicrobial usages form yet another promising area of application in biomedical sciences. Iron oxide nanoparticles generated from plant extracts have studied for their strong antibacterial properties. Nanoparticles prepared from Tridax procumbens are able to kill a lot of gram-negative bacteria, like *Pseudomonas aeruginosa*, while *Dodonaea viscosa* leaf extract-based nanoparticles are excellent antimicrobials against a range of human pathogens. All these findings suggest their possible application in the development of novel antimicrobials and therapies for infectious disease processes (Saif et al., 2016).

For environmental uses, NPs have been potent agents for pollution management and remediation. In the interior manifestation, they are being documented to facilitate treatment of waste. The broad range contaminants, such as organic compounds & heavy metals, capable of being removed from wastewater treatment using these nanoparticles has already been documented (Salem et al., 2019; Kolya & Kang, 2022). Iron oxide nanoparticles are known to be among the best adsorbents. They are very effective in removing heavy metals from an aqueous environment, particularly Fe_3O_4 (Samuel et al., 2021). One demonstration of this capability becomes obvious when nanoparticles having zero valence are prepared from eucalyptus leaf extract. In this case, nanoparticles exhibited chromium adsorption achievement level of 98.1% within just 30 minutes (Saif et al., 2016). Iron nanoparticles, green synthesized, are also expected to have catalytic properties through which they act in degradation of organic-coloured chemicals, which are one of the severest environmental problems found in textile and other industries. Iron nanoparticles derived from green tea showed much higher catalytic activity than other synthesized methods in preparation while degrading methylene blue (Saif et al., 2016). Though still in the infancy of engineering, iron nanoparticles have found their way into the field of biotechnology, primarily in the control of plant diseases. With increasing concerns about the environmental effects of chemical pesticides, iron oxide nanoparticles are seen as promising alternatives. For example, Fe_2O_3 nanoparticles synthesized from *Skimmia laureola* extract exhibit considerable antibacterial activity against *Ralstonia solanacearum*, the pathogen responsible for tomato wilt disease (Alam et al., 2018).

This application shows the potential for exploiting biogenic iron nanoparticles toward developing eco-friendly and effective strategies for crop protection.

In the field of catalysis, iron nanoparticles, especially those developed by green methods, show more activity compared to the conventional iron compounds. Their abilities to act as catalysts in the formation of free radicals from hydrogen peroxide (H_2O_2) raise hopes for industrial and environmental processes. The enhanced photocatalytic property of Fe_2O_3 NPs derived through hydrothermal treatment using biogenic (plant) extracts as opposed to their commercially available counterparts reinforces their promising applications in catalysis (Saif et al., 2016). The unique magnetic properties of iron-based nanoparticles have sounded an alert in recent years in information storage technologies. These nanoparticles are being explored for various uses in the electronics and optoelectronics industries, based on their unique electronic and magnetic properties (Naseem & Farrukh, 2015). The petroleum industry has also found applications of iron-based nanoparticles, using them to enhance drilling fluids, improve oil recovery, and stimulate wells because of their size and reactive capability. In the energy sector, nanoparticles of iron drive innovation in field of renewable energy. They help to increase in efficiency of solar cells and are being studied for their applications in producing fuels (Mughal et al., 2021). With the global push for more sustainable sources of energy, these nanoparticles contributions to development of energy conversion & storage technologies will continue to expand.

6. SAFETY AND TOXICITY OF GREEN-SYNTHESIZED FE NPS

Using biogenic techniques to produce Fe NPs is safer for the environment than the traditional ways of producing them. Yet, their manufacturing methods and the amount, nature of the coating and the situation they are applied in can make a huge difference in how harmful or safe they are. Even small amounts of certain chemicals can still be harmful, especially when we consider how the chemical behaves. Usually, low doses are safe (between 1 mg/kg and 1 mg/L), as the liver, kidneys and brain are unharmed. However, higher doses over 10 mg/kilogramme typically cause harm to the liver, kidneys and brain (Kulkarni et al., 2020; Köktürk et al., 2023). Feeding zebrafish embryos with water that contained 50 mg/L of this substance caused severe consequences, with 60% deceasing and the others developing swelling around the heart and abnormalities in their spines (Köktürk et al., 2023).

Applying a coating to FeNPs could influence how biocompatible and toxic they are. Covering nanoparticles with plant-based coatings such as from *Spinacia oleracea* and *Ficus palmata*, keeps their compatibility with the body high, helping lower hemolytic activity to 5% or less and making ROS generation less common (Kulkarni et al., 2020; Kebede et al., 2012). In some cases, uncoated iron nanoparticles may cause cell death on their own, especially when used at a dose of more than 1000 particles per millilitre (Kebede et al., 2012). Many different responses from organisms to FeONPs demonstrate that these particles are not easy to understand from a toxic viewpoint. FeO NPs at a dose of 100 mg/kg seem to cause liver stress in rats, as indicated by the increase of SGPT and bilirubin in their blood (Kulkarni et al., 2020). Exposing zebrafish larvae to 10 mg/L of FeNPs caused them to move around more, behave nervously and disrupt their daily rhythms. Tests done with red blood cells showed that, even after treating them with the highest dose of FeONPs, only a slight hemolysis was recorded, with the rate always staying below 5%. Moreover, high levels of antioxidants in plant extracts including Sageretia thea made it very effective, showing DPPH scavenging activity of 78.06% at 200 µg/mL (Israeel et al., 2024). Still, many are concerned about the impact on the environment and development in water environments. Studies found that feeding zebrafish larvae 10 mg/L of FeONPs leads to them being less active and drawn to the walls, where they prefer to hide (Köktürk et al., 2023). Accordingly, it is necessary to control nanoparticle levels at less than 1 mg/L to avoid harm to the environment from repeated exposure.

Various suggestions have been made to manage these toxicity challenges. Favourable results were obtained when peptides were used at 1–25 µg/mL, because at this range their antibacterial and biocompatible properties were balanced (Kebede et al., 2012; Mohamed et al., 2023). In addition, adding plant materials to nanoparticles helps them remain stable and reduces their likelihood of producing dangerous reactive oxygen species (Kulkarni et al., 2020; Kebede et al., 2012). In particular, studies show that FeONPs fight several bacteria such as Salmonella typhi and Escherichia coli, although their success depends on the concentrations applied; often, they add to effects of toxicity (Kebede et al., 2012; Mohamed et al., 2023). FeONP appears to be more successful in controlling Staphylococcus aureus than Gramme-negative bacteria which could highlight a strategy for focusing on particular strains in the future (Mohamed et al., 2023). Typically, medicines made from FeNPs produced in sustainable ways show great promise and are safe for humans at lower levels, but proper supervision and dressing the nanoparticles are required to avoid any effects on our organs or the environment. Further studies are necessary to ensure that all biological and ecological models are protected against chemical exposure (Kebede et al., 2012; Mohamed et al., 2023).

7. CONCLUSION

Biogenic synthesis of metallic Fe NPs is important milestone in nanotechnology, which provides an economically viable and green alternative to the present physical or chemical pathways. This helps to lessen the whole harmful chemical usages and strengthens the compatibility and functionality of the produced nanoparticles through use of natural materials like plant component, algae, and micro-organisms as the articles of use. These Fe NPs have wide range of applications in biomedicine, environmental remediation, & agriculture highlighting their versatility and potential to address pressing global challenges. However, the full potential of synthetic polymers can only be realized through continuous research, which aims to refine synthesis methods, enhance scalability, and discover new applications. The future of green nanotechnology will be to take the method of sustainable development with innovative solutions to advances in all fields of science and industry.

REFERENCES

Aisida, S. O., Madubuonu, N., Alnasir, M. H., Ahmad, I., Botha, S., Maaza, M., & Ezema, F. I. (2019). Biogenic synthesis of iron oxide nanorods using Moringa oleifera leaf extract for antibacterial applications. *Applied Nanoscience*, *10*(1), 305–315. DOI: 10.1007/s13204-019-01099-x

Alam, T., Khan, R. A., Ali, A., Sher, H., Ullah, Z., & Ali, M. (2018). Biogenic synthesis of iron oxide nanoparticles via Skimmia laureola and their antibacterial efficacy against bacterial wilt pathogen Ralstonia solanacearum. *Materials Science and Engineering C*, *98*, 101–108. DOI: 10.1016/j.msec.2018.12.117 PMID: 30812984

Alphandéry, E. (2020). Bio-synthesized iron oxide nanoparticles for cancer treatment. *International Journal of Pharmaceutics*, *586*, 119472. DOI: 10.1016/j.ijpharm.2020.119472 PMID: 32590095

Ashrafi-Saiedlou, S., Rasouli-Sadaghiani, M., Fattahi, M., & Ghosta, Y. (2025). Biosynthesis and characterization of iron oxide nanoparticles fabricated using cell-free supernatant of Pseudomonas fluorescens for antibacterial, antifungal, antioxidant, and photocatalytic applications. *Scientific Reports*, *15*(1). Advance online publication. DOI: 10.1038/s41598-024-84974-0 PMID: 39762412

Batool, F., Iqbal, M. S., Khan, S., Khan, J., Ahmed, B., & Qadir, M. I. (2021). Biologically synthesized iron nanoparticles (FeNPs) from Phoenix dactylifera have anti-bacterial activities. *Scientific Reports*, *11*(1). Advance online publication. DOI: 10.1038/s41598-021-01374-4 PMID: 34764312

Beheshtkhoo, N., Kouhbanani, M. J., Savardashtaki, A., Amani, A. M., & Taghizadeh, S. (2018). Green synthesis of iron oxide nanoparticles by aqueous leaf extract of Daphne mezereum as a novel dye removing material. *Applied Physics. A, Materials Science & Processing*, *124*(5). Advance online publication. DOI: 10.1007/s00339-018-1782-3

Bensy, A. D., Christobel, G. J., Muthusamy, K., Alfarhan, A., & Anantharaman, P. (2022). Green synthesis of iron nanoparticles from Ulva lactuca and bactericidal activity against enteropathogens. *Journal of King Saud University. Science*, *34*(3), 101888. DOI: 10.1016/j.jksus.2022.101888

Biswas, A., Vanlalveni, C., Lalfakzuala, R., Nath, S., & Rokhum, S. L. (2021). Mikania mikrantha leaf extract mediated biogenic synthesis of magnetic iron oxide nanoparticles: Characterization and its antimicrobial activity study. *Materials Today: Proceedings*, *42*, 1366–1373. DOI: 10.1016/j.matpr.2021.01.108

Buarki, F., AbuHassan, H., Hannan, F. A., & Henari, F. Z. (2022). Green Synthesis of Iron Oxide Nanoparticles Using Hibiscus rosa sinensis Flowers and Their Antibacterial Activity. *Journal of Nanotechnology*, *2022*, 1–6. DOI: 10.1155/2022/5474645

Butt, B. Z., Sumera Javad, S. J., & Saba Zia, S. Z. (2018). Biogenic synthesis of iron nanoparticles from Catharanthus roseus. *Asian Journal of Agriculture and Biology*, *6*(1), 83–89.

Caf, F. (2022). Biogenic Synthesis of Iron Oxide Nanoparticle Using Padina Pavonica Extract: Application for Photocatalytic Degradation of Congo Red Dye, Neurotoxicity and Antioxidant Activity. *Turkish Journal of Fisheries and Aquatic Sciences*, *23*(2). Advance online publication. DOI: 10.4194/trjfas21398

Da'na, E., Taha, A., & Afkar, E. (2018). Green Synthesis of Iron Nanoparticles by Acacia nilotica Pods Extract and Its Catalytic, Adsorption, and Antibacterial Activities. *Applied Sciences (Basel, Switzerland)*, *8*(10), 1922. DOI: 10.3390/app8101922

Devi, H. S., Boda, M. A., Shah, M. A., Parveen, S., & Wani, A. H. (2018). Green synthesis of iron oxide nanoparticles using Platanus orientalis leaf extract for antifungal activity. *Green Processing and Synthesis*, *8*(1), 38–45. DOI: 10.1515/gps-2017-0145

Diaz-Puertas, R., Alvarez-Martinez, F. J., Falco, A., Barrajon-Catalan, E., & Mallavia, R. (2023). Phytochemical-Based Nanomaterials against Antibiotic-Resistant Bacteria: An Updated Review. *Polymers*, *15*(6), 1392. DOI: 10.3390/polym15061392 PMID: 36987172

Herlekar, M., Barve, S., & Kumar, R. (2014). Plant-Mediated Green Synthesis of Iron Nanoparticles. *Journal of Nanoparticles*, *2014*, 1–9. DOI: 10.1155/2014/140614

Israeel, M., Iqbal, J., Abbasi, B. A., Ijaz, S., Ullah, R., Zarshan, F., Yaseen, T., Khan, G., Murtaza, G., Ali, I., Alarjani, K. M., Elshikh, M. S., Rizwan, M., Khan, S., & Iqbal, R. (2024). Potential biological applications of environment friendly synthesized iron oxide nanoparticles using Sageretia thea root extract. *Scientific Reports*, *14*(1). Advance online publication. DOI: 10.1038/s41598-024-79953-4 PMID: 39550505

Jamzad, M., & Bidkorpeh, M. K. (2020). Green synthesis of iron oxide nanoparticles by the aqueous extract of Laurus nobilis L. leaves and evaluation of the antimicrobial activity. *Journal of Nanostructure in Chemistry*, *10*(3), 193–201. DOI: 10.1007/s40097-020-00341-1

Kang, Y., Kim, M., Shim, C., Bae, S., & Jang, S. (2021). Potential of Algae–Bacteria Synergistic Effects on Vegetable Production. *Frontiers in Plant Science*, *12*. Advance online publication. DOI: 10.3389/fpls.2021.656662 PMID: 33912211

Kebede, A., Singh, A. K., Rai, P. K., Giri, N. K., Rai, A. K., Watal, G., & Gholap, A. V. (2012). Controlled synthesis, characterization, and application of iron oxide nanoparticles for oral delivery of insulin. *Lasers in Medical Science*, *28*(2), 579–587. DOI: 10.1007/s10103-012-1106-3 PMID: 22581389

Köktürk, M., YiĞiT, A., & Sulukan, E. (2023). Green synthesis iron oxide nanoparticles (FE@AV NPs) induce developmental toxicity and Anxiety-Like behavior in zebrafish Embryo-Larvae. *Marine Science and Technology Bulletin*, *12*(1), 39–50. DOI: 10.33714/masteb.1220668

Kolya, H., & Kang, C. (2022). Biogenic Synthesis of Silver-Iron Oxide Nanoparticles Using Kulekhara Leaves Extract for Removing Crystal Violet and Malachite Green Dyes from Water. *Sustainability*, *14*(23), 15800. DOI: 10.3390/su142315800

Kulkarni, S., Mohanty, N., Kadam, N. N., Swain, N., & Thakur, M. (2020). Green synthesis to develop Iron-Nano formulations and its toxicity assays. *Journal of Pharmacopuncture*, *23*(3), 165–172. DOI: 10.3831/kpi.2020.23.3.165 PMID: 33072414

Kumar, A. P. N., Kumar, M., Jose, A., Tomer, V., Oz, E., Proestos, C., Zeng, M., Elobeid, T. K. S., & Oz, F. (2023). Major Phytochemicals: Recent Advances in Health Benefits and Extraction Method. *Molecules (Basel, Switzerland)*, *28*(2), 887. DOI: 10.3390/molecules28020887 PMID: 36677944

Lakshminarayanan, S., Shereen, M. F., Niraimathi, K. L., Brindha, P., & Arumugam, A. (2021). One-pot green synthesis of iron oxide nanoparticles from Bauhinia tomentosa: Characterization and application towards synthesis of 1, 3 diolein. *Scientific Reports*, *11*(1). Advance online publication. DOI: 10.1038/s41598-021-87960-y PMID: 33883589

Lee, S., & Ryu, C. (2021). Algae as New Kids in the Beneficial Plant Microbiome. *Frontiers in Plant Science*, *12*. Advance online publication. DOI: 10.3389/fpls.2021.599742 PMID: 33613596

Mohamed, A., Atta, R. R., Kotp, A. A., El-Ela, F. I. A., El-Raheem, H. A., Farghali, A., Alkhalifah, D. H. M., Hozzein, W. N., & Mahmoud, R. (2023). Green synthesis and characterization of iron oxide nanoparticles for the removal of heavy metals (Cd2+ and Ni2+) from aqueous solutions with Antimicrobial Investigation. *Scientific Reports*, *13*(1). Advance online publication. DOI: 10.1038/s41598-023-31704-7 PMID: 37142660

Mughal, B., Zaidi, S. Z. J., Zhang, X., & Hassan, S. U. (2021). Biogenic Nanoparticles: Synthesis, Characterisation and Applications. *Applied Sciences (Basel, Switzerland)*, *11*(6), 2598. DOI: 10.3390/app11062598

Nahari, M. H., Ali, A. A., Asiri, A., Mahnashi, M. H., Shaikh, I. A., Shettar, A. K., & Hoskeri, J. (2022). Green Synthesis and Characterization of Iron Nanoparticles Synthesized from Aqueous Leaf Extract of Vitex leucoxylon and Its Biomedical Applications. *Nanomaterials (Basel, Switzerland)*, *12*(14), 2404. DOI: 10.3390/nano12142404 PMID: 35889627

Naseem, T., & Farrukh, M. A. (2015). Antibacterial Activity of Green Synthesis of Iron Nanoparticles Using Lawsonia inermis and Gardenia jasminoides Leaves Extract. *Journal of Chemistry*, *2015*, 1–7. DOI: 10.1155/2015/912342

Saif, S., Tahir, A., & Chen, Y. (2016). Green Synthesis of Iron Nanoparticles and Their Environmental Applications and Implications. *Nanomaterials (Basel, Switzerland)*, *6*(11), 209. DOI: 10.3390/nano6110209 PMID: 28335338

Salem, D. M., Ismail, M. M., & Aly-Eldeen, M. A. (2019). Biogenic synthesis and antimicrobial potency of iron oxide (Fe3O4) nanoparticles using algae harvested from the Mediterranean Sea, Egypt. *Egyptian Journal of Aquatic Research*, *45*(3), 197–204. DOI: 10.1016/j.ejar.2019.07.002

Samuel, M. S., Datta, S., Chandrasekar, N., Balaji, R., Selvarajan, E., & Vuppala, S. (2021). Biogenic Synthesis of Iron Oxide Nanoparticles Using Enterococcus faecalis: Adsorption of Hexavalent Chromium from Aqueous Solution and In Vitro Cytotoxicity Analysis. *Nanomaterials, 11*(12), 3290. https://doi.org/ (Creative Commons Attribution 4.0 International (CC BY 4.0)) (https://creativecommons.org/licenses/by/4.0/deed.en)DOI: 10.3390/nano11123290

Sandhya, J., & Kalaiselvam, S. (2019). Biogenic synthesis of magnetic iron oxide nanoparticles using inedible borassus flabellifer seed coat: characterization, antimicrobial, antioxidant activity and in vitro cytotoxicity analysis. *Materials Research Express, 7*(1), 015045. https://doi.org/ (Creative Commons Attribution 4.0 International (CC BY 4.0)) (https://creativecommons.org/licenses/by/4.0/deed.en)DOI: 10.1088/2053-1591/ab6642

Saod, W. M., Al-Janaby, M. S., Gayadh, E. W., Ramizy, A., & Hamid, L. L. (2024). Biogenic synthesis of iron oxide nanoparticles using Hibiscus sabdariffa extract: Potential for antibiotic development and antibacterial activity against multidrug-resistant bacteria. *Current Research in Green and Sustainable Chemistry*, *8*, 100397. DOI: 10.1016/j.crgsc.2024.100397

Shabbir, M. A., Naveed, M., Rehman, S. U., Ain, N. U., Aziz, T., Alharbi, M., Alsahammari, A., & Alasmari, A. F. (2023). Synthesis of Iron Oxide Nanoparticles from Madhuca indica Plant Extract and Assessment of Their Cytotoxic, Antioxidant, Anti-Inflammatory, and Anti-Diabetic Properties via Different Nanoinformatics Approaches. *ACS Omega*, *8*(37), 33358–33366. DOI: 10.1021/acsomega.3c02744 PMID: 37744851

Tan, H. L., Lim, Y. C., Ng, L. Y., & Lim, Y. P. (2023). Plant-mediated synthesis of iron nanoparticles for environmental application: Mini review. *Materials Today: Proceedings*, *87*, 64–69. DOI: 10.1016/j.matpr.2023.02.101

Tyagi, P. K., Gupta, S., Tyagi, S., Kumar, M., Pandiselvam, R., Daştan, S. D., Sharifi-Rad, J., Gola, D., & Arya, A. (2021). Green Synthesis of Iron Nanoparticles from Spinach Leaf and Banana Peel Aqueous Extracts and Evaluation of Antibacterial Potential. *Journal of Nanomaterials*, *2021*, 1–11. DOI: 10.1155/2021/4871453

ADDITIONAL READING

Abdeen, M., Sabry, S., Ghozlan, H., El-Gendy, A. A., & Carpenter, E. E. (2016). Microbial-Physical Synthesis of Fe and Fe3O4 Magnetic Nanoparticles Using Aspergillus niger YESM1 and Supercritical Condition of Ethanol. *Journal of Nanomaterials*, *2016*, 1–7. DOI: 10.1155/2016/9174891

Abdullah, J. A., Eddine, L. S., Abderrhmane, B., Alonso-González, M., Guerrero, A., & Romero, A. (2020). Green synthesis and characterization of iron oxide nanoparticles by pheonix dactylifera leaf extract and evaluation of their antioxidant activity. *Sustainable Chemistry and Pharmacy*, *17*, 100280. DOI: 10.1016/j.scp.2020.100280

Ajinkya, N., Yu, X., Kaithal, P., Luo, H., Somani, P., & Ramakrishna, S. (2020). Magnetic Iron Oxide Nanoparticle (IONP) Synthesis to Applications: Present and Future. *Materials (Basel)*, *13*(20), 4644. DOI: 10.3390/ma13204644 PMID: 33080937

Akhtar, M. S., Panwar, J., & Yun, Y. (2013). Biogenic synthesis of metallic nanoparticles by plant extracts. *ACS Sustainable Chemistry & Engineering*, *1*(6), 591–602. DOI: 10.1021/sc300118u

Priya, N., Naveen, N., Kaur, K., & Sidhu, A. K. (2021). Green Synthesis: An eco-friendly route for the synthesis of iron oxide nanoparticles. *Frontiers in Nanotechnology*, *3*. Advance online publication. DOI: 10.3389/fnano.2021.655062

KEY TERMS AND DEFINITIONS

Antibacterial Activity: The power of iron-based nanoparticles to block or destroy harmful bacteria. The effect of this activity varies according to the dosage and is usually stronger against Gram-positive bacteria.

Biocompatibility: The ability that permits material such as green-synthesised FeNPs, to be easily accepted by biological systems without prompting harmful reactions or alerting the immune system.

Dose Optimization: The discovery of the minimum amount of FeNPs that brings therapeutic or antimicrobial benefits with only minor consequences.

Hemolytic Activity: The power of a substance to trigger the destruction of red blood cells (hemolysis). When the hemolysis rate is under 5%, it means the nanoparticles have a good match with blood.

Reactive Oxygen Species (ROS): Molecules formed with oxygen that react easily and can damage cells in the body. If they are not controlled well, FeNPs may produce ROS, causing oxidative stress, inflammation and leading to toxicity.

Surface Functionalization: Making changes to nanoparticles using biomolecules or coatings from plants to stabilise them, make them less toxic and improve their ability to target specific cells.

Toxicity Assessment: Evaluating how substances, like FeNPs, influence biological systems by means of biological models for both in vitro and in vivo tests.

Chapter 5
Role of Doped Nanomaterials in the Synthesis and Characterization of Bio-Organic-Inorganic Composites

Sumana Dutta
Asutosh College, India

Sayanti Kar
Asutosh College, India

Gaurav Pant
Graphic Era University, India

Souvik Paul
https://orcid.org/0009-0007-9512-5752
Asutosh College, India

Riya Mukherjee
Graduate Institute of Biomedical Science, Taiwan

Arjun Sarkar
Leibniz Institute for Natural Product Research and Infection Biology, Germany

Ayantika Banerjee
https://orcid.org/0000-0003-0756-8005
Federal Institute for Material Research and Testing, Germany

ABSTRACT

The bio-inspired synthesis of organic and inorganic nanomaterials has gained significant attention for creating hierarchical structures. Blending these materials enhances nanocomposite membranes for filtration and sensing, with elements like Co^{2+} *and* Mn^{2+} *improving their properties. Many organisms, such as mollusks and sponges, naturally produce organic-inorganic composites through biomineraliza-*

DOI: 10.4018/979-8-3373-2180-6.ch005

tion. Biomolecules—proteins, peptides, DNA, RNA, and polysaccharides—guide nucleation, stabilization, and assembly of inorganic frameworks. Proteins aid in forming organized nanostructures, while peptides offer customizable designs for tissue engineering and nanotechnology. Doping is widely used to tailor nanoparticles' electrical, optical, and morphological properties. Doped nanomaterials, like Ag-N-TiO_2 and ZnO composites, exhibit antimicrobial efficacy, while co-doping strategies further enhance microbial inhibition. This review explores biomolecules' synergistic role in nanotechnology for sustainable, multifunctional materials in industry and biomedicine.

INTRODUCTION

Doping is a common method for customizing nanoparticles to continue improving their electrical, optical, and morphological properties. Doping may help boost the antimicrobial effect, according to a few earlier studies (Wei et al., 2011). The doping of nanomaterials provides a multi-functional method of fine-tuning the characteristics of the materials while retaining their high surface areas. Through the doping of various elements, the electronic, optical, photochemical, photo-electrochemical, photocatalytic, and photo-excited relaxation properties can be tuned in the desired direction (Bayne et al., 1998). Semiconductors for particularly sensitive sensing applications can be created by doping with nanoparticles. Now, semiconductor sensors can be made with extremely high accuracy, reliability, and precision by doping with nanoparticles. This article will review the position of several dopant contaminants, including transition metals, rare earth minerals, Co^{2+}, Ni^{2+}, and Mn^{2+} (Drummond & Bapna, 2003).

There are methods by which microorganisms regulate self-gathered hierarchical organic or inorganic frameworks. The protoplasmic surface boundary layer is where cells normally find inorganic materials. Subsequently, the interfacial stress between the cell lines, vesicles, and the ongoing mineralization process helps regulate biomineral formation (Ferracane, 2011). Inorganic materials can obtain structural information directly or indirectly from cellular composition or by modification of their inorganic framework. The use of biomimetic mineralization is broadening into previously unavailable markets, such as the innovation of lithium-ion battery component materials and exquisitely structured composite materials utilizing carbon dioxide gas.

The organisms that generate organic and inorganic bio-composites are equivalent to mollusks that generate nacres or shells with a crystalline state of calcium carbonates (Hosoya et al., 2011). The silica spicules manufactured by marine sponges can grow up to 3 meters in length and have been shown to exhibit light-guiding properties (Sevinç & Hanley, 2010). Specialized biomolecules - including peptides, RNA, polysaccharides, DNA, and proteins - are required to be unearthed to take part, most importantly, in trying to direct the structure of such inorganic frameworks that are being gathered in a hierarchy (Kawashita et al., 2000) (Table 1).

Table 1. Role of Organisms and Biomolecules in Bio-Composite Formation

Organisms	Bio-Composite Produced	Material Type	Key Features	Role of Biomolecules
Mollusks	Nacres, Shells	Crystalline Calcium Carbonate	Durable, protective, structured	Proteins and polysaccharides guide nucleation and crystalline growth.
Marine Sponges	Silica Spicules	Silica	Light-guiding properties, structural support	Specialized biomolecules like peptides and RNA direct hierarchical assembly.
Various Organisms	Organic/Inorganic Composites	Calcium carbonate, silica	Hierarchical structures	DNA, RNA, peptides, and proteins organize and stabilize the frameworks.

Many scientists have concentrated on the influence that biomolecules play in the initiation and stability of crystals. Most relevantly, the proteins that regulate how these biomaterials are formed regularly have emerged to acknowledge and attach, particularly to a single or even a lot of faces related to the developing crystal. The biomimetic method could perhaps result in the formation of organized composite and ceramic materials, controlled inorganic nanophase generation, and massive crystalline solid engineering (Zhang et al., 2014). The Importance of bio-inspired synthesis of doped nanomaterials is explained in (Figure 1).

Figure 1. Importance of bio-inspired synthesis of doped nanomaterials

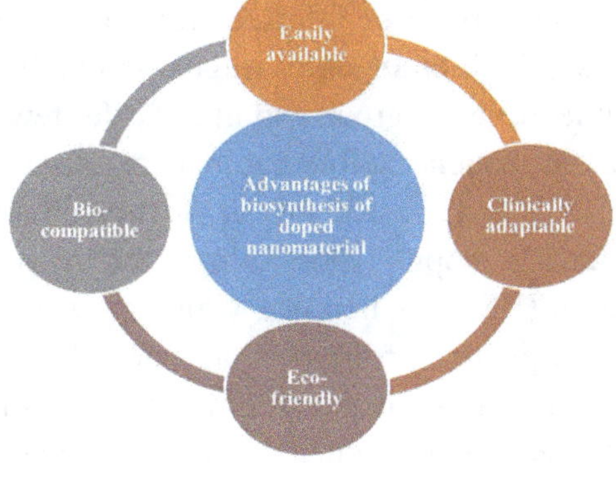

ADVANTAGES OF USING BIOMOLECULES FOR BIO-INSPIRED SYNTHESIS AND THEIR SOURCES

Using biomaterials to generate inorganic materials in the laboratory offers many advantages. One key benefit is the ability to create these materials under mild conditions, such as in water, at room temperature, and with a neutral pH, unlike traditional methods that require harsh experimental settings. Bio-inspired synthesis is inherently "green" processing because it uses less energy and doesn't use hazardous solvents. These biomolecular influences frequently have a strong impact on the properties of the synthesized material. The ability to make materials with innumerable or highly specific functions is supplied by biomolecules. A biomineral can acknowledge, converse, or impact the creation of an inorganic mineral, which is extremely likely to appear owing to the extensive diversity of natural and man-made biomolecules (Kasraei et al., 2014). Biomolecules influence the kinetics of inorganic phase nucleation and growth in molluscan shells in various ways. They provide diverse nucleation sites, bind to specific crystal planes, regulate the growth of calcium carbonate crystals, control crystal formation in mineralized tissues, and affect polymorph selection, crystallization pathways, and the overall growth of mineral phases (Schoeppler et al., 2019). Additionally, biomolecules can alter the mechanical stiffness and nanostructure of inorganic materials during synthesis. Research has shown that proteins and amino acids significantly influence the nanostructure and nanomechanical properties of calcium-silicate-hydrate systems, which are vital for numerous biological and therapeutic applications (Kamali & Ghahremaninezhad, 2018).

The main origins of the biomolecules used for the bio-inspired synthesis of materials are the biominerals separated from or acquired from bio-mineralizing things and the peptides noted in the case of bio-mineralization (Uskoković & Bertassoni, 2010). The biomimetic synthesis of inorganic compounds has made extensive use of biomolecules isolated from or inferred from bio-mineralizing organisms. For example, in this case, the biomolecules might well be hard to get or produce, might need specialized facilities to grow, and might offer few or no possibilities to change or engineer protein sequences. Recombinant expression of remineralization proteins from bacterial cells and subsequent purification could solve many of these troubles (Melo et al., 2013). Popular possibilities for bioinspired research, such as bovine serum albumin (BSA) or hen egg white lysozyme (HEWL), may also possess a portion of the sequential properties intrinsic to bio-mineralizing peptides (Ferracane, 1995). Analogs of biomolecules competent in bio-mineralization can be developed based on these recognized sequences. Synthetic materials, for example, are manufactured as substrates for inorganic polymerization in addition to biologically relevant analogs that are inspired by the way biological processes (Jin et al., 2015).

More opportunities to create the blueprints for biosynthetic pathways of bio-inspired inorganic materials have emerged because of progressions in RNA, peptide, and DNA recognition, isolation, and synthesis. Phage display is one such technique using methodical ligand amplification by incredibly rapid concentration (SELEX) to acknowledge RNA as well as peptide recognition (Tavassoli Hojati et al., 2013).

Nucleic acids, polysaccharides, proteins, and peptides are four broad categories that can be used to cluster numerous biomolecules used in bio-inspired synthesis (Table 2).

Table 2. Biomolecules Used in Bio-Inspired Synthesis and Their Sources and Advantages

Biomolecule Type	Sources	Advantages
Proteins	Biomolecules from organisms, recombinant expression	Hierarchical organization, molecular recognition, specific target binding
Peptides	Synthesized or derived from proteins	Tissue engineering, gene delivery, nanofabrication, self-assembly into supramolecular structures
DNA and RNA	Self-assembled nanoscale structures	Nanoparticle templates, nanophotonic and nanoelectronic devices, genome delivery systems
Polysaccharides	Natural sources like cellulose, chitin	Roles in bio-mineralization, specific chemical functionalities, potential copolymer formation with chitosan

Protein-mediated bio-inspired synthesis

Proteins serve as the operational building blocks for the formation of materials with numerous purposes (M. Fang et al., 2006). Proteins' capacity to self-assemble would facilitate a hierarchical organization with favorable characteristics at multiple large scales based on the molecular recognition at the organic/inorganic interface. Proteins with advanced molecular recognition mechanisms have superior target binding specificity (Adams et al., 2006).

Peptide-mediated Bio-inspired synthesis

Proteins have more complex functionalities than peptides, which are composed of short chains of amino acid. Although peptides might not even have as precise a function as proteins, well-established biochemical as well as genetic engineering procedures make it simpler to create peptides with desired amino acid sequences. As a result, they are commonly used in applications such as tissue engineering and three-dimensional cellular culture, governed pharmaceutical and gene delivery, biomineralization, peptide retention in membranes, and nanofabrication. Peptides are created to compress into morphologies, such as helix, sheet, or another shape. These 3D building blocks might self-assemble into supramolecular frameworks. Peptides are flexible smart resources for systematic component layouts because the supramolecular frameworks can be governed by modifying the physical and chemical elements of the environment, including pH, temperature, and salt concentration (Allaker, 2010).

DNA and RNA-mediated bio-inspired synthesis

At the nanoscale, RNA and DNA can self-assemble into distinct secondary and tertiary structures, attempting to make them an excellent blueprint for the advancement of nanoparticles (Peng et al., 2017). Nanoparticles of gold (Au NPs) assessing 5 nm were developed into periodic square lattices using a self-assembled two-dimensional (2D) DNA nano grid as a template, as shown in Figure. 9. Approximately 38 nm separates the centers of adjacent particles. Implementations for these exact governed Au NP distributions usually involve Nanophotonic and nanoelectronic devices (Takahashi et al., 2017). In recent times, a nanohybrid substance with such an inorganic shell, as well as a chemically distinct bio-core, was noted. It comprised a 100 nm DNA biocore surrounded by an inorganic nanoshell—constructed from a highly porous, stacked metal hydroxide—with an overall thickness of 10 nm. The potential to encompass and release DNA is made feasible by the MH nanoshell's pH-dependent solubility, which is essential for maximizing the solidity of foundation sequence-manipulated and probe-functionalized DNA molecules with crafted data. Consequently, these conglomerate composites could be employed as sophisticated genome distribution systems and biological diagnostic systems involving DNA-derived factors (Ghosh et al., 2012).

Polysaccharide-mediated bio-inspired synthesis

Although polysaccharides have existed since the start of evolvement, there seems to be a slow but rising interest in investigating their role in bio-mineralization. Generally, bio-minerals might not have a specific type of polysaccharide associated with them. In bio-minerals, only polysaccharides that are hydroxylated, carboxylated, sulfated, or encompass a blend of these active components are prevalent (Jiménez et al., 2015). Cellulose and Chitin share a very similar chemical structure, except that chitin contains an acetamido group in place of a hydroxyl group. There is no such idea as 100% acetylated pure chitin in nature. When taken in conjunction with chitosan, its N-deacetylated derivative, chitin frequently forms a copolymer. (1-4)-2-amino-2-deoxy-d-glucose polymeric component is said to be chitosan.

Due to their low danger of foreign body responses, inherent antibacterial properties, chemical stability, biocompatibility, and adaptability to be shaped into numerous geometrical structures, a polymeric form of chitosan nanocomposites has generated a great deal of significance in research in tissue engineering. In recent times, carbon nanotubes have always been transplanted onto chitosan natural polymer to boost the mechanical properties of synthetic bone (Cheng et al., 2015). This combination brings a different point of view for generating biomimetic nanocomposites since it can be implemented in other HAp-biopolymer systems (Cavassin et al., 2015). Bio-inspired synthesis processes for industrial applications face challenges in maintaining reproducibility and quality due to their complexity, which involves intricate molecular structures and precise functionalization. Maintaining quality and preserving the intended properties while scaling up is also a challenge. Reproducibility is another major hurdle, as ensuring consistent results across larger batches is a major hurdle; another difficulty is cost-effectiveness. To address these issues, a comprehensive understanding of biological systems, optimization of synthesis parameters, advanced manufacturing techniques, standardization of protocols and quality control measures, interdisciplinary collaboration between biologists, chemists, and material scientists, and computational modelling to simulate and predict nanomaterial behaviour in complex biological environments (Harun-Ur-Rashid et al., 2023). Scalability and reproducibility can be increased by using environmentally friendly synthesis techniques. For instance, bio-inspired synthesis methods that use renewable resources instead of dangerous chemicals may be easier to apply in industry (Kulkarni et al., 2023).

DOPED NANOMATERIALS

Doped nanoparticles use several significant mechanisms for demonstrating improved antibacterial activities. Doping nanomaterials can produce defects or change their electrical structures, which makes it easier for Reactive Oxygen Species (ROS) such as superoxide anions and hydroxyl radicals to form. These ROS cause bacterial cells to experience oxidative stress, which damages DNA, lipids, and proteins and finally results in cell death. It has been found that doping V_6O_{13} with nickel ions modulates oxygen defects, boosting ROS production and improving antibacterial activity (Liu et al., 2024). The efficiency and kinds of ROS produced can vary depending on the dopant and host material doping, even though many doped nanomaterials increase ROS generation. Certain elements in ZnO nanoparticles can greatly increase their antibacterial activity mediated by ROS (Abebe et al., 2020).

Doped nanomaterials can give off harmful metal ions, which can interfere with physiological functions such as membrane integrity and enzyme performance. The type of ions produced is dependent on the dopant utilized; for instance, elements doped with silver release Ag^+ ions to have antibacterial properties (Wang et al., 2017). Certain nanomaterials can interact with bacterial cell membranes, causing structural damage and increased permeability, which can significantly increase their antibacterial efficacy (Jiang et al., 2020).

Significant antimicrobial resistance against *S. aureus*, *E. coli*, and *B. subtilis* is accomplished by injecting Ag with the polymer chitosan and iron oxide (Leite et al., 2004). The study demonstrates that even under fluorescent light illumination, N-doped and Ag-TiO_2 both strengthen the anti-bacterial characteristics of TiO_2 nanoparticles. After 24 hours of cultivation, a 1% Ag-N-TiO_2 sample exhibited a strong anti-bacterial efficacy, with a clear anti-bacterial perimeter of 33.0 mm against *E. coli* and 22.8 mm against *B. subtilis* (Jia et al., 2013). Interestingly, through leveraging the simultaneous Ag and TiO_2 exhibit antibacterial action, TiO_2-Ag nanomaterials had been found to have significant light-independent antibacterial effects against both *E. coli* as well as *B. subtilis* particles (das Neves et al., 2014). Ag-doped TiO_2 nanoparticles also demonstrated effective lethality against *E. coli* in a different investigation (Kelly et al., 2003). Decontamination moderated by TiO_2 is mostly beneficial in improving the countryside where there is no availability of electricity for disinfection owing to the possible stimulation by visible light whenever treated along with new elements and to the photocatalytic performance UV-A (Chung, 2010; Sing et al., 1985). The antimicrobial efficacy of two geometrically isomeric ferrocene-carborane compounds in association with nanoscale titania as opposed to five MDR strains of *A. baumannii*. Various research subjects reveal that the targeted microorganisms are being eliminated by the synergistic antimicrobial property of ferrocene-carborane derivatives. In an aqueous growth media with *Shewanella oneidensis* MR-1 as well as *Mycobacterium smegmatis*, the research focuses on the antibacterial properties of Cu-doped TiO_2 nanoparticles and TiO_2 nanoparticles. The key conclusions were: 20 mg/l of Cu-doped TiO_2 nanoparticles substantially lessened rates of growth of *M. smegmatis* through three multitudes; 100 mg/l of TiO_2 particles did not seem to impede the performance of such two organisms in the presence of the aqueous culture. Such types of nanomaterials are harmful to *M. smegmatis* since TiO_2 nanoparticles doped in Cu should emit Cu^{2+} in an adulterated medium (although it is not in the case of metal-reducing bacterium *Shewanella*) (Pechini, 1967). In an investigation, titanium, as well as iron oxide nanoparticles with inactive succinic anhydride accompanied by the linker of dopamine on the superficial part of the nanomaterials as well as successive inactivation of IgG (IgG-Fe3O4@TiO_2 magnetic nanomaterials) had been satisfactorily employed as opposed to antibiotic-resistant bacteria, *S. pyogenes*, multidrug-resistant *S. pyogene*, and *Staphylococcus saprophyticus* (Razavi et al., 2012). Furthermore, it has been already shown by scientists that nanoparticles can inhibit bacterial growth while exposed to low-power UV light throughout short stops. Significant photocatalytic performance, as well as oxidizing ability, is present in gold-capped TiO_2 nanocomposites (Dizaj et al., 2014) (Table 3).

Table 3. Summary of Antimicrobial Efficacy of Various Nanomaterials Against Different Microorganisms

Nanomaterial	Target Microorganisms	Key Findings
Ag with Chitosan and Iron Oxide	*S. aureus, E. coli, B. subtilis*	Effective antimicrobial resistance.
N-doped and Ag-TiO_2 Nanoparticles	*E. coli, B. subtilis*	Strengthen antibacterial characteristics under fluorescent light. 1% Ag-N-TiO_2: 33.0 mm (E. coli), 22.8 mm (B. subtilis).
TiO_2-Ag Nanomaterials	*E. coli, B. subtilis*	Significant light-independent antibacterial effects.
Ag-doped TiO_2 Nanoparticles	*E. coli*	Effective lethality against E. coli.
Ferrocene-Carborane Compounds	*A. baumannii (MDR strains)*	Antimicrobial efficacy against MDR strains of A. baumannii.
Cu-doped TiO_2 Nanoparticles	*M. smegmatis, Shewanella oneidensis MR-1*	20 mg/l Cu-doped TiO_2 reduced the growth of M. smegmatis. No effect on *Shewanella oneidensis* MR-1.
IgG-Fe_3O_4@ TiO_2 Magnetic Nanomaterials	*S. pyogenes, S. saprophyticus*	Effective against antibiotic-resistant bacteria such as *S. pyogenes* and *S. saprophyticus.*
Ag-enhanced CuO Nanoparticles	*Gram-positive & Gram-negative bacteria*	Enhanced microbial cell destruction due to the inclusion of silver.
MgO Nanoparticles with Halogens	*Gram-positive & Gram-negative bacteria, spores*	Immune to the antimicrobial activity of MgO nanoparticles treated with halogens (Cl_2, Br_2)
Ag-doped MnO_2 Nanoparticles	*Methicillin-resistant Staphylococcus aureus* (MRSA)	Demonstrated antimicrobial properties against MRSA.
Ag-doped Bi_2O_3 Nanoparticles	*E. coli (antibacterial), Aspergillus niger (antifungal)*	Enhanced antibacterial activity against E. coli and antifungal activity against Aspergillus niger.
Zn-Cu Nanocomposite (Zn-doped CuO)	*E. coli, S. aureus*	10,000-fold increase in sterilization capacity compared to untreated ZnO and Cu nanoparticles.
Se-doped ZnO Nanoparticles	*E. aerogenes, P. aeruginosa, K. pneumonia, E. faecalis, Alcaligenes faecalis, Fusarium, Aspergillus*	Se-doped ZnO showed greater antibacterial and antifungal actions compared to un-doped ZnO. It has the best inhibition against E. faecalis, K. pneumoniae, P. aeruginosa, and fungi.

continued on following page

Table 3. Continued

Nanomaterial	Target Microorganisms	Key Findings
ZnO Doped with Silver	*Aspergillus*	Nano-ZnO was twice as effective as its non-nano counterparts in destroying Aspergillus. 5% nano-Pd increased efficacy by four times.
Sn-doped ZnO	*S. aureus*	Enhanced effectiveness in inhibiting S. aureus.
Co-doped ZnO Nanoparticles	*S. aureus, B. subtilis, Bacillus atrophaeus, K. pneumoniae, E. coli, Shigella, P. aeruginosa*	Enhanced antibacterial activity against both Gram-positive and Gram-negative bacteria
Mn-doped ZnO Nanoparticles	*S. typhi, S. aureus, B. subtilis, E. coli*	Greater antimicrobial effect compared to undoped ZnO. Enhanced activity against both Gram-negative and Gram-positive bacteria.
Au-doped ZnO Nanoparticles	*S. typhi, K. pneumoniae, S. aureus, Enterococci sp.*	Increased antibacterial activities against a range of bacteria compared to gold and ZnO nanoparticles.

Likewise, the inclusion of silver enhanced the capability of CuO nanoparticles to destroy microbial cells (Sun et al., 2011). Gram-positive as well as Gram-negative bacteria, as well as spores, exhibited immunity to the antimicrobial activity of MgO nanoparticles treated with halogens (Cl_2, Br_2) (Hussain et al., 2006). Silver-doped MnO2 nanoparticles generated chemically have always had an antimicrobial property on *Methicillin-resistant Staphylococcus aureus* (MRSA). Silver-doped bismuth oxide (Bi_2O_3) was known to effectively enhance antibacterial activity (*E. coli*) as well as antifungal activity (*Aspergillus niger*) (Rhim et al., 2013).

Evaluating both Gram-positive and Gram-negative bacteria, *E. coli* and *S. aureus*, the CuO doped in zinc into a tincture solution or settled onto the substance is resistant to bacteria. After just being subjected to microorganisms for 10 minutes, a remarkable 10,000-fold boost (Duncan, 2011) in the Zn-Cu nanocomposite's capacity to sterilize was seen in comparison to the untreated ZnO and Cu nanoparticles. Similar results are shown when treating multidrug-resistant microorganisms (MDR), demonstrating the effectiveness of such a composite (John & Thomas, 2008).

Lately, it has been claimed that the ZnO nanostructures, having doping with several other compounds, can expand the range of its characteristics. (Nguong et al., 2013). ZnO nanoparticles with Se-doped, as well as un-doped ZnO, have both been trialed for their antibacterial activities against the bacteria *E. aerogenes*, *P. aeruginosa*, *K. pneumonia*, and *E. faecalis*, *Alcaligenes faecalis*, *Fusarium*, and *Aspergillus*. Likewise, ZnO doped with Se showed greater actions, yet it was Se-doped ZnO that showed the greatest inhibition of the growth of *E. faecalis*, followed by *K. pneumoniae* and *P. aeruginosa*. Again, Se-doped ZnO showed an excellent inhibition against the fungi Aspergillus and Fusarium, so even though nanoparticles of zinc oxide just have shown a reasonable zone of inhibition (Nguong et al., 2013). Throughout the study of ZnO doped with silver thin films, it must have been acknowledged that having to raise the silver content lowers the E. Even before compared to its non-nano equivalents, nano-ZnO was twice as efficient at destroying Aspergillus, and then when it was filled with 5 percent nano-Pd, it became four times more potent (Othman, 2014). It's significant to observe that doped Sn raised the effectiveness of ZnO's inhibition effect as opposed to *S. aureus* (Roy et al., 2014). Co-doped ZnO nanoparticles have been demonstrated to have enhanced antibacterial activity against Gram-positive *S. aureus*, *B. subtilis*, and *Bacillus atrophaeus*, and Gram-negative *K. pneumonia, E. coli*, *Shigella*, as well as *P. aeruginosa.* [41]. Furthermore, significant growth related to the ZnO nanoparticles doped with Co has been noted as opposed to *S. typhi*, *S. aureus, B. subtilis*, and *E. coli* (2011). Desselberger (2000) created ZnO nanoparticles doped with Mn and established the fact that they displayed a greater antimicrobial effect than undoped ZnO nanoparticles against both Gram-negative and Gram-positive bacteria. Mn-doped ZnO has been shown to possess enhanced antibacterial activity against *S. typhi, S. aureus, B. subtilis, and E. coli* (Ruiz-Hitzky et al., 2013). Another study also showed that Mn doping raised the zone of inhibition against *Serratia marcescens, E. coli*, *B. subtilis*, *S. aureus*, and *B. subtilis* (Zafar et al., 2016). Particularly in comparison to gold nanoparticles or ZnO nanoparticles themselves, Au-doped ZnO exhibited increased antibacterial activities against *S. Typhi, K. pneumoniae*, *S. aureus,* and *Enterococci sp.* (Giri et al., 2013). Ag ZnO nanoparticles were formed by using sol-gel, sonochemical, and microwave methods. In contrast to ZnO nanoparticles, the results revealed that Ag-doped ZnO nanoparticles had stronger antibacterial activity. Furthermore, it has been reported that adding Ag to ZnO nanoparticles with tyrosine support managed to improve the photo-degradation of biological dye contaminants and bacterial elimination (Jha, 2020). ZnO nanoparticles coated with polyvinyl alcohol (PVA) were also found to have modified intracellular structural features, increased membrane permeability, and cellular internalization (Damm et al., 2008). Based on the rough surface structure of nanomaterials, one interpretation of the antimicrobial effect of doped nanoparticles is possible. It is presumed that integrating oxides of metal or transition metals into a compound enhances the list of obvious flaws (Eichhorn et al., 2010) (Table 3).

To increase the mechanical strength, electrical conductivity, and biocompatibility of bio-organic-inorganic composites, doped nanomaterials are essential. But for them to be successfully incorporated into various applications, it is essential to evaluate their long-term stability and biocompatibility under actual working settings. Various studies and research have been conducted on the long-term stability and biocompatibility of doped nanomaterials. The significance of surface functionalization for biological applications was highlighted by a study that used thermal decomposition to create ultrafine, monodisperse nanoparticles in organic solvents. These nanoparticles, which were initially stabilized by hydrophobic coatings, showed difficulties dispersing in aqueous solutions (C. Fang et al., 2009). The study on nanoparticle toxicity in biomedical applications highlights that size and shape significantly influence biocompatibility, with smaller nanoparticles exhibiting higher cellular uptake but potentially increased toxicity, while spherical shapes tend to be more biocompatible. Surface charge and coating also play a crucial role, as positively charged nanoparticles interact more with cell membranes, increasing cytotoxicity, whereas surface modifications like PEGylation improve stability and reduce harmful effects (Saikia, 2024). While bioactive coatings and hydrogels reduce immunological reactions, the study emphasizes advanced coatings, conductive polymers, and nanocomposites to stop degradation. More dependable implantable sensors and brain interfaces are made possible by flexible, adaptable materials that enhance mechanical compatibility and signal transfer (Park et al., 2022). The properties of bio-organic-inorganic composites are greatly improved by doped nanomaterials, which may provide benefits over conventionally synthesized materials in terms of effectiveness, affordability, and environmental impact. When compared to traditional materials, bio-inspired materials frequently have distinctive hierarchical structures that enhance their mechanical qualities, such as greater strength and toughness (Sonia et al., 2024). To create materials that are both biodegradable and maybe more affordable since they utilize renewable resources, bio-nanocomposites are created by combining nanoscale fillers with biopolymers. This strategy can lessen dependency on pricey, non-renewable raw materials (Nandhini et al., 2024). These biomaterials have low adverse environmental effects and great biodegradability.

It is generally accepted that the progression of most contemporary technologies tends to depend on the creation of novel, nanoscale, self-organized materials with novel architectures and properties. Here, we focus on hybrid metal/ceramic nanofiber (nf) materials. As catalysts (such as Pt/TiO_2-nf, photocatalysts (such as Nd doped TiO_2-nanotubes2), and electro-catalysts (such as Pd/TiO_2-nanotubes), these have sparked a great deal of interest. Other usages for metal/organic nanohybrids include photovoltaic cells, lithium batteries, H_2 storage and sensing materials, and low-temperature fuel cell components (Cano et al., 2017). The family of carbon nanotubes (CNTs) that have been augmented and adorned with metals is a related group of hybrid materials that is also relevant to this introduction. This study's novel discovery is that now the metallic nanoparticles used to decorate the ceramic nanofibers (nf) were doped with organic material, likely to result in triply-hybrid materials that include members of the three main families of organics, metals, and ceramics. This was achieved by using a newly invented method of analysis for this kind of metal doping with small molecules or polymers, generating a range of novel organics-metal composite materials. These composites, which are premised on copper, silver, gold, palladium, or platinum and some of their alloys, have already shown promise for several practical uses and attributes, including those attempting to influence the mechanical and conductive properties of metals, forming new biomaterials, and forming new, improved catalysts.

We provide a brief description of this methodology in this introduction because doping metals with an organic substance is a recent method of creating new materials. The basic aspect has been to lessen a metal cation in the mere existence of the molecules to be trapped, resulting in tiny aggregates that the trapped molecules are enclosed within.

Biomedical and biological courses are needed to encounter extraordinary advancements thanks to nanotechnology (Cheung et al., 2008). Likewise, it is expected that nanotechnologies will revolutionize the sensitivity of tools and devices within a wide range of domains of science and technology in addition to the regulation of material characteristics at ultrafine scales. It has been found that the use of metal oxide nanoparticles (NPs) is way more efficient than using bigger particles to counteract both gram-positive and gram-negative bacteria. Recent studies have demonstrated the efficacy of using silver-decorated materials to treat biofilm infections, including those brought on by oral pathogens like *Lactobacillus acidophilus*, *Streptococcus mutans*, *Candida albicans*, and *Aggregatibacter actionmycetemcomitans*. In contrast to their components, the synergistic antibacterial action of nano-hybrid or nanocomposites has been discovered to be more advantageous as opposed to Gram-negative and Gram-positive bacteria (Sharabi et al., 2014). Bacterial activity and Reactive oxygen species (ROS) production can both be elevated by incorporating ZnO and Ag nanoparticles. Integrating an antibacterial nano-hybrid material into a resin-based restorative material may prevent the creation of oral biofilm on the surface of restorations, thus further lowering the probability of risk of secondary caries.

CONCLUSION

In conclusion, there have been substantial advances in the layout of biologically inspired methodologies for controlling the nucleation and growth of inorganic crystals, especially at the nanometer scale. Under relatively uncomplicated situations, a wide range of self-assembled structured inorganic materials have been effectively built utilizing biomolecules. It has been revealed that biomolecules have the power to manage or alter respective sizes, forms, crystal configurations, and certain other characteristics related to synthetic artificial substances. These inorganic nanostructures, which also seem to be used in the biomedical, biosensor, bio-ceramic, and other fields, including hydroxyapatite (HAp), silver, SiO_2, TiO_2, ZrO_2, gold, CdS, and Fe_3O_4, as examples. Furthermore, chimeric biomolecules with desired characteristics can still be formed utilizing modern biotechnology. These materials could be used to construct hierarchical structures gathered and enhanced composite materials.

Even a significant number of biomolecules promoting the formulation of materials have indeed been discovered. Besides that, there are now more inorganic materials available for bioinspired synthesis. To employ the possibility of biomolecules for synthesizing nanoparticles, however, requires a thorough understanding of these already-established subjects. Several fascinating as well as dynamic fresh notions have already been given a lot of recognition or have still not been fully realized. It is also essential to formulate more self-assembled grouping manner biological molecules that might template as well as mentor the artificial establishment. The prospect of biomimetic chemical synthesis guarantees to be thrilling, vibrant, and filled with applications with the renewed emphasis and innovation of researchers from different fields of study.

REFERENCES

Abebe, B., Zereffa, E. A., Tadesse, A., & Murthy, H. C. A. (2020). A Review on Enhancing the Antibacterial Activity of ZnO: Mechanisms and Microscopic Investigation. *Nanoscale Research Letters*, *15*(1), 1–19. DOI: 10.1186/S11671-020-03418-6/TABLES/2 PMID: 33001404

Adams, L. K., Lyon, D. Y., & Alvarez, P. J. J. (2006). Comparative eco-toxicity of nanoscale TiO2, SiO2, and ZnO water suspensions. *Water Research*, *40*(19), 3527–3532. DOI: 10.1016/J.WATRES.2006.08.004 PMID: 17011015

Allaker, R. P. (2010). The use of nanoparticles to control oral biofilm formation. *Journal of Dental Research*, *89*(11), 1175–1186. DOI: 10.1177/0022034510377794 PMID: 20739694

Bayne, S. C., Thompson, J. Y., Swift, E. J., Stamatiades, P., & Wilkerson, M. (1998). A CHARACTERIZATION OF FIRST-GENERATION FLOWABLE COMPOSITES. *The Journal of the American Dental Association*, *129*(5), 567–577. DOI: 10.14219/JADA.ARCHIVE.1998.0274 PMID: 9601169

Cano, L., Pollet, E., Avérous, L., & Tercjak, A. (2017). Effect of TiO2 nanoparticles on the properties of thermoplastic chitosan-based nano-biocomposites obtained by mechanical kneading. *Composites. Part A, Applied Science and Manufacturing*, *93*, 33–40. DOI: 10.1016/J.COMPOSITESA.2016.11.012

Cavassin, E. D., de Figueiredo, L. F. P., Otoch, J. P., Seckler, M. M., de Oliveira, R. A., Franco, F. F., Marangoni, V. S., Zucolotto, V., Levin, A. S. S., & Costa, S. F. (2015). Comparison of methods to detect the in vitro activity of silver nanoparticles (AgNP) against multidrug resistant bacteria. *Journal of Nanobiotechnology*, *13*(1). Advance online publication. DOI: 10.1186/S12951-015-0120-6 PMID: 26438142

Cheng, L., Zhang, K., Weir, M. D., Melo, M. A. S., Zhou, X., & Xu, H. H. K. (2015). Nanotechnology strategies for antibacterial and remineralizing composites and adhesives to tackle dental caries. *Nanomedicine (London, England)*, *10*(4), 627–641. DOI: 10.2217/NNM.14.191 PMID: 25723095

Cheung, H. Y., Lau, K. T., Tao, X. M., & Hui, D. (2008). A potential material for tissue engineering: Silkworm silk/PLA biocomposite. *Composites. Part B, Engineering*, *39*(6), 1026–1033. DOI: 10.1016/J.COMPOSITESB.2007.11.009

Chung, D. D. L. (2010). *Composite Materials*. https://doi.org/DOI: 10.1007/978-1-84882-831-5

Damm, C., Münstedt, H., & Rösch, A. (2008). The antimicrobial efficacy of polyamide 6/silver-nano- and microcomposites. *Materials Chemistry and Physics, 108*(1), 61–66. DOI: 10.1016/J.MATCHEMPHYS.2007.09.002

das Neves. P. B. A., Agnelli, J. A. M., Kurachi, C., & de Souza, C. W. O. (2014). Addition of silver nanoparticles to composite resin: effect on physical and bactericidal properties in vitro. *Brazilian Dental Journal, 25*(2), 141–145. https://doi.org/ DOI: 10.1590/0103-6440201302398

Desselberger, U. (2000). Emerging and Re-emerging Infectious Diseases. *The Journal of Infection, 40*(1), 3–15. DOI: 10.1053/JINF.1999.0624 PMID: 10762105

Dizaj, S. M., Lotfipour, F., Barzegar-Jalali, M., Zarrintan, M. H., & Adibkia, K. (2014). Antimicrobial activity of the metals and metal oxide nanoparticles. *Materials Science and Engineering C, 44*, 278–284. DOI: 10.1016/J.MSEC.2014.08.031 PMID: 25280707

Drummond, J. L., & Bapna, M. S. (2003). Static and cyclic loading of fiber-reinforced dental resin. *Dental Materials, 19*(3), 226–231. DOI: 10.1016/S0109-5641(02)00034-9 PMID: 12628435

Duncan, T. V. (2011). Applications of nanotechnology in food packaging and food safety: Barrier materials, antimicrobials and sensors. *Journal of Colloid and Interface Science, 363*(1), 1–24. DOI: 10.1016/J.JCIS.2011.07.017 PMID: 21824625

Eichhorn, S. J., Dufresne, A., Aranguren, M., Marcovich, N. E., Capadona, J. R., Rowan, S. J., Weder, C., Thielemans, W., Roman, M., Renneckar, S., Gindl, W., Veigel, S., Keckes, J., Yano, H., Abe, K., Nogi, M., Nakagaito, A. N., Mangalam, A., Simonsen, J., … Peijs, T. (2010). Review: current international research into cellulose nanofibres and nanocomposites. *Journal of Materials Science 2009 45:1, 45*(1), 1–33. https://doi.org/DOI: 10.1007/S10853-009-3874-0

Fang, C., Bhattarai, N., Sun, C., & Zhang, M. (2009). Functionalized Nanoparticles with Long-Term Stability in Biological Media. *Small (Weinheim an Der Bergstrasse, Germany), 5*(14), 1637. https://doi.org/DOI: 10.1002/SMLL.200801647

Fang, M., Chen, J. H., Xu, X. L., Yang, P. H., & Hildebrand, H. F. (2006). Antibacterial activities of inorganic agents on six bacteria associated with oral infections by two susceptibility tests. *International Journal of Antimicrobial Agents, 27*(6), 513–517. DOI: 10.1016/J.IJANTIMICAG.2006.01.008 PMID: 16713190

Ferracane, J. L. (1995). Current trends in dental composites. *Critical Reviews in Oral Biology and Medicine : An Official Publication of the American Association of Oral Biologists, 6*(4), 302–318. DOI: 10.1177/10454411950060040301 PMID: 8664421

Ferracane, J. L. (2011). Resin composite—State of the art. *Dental Materials*, *27*(1), 29–38. DOI: 10.1016/J.DENTAL.2010.10.020 PMID: 21093034

Ghosh, S., Goudar, V. S., Padmalekha, K. G., Bhat, S. V., Indi, S. S., & Vasan, H. N. (2012). ZnO/Ag nanohybrid: Synthesis, characterization, synergistic antibacterial activity and its mechanism. *RSC Advances*, *2*(3), 930–940. DOI: 10.1039/C1RA00815C

Giri, A., Goswami, N., Sarkar, S., & Kumar Pal, S. (2013). Bio-Nanomaterials: Understanding Key Biophysics and Their Applications. *Nanotechnology*, *11*, 41–110.

Harun-Ur-Rashid, M., Jahan, I., Foyez, T., & Imran, A. Bin. (2023). Bio-Inspired Nanomaterials for Micro/Nanodevices: A New Era in Biomedical Applications. *Micromachines 2023, Vol. 14, Page 1786*, *14*(9), 1786. https://doi.org/DOI: 10.3390/MI14091786

Hosoya, Y., Shiraishi, T., Odatsu, T., Nagafuji, J., Kotaku, M., Miyazaki, M., & Powers, J. M. (2011). Effects of polishing on surface roughness, gloss, and color of resin composites. *Journal of Oral Science*, *53*(3), 283–291. DOI: 10.2334/JOSNUSD.53.283 PMID: 21959654

Hussain, F., Hojjati, M., Okamoto, M., & Gorga, R. E. (2006). Review article: Polymer-matrix Nanocomposites, Processing, Manufacturing, and Application: An Overview. *Journal of Composite Materials*, *40*(17), 1511–1575. DOI: 10.1177/0021998306067321

Jha, P. (2020). Effect of plasticizer and antimicrobial agents on functional properties of bionanocomposite films based on corn starch-chitosan for food packaging applications. *International Journal of Biological Macromolecules*, *160*, 571–582. DOI: 10.1016/J.IJBIOMAC.2020.05.242 PMID: 32485256

Jia, Y., Yu, X. Y., Luo, T., Zhang, M. Y., Liu, J. H., & Huang, X. J. (2013). PEG aggregation templated porous ZnO nanostructure: Room temperature solution synthesis, pore formation mechanism, and their photoluminescence properties. *CrystEngComm*, *15*(18), 3647–3653. DOI: 10.1039/C3CE27091B

Jiang, S., Lin, K., & Cai, M. (2020). ZnO Nanomaterials: Current Advancements in Antibacterial Mechanisms and Applications. *Frontiers in Chemistry*, *8*, 552221. DOI: 10.3389/FCHEM.2020.00580/BIBTEX PMID: 32793554

Jiménez, A. B. P., Aguilar, C. A. H., Ramos, J. M. V., & Thangarasu, P. (2015). Synergistic antibacterial activity of nanohybrid materials ZnO-Ag and ZnO-Au: Synthesis, characterization, and comparative analysis of undoped and doped ZnO nanoparticles. *Australian Journal of Chemistry*, *68*(2), 288–297. DOI: 10.1071/CH14123

Jin, J., Wei, G., Hao, A., Zhang, W., Zhang, Y., Wang, Q., & Chen, Q. (2015). ZnO-AlBw-reinforced dental resin composites: The effect of pH level on mechanical properties. *RSC Advances*, *5*(33), 26430–26437. DOI: 10.1039/C5RA01413A

John, M. J., & Thomas, S. (2008). Biofibres and biocomposites. *Carbohydrate Polymers*, *71*(3), 343–364. DOI: 10.1016/J.CARBPOL.2007.05.040

Kamali, M., & Ghahremaninezhad, A. (2018). Effect of Biomolecules on the Nano-structure and Nanomechanical Property of Calcium-Silicate-Hydrate. *Scientific Reports*, *8*(1). Advance online publication. DOI: 10.1038/S41598-018-27746-X PMID: 29934541

Kasraei, S., Sami, L., & Hendi, S., AliKhani, M.-Y., Rezaei-Soufi, L., & Khamverdi, Z. (2014). Antibacterial properties of composite resins incorporating silver and zinc oxide nanoparticles on Streptococcus mutans and Lactobacillus. *Restorative Dentistry & Endodontics*, *39*(2), 109. DOI: 10.5395/RDE.2014.39.2.109 PMID: 24790923

Kawashita, M., Tsuneyama, S., Miyaji, F., Kokubo, T., Kozuka, H., & Yamamoto, K. (2000). Antibacterial silver-containing silica glass prepared by sol–gel method. *Biomaterials*, *21*(4), 393–398. DOI: 10.1016/S0142-9612(99)00201-X PMID: 10656321

Kelly, K. L., Coronado, E., Zhao, L. L., & Schatz, G. C. (2003). The optical properties of metal nanoparticles: The influence of size, shape, and dielectric environment. *The Journal of Physical Chemistry B*, *107*(3), 668–677. DOI: 10.1021/JP026731Y/ASSET/IMAGES/LARGE/JP026731YF00014.JPEG

Kulkarni, D., Sherkar, R., Shirsathe, C., Sonwane, R., Varpe, N., Shelke, S., More, M. P., Pardeshi, S. R., Dhaneshwar, G., Junnuthula, V., & Dyawanapelly, S. (2023). Biofabrication of nanoparticles: Sources, synthesis, and biomedical applications. *Frontiers in Bioengineering and Biotechnology*, *11*, 1159193. DOI: 10.3389/FBIOE.2023.1159193/PDF PMID: 37200842

Leite, E. R., Bernardi, M. I. B., Longo, E., Varela, J. A., & Paskocimas, C. A. (2004). Enhanced electrical property of nanostructured Sb-doped SnO2 thin film processed by soft chemical method. *Thin Solid Films*, *449*(1–2), 67–72. DOI: 10.1016/J.TSF.2003.10.101

Liu, L., Li, S., Zhu, W., Bao, Q., Liang, Y., Zhao, T., Li, X., & Zhou, J. (2024). Study on the mechanism of ROS-induced oxidative stress injury and the broad-spectrum antimicrobial performance of nickel ion-doped V6O13 powder. *Scientific Reports*, *14*(1), 22374. DOI: 10.1038/S41598-024-71959-2 PMID: 39333647

Melo, M. A. S., Guedes, S. F. F., Xu, H. H. K., & Rodrigues, L. K. A. (2013). Nanotechnology-based restorative materials for dental caries management. *Trends in Biotechnology*, *31*(8), 459–467. DOI: 10.1016/J.TIBTECH.2013.05.010 PMID: 23810638

Nandhini, J., Karthikeyan, E., & Rajeshkumar, S. (2024). Eco-friendly bio-nanocomposites: pioneering sustainable biomedical advancements in engineering. *Discover Nano 2024 19:1*, *19*(1), 1–31. https://doi.org/DOI: 10.1186/S11671-024-04007-7

Nguong, C. W., Lee, S. N. B., & Sujan, D. (2013). A Review On Natural Fibre Reinforced Polymer Composites. *World Academy of Science, Engineering and Technology, International Journal of Chemical, Molecular, Nuclear, Materials and Metallurgical Engineering*, *7*(1), 52–59. https://publications.waset.org/6783/pdf

Othman, S. H. (2014). Bio-nanocomposite Materials for Food Packaging Applications: Types of Biopolymer and Nano-sized Filler. *Agriculture and Agricultural Science Procedia*, *2*, 296–303. DOI: 10.1016/J.AASPRO.2014.11.042

Park, J., Lee, Y., Kim, T. Y., Hwang, S., & Seo, J. (2022). Functional Bioelectronic Materials for Long-Term Biocompatibility and Functionality. *ACS Applied Electronic Materials*, *4*(4), 1449–1468. DOI: 10.1021/ACSAELM.1C01212/ASSET/IMAGES/LARGE/EL1C01212_0007.JPEG

Pechini, M. P. (1967). *Method of preparing lead and alkaline earth titanates and niobates and coating method using the same to form a capacitor*.

Peng, J. min, Lin, J. cheng, Chen, Z. yu, Wei, M. chao, Fu, Y. xiang, Lu, S. shen, Yu, D. sheng, & Zhao, W. (2017). Enhanced antimicrobial activities of silver-nanoparticle-decorated reduced graphene nanocomposites against oral pathogens. *Materials Science and Engineering C*, *71*, 10–16. DOI: 10.1016/J.MSEC.2016.09.070 PMID: 27987652

Razavi, R. S., Loghman-Estarki, M. R., & Farhadi-Khouzani, M. (2012). Synthesis and Characterization of ZnO Nanostructures by Polymeric Precursor Route. *Acta Physica Polonica A*, *121*(1), 98–100. DOI: 10.12693/APHYSPOLA.121.98

Rhim, J. W., Park, H. M., & Ha, C. S. (2013). Bio-nanocomposites for food packaging applications. *Progress in Polymer Science, 38*(10–11), 1629–1652. DOI: 10.1016/J.PROGPOLYMSCI.2013.05.008

Roy, S. B., Shit, S. C., Sengupta, R. A., & Shukla, P. R. (2014). A Review on Bio-Composites: Fabrication, Properties and Applications. *International Journal of Innovative Research in Science, Engineering and Technology, 03*(10), 16814–16824. DOI: 10.15680/ijirset.2014.0310058

Ruiz-Hitzky, E., Darder, M., Fernandes, F. M., Wicklein, B., Alcântara, A. C. S., & Aranda, P. (2013). Fibrous clays based bionanocomposites. *Progress in Polymer Science, 38*(10–11), 1392–1414. DOI: 10.1016/J.PROGPOLYMSCI.2013.05.004

Saikia, N. (2024). Inorganic-Based Nanoparticles and Biomaterials as Biocompatible Scaffolds for Regenerative Medicine and Tissue Engineering: Current Advances and Trends of Development. *Inorganics 2024, Vol. 12, Page 292, 12*(11), 292. https://doi.org/DOI: 10.3390/INORGANICS12110292

Schoeppler, V., Lemanis, R., Reich, E., Pusztai, T., Gránásy, L., & Zlotnikov, I. (2019). Crystal growth kinetics as an architectural constraint on the evolution of molluscan shells. *Proceedings of the National Academy of Sciences of the United States of America, 116*(41), 20388–20397. DOI: 10.1073/PNAS.1907229116/SUPPL_FILE/PNAS.1907229116.SM03.MP4 PMID: 31551265

Sevinç, B. A., & Hanley, L. (2010). Antibacterial activity of dental composites containing zinc oxide nanoparticles. *Journal of Biomedical Materials Research. Part B, Applied Biomaterials, 94*(1), 22–31. DOI: 10.1002/JBM.B.31620 PMID: 20225252

Sharabi, M., Mandelberg, Y., Benayahu, D., Benayahu, Y., Azem, A., & Haj-Ali, R. (2014). A new class of bio-composite materials of unique collagen fibers. *Journal of the Mechanical Behavior of Biomedical Materials, 36*, 71–81. DOI: 10.1016/J.JMBBM.2014.04.008 PMID: 24801102

Sing, K. S. W., Everett, D. H., Haul, R. A. W., Moscou, L., Pierotti, R. A., Rouquerol, J., & Siemieniewska, T. (1985). Reporting Physisorption Data for Gas/Solid Systems with Special Reference to the Determination of Surface Area and Porosity. *Pure and Applied Chemistry, 57*(4), 603–619. DOI: 10.1351/PAC198557040603/MACHINEREADABLECITATION/RIS

Sonia, P., Srinivas, R., Kansal, L., Abdul-Zahra, D. S., Reddy, U., & Kumari, V. (2024). Bioinspired Composites a Review: Lessons from Nature for Materials Design and Performance. *E3S Web of Conferences, 505*, 01024. https://doi.org/DOI: 10.1051/E3SCONF/202450501024

Sun, J., Forster, A. M., Johnson, P. M., Eidelman, N., Quinn, G., Schumacher, G., Zhang, X., & Wu, W. L. (2011). Improving performance of dental resins by adding titanium dioxide nanoparticles. *Dental Materials*, *27*(10), 972–982. DOI: 10.1016/J.DENTAL.2011.06.003 PMID: 21774976

Takahashi, C., Matsubara, N., Akachi, Y., Ogawa, N., Kalita, G., Asaka, T., Tanemura, M., Kawashima, Y., & Yamamoto, H. (2017). Visualization of silver-decorated poly (DL-lactide-co-glycolide) nanoparticles and their efficacy against Staphylococcus epidermidis. *Materials Science and Engineering C*, *72*, 143–149. DOI: 10.1016/J.MSEC.2016.11.051 PMID: 28024570

Tavassoli Hojati, S., Alaghemand, H., Hamze, F., Ahmadian Babaki, F., Rajab-Nia, R., Rezvani, M. B., Kaviani, M., & Atai, M. (2013). Antibacterial, physical and mechanical properties of flowable resin composites containing zinc oxide nanoparticles. *Dental Materials*, *29*(5), 495–505. DOI: 10.1016/J.DENTAL.2013.03.011 PMID: 23562673

Uskoković, V., & Bertassoni, L. E. (2010). Nanotechnology in dental sciences: Moving towards a finer way of doing dentistry. *Materials (Basel)*, *3*(3), 1674–1691. DOI: 10.3390/MA3031674 PMID: 27103959

Wang, L., Hu, C., & Shao, L. (2017). The antimicrobial activity of nanoparticles: Present situation and prospects for the future. *International Journal of Nanomedicine*, *12*, 1227–1249. DOI: 10.2147/IJN.S121956 PMID: 28243086

Wei, Y. J., Silikas, N., Zhang, Z. T., & Watts, D. C. (2011). Hygroscopic dimensional changes of self-adhering and new resin-matrix composites during water sorption/desorption cycles. *Dental Materials*, *27*(3), 259–266. DOI: 10.1016/J.DENTAL.2010.10.015 PMID: 21111476

Zafar, R., Zia, K. M., Tabasum, S., Jabeen, F., Noreen, A., & Zuber, M. (2016). Polysaccharide based bionanocomposites, properties and applications: A review. *International Journal of Biological Macromolecules*, *92*, 1012–1024. DOI: 10.1016/J.IJBIOMAC.2016.07.102 PMID: 27481340

Zhang, Y., Gao, X., Zhi, L., Liu, X., Jiang, W., Sun, Y., & Yang, J. (2014). The synergetic antibacterial activity of Ag islands on ZnO (Ag/ZnO) heterostructure nanoparticles and its mode of action. *Journal of Inorganic Biochemistry*, *130*(1), 74–83. DOI: 10.1016/J.JINORGBIO.2013.10.004 PMID: 24176922

Chapter 6
Sustainable Bio-Based Nanocomposites:
Engineering Regenerable Materials for Tomorrow

Chaitanya Surya Chandra Nadimpalli
https://orcid.org/0000-0003-1070-3079
Raghu Engineering College, India

Murali Monangi
https://orcid.org/0009-0001-7832-2216
Raghu Engineering College, India

Kalyani Teku
https://orcid.org/0000-0002-3844-2483
Raghu Engineering College, India

V. Subrahmanyam Surya
Raghu Engineering College, India

ABSTRACT

Bio-based nanocomposites represent a transformative frontier in materials science, offering sustainable alternatives to conventional petroleum-derived composites. This chapter examines sustainable bio-based nanocomposites with emphasis on their regenerability across multiple life cycles. Beginning with fundamental principles governing synthesis and structure-property relationships, the work explores interactions between bio-based matrices and nanoscale reinforcements. Processing techniques balancing performance with sustainability are analyzed, establishing foundations for understanding degradation mechanisms and regeneration strategies. Both chemical and physical aging processes are investigated, connecting these insights

DOI: 10.4018/979-8-3373-2180-6.ch006

to effective regeneration approaches. The chapter addresses lifecycle assessment through environmental impact evaluations spanning multiple regeneration cycles. Advanced characterization techniques and economic considerations are discussed alongside emerging trends, serving as a foundational resource and catalyst for future innovations in sustainable materials.

1. INTRODUCTION

The emergence of bio-based nanocomposites marks a pivotal transformation in materials science and engineering, representing a fundamental shift toward developing truly sustainable material alternatives that address the growing environmental imperatives of the 21st century (Thomas et al., 2022). In an era where environmental concerns and resource scarcity increasingly influence scientific inquiry and industrial practices, these advanced materials offer promising pathways toward addressing the complex challenges of sustainability while maintaining the demanding performance requirements of modern technological applications (Thakur, 2023).

1.1 Scope and Significance of the Work

This chapter aims to provide a comprehensive examination of sustainable bio-based nanocomposites with specific focus on their regenerability characteristics and circular economy integration. The scope encompasses fundamental scientific principles underlying material behavior, advanced processing methodologies enabling sustainable manufacturing, comprehensive performance characteristics across multiple application domains, systematic environmental impact assessment throughout complete lifecycles, and detailed economic considerations spanning initial development through commercial implementation (Gopi et al., 2023).

The work specifically addresses materials comprising nanoscale reinforcements embedded within matrices derived from renewable biological sources, with particular emphasis on systems designed for multiple functional lifecycles through controlled regeneration processes. This includes natural polymer matrices such as cellulose, starch, and protein-based systems, bio-derived synthetic polymers including polylactic acid (PLA) and polyhydroxyalkanoates (PHAs), and various nanoscale reinforcements ranging from cellulose nanocrystals to bio-derived carbon nanomaterials (Klemm et al., 2023).

The significance of this research lies in its potential to address critical global challenges including plastic pollution, resource depletion, climate change mitigation, and the urgent need for circular economy implementation. With global plastic production exceeding 380 million tonnes annually and less than 10% being effectively recycled, the development of regenerable bio-based alternatives represents a crucial technological imperative that could fundamentally transform material usage patterns across multiple industries (Vatansever et al., 2023).

Table 1. Market Growth Projections for Bio-based Nanocomposites

Market Segment	2023 Value (USD Million)	2034 Projection (USD Million)	CAGR (%)
Nanocellulose Market	560.49	3,162.42	20.14
Biodegradable Plastics	42,760	89,400	9.5
Bio-based Chemicals	85,200	127,600	4.8

Source: Polaris Market Research (2024), Expert Market Research (2024)

The economic significance is equally compelling, with the global bio-based chemicals market projected to reach USD 127.6 billion by 2030, driven by increasing environmental regulations, consumer demand for sustainable products, and technological advancements enabling commercial viability (Expert Market Research, 2024). The biodegradable plastics market demonstrates exceptional growth potential, expanding from $42.76 billion in 2020 to an estimated $89.40 billion by 2028, reflecting a compound annual growth rate of 9.5% that highlights the accelerating market acceptance of sustainable alternatives (Mishra et al., 2024).

Primary Objectives:

First, analyzing the fundamental principles governing bio-based nanocomposite synthesis and regeneration, providing readers with essential scientific understanding of these complex material systems including molecular-level interactions, processing-structure-property relationships, and the mechanisms enabling multiple lifecycle functionality (Jawaid et al., 2023).

Second, evaluating current processing approaches and their effectiveness for regenerable material development, offering practical insights for material scientists and engineers including comparative analysis of different synthesis methods, scalability considerations, and optimization strategies for commercial implementation (Kargarzadeh et al., 2022).

Third, examining degradation mechanisms and their implications for regeneration strategies, enabling better design of durable regenerable systems through understanding of chemical, physical, and environmental factors affecting material performance over extended periods (Nikolaidis et al., 2023).

Fourth, assessing environmental and economic impacts across complete material lifecycles, providing comprehensive sustainability analysis including carbon footprint quantification, resource utilization efficiency, and economic competitiveness relative to conventional alternatives (Youssef & El-Sayed, 2022).

Finally, identifying emerging trends and future research directions that will shape the field's development over the coming decades, including technological convergence opportunities, novel application domains, and regulatory framework evolution (StartUs Insights, 2025).

1.2 Evolution and Current State

The evolution from conventional composite materials to bio-based nanocomposites reflects a broader transition in materials science toward sustainability-driven innovation (Faruk et al., 2021). Traditional composites, typically comprising synthetic polymers reinforced with various fillers, have revolutionized numerous industries due to their exceptional mechanical properties, lightweight characteristics, and design flexibility. However, these materials present significant environmental challenges, particularly regarding their petroleum-based origins, energy-intensive manufacturing processes, and limited end-of-life options (Bledzki & Jaszkiewicz, 2020).

Early bio-composites often suffered from performance limitations compared to their synthetic counterparts, particularly in terms of mechanical properties, thermal stability, and moisture resistance (Mohanty et al., 2022). The incorporation of nanotechnology into bio-based composites represented a fundamental breakthrough, enabling significant performance enhancements while maintaining environmental benefits (Dufresne, 2022). Recent advances demonstrate that functional biopolymer-based nanocomposites exhibit remarkably improved properties, including enhanced barrier characteristics, oxygen impermeability, solvent resistance, moisture control, thermal stability, and antimicrobial functionality (Lagarón et al., 2022).

Despite significant advances, several critical challenges persist in the development and widespread adoption of bio-based nanocomposites. Primary among these is the "end-of-life" question concerning optimal resource utilization after initial use phases. While biodegradability offers one solution pathway, this approach essentially represents a single-use strategy, failing to optimize resource utilization across multiple lifecycles (Pöhlein et al., 2022). The continuous disposal of conventional plastics results in accumulation of persistent waste, leading to significant pollution of both marine and terrestrial habitats (Wang et al., 2023).

1.3 The Regenerability Paradigm

The concept of regenerability introduces a fundamentally different approach to sustainable materials development that transcends traditional recycling limitations. Rather than focusing solely on biodegradability or conventional recyclability, regenerability encompasses the ability of materials to be recovered and restored to their original functionality through multiple lifecycles while maintaining critical performance characteristics (Misra et al., 2023).

This approach aligns with circular economy principles by maintaining the value of materials and resources for extended periods, potentially achieving resource utilization efficiencies several times greater than conventional linear material flows (Avérous & Pollet, 2023). For bio-based nanocomposites, regenerability represents a particularly promising avenue that builds upon the inherent advantages of renewable feedstocks, potentially lower environmental impacts, and biodegradable end-of-life options.

Implementing regenerability requires sophisticated understanding of fundamental degradation mechanisms and development of effective regeneration strategies that preserve critical performance characteristics through multiple use cycles (Zhang et al., 2022). This necessitates innovation in material design, processing technologies, collection systems, and quality control methodologies specifically adapted for regenerable material management.

2. FUNDAMENTAL PRINCIPLES OF BIO-BASED NANOCOMPOSITES

2.1 Bio-based Matrices

The matrix phase in bio-based nanocomposites provides the continuous medium that binds reinforcement elements, determines many bulk properties, and often represents the largest volume fraction of the material (Hubbe et al., 2022). Bio-based matrices encompass a diverse range of polymers derived from renewable resources, each offering distinct advantages and challenges for regenerable applications.

Natural Polymers

Natural polymers extracted directly from biomass represent some of the most abundant renewable resources available. Cellulose, the most prevalent natural polymer on Earth, offers excellent mechanical properties and thermal stability, though processing challenges stem from limited solubility in conventional solvents and high crystallinity that resists modification (Moon et al., 2022). Scientific pursuit to seek environmentally friendly materials from renewable sources has increasingly focused on cellulose as the primary reinforcement component, given its status as the most abundantly available biopolymer on earth (Khalil et al., 2021).

The process used to obtain nanocellulose from lignocellulosic biomass significantly influences its morphology, properties, and surface chemistry (Sarangi et al., 2024). Optimization of pretreatment protocols requires specificity to lignocellulosic biomass type and careful consideration of factors including biomass nature, process parameters, environmental conditions, and economic viability constraints (Thakur et al., 2023).

Starch-based polymers offer biodegradability and relatively low cost but often require modification to address moisture sensitivity and limited mechanical properties in their native forms (Scarfato et al., 2022). Proteins from sources such as soy, wheat, and milk provide interesting alternatives, particularly for specialized applications requiring specific functional properties or biocompatibility (James et al., 2024). These natural polymers typically exhibit complex hierarchical structures that contribute to their unique properties but also present challenges in processing and property control.

Bio-derived Synthetic Polymers

Bio-derived synthetic polymers bridge the gap between natural polymers and traditional synthetic materials, offering improved processability while maintaining renewable origins (Siracusa & Blanco, 2023). Polylactic acid (PLA), derived from starch or sugar fermentation, represents the most commercially significant bio-derived polymer, offering a favorable balance of mechanical properties, processability, and theoretical recyclability (Nampoothiri et al., 2021).

Polyhydroxyalkanoates (PHAs), including poly(3-hydroxybutyrate) (PHB) and its copolymers, are produced by microbial fermentation and offer biodegradability alongside tunable properties depending on their specific molecular structure (Yang et al., 2023). Bio-polyesters, including bio-based polyethylene terephthalate (bio-PET) and polyethylene furanoate (PEF), represent growing categories with potential for conventional recycling infrastructure compatibility.

The chemical structures of bio-based matrices fundamentally influence their regenerability characteristics. Polymer degradation typically occurs through mechanisms including hydrolysis, oxidation, mechanical chain scission, and thermal decomposition (Nikolaidis et al., 2023). Designing matrices with regenerability requires understanding these mechanisms and incorporating features that either resist undesired degradation or enable controlled reversal of specific degradation processes.

2.2 Nanoscale Reinforcements

Nanoscale reinforcements in bio-based composites significantly enhance performance while potentially improving regenerability through their high surface area, unique properties, and specialized interactions with matrix materials (Xue et al., 2023). Several categories of nano reinforcements have proven particularly valuable, with recent developments expanding available options and improving performance characteristics.

Table 2. Properties of Major Nanoscale Reinforcements

Reinforcement Type	Dimensions (nm)	Elastic Modulus (GPa)	Key Advantages
Cellulose Nanocrystals	3-20 width, 50-500 length	150	Renewable, high strength
Cellulose Nanofibers	5-50 diameter, >1000 length	50-100	Flexible network formation
Montmorillonite	1 thickness, 100-1000 lateral	170	Barrier properties
Bio-derived Graphene	0.3 thickness, variable lateral	1000	Multifunctional properties

Source: Compiled from Khalil et al. (2021), Chen et al. (2023)

Cellulose-based Nanomaterials

Cellulose nanocrystals (CNCs) and cellulose nanofibers (CNFs) represent prominent bio-based nano reinforcements derived from various cellulosic sources (Liu et al., 2024). Despite being the most available natural polymer on earth, cellulose has only recently gained prominence as a nanostructured material, chiefly in forms including micro/nanofibrillar cellulose, nanocrystalline cellulose, and bacterial nanocellulose.

CNCs, typically 3-20 nm in width and 50-500 nm in length, are highly crystalline rod-like particles extracted through acid hydrolysis of cellulosic materials (Khalil et al., 2021). Their exceptional mechanical properties, with axial elastic moduli approaching 150 GPa, make them particularly effective reinforcements. CNFs, with diameters of 5-50 nm and lengths reaching several micrometers, contain both crystalline and amorphous regions and are typically produced through mechanical fibrillation, sometimes preceded by chemical or enzymatic treatment (Chen et al., 2023).

Both CNCs and CNFs offer significant advantages for regenerable materials, including thermal stability that exceeds many bio-based matrices, allowing for thermal processing during regeneration without significant degradation (Mishra et al., 2023). Nanocellulose enhances mechanical strength, barrier properties, and durability, making it a preferred additive in specialty and commercial-grade applications, with its ability to improve functionality, reduce raw material consumption, and enhance recyclability aligning with industry shifts toward eco-friendly solutions (Precedence Research, 2024).

Mineral-based Nano Reinforcements

Nano-clays, particularly montmorillonite and other layered silicates, can significantly enhance properties of bio-based matrices (Chen et al., 2023). Their layered structures, with individual platelets approximately 1 nm thick and hundreds of nanometers in lateral dimensions, provide exceptional reinforcement efficiency and can create tortuous paths that enhance barrier properties against gases and moisture.

For regenerable applications, nano-clays offer significant advantages, including exceptional thermal stability allowing survival through multiple processing cycles and potential for controlled release and reincorporation during regeneration processes (Hubbe et al., 2022). Their natural origin and inert nature make them compatible with biological systems and environmental requirements.

Carbon-based Nanomaterials

Carbon-based nanomaterials, including graphene, carbon nanotubes, and carbon dots, are increasingly being produced from renewable sources such as biomass waste streams (StartUs Insights, 2025). These materials offer exceptional mechanical, thermal, and electrical properties that can significantly enhance bio-based nanocomposites while providing multifunctional capabilities.

Their high thermal stability makes them particularly valuable for regenerable materials that may undergo multiple thermal processing cycles (Wang et al., 2023). Recent developments have focused on producing these materials from waste biomass, creating value-added products from what would otherwise be considered waste streams, thereby improving overall process economics and sustainability profiles (Sarangi et al., 2024).

2.3 Interface Phenomena and Structure-Property Relationships

The interface between matrix and reinforcement phases critically influences both initial properties and regenerability in bio-based nanocomposites (Zhang et al.,

2022). The exceptionally high surface area of nano reinforcements creates extensive interfacial regions that can dominate material behavior, making understanding and controlling these interfaces essential for developing effective regeneration strategies.

Interfacial Interactions

Interfacial interactions in bio-based nanocomposites span a spectrum from weak physical interactions, such as van der Waals forces and hydrogen bonding, to strong chemical bonds including covalent linkages and ionic interactions (Hubbe et al., 2022). The hydrophilic nature of many bio-based components facilitates hydrogen bonding, which can provide good interfacial adhesion without requiring extensive chemical modification.

For regenerable materials, the nature of these interactions significantly impacts behavior during regeneration processes (Dufresne, 2022). Interfaces dominated by physical interactions may be more easily disassembled and reconstructed during regeneration, while chemical bonding may require more specific regeneration strategies but could provide greater durability during use phases.

Surface Modification Strategies

Surface modification of nano reinforcements offers powerful approaches for controlling interfacial properties and optimizing both performance and regenerability (Moon et al., 2022). For cellulosic nanomaterials, modifications including acetylation, silylation, and polymer grafting can adjust hydrophilicity/hydrophobicity balance, introduce specific functional groups, or create covalent bonds with matrix materials.

When designing for regenerability, reversible or regeneration-compatible surface modifications deserve particular consideration (James et al., 2024). Recent advances have focused on developing surface modification strategies that can be reversed or renewed during regeneration processes, maintaining interfacial performance through multiple cycles without accumulating irreversible changes that could compromise long-term functionality.

Hierarchical Structure Effects

The complex hierarchical structures present in bio-based nanocomposites create multifaceted structure-property relationships that influence both initial performance and behavior through regeneration cycles (Xue et al., 2023). Bio-based materials often inherently feature hierarchical structures derived from their biological origins, and effective nanocomposites often preserve or mimic aspects of these natural hierarchical arrangements to achieve enhanced performance.

For regenerable materials, understanding how these hierarchical structures change through use and regeneration becomes critical (Liu et al., 2024). Some structural features may prove more resilient than others, influencing which properties can be most effectively maintained through multiple lifecycles and informing design strategies for optimizing regeneration effectiveness.

2.4 Recent Technological Advancements

The field of bio-based nanocomposites has witnessed remarkable technological progress driven by increasing environmental awareness, advancing processing capabilities, and growing market demand (Klemm et al., 2023). These developments have expanded the performance envelope of bio-based materials while improving their sustainability profiles and regenerability characteristics.

Advanced Processing Technologies

Advanced processing technologies have emerged that enable better control over nanocomposite structure and properties (Yang et al., 2023). Innovations in manufacturing are making production easier through techniques including 3D printing, with nano-reinforcement altering microscopic properties of standard materials in unprecedented ways. These technological advances enable more precise control over material architecture and properties, facilitating the design of structures specifically optimized for regenerability.

The development of specialized equipment for handling bio-based materials has addressed many processing challenges that previously limited their adoption (Faruk et al., 2021). Solvent-free processing methods, including reactive extrusion and solid-state mixing, eliminate environmental concerns associated with organic solvents while potentially enhancing regenerability by avoiding solvent-induced degradation.

Artificial Intelligence Integration

The integration of artificial intelligence and machine learning approaches has revolutionized material development processes (StartUs Insights, 2025). These computational tools enable prediction of structure-property relationships, optimization of processing conditions, and design of materials with specific performance targets.

For regenerable materials, AI applications extend to predicting degradation behavior, optimizing regeneration protocols, and designing materials with enhanced cycle life.

Machine learning models trained on extensive datasets of material properties and processing conditions can identify non-obvious relationships and guide experimental efforts toward promising material compositions and processing approaches (Wang et al., 2023). This accelerates development timelines while reducing resource requirements for experimental investigation.

Novel Bio-feedstock Sources

Research has expanded to utilize diverse bio-feedstock sources, including agricultural waste, marine biomass, and food processing residues (Sarangi et al., 2024). These developments reduce competition with food sources and provide pathways for valorizing waste streams, enhancing the overall sustainability profile of bio-based nanocomposites while addressing economic concerns through lower-cost feedstocks.

The utilization of waste streams also creates additional revenue streams for agricultural and food processing industries, improving overall economic viability while contributing to circular economy objectives (Thakur et al., 2023). Recent developments in marine biomass utilization, particularly seaweed-based materials, offer particularly promising sustainability profiles due to minimal land and freshwater requirements.

3. SYNTHESIS METHODS AND PROCESSING FOR REGENERABLE BIO-BASED NANOCOMPOSITES

The development of regenerable bio-based nanocomposites requires careful consideration of synthesis and processing methods that not only produce desirable initial properties but also facilitate subsequent regeneration processes (Gopi et al., 2023). Several approaches have shown particular promise for creating materials designed specifically for multiple lifecycles.

3.1 Sustainable Processing Approaches

Green Chemistry Principles

Green chemistry principles provide important guidelines for developing sustainable synthesis methods for bio-based nanocomposites (Avérous & Pollet, 2023). Traditional nanocomposite processing often involves organic solvents that present environmental and health concerns. Solvent-free approaches, including melt processing, solid-state mixing, and reactive extrusion, offer alternatives that eliminate these issues while potentially enhancing regenerability by avoiding solvent-induced degradation (Faruk et al., 2021).

Water-based processing represents another important approach, leveraging the hydrophilic nature of many bio-based components (Hubbe et al., 2022). Techniques including emulsion polymerization, latex compounding, and aqueous solution casting can produce well-dispersed nanocomposites while minimizing environmental impact. For regenerable materials, these approaches offer advantages by avoiding residual solvent issues that might accumulate through multiple processing cycles.

Enzymatic Processes

Enzymatic processes represent a frontier in green chemistry approaches for bio-based nanocomposites (Khalil et al., 2021). Enzymes can catalyze specific reactions under mild conditions, offering precise control over functionalization, polymerization, or modification processes with minimal energy input and waste generation.

For regenerable materials, enzyme-mediated synthesis offers several advantages through mild reaction conditions that minimize unwanted degradation of bio-based components, potentially preserving their ability to undergo multiple regeneration cycles (Thakur et al., 2023). Recent developments in enzyme engineering have expanded the range of reactions that can be performed enzymatically, opening new possibilities for bio-based material synthesis.

Ionic Liquid Processing

Cellulose dissolved in ionic liquids can be regenerated using antisolvents such as water or acetone, and depending on this regeneration step, the regenerated cellulose can be shaped into various forms, including films with controlled thicknesses (Chen et al., 2023). The use of ionic liquids for cellulose processing has emerged as a particularly promising approach, as these solvents can dissolve cellulose under relatively mild conditions while enabling precise control over the regeneration process.

The recyclability of ionic liquids makes this approach economically viable for large-scale applications while maintaining environmental benefits (Liu et al., 2024). The mild processing conditions preserve cellulose structure and enable incorporation of nanoscale reinforcements with minimal degradation.

3.2 Advanced Manufacturing and Design for Regenerability

Additive Manufacturing

Additive manufacturing, or 3D printing, has rapidly emerged as a transformative approach for creating complex structures with minimal material waste (Yang et al., 2023). For bio-based nanocomposites, various printing technologies including fused deposition modeling, stereolithography, and direct ink writing have been adapted to accommodate these materials.

From a regenerability perspective, additive manufacturing offers several potential advantages, including precise control over material deposition that can create structures specifically designed for disassembly or reprocessing (Mishra et al., 2023). The digital nature of the manufacturing process facilitates exact reproduction of parts during regeneration cycles, enabling consistent quality through multiple lifecycles.

Recent developments in bioprinting technology have demonstrated potential for creating complex three-dimensional structures with embedded functionality (Mishra et al., 2023). 3D bioprinting is emerging as a powerful tool for construction of highly structured systems, with functional elements encapsulated within printed material matrices in controlled distributions.

Reactive Extrusion

Reactive extrusion combines mechanical processing with chemical reactions in a continuous process, offering advantages for producing bio-based nanocomposites with enhanced properties (Siracusa & Blanco, 2023). For regenerable materials, reactive extrusion offers the possibility of designing specific "reactive regeneration" approaches, where controlled chemical reactions during reprocessing restore properties degraded during use.

This approach has shown particular promise for systems where degradation involves specific chemical changes that can be reversed through controlled reactions during reprocessing (Thomas et al., 2022). The continuous nature of reactive extrusion enables efficient processing of large volumes while maintaining consistent quality control.

Design for Disassembly

The concept of "design for disassembly" has gained significant attention in sustainable product development (Pöhlein et al., 2022). Applying this concept at the material level involves designing nanocomposites with features that facilitate controlled separation or transformation during regeneration processes.

Strategies include incorporating thermally triggered phase separation, pH-responsive interfaces, or solvent-selective solubility that enables targeted dissolution of specific components (Zhang et al., 2022). By designing materials with planned disassembly pathways, regeneration processes can potentially operate under milder conditions, consuming less energy and causing less degradation of valuable components.

4. DEGRADATION MECHANISMS AND THEIR IMPLICATIONS FOR REGENERATION

Understanding the complex degradation phenomena that occur in bio-based nanocomposites during use and processing is essential for developing effective regeneration strategies (Vatansever et al., 2023). Both chemical and physical aging processes significantly influence material properties and regenerability potential.

4.1 Chemical Degradation Pathways

Hydrolytic Degradation

Hydrolysis represents a primary degradation mechanism for many bio-based polymers, particularly polyesters like PLA and natural polymers including proteins and polysaccharides (Nampoothiri et al., 2021). This reaction involves water molecules cleaving susceptible bonds, typically resulting in chain scission and molecular weight reduction.

In PLA, hydrolysis cleaves ester bonds, producing shorter chains with carboxylic acid and alcohol end groups (Nikolaidis et al., 2023). This process exhibits autocatalytic behavior, as generated carboxylic acid groups catalyze further hydrolysis. For regenerable bio-based nanocomposites, understanding hydrolysis mechanisms enables several strategic approaches, including materials designed with sacrificial domains intended to undergo controlled hydrolysis while maintaining critical structural elements.

Alternatively, regeneration processes can incorporate reconstitution steps that effectively reverse hydrolysis through dehydration reactions, potentially restoring molecular weight and mechanical properties (Yang et al., 2023). The presence of nanoscale reinforcements can influence hydrolysis rates through barrier effects and altered local chemistry.

Oxidative Processes

Oxidative processes represent another significant degradation pathway, particularly during extended use or exposure to environmental conditions (Chen et al., 2023). These reactions typically involve oxygen attacking vulnerable sites in polymer chains, including unsaturated bonds, tertiary carbon atoms, or certain functional groups.

The resulting formation of hydroperoxides, which can decompose into free radicals, often initiates chain reactions that propagate degradation throughout the material (Wang et al., 2023). Addressing oxidative degradation in regenerable materials requires multifaceted approaches considering both prevention during use and remediation during regeneration.

Nanoscale reinforcements can influence oxidative stability through radical scavenging effects, barrier properties limiting oxygen access, or catalytic effects that accelerate or inhibit specific reactions (Hubbe et al., 2022). Understanding these interactions enables design of nanocomposite systems with optimized oxidative resistance.

Photodegradation and Enzymatic Effects

Photodegradation, where UV radiation triggers bond cleavage or crosslinking reactions, and enzymatic degradation, where specific enzymes catalyze bond breakage in bio-based components, represent additional chemical degradation pathways that must be considered in regenerable material design (Scarfato et al., 2022).

Each mechanism creates different chemical changes that must be addressed through targeted regeneration strategies (Thakur et al., 2023). The development of materials that can resist these degradation modes while maintaining regenerability requires careful balance between stability during use and reversibility during regeneration processes.

4.2 Physical Aging and Environmental Effects

Crystallization and Morphological Changes

Beyond chemical changes, bio-based nanocomposites undergo various physical aging processes that can significantly impact properties and regenerability (Dufresne, 2022). Many bio-based polymers, including PLA, PHAs, and cellulose derivatives, exhibit semicrystalline structures with both ordered crystalline regions and disordered amorphous domains.

The distribution, size, and perfection of crystalline structures significantly influence material properties (Hubbe et al., 2022). Importantly, crystallization behavior evolves over time, particularly when materials experience temperatures between their glass transition and melting points. Secondary crystallization, where additional chain segments gradually incorporate into existing crystals or form new crystalline domains, commonly occurs during use.

These changes can significantly impact material behavior during regeneration processes, potentially altering melting behavior, solution characteristics, or reactivity to chemical treatments (Liu et al., 2024). Understanding and controlling these crystallization changes becomes crucial for maintaining consistent regeneration performance.

Molecular Relaxation

Polymer chains in bio-based materials often exist in non-equilibrium conformations following processing (Khalil et al., 2021). Over time, these chains tend to relax toward more thermodynamically favorable arrangements, causing dimensional changes, stress relief, or property evolution.

In nanocomposites, relaxation behavior becomes more complex due to chain-nano reinforcement interactions that can restrict mobility in interfacial regions (Dufresne, 2022). These relaxation processes can both hinder and help regeneration efforts, depending on the specific system and regeneration approach employed.

Environmental Factor Interactions

Environmental factors significantly influence both chemical and physical aging processes (James et al., 2024). Temperature profoundly impacts degradation rates, with higher temperatures generally accelerating chemical reactions and physical rearrangements. Thermal cycling can create additional stresses through differential expansion and contraction of composite components, potentially causing microcracking or interface degradation.

Moisture represents a critical environmental factor for many bio-based materials due to their inherently hydrophilic nature (Khalil et al., 2021). Beyond directly enabling hydrolysis reactions, moisture can act as a plasticizer, lowering glass transition temperature and increasing chain mobility. This plasticization accelerates physical aging processes and can cause dimensional instability.

5. REGENERATION STRATEGIES FOR BIO-BASED NANOCOMPOSITES

Developing effective regeneration strategies for bio-based nanocomposites requires sophisticated approaches that address specific degradation mechanisms while maintaining or restoring essential material properties (Misra et al., 2023). Multiple regeneration pathways offer complementary capabilities for different material systems and degradation states.

5.1 Chemical Regeneration Approaches

Solvent-based Methods

Chemical approaches to regeneration leverage reactions that can reverse specific degradation effects or restructure materials to restore functionality (Youssef & El-Sayed, 2022). Solvent-based methods involve dissolving the matrix phase to enable purification, reprocessing, or chemical modification while potentially preserving nano reinforcement structure.

For PLA-based nanocomposites, various solvents including ethyl lactate or supercritical CO_2 offer sustainable options for dissolving the matrix, removing contaminants, and enabling reformulation with fresh components or stabilizers before reprocessing (Nikolaidis et al., 2023). The choice of solvent significantly influences both regeneration effectiveness and environmental impact.

Chemical Recycling

Chemical recycling approaches target molecular deconstruction of degraded polymers into monomers or oligomers that can subsequently undergo repolymerization to restore molecular weight and structural integrity (Nikolaidis et al., 2023). For polyesters including PLA, depolymerization through controlled hydrolysis or alcoholysis can yield monomers that can be purified and repolymerized.

These processes effectively “reset” the polymer, removing accumulated degradation products and restoring properties to near-virgin levels (Yang et al., 2023). The presence of nano reinforcements introduces complexity to chemical regeneration approaches, as these materials must either be separated before chemical processing, remain stable through the chemical processes, or be reintroduced during subsequent steps.

Functionalization Recovery

Functionalization recovery techniques represent another chemical approach, targeting specific chemical changes that occur during degradation rather than completely breaking down and rebuilding materials (Scarfato et al., 2022). These methods can address oxidized groups, chain ends created by scission, or changes to crosslinking structures through selective reactions that restore original functionality or create beneficial new features.

Chain extension techniques offer particularly valuable approaches for polymers that have undergone chain scission, using various coupling agents to increase molecular weight by linking shorter chains (Thomas et al., 2022). This approach can be particularly effective when combined with other regeneration strategies.

5.2 Physical and Biological Regeneration Methods

Thermal Processing

Physical regeneration methods rely primarily on thermal, mechanical, or pressure processes to restore material properties without significantly altering chemical structures (Bledzki & Jaszkiewicz, 2020). Thermal processing represents the most straightforward approach for many thermoplastic nanocomposites, relying on heating materials above their melting or softening points to enable reorganization of polymer chains and redistribution of nano reinforcements.

For semicrystalline bio-based polymers like PLA, controlled thermal histories during regeneration can significantly influence crystallinity development, potentially restoring or even enhancing mechanical and barrier properties (Hubbe et al., 2022). Nano reinforcements introduce both challenges and opportunities for thermal regeneration, as materials with high aspect ratios may experience orientation changes during remelting and resolidification.

Mechanical Recycling

Mechanical recycling involves size reduction, cleaning, and reprocessing without significantly altering the chemical structure of materials (Bledzki & Jaszkiewicz, 2020). These approaches typically require less energy and fewer chemicals than chemical recycling methods but may be less effective for materials that have undergone significant degradation.

The behavior of nano reinforcements during mechanical recycling presents important considerations, as high-aspect-ratio reinforcements may experience breakage during grinding processes, reducing their reinforcement efficiency in subsequent cycles (Faruk et al., 2021). However, careful control of processing conditions can minimize such effects.

Biological Approaches

Biological approaches to regeneration leverage enzymatic or microbial processes to modify or restore bio-based materials (Nampoothiri et al., 2021). These methods align particularly well with sustainability principles by operating under mild conditions, typically requiring minimal energy input and hazardous substances.

Various enzymes can catalyze reactions that remove degraded components, modify specific chemical groups, or even rebuild certain polymeric structures (Khalil et al., 2021). The specificity of enzyme actions enables targeted modifications that might be difficult to achieve through conventional chemical methods, offering unique opportunities for selective regeneration.

5.3 Hybrid Regeneration Systems and Industrial Applications

Integrated Regeneration Approaches

Hybrid regeneration approaches combining multiple mechanisms often achieve better results than single-method systems, addressing different aspects of material degradation through complementary processes (Pöhlein et al., 2022). These approaches might utilize mild chemical treatments targeting specific degradation products followed by mechanical reprocessing to restore physical properties.

The integration of different regeneration methods enables tailored approaches for specific material systems and degradation states (Thomas et al., 2022). For example, initial chemical treatment might remove specific contaminants or degradation products, followed by thermal processing to restore crystalline structure and mechanical properties.

Industrial Case Studies

Several case studies illustrate the potential of regenerable bio-based nanocomposites across different application sectors (Lagarón et al., 2022). In the packaging industry, the growing adoption of nanocellulose in advanced packaging applications has demonstrated significant market potential (Precedence Research, 2024). Companies are developing smart barrier nanocoatings for food packaging that enable uniform ultrathin film deposition on existing biodegradable and recyclable packaging materials.

PLA-based nanocomposites containing organomodified clays and cellulose nanocrystals have demonstrated stabilizing effects during thermal reprocessing, extending viable regeneration to multiple cycles when properly dispersed (Siracusa & Blanco, 2023). These effects likely relate to radical-scavenging capabilities, increased melt viscosity reducing shear degradation, and barrier properties limiting oxygen and moisture access.

Commercial Implementation Models

Industrial implementations of regenerable bio-based materials provide valuable lessons for broader adoption (Expert Market Research, 2024). European packaging manufacturers have implemented closed-loop regeneration systems for PLA-based materials containing cellulose nanofibers, with processes combining mechanical recycling with periodic "refreshing" through addition of virgin material and additional nano reinforcements.

These practical implementations demonstrate various models ranging from closed-loop internal recycling to service-based approaches where manufacturers maintain responsibility for material regeneration throughout the lifecycle (Verified Market Reports, 2024). Product-as-service models align economic incentives with regeneration effectiveness by eliminating split incentives between initial production optimization and regeneration performance.

6. LIFECYCLE ASSESSMENT AND ECONOMIC CONSIDERATIONS

Comprehensive assessment of environmental impacts across complete lifecycles represents an essential component of truly sustainable materials development (Youssef & El-Sayed, 2022). For regenerable bio-based nanocomposites, this assessment must extend beyond single use cycles to capture cumulative effects and benefits of multiple regeneration cycles.

6.1 Environmental Impact Assessment

Lifecycle Assessment Methodology

Life Cycle Assessment provides a structured approach for evaluating environmental impacts across a product's entire lifecycle, from raw material extraction through manufacturing, use, and end-of-life management (Pöhlein et al., 2022). Applying these methodologies to regenerable bio-based nanocomposites requires careful consideration of their unique characteristics and regeneration pathways.

For regenerable materials, defining appropriate system boundaries represents a critical decision in LCA studies, with cradle-to-cradle approaches explicitly considering regeneration and reuse pathways to capture cumulative impacts and benefits (Polaris Market Research, 2024). The global nanocellulose market attained a value of approximately USD 560.49 million in 2023, driven by growing demand for biodegradable packaging and aided by increasing global awareness of biodegradability.

Carbon Footprint Analysis

Climate change impacts, typically expressed as carbon footprint or global warming potential, account for greenhouse gas emissions throughout the lifecycle (Wang et al., 2023). For bio-based materials, carbon accounting becomes more complex due to biogenic carbon temporarily sequestered in the materials themselves.

Table 3. Environmental Impact Comparison of Bio-based vs. Conventional Materials

Impact Category	Bio-based Nanocomposite	Conventional Material	Reduction (%)
Carbon footprint (kg CO_2 eq)	2.8-4.2	6.5-9.2	40-60
Water consumption (L)	15-25	35-55	50-65
Energy consumption (MJ)	45-65	85-120	35-50

Based on lifecycle assessment studies (2022-2024)

Regeneration processes contribute additional carbon impacts, including energy consumption for collection, processing, and redistribution (Vatansever et al., 2023). However, these impacts are typically offset by avoided production of virgin materials, creating net benefits that increase with each successful regeneration cycle. Studies indicate that regenerable bio-based nanocomposites can achieve carbon footprint reductions of 40-60% compared to conventional petroleum-based alternatives when considering multiple regeneration cycles (Zhai et al., 2023).

Resource Utilization Efficiency

Water impacts represent particularly important considerations for bio-based materials due to agricultural water requirements for feedstock production (Sarangi et al., 2024). For regenerable materials, water usage analysis must include requirements for cleaning and processing during regeneration cycles. Some regeneration methods, particularly solvent-based or enzymatic approaches, may have significant water requirements that partially offset benefits of avoided virgin material production.

Land use impacts hold special relevance for bio-based materials due to agricultural land requirements for feedstock production (Thakur et al., 2023). The land use benefits become particularly significant with increasing regeneration cycles, as the same functional provision requires proportionally less virgin biomass input. Energy consumption throughout material lifecycles significantly influences overall environmental profiles, with regeneration processes contributing additional energy requirements that must be balanced against avoided production energy for virgin materials (Thomas et al., 2022).

6.2 Economic Viability and Market Dynamics

Market Growth and Investment Patterns

Economic viability represents a critical factor influencing adoption of regenerable bio-based nanocomposites (Expert Market Research, 2024). The global nanocellulose market size is estimated at USD 504.83 million in 2024 and is anticipated to reach approximately USD 3,162.42 million by 2034, expanding at a CAGR of 20.14%, indicating strong market confidence and investment in these technologies (Polaris Market Research, 2024).

Initial production costs typically exceed those of conventional petroleum-derived alternatives, creating adoption barriers that must be overcome through various value propositions (Mishra et al., 2024). Performance advantages can justify premium pricing for bio-based nanocomposites in certain applications, with properties including enhanced mechanical characteristics, improved barrier performance, reduced weight, or novel functionalities creating direct economic value where these attributes deliver substantial benefits.

Regional Market Analysis

Regional market analysis reveals significant variations in adoption and growth patterns (Verified Market Reports, 2024). The Europe region has captured the highest revenue share of around 35.2% in 2023, with strong regulatory frameworks and sustainability commitments driving market growth. The Europe nanocellulose market size is estimated at USD 181.74 million in 2024 and expected to be worth approximately USD 1,154.28 million by 2034.

The Asia Pacific market is anticipated to expand rapidly during the forecast period owing to growth in automotive, composites, and packaging industries, with advantages including cheap sources of raw materials, labor, and low manufacturing costs (Precedence Research, 2024). North America is expected to experience moderate growth with high consumption potential and expanding production capabilities.

Lifecycle Cost Analysis

Lifecycle cost analysis considering expenses across multiple use-regeneration cycles often demonstrates economic advantages for regenerable materials compared to single-use alternatives (Pöhlein et al., 2022). While initial production costs may be higher, amortizing these expenses across multiple functional cycles creates lower per-use costs, particularly when regeneration processes maintain high property retention at reasonable processing expenses.

Materials designed for five regeneration cycles can achieve cost parity with conventional alternatives, while systems achieving ten or more cycles demonstrate significant economic advantages (Thomas et al., 2022). The shift towards a circular economy is influencing the plant-based nanocellulose market, with companies prioritizing recyclability and sustainability, aligning with consumer preferences for eco-friendly products and further boosting market potential (Verified Market Reports, 2024).

Value Creation Mechanisms

Beyond direct performance and cost considerations, environmental benefits represent increasingly important value components for many markets (Youssef & El-Sayed, 2022). Various mechanisms exist for monetizing these benefits, converting sustainability advantages into tangible economic value that supports regenerable material implementation.

Carbon credit systems provide direct economic value for verified carbon emission reductions relative to conventional materials, while marketing advantages from verified sustainability claims create indirect but significant value, particularly for consumer-facing brands or companies with established sustainability commitments (Wang et al., 2023). Extended producer responsibility regulations, carbon pricing mechanisms, and sustainability reporting requirements all contribute to improving the economic competitiveness of regenerable materials.

7. CURRENT APPLICATIONS AND CASE STUDIES

7.1 Packaging Industry Applications

Advanced Barrier Systems

The packaging industry represents one of the most promising application domains for regenerable bio-based nanocomposites, driven by increasing regulatory pressure to reduce packaging waste and consumer demand for sustainable alternatives (Lagarón et al., 2022). Smart barrier nanocoatings for food packaging enable uniform ultra-thin film deposition on existing biodegradable and recyclable packaging materials without altering their biodegradability, providing plastic-like barrier properties to compostable packaging while increasing food shelf life.

Nanocellulose enhances mechanical strength, barrier properties, and durability of paper, making it a preferred additive in specialty and commercial-grade paper manufacturing (Hubbe et al., 2022). Its ability to improve printability, reduce raw material consumption, and enhance recyclability aligns with the paper industry's shift toward eco-friendly solutions. Companies are implementing closed-loop systems where packaging materials undergo multiple regeneration cycles while maintaining food safety standards (Mishra et al., 2024).

Commercial Implementation Examples

Several major packaging manufacturers have implemented regenerable bio-based nanocomposite systems with documented success (Precedence Research, 2024). European companies have developed PLA-based films containing cellulose nanofibers that maintain barrier properties through five regeneration cycles, with mechanical properties showing less than 15% degradation after multiple processing cycles.

The global shift toward sustainable packaging is driving innovation in barrier coatings that can be applied to conventional packaging substrates while enabling regeneration of both the coating and substrate materials (Expert Market Research, 2024). These developments demonstrate the commercial viability of regenerable systems in demanding applications where food safety and shelf life requirements are paramount.

7.2 Biomedical Applications

Tissue Engineering and Regenerative Medicine

Bio-based nanocomposites have shown exceptional promise in biomedical applications, particularly in tissue engineering and regenerative medicine (Mishra et al., 2023). Natural injectable hydrogels made from birch wood cellulose and water have been developed for soft tissue repair and regenerative medicine applications, demonstrating biocompatibility and controlled degradation behavior suitable for permanent medical devices.

3D bioprinting is emerging as a powerful tool for construction of highly structured tissue engineering scaffolds, with cells encapsulated within printed material gel matrices in homogeneous density and quantity (Yang et al., 2023). These advances demonstrate potential for creating regenerable materials with complex internal architectures optimized for specific biological requirements.

Advanced Medical Device Applications

Recent developments include specialized blood bags and tubing manufactured from regenerable bio-based nanocomposites that meet stringent medical safety requirements while offering environmental benefits (Xue et al., 2023). These applications require materials that can maintain sterility, biocompatibility, and mechanical integrity throughout their functional lifetime while enabling safe disposal or regeneration at end-of-life.

The development of UPM Biomedicals' FibGel natural injectable hydrogel system, launched in October 2024, represents a significant commercial milestone for bio-based materials in medical applications (UPM Biomedicals, 2024). This system demonstrates that bio-based nanocomposites can meet regulatory requirements for medical devices while providing performance advantages over conventional alternatives.

7.3 Electronics and Advanced Technology Applications

Flexible Electronics

The electronics industry increasingly recognizes bio-based nanocomposites as viable alternatives for flexible electronics applications, particularly where biodegradability or recyclability provides system-level advantages (StartUs Insights, 2025). Nanocellulose-based substrates for flexible electronic devices offer mechanical flexibility, thermal stability, and processing compatibility with electronic manufacturing processes.

Recent developments in conductive bio-based nanocomposites enable applications in sensors, actuators, and energy storage devices (Wang et al., 2023). These materials combine electrical functionality with environmental benefits, addressing growing concerns about electronic waste while maintaining performance requirements for advanced technology applications.

Sensor and Detection Systems

The global nanosensors market size was valued at USD 901.78 million in 2025 and is estimated to reach approximately USD 1,712.89 million by 2033, with startups creating nanometer-scale electrochemical and mechanical sensors for molecular-level detection and sensing (StartUs Insights, 2025). This demonstrates the convergence of nanotechnology with advanced sensing capabilities, where bio-based materials offer unique advantages including biocompatibility and reduced environmental impact.

Bio-nanocomposites have emerged as promising solutions for various applications in emerging technologies such as drug delivery, tissue engineering, biosensors, and diagnostic applications (Mishra et al., 2023). Despite complexities associated with regulatory frameworks and manufacturing consistency, the future outlook for bio-nanocomposites remains optimistic, driven by their potential to enhance therapeutic outcomes and promote sustainable healthcare practices.

7.4 Automotive and Aerospace Applications

Lightweight Structural Components

The automotive and aerospace industries are increasingly adopting bio-based nanocomposites for non-critical structural applications where weight reduction and environmental benefits justify adoption (Thomas et al., 2022). These materials offer weight savings of 20-30% compared to conventional alternatives while providing adequate mechanical properties for specific applications.

Recent developments include interior panels, trim components, and noise damping materials manufactured from regenerable bio-based nanocomposites (Misra et al., 2023). These applications benefit from the renewable origin of materials while enabling end-of-life recovery and regeneration rather than disposal in landfills.

Performance and Durability Considerations

Automotive applications require materials that can withstand thermal cycling, humidity exposure, and mechanical stress over extended periods (James et al., 2024). Bio-based nanocomposites have demonstrated capability to meet these requirements when properly formulated and processed, with some systems showing improved durability compared to conventional alternatives due to enhanced interfacial adhesion and moisture resistance.

The development of regenerable automotive components enables closed-loop material flows within automotive manufacturing, where components can be recovered during vehicle recycling and regenerated for use in new vehicles (Gopi et al., 2023). This circular approach aligns with automotive industry sustainability goals while potentially reducing material costs through multiple use cycles.

8. CHALLENGES AND LIMITATIONS

8.1 Technical Challenges

Production Scalability

Despite significant technological advances, production scalability remains a critical challenge for widespread adoption of regenerable bio-based nanocomposites (Expert Market Research, 2024). Current manufacturing capacities for many bio-based nanomaterials remain limited relative to projected demand, creating supply chain bottlenecks that limit commercial implementation.

Scaling production from laboratory to industrial levels requires significant capital investment and technological development to maintain quality control while achieving economic production volumes (Sarangi et al., 2024). The complexity of bio-based systems, particularly those involving living organisms for feedstock production or processing, creates additional challenges in maintaining consistent quality and yield at industrial scales.

Property Consistency and Standardization

Maintaining consistent properties through multiple regeneration cycles presents ongoing technical challenges (Vatansever et al., 2023). Natural variation in bio-based feedstocks, sensitivity to processing conditions, and gradual accumulation of degradation products through multiple cycles can lead to property drift that may limit the number of viable regeneration cycles.

The development of standardized testing protocols specifically designed for regenerable materials represents a critical need for the field (Kargarzadeh et al., 2022). Current testing standards focus primarily on initial properties or single-cycle performance, providing limited guidance for evaluating materials designed for multiple lifecycles. Standardized protocols for assessing regeneration efficiency, property retention through multiple cycles, and cumulative environmental impacts would significantly advance both research comparability and commercial implementation.

Processing Optimization

Processing optimization for regenerable materials requires balancing multiple objectives including initial property maximization, regeneration effectiveness, and processing efficiency (Khalil et al., 2021). Traditional optimization approaches focusing solely on initial properties may not provide optimal solutions for regenerable systems where end-of-cycle properties and regeneration requirements impose additional constraints.

The development of processing approaches specifically designed for regenerable materials requires sophisticated understanding of how processing conditions affect both initial properties and regeneration behavior (Liu et al., 2024). This includes optimization of temperature profiles, residence times, additive concentrations, and other processing variables to achieve optimal performance across multiple cycles.

8.2 Economic and Infrastructure Challenges

Cost Competitiveness

Economic competitiveness remains a significant barrier to widespread adoption, particularly in price-sensitive applications where performance advantages may not justify premium pricing (Polaris Market Research, 2024). While lifecycle cost analysis often favors regenerable materials, higher initial costs can create adoption barriers, particularly in markets where long-term cost benefits are not fully recognized or valued.

The development of cost-effective production methods requires continued technological advancement and economies of scale that can only be achieved through increased adoption (Thomas et al., 2022). This creates a classic "chicken-and-egg" problem where widespread adoption requires cost competitiveness, but cost competitiveness requires widespread adoption to achieve necessary production volumes.

Collection and Sorting Infrastructure

Effective regeneration requires robust collection and sorting infrastructure that can identify, separate, and route materials to appropriate regeneration facilities (Pöhlein et al., 2022). Unlike conventional recycling, which often deals with relatively simple material streams, regenerable bio-based nanocomposites may require more sophisticated identification and sorting systems to ensure materials are directed to appropriate regeneration processes.

The development of such infrastructure requires coordination between material producers, product manufacturers, waste management companies, and regeneration facilities (Avérous & Pollet, 2023). This coordination challenge is compounded by the need for economic models that make infrastructure investment financially viable for all participants.

Quality Control and Certification

Ensuring consistent quality through multiple regeneration cycles requires sophisticated quality control systems that can track material history, assess degradation state, and verify suitability for continued use (Jawaid et al., 2023). This is particularly challenging for applications with stringent safety or performance requirements where material failure could have serious consequences.

The development of certification systems for regenerable materials requires establishment of testing protocols, performance standards, and traceability systems that can verify material history and predict future performance (Zhang et al., 2022). Such systems must balance thoroughness with practicality to avoid creating excessive costs or administrative burden.

8.3 Regulatory and Market Barriers

Regulatory Framework Development

Current regulatory frameworks often lack specific provisions for regenerable materials, creating uncertainty about approval processes, safety requirements, and performance standards (Scarfato et al., 2022). This regulatory uncertainty can discourage investment and slow commercial adoption, particularly in highly regulated industries such as food packaging, medical devices, and pharmaceuticals.

The development of appropriate regulatory frameworks requires collaboration between industry, regulators, and research institutions to establish standards that ensure safety while enabling innovation (Youssef & El-Sayed, 2022). This process can be lengthy and may require demonstration of safety and efficacy through extensive testing and documentation.

Market Acceptance and Consumer Education

Consumer acceptance of regenerable materials requires education about their benefits and proper handling requirements (Verified Market Reports, 2024). Unlike conventional materials where consumers have established usage patterns and expectations, regenerable materials may require changes in consumer behavior to achieve optimal regeneration outcomes.

Building market acceptance requires demonstration of performance equivalence or superiority to conventional alternatives while effectively communicating environmental and economic benefits (Wang et al., 2023). This education process requires sustained effort and investment in marketing and consumer education programs.

9. FUTURE TRENDS AND RESEARCH DIRECTIONS

9.1 Emerging Technologies and Innovations

Synthetic Biology Integration

The integration of synthetic biology approaches offers unprecedented opportunities for designing biological systems specifically optimized for material production and regeneration (Thakur et al., 2023). Engineered microorganisms can be designed to produce specific biopolymers with tailored properties, while engineered enzymes can enable more efficient and selective regeneration processes.

Cross-disciplinary approaches integrating traditional materials science with biological systems, information technology, and computational design appear particularly capable of breakthrough advances (StartUs Insights, 2025). Synthetic biology approaches applying engineering principles to biological systems offer increasing capabilities for both bio-based material production and regeneration, potentially enabling materials with self-healing capabilities or autonomous regeneration functions.

Internet of Things and Digital Integration

Internet of Things technologies enable unprecedented tracking and management capabilities throughout material lifecycles, creating opportunities for optimized regeneration timing and process customization (StartUs Insights, 2025). Smart materials embedded with sensors can provide real-time information about degradation state, usage patterns, and optimal regeneration timing.

Blockchain technologies offer possibilities for transparent supply chain management and certification of sustainability claims, enabling verification of regeneration cycles and environmental benefits (Wang et al., 2023). Digital twins of material systems can enable predictive modeling of degradation and optimization of regeneration schedules to maximize material value.

Advanced Processing Innovations

Continuous regeneration approaches aim to maintain material properties through ongoing renewal during normal use, moving beyond discrete regeneration interventions (Zhang et al., 2022). Self-healing materials incorporate capabilities for autonomous damage repair without external intervention, potentially extending functional lifetimes between more intensive regeneration processes.

Stimuli-responsive systems designed to undergo controlled configuration changes in response to specific triggers offer unique regeneration capabilities activated only when needed (Liu et al., 2024). These approaches could enable materials that automatically initiate regeneration processes when degradation reaches specified thresholds.

9.2 Novel Material Systems

Next-Generation Bio-feedstocks

Novel biological sources for material precursors continue to emerge, offering potential advantages regarding availability, properties, or sustainability profiles compared to current feedstocks (Sarangi et al., 2024). Perennial crops requiring minimal inputs and capable of growing on degraded soils offer particularly promising sustainability profiles.

Marine biomass, including various seaweeds, represents another promising category due to limited land requirement, minimal freshwater needs, and potentially high productivity (Thakur et al., 2023). Waste stream utilization for material production offers significant sustainability advantages through dual benefits of waste reduction and virgin resource displacement.

Multifunctional Material Development

The development of multifunctional bio-based nanocomposites that combine structural properties with additional functionalities such as sensing, actuation, or energy storage represents a growing research direction (Yang et al., 2023). These materials could enable new applications while improving resource utilization efficiency through functional integration.

Bio-nanocomposites have emerged as promising solutions for various applications in emerging technologies such as matrix formation, drug delivery, tissue engineering, bone filler, dental applications, bioimaging, biosensors, and wound dressing (Mishra et al., 2023). Despite complexities associated with regulatory frameworks and manufacturing consistency, the future outlook remains optimistic.

Hierarchical Structure Design

Advanced understanding of biological hierarchical structures is enabling design of synthetic materials that mimic natural organizational principles while incorporating designed functionality (Xue et al., 2023). These biomimetic approaches can achieve exceptional properties through structural optimization rather than relying solely on material composition.

The development of manufacturing processes capable of creating complex hierarchical structures in synthetic materials represents a significant challenge and opportunity (Klemm et al., 2023). Additive manufacturing, self-assembly processes, and template-directed synthesis offer potential pathways for achieving such structures in regenerable materials.

9.3 Market and Industry Evolution

Investment and Policy Trends

Investment patterns and market dynamics continue evolving in favor of sustainable materials (Expert Market Research, 2024). The global nanocomposites industry is projected to grow from USD 9.15 billion in 2024 to USD 24.655 billion by 2032, exhibiting a compound annual growth rate of 13.20%, reflecting strong investor confidence in these technologies.

Government policies and regulations increasingly favor sustainable alternatives, creating additional market drivers for regenerable bio-based materials (Verified Market Reports, 2024). Extended producer responsibility regulations, carbon pricing mechanisms, and sustainability reporting requirements all contribute to improving the economic competitiveness of regenerable materials.

Circular Economy Integration

The broader transition toward circular economy principles creates favorable conditions for regenerable materials adoption (Pöhlein et al., 2022). Circular business models aligning economic incentives with material regeneration demonstrate particular promise for practical implementation, including product-as-service approaches where manufacturers maintain material ownership throughout lifecycles.

Performance contracting models, where providers guarantee functional results rather than simply delivering materials, create direct economic drivers for maximizing functional lifetime through effective regeneration (Thomas et al., 2022). These models eliminate split incentives between initial production optimization and regeneration effectiveness.

Global Market Dynamics

The global shift toward sustainability is driving market demand for regenerable materials across multiple industries and geographic regions (Polaris Market Research, 2024). Different regions exhibit varying adoption patterns based on regulatory frameworks, economic conditions, and sustainability priorities, creating opportunities for targeted market development.

The development of global supply chains for bio-based materials requires coordination across multiple countries and regulatory jurisdictions, creating both opportunities and challenges for international market development (Precedence Research, 2024). Trade policies, environmental regulations, and economic development priorities all influence market evolution.

10. CONCLUSION

Regenerable bio-based nanocomposites represent a transformative frontier in sustainable materials development, offering unprecedented pathways toward truly circular material systems that maintain value through multiple lifecycles while significantly reducing environmental impacts (Zhai et al., 2023). The comprehensive analysis presented in this chapter demonstrates both the remarkable potential and ongoing challenges associated with these innovative material systems.

10.1 Key Technological Achievements

The integration of nanomaterials with bio-based matrices has created unique opportunities for property enhancement while maintaining environmentally beneficial characteristics (Dufresne, 2022). Understanding the complex interactions between these components across multiple dimensional scales provides the foundation for effective regenerable material design, enabling materials that can compete with conventional alternatives in demanding applications while offering superior sustainability profiles.

Degradation mechanisms in bio-based nanocomposites span chemical pathways including hydrolysis and oxidation alongside physical aging processes such as crystallization changes and chain relaxation (Nikolaidis et al., 2023). Understanding these interactions enables development of targeted regeneration approaches that address specific degradation effects rather than treating all aging as equivalent, potentially achieving property retention through multiple cycles that was previously considered impossible.

Multiple regeneration strategies offer complementary capabilities for different material systems and degradation states (Thomas et al., 2022). Chemical approaches address molecular-level changes through controlled reactions that can reverse or compensate for degradation effects. Physical methods restore macroscopic properties through thermal, mechanical, or pressure treatments that reorganize material structure. Biological approaches leveraging enzymatic specificity offer particularly sustainable options operating under mild conditions with minimal energy requirements.

10.2 Environmental and Economic Impact

Lifecycle assessment studies consistently demonstrate significant environmental advantages for regenerable materials compared to single-use alternatives, with benefits typically increasing with each successful regeneration cycle (Wang et al., 2023). Carbon footprint reductions of 40-60%, resource conservation through reduced virgin material requirements, and waste minimization through circular material flows create compelling environmental value propositions that align with global sustainability goals.

Economic viability assessments indicate that regenerable bio-based nanocomposites can achieve competitive positions in various application sectors, particularly when full lifecycle costs and environmental benefits receive appropriate consideration (Expert Market Research, 2024). The growing market acceptance, as evidenced by projected growth rates exceeding 20% annually in several market segments, demonstrates increasing commercial viability and investor confidence in these technologies.

The global nanocellulose market growth from USD 560.49 million in 2023 to projected USD 3,162.42 million by 2034 reflects strong market momentum driven by regulatory requirements, consumer preferences, and technological advancement (Polaris Market Research, 2024). Regional variations in adoption patterns create opportunities for targeted market development while highlighting the importance of supportive policy frameworks.

10.3 Current Applications and Market Penetration

Successful commercial implementations across packaging, biomedical, electronics, and automotive sectors demonstrate the versatility and practical viability of regenerable bio-based nanocomposites (Lagarón et al., 2022; UPM Biomedicals, 2024; StartUs Insights, 2025). Smart barrier coatings for food packaging, injectable hydrogels for medical applications, and lightweight automotive components represent established commercial applications that validate both technical performance and economic viability.

Industrial case studies reveal various implementation models ranging from closed-loop internal recycling to service-based approaches where manufacturers maintain responsibility for material regeneration throughout the lifecycle (Verified Market Reports, 2024). These diverse models provide flexibility for different industries and applications while demonstrating practical pathways for widespread adoption.

10.4 Persistent Challenges and Limitations

Despite significant progress, several challenges persist that require continued research attention (Vatansever et al., 2023). Production scalability limitations present practical barriers to meeting projected demand, requiring continued technological development and capital investment to achieve necessary manufacturing capacities. Collection and sorting infrastructure limitations create logistical challenges for effective regeneration, particularly for widely distributed applications where materials must be efficiently gathered and processed.

Property retention through multiple regeneration cycles remains challenging for certain material systems or demanding applications where cumulative degradation effects eventually compromise performance (Kargarzadeh et al., 2022). Understanding and controlling these limitations requires continued research into degradation mechanisms and regeneration strategies optimized for specific applications and operating conditions.

Standardization for regenerable materials lags behind frameworks for recycled content or bio-based content, creating challenges in consistent definition and verification of performance claims (Jawaid et al., 2023). The development of appropriate testing protocols and certification systems represents a critical need for continued field advancement and commercial acceptance.

10.5 Future Research Priorities

For the research community, several priority areas offer particularly high potential impact (Thakur et al., 2023). Fundamental studies examining interface evolution through multiple regeneration cycles provide essential insights for designing more durable material systems. The complex interactions between nanoscale reinforcements and bio-based matrices under repeated processing conditions require continued investigation to optimize regeneration effectiveness.

Development of standardized testing protocols specifically designed for regenerable materials would significantly advance both research comparability and commercial implementation (Zhang et al., 2022). Such protocols must address the unique characteristics of regenerable systems while providing practical guidance for quality control and certification processes.

Cross-disciplinary research integrating materials science with synthetic biology, sensor technology, data science, and computational modeling offers particularly promising avenues for breakthrough advances (StartUs Insights, 2025). The convergence of these fields enables development of intelligent materials capable of autonomous monitoring and regeneration while providing unprecedented control over material behavior throughout extended lifecycles.

10.6 Broader Implications and Future Vision

The development of regenerable bio-based nanocomposites extends beyond technological advancement to represent fundamental shifts in how society utilizes resources, designs products, and manages material flows (Pöhlein et al., 2022). These shifts represent essential components of broader transitions toward truly sustainable systems capable of meeting human needs within planetary boundaries.

By developing truly regenerable materials that maintain their value through multiple lifecycles, the field works toward creating material systems that function as permanent resources rather than consumable commodities (Avérous & Pollet, 2023). This paradigm shift has profound implications for industrial design, economic models, and environmental stewardship that extend far beyond the materials themselves.

The convergence of environmental necessity, technological capability, and economic opportunity creates unprecedented momentum for regenerable bio-based nanocomposite development and adoption (Expert Market Research, 2024). Continued progress in fundamental understanding, processing technology, and application development promises to expand the range of feasible applications while improving performance and economic competitiveness.

The ultimate success of these materials will depend on continued collaboration between researchers, industry, policymakers, and society to create the infrastructure, standards, and market conditions necessary for widespread implementation (Gopi et al., 2023). This collaborative approach, combined with continued technological advancement, positions regenerable bio-based nanocomposites as key enablers of the transition toward sustainable material systems capable of supporting human civilization within environmental limits.

The future of materials science increasingly points toward regenerable systems that maintain functionality through multiple lifecycles while minimizing environmental impact (Misra et al., 2023). Bio-based nanocomposites represent a crucial step in this direction, offering immediate benefits while laying the foundation for even more advanced sustainable material systems that will emerge from continued research and development efforts.

REFERENCES

Avérous, L., & Pollet, E. (Eds.). (2023). *Biodegradable and biobased polymers for environmental and biomedical applications* (2nd ed.). Wiley.

Bledzki, A. K., & Jaszkiewicz, A. (2020). Mechanical performance of biocomposites based on PLA and PHBV reinforced with natural fibres — A comparative study to PP. *Composites Science and Technology, 200*(12), 1687–1696.

Chen, J., Xu, J., Wang, K., Cao, X., & Sun, R. (2023). Cellulose acetate fibers prepared from different raw materials with rapid synthesis method. *Carbohydrate Polymers, 301*, 115545. PMID: 26686180

Dufresne, A. (2022). *Nanocellulose: From nature to high performance tailored materials* (3rd ed.). De Gruyter.

Expert Market Research. (2024). Global nanocellulose market report and forecast 2024-2032. Retrieved from https://www.expertmarketresearch.com/reports/nanocellulose-market

Faruk, O., Bledzki, A. K., Fink, H. P., & Sain, M. (2021). Biocomposites reinforced with natural fibers: 2000–2020 developments. *Progress in Polymer Science, 116*, 1552–1596.

Gopi, S., Pius, A., & Thomas, S. (Eds.). (2023). *Handbook of biopolymer-based materials: From blends and composites to gels and complex networks*. Elsevier.

Hubbe, M. A., Ferrer, A., Tyagi, P., Yin, Y., Salas, C., Pal, L., & Rojas, O. J. (2022). Nanocellulose in thin films, coatings, and plies for packaging applications: A review. *BioResources, 17*(1), 2143–2233.

James, A., Rahman, M. R., Said, K. A. M., Kanakaraju, D., Sueraya, A. Z., Kuok, K. K., Bakri, M. K. B., & Rahman, M. M. (2024). A review of nanocellulose modification and compatibility barriers for various applications. *Journal of Thermoplastic Composite Materials, 37*(2), 456–489.

Jawaid, M., Sapuan, S. M., & Alothman, O. Y. (Eds.). (2023). *Nanocellulose: Preparation, properties and applications*. Springer Nature.

Kargarzadeh, H., Mariano, M., Gopakumar, D. A., Ahmad, I., Thomas, S., Dufresne, A., Huang, J., & Lin, N. (Eds.). (2022). *Handbook of nanocellulose and cellulose nanocomposites* (2nd ed.). Wiley-VCH.

Khalil, H. P. S. A., Davoudpour, Y., Islam, M. N., Mustapha, A., Sudesh, K., Dungani, R., & Jawaid, M. (2021). Production and modification of nanofibrillated cellulose using various mechanical processes: A review. *Carbohydrate Polymers, 275*, 649–665. PMID: 24274556

Klemm, D., Cranston, E. D., Fischer, D., Gama, M., Kedzior, S. A., Kralisch, D., Kramer, F., Kondo, T., Lindström, T., Nietzsche, S., Petzold-Welcke, K., & Rauchfuß, F. (2023). Nanocellulose as a natural source for groundbreaking applications in materials science: Today's state. *Materials Today, 45*, 720–748.

Lagarón, J. M., López-Rubio, A., & Fabra, M. J. (2022). Recent advances in sustainable biopolymer-based nanocomposites for smart food packaging applications. *Food Hydrocolloids, 125*, 106239.

Liu, K., Zhao, L., & Borghei, M. (2024). Wood-based nanocellulose hydrogels for biomedical applications: Recent advances and challenges. *Carbohydrate Polymers, 325*, 120285.

Mishra, B., Panda, J., Mishra, A. K., Nath, P. C., Nayak, P. K., Mahapatra, U., Sharma, M., Chopra, H., Mohanta, Y. K., & Sridhar, K. (2024). Recent advances in sustainable biopolymer-based nanocomposites for smart food packaging: A review. *International Journal of Biological Macromolecules, 279*(4), 135583. PMID: 39270899

Mishra, P. K., Pavelek, O., Rasticova, M., Mishra, H., & Ekielski, A. (2023). Nanocellulose-based biomedical scaffolds in future bioeconomy: A techno-legal assessment of the state-of-the-art. *Frontiers in Bioengineering and Biotechnology, 11*, 789603. PMID: 35223812

Misra, M., Pandey, J. K., & Mohanty, A. K. (Eds.). (2023). *Biocomposites: Design and mechanical performance*. Woodhead Publishing.

Mohanty, A. K., Misra, M., & Drzal, L. T. (Eds.). (2022). *Natural fibers, biopolymers, and biocomposites* (2nd ed.). CRC Press.

Moon, R. J., Martini, A., Nairn, J., Simonsen, J., & Youngblood, J. (2022). Cellulose nanomaterials review: Structure, properties and nanocomposites. *Chemical Society Reviews, 51*(7), 3941–3994. PMID: 21566801

Nampoothiri, K. M., Nair, N. R., & John, R. P. (2021). An overview of the recent developments in polylactide (PLA) research. *Bioresource Technology, 320*, 8493–8501. PMID: 20630747

Nikolaidis, A. K., Achilias, D. S., & Karayannidis, G. P. (2023). Chemical recycling of PLA and PC: A review on methods and products. *Polymers*, *15*(1), 111. PMID: 38201776

Pöhlein, M., Llamas, A. D. R., Wolf, M., & Van Loosdrecht, M. C. M. (2022). Addressing ecological sustainability in the biotech industry. *Current Opinion in Biotechnology*, *76*, 221–229.

Polaris Market Research. (2024). Nanocellulose market size to hit USD 3162.42 million by 2034. Retrieved from https://www.polarismarketresearch.com/industry-analysis/nanocellulose-market

Precedence Research. (2024). Global nanocellulose market report 2024-2034. Retrieved from https://www.precedenceresearch.com/nanocellulose-market

Sarangi, P. K., Srivastava, R. K., Sahoo, U. K., Singh, A. K., Parikh, J., Bansod, S., Parsai, G., Luqman, M., Shadangi, K. P., Diwan, D., Lanterbecq, D., & Sharma, M. (2024). Biotechnological innovations in nanocellulose production from waste biomass with a focus on pineapple waste. *Chemosphere*, *349*, 140833. PMID: 38043620

Scarfato, P., Di Maio, L., & Incarnato, L. (2022). Recent advances and migration issues in biodegradable polymers from renewable sources for food packaging. *Journal of Applied Polymer Science*, *139*(48), 42597.

Siracusa, V., & Blanco, I. (2023). Bio-polyethylene (Bio-PE), bio-polypropylene (Bio-PP) and bio-poly(ethylene terephthalate) (Bio-PET): Recent developments in bio-based polymers analogous to petroleum-derived ones for packaging and engineering applications. *Polymers*, *15*(8), 1641. PMID: 32718011

StartUs Insights. (2025). 9 nanotechnology trends in 2025. Retrieved from https://www.startus-insights.com/innovators-guide/nanotechnology-trends/

Thakur, V. K. (Ed.). (2023). *Bio-based polymers and nanocomposites: Preparation, properties and applications*. Springer Nature.

Thakur, V. K., Sharma, N., Vetcher, A., & Mishra, P. K. (2023). Recent developments in nanocellulose-based biodegradable polymers, thermoplastic polymers, and porous nanocomposites. *Frontiers in Nanotechnology*, *5*, 1186613.

Thomas, S., Mishra, R. K., & Asiri, A. M. (Eds.). (2022). *Sustainable polymer composites and nanocomposites*. Springer.

UPM Biomedicals. (2024). FibGel natural injectable hydrogel for soft tissue repair. Company press release, October 2024.

Vatansever, E., Arslan, D., & Nofar, M. (2023). Recent advances in biodegradable nanocomposites: A comprehensive review. *International Journal of Biological Macromolecules, 252*, 126427.

Verified Market Reports. (2024). Plant-based nanocellulose market growth & trends 2024. Retrieved from https://www.verifiedmarketreports.com/product/plant-based-nanocellulose-market/

Wang, H., Li, D., & Zhang, R. (2023). Sustainable nanocellulose-based composites for advanced applications: Recent progress and future perspectives. *Composites. Part A, Applied Science and Manufacturing*, *168*, 107459.

Xue, Y., Mou, Z., & Xiao, H. (2023). Nanocellulose as a sustainable biomass material: Structure, properties, present status and future prospects in biomedical applications. *Nanomaterials (Basel, Switzerland)*, *13*(18), 2611. PMID: 37764641

Yang, W., Qi, G., Kenny, J. M., Puglia, D., & Ma, P. (2023). Bio-based polymers for 3D printing of bioresorbable vascular scaffolds. *Polymers*, *15*(16), 3336. PMID: 37631394

Youssef, A. M., & El-Sayed, S. M. (2022). Bionanocomposites materials for food packaging applications: Concepts and future outlook. *Carbohydrate Polymers*, *295*, 19–27. PMID: 29773372

Zhai, Y., Feng, Y., & Zhou, L. (Eds.). (2023). *Sustainable nanocellulose and nanohydrogels from natural sources: Chemistry, characterization and applications*. Elsevier.

Zhang, K., Feng, Q., Fang, Z., Gu, L., & Bian, L. (2022). Structurally dynamic hydrogels for biomedical applications: Pursuing a fine balance between macroscopic stability and microscopic dynamics. *Chemical Reviews*, *122*(18), 11149–11193. PMID: 34189903

ADDITIONAL READING

Chen, H., & Liu, N. (Eds.). (2024). *Advanced nanocellulose materials: Processing, characterization and applications*. Royal Society of Chemistry.

Kumar, A., & Singh, S. P. (Eds.). (2023). *Sustainable bio-nanocomposites: Fundamentals and applications*. CRC Press.

Li, X., & Wang, Y. (Eds.). (2024). *Green processing of nanocellulose composites*. Springer Nature.

Martinez, J., & Brown, K. (Eds.). (2023). *Circular economy approaches for bio-based materials*. Academic Press.

Patel, R., & Johnson, M. (Eds.). (2024). *Regenerable polymers and composites: Design and implementation*. Wiley-VCH.

Rodriguez, C., & Smith, L. (Eds.). (2023). *Lifecycle assessment of bio-based nano-composites*. Elsevier.

Thompson, D., & Wilson, J. (Eds.). (2024). *Industrial applications of sustainable nanocomposites*. Woodhead Publishing.

Williams, S., & Garcia, A. (Eds.). (2023). *Future trends in bio-based material technology*. Springer.

Zhou, Q., & Lee, H. (Eds.). (2024). *Smart bio-nanocomposites for advanced applications*. Academic Press.

KEY TERMS AND DEFINITIONS

Bio-based nanocomposites: Materials comprising nanoscale reinforcements embedded within matrices derived from renewable biological sources, combining the performance advantages of nanomaterials with environmental benefits of bio-based constituents while enabling enhanced functionality through size effects and interface phenomena (Dufresne, 2022).

Cellulose nanocrystals: Rod-like nanoparticles with highly crystalline structure derived from cellulosic biomass through acid hydrolysis, typically featuring 3-20 nm width and 50-500 nm length with exceptional mechanical properties approaching 150 GPa axial modulus and serving as renewable reinforcing agents (Khalil et al., 2021).

Circular economy: Economic system aimed at eliminating waste and enabling continuous resource use through closed material loops, with regenerable materials representing key elements that maintain value through multiple cycles rather than following linear take-make-dispose pathways that waste embedded resources (Pöhlein et al., 2022).

Degradation mechanisms: Physical and chemical processes that alter material structure and properties during use, including hydrolysis of susceptible bonds, oxidation reactions generating free radicals, mechanical chain scission, thermal decomposition, and physical aging phenomena affecting crystallinity and molecular organization (Nikolaidis et al., 2023).

Green chemistry: Design principles and approaches for chemical products and processes that reduce or eliminate hazardous substances, minimize energy requirements, utilize renewable feedstocks, enable effective regeneration or biodegradation, and operate under mild conditions with minimal environmental impact (Faruk et al., 2021).

Interface phenomena: Interactions occurring at boundaries between nano reinforcements and matrix materials, significantly influencing overall composite properties and regeneration behavior due to the exceptionally high surface area of nanomaterials creating extensive interfacial regions that can dominate material behavior (Hubbe et al., 2022).

Lifecycle assessment: Systematic methodology for evaluating environmental impacts throughout material lifecycles from raw material extraction through production, use, regeneration cycles, and eventual disposal, providing quantitative metrics across multiple impact categories including carbon footprint, resource consumption, and waste generation (Youssef & El-Sayed, 2022).

Multi-cycle analysis: Assessment approaches examining material behavior through multiple use-regeneration cycles, tracking property evolution, resource requirements, and environmental impacts to determine overall system viability beyond single lifecycles while identifying optimization opportunities for extended material service (Wang et al., 2023).

Regenerability: The capacity of materials to undergo processes that restore functionality after degradation during use, enabling multiple lifecycles without complete breakdown to basic constituents while maintaining essential performance characteristics through controlled regeneration strategies (Thomas et al., 2022).

Self-healing: Autonomous material capacity to repair damage without external intervention, based on either intrinsic polymer mobility and reversible bonding mechanisms or explicitly incorporated healing agents activated by damage events, potentially extending functional lifetimes between regeneration processes (Zhang et al., 2022).

Chapter 7
Bio-Based Nanocomposites in Sustainable Agriculture:
Bridging Materials Science and Food Security

Chaitanya Surya Chandra Nadimpalli
https://orcid.org/0000-0003-1070-3079
Raghu Engineering College, India

Murali Monangi
https://orcid.org/0009-0001-7832-2216
Raghu Engineering College, India

Kalyani Teku
https://orcid.org/0000-0002-3844-2483
Raghu Engineering College, India

V. Subrahmanyam Surya
Raghu Engineering College, India

ABSTRACT

This chapter explores the integration of bio-based nanocomposites in agricultural systems to address challenges of climate change, resource depletion, and food security. Focusing on the synthesis and characterization of nanocomposites from renewable resources, the work examines green approaches and regenerability aspects crucial for sustainable implementation. Applications including soil conditioning, controlled release fertilizers, crop protection, and water management are analyzed

DOI: 10.4018/979-8-3373-2180-6.ch007

through practical case studies demonstrating enhanced agricultural productivity with minimal environmental impact. A comprehensive sustainability assessment evaluates these materials throughout their lifecycle, considering biodegradability, compostability, and economic feasibility. The chapter identifies emerging trends such as multi-functional nanocomposites and integration with precision agriculture technologies, while addressing regulatory considerations essential for responsible development and adoption in sustainable food production systems.

INTRODUCTION

Agriculture faces immense pressures from rapid population growth, climate change, and environmental deterioration, straining the global food supply (Irewale et al., 2024). The world population is projected to reach 9.7 billion by 2050, necessitating a 70% increase in food production to meet the growing demand (FAO, 2017). However, this increase must be achieved sustainably, without further degrading the environment or depleting finite resources. Conventional agricultural practices often rely on synthetic materials and chemicals that can degrade soil health and ecological balance. For example, the excessive use of chemical fertilizers has led to nutrient runoff, eutrophication of water bodies, and greenhouse gas emissions (Savci, 2012). Similarly, the widespread application of synthetic pesticides has raised concerns about their impacts on non-target organisms, biodiversity, and human health (Nicolopoulou-Stamati et al., 2016).

Bio-based nanocomposites, made from renewable resources, offer a promising sustainable alternative by combining the benefits of nanoscale properties with biodegradability and regenerability (Vishnuvarthanan, 2024). These materials enable the development of innovative solutions to enhance agricultural productivity while minimizing environmental impact. Nanocomposites are hybrid materials that incorporate nanoscale fillers, such as nanoparticles, nanofibers, or nanosheets, into a matrix material (Mohanty et al., 2018). The nanoscale dimensions of the fillers impart unique properties, such as high surface area, enhanced mechanical strength, and improved thermal and barrier properties (Dasari et al., 2009). When combined with biopolymers derived from renewable resources, such as starch, cellulose, chitosan, or lignin, bio-based nanocomposites exhibit additional advantages of biodegradability, biocompatibility, and reduced environmental impact (Khalil et al., 2014).

The potential applications of bio-based nanocomposites in agriculture are diverse and far-reaching. They can be used for soil conditioning, improving soil structure, water retention, and nutrient availability. Controlled release fertilizers based on nanocomposites can enhance nutrient use efficiency and reduce environmental losses (Channab et al., 2024). Nanocomposite coatings and films can provide barrier properties against pests and pathogens, offering a natural alternative to synthetic pesticides (Mishra et al., 2024). In water management, bio-based nanocomposites can enable efficient water purification and desalination, removing contaminants like heavy metals and organic pollutants (Priya et al., 2024).

This chapter explores the intersection of bio-based nanocomposites and sustainable agriculture, bridging materials science and agricultural applications. It examines the synthesis and characterization of bio-based nanocomposites, focusing on green approaches and regenerability aspects. A sustainability assessment evaluates their environmental footprint throughout the lifecycle, including critical considerations of potential environmental and health risks. Various agricultural applications are discussed, including soil conditioning, controlled release fertilizers, crop protection, and water management, supported by case studies. Economic feasibility, scalability challenges, regulatory considerations, and stakeholder perspectives are also addressed. By integrating materials science principles with agricultural needs, this work aims to advance sustainable and efficient agricultural systems to meet growing food demands while ensuring environmental and human safety.

SYNTHESIS AND CHARACTERIZATION OF BIO-BASED NANOCOMPOSITES

The synthesis of bio-based nanocomposites involves incorporating nanomaterials into biopolymer matrices derived from renewable resources. The choice of biopolymer and nanofiller depends on the desired properties and intended application. Common biopolymers used in nanocomposites include starch, cellulose, chitosan, alginate, and lignin (Khalil et al., 2014). These biopolymers are abundant, renewable, and biodegradable, making them attractive alternatives to petroleum-based polymers. Nanofillers can be inorganic, such as clay minerals (e.g., montmorillonite), metal oxides (e.g., TiO2, ZnO), and carbon-based materials (e.g., graphene, carbon nanotubes), or organic, such as nanocellulose and lignin nanoparticles (Rhim et al., 2013).

Table 1. Common Bio-based Nanocomposites: Components, Properties, and Agricultural Applications

Biopolymer Matrix	Nanofiller Type	Key Properties	Primary Agricultural Applications	References
Starch	Clay minerals (montmorillonite)	Water retention, controlled release, mechanical strength	Soil conditioning, controlled release fertilizers, hydrogels	Wang et al., 2016; Channab et al., 2024
Chitosan	Metal oxide nanoparticles (ZnO, Ag)	Antimicrobial activity, barrier properties, film-forming	Crop protection, post-harvest preservation, biopesticide delivery	Mishra et al., 2024; Khalil et al., 2014
Cellulose/Nanocellulose	Graphene oxide, carbon nanotubes	High mechanical strength, biodegradability, large surface area	Mulch films, water purification membranes, reinforced coatings	Priya et al., 2024; Abraham et al., 2011
Alginate	Clay nanoparticles	Biocompatibility, gel formation, nutrient encapsulation	Fertilizer encapsulation, seed coating, soil amendments	Irewale et al., 2024
Lignin	Graphene oxide	Antioxidant properties, UV protection, mechanical reinforcement	Controlled release coatings, UV-protective films	Channab et al., 2024

Green synthesis approaches that minimize environmental impact while optimizing material properties are emphasized in the production of bio-based nanocomposites (Dwivedi et al., 2025). These methods capitalize on the natural reducing and stabilizing properties of biomolecules present in plant extracts and agricultural waste (Vandana et al., 2024). For example, Channab et al. (2024) synthesized a starch-lignin-graphene oxide nanocomposite using kraft lignin derived from pulp and paper industry waste. The lignin acted as a natural reducing agent, facilitating the formation of graphene oxide nanosheets without the need for harsh chemicals. Similarly, Dwivedi et al. (2025) used calyx leaves extract of Physalis peruviana fruits to synthesize CeO2-TiO2 nanocomposites, harnessing the antioxidant and antimicrobial properties of the plant extract.

The green synthesis of bio-based nanocomposites often involves solution mixing, melt compounding, or in situ polymerization methods (Abraham et al., 2011). Solution mixing is a simple and versatile technique where the biopolymer and nanofiller are dispersed in a common solvent, followed by solvent evaporation to obtain the nanocomposite film or coating. Melt compounding uses high temperature and shear forces to mix the biopolymer and nanofiller in a molten state, resulting in a homogeneous dispersion. In situ polymerization involves the synthesis of the biopolymer matrix in the presence of the nanofiller, leading to strong interfacial interactions and improved dispersion.

Various characterization techniques are essential to understand the properties and behavior of bio-based nanocomposites. Advanced microscopy methods like scanning electron microscopy (SEM) and transmission electron microscopy (TEM) provide insights into the morphology and dispersion of nanofillers within the biopolymer matrix. SEM images reveal the surface topography and interfacial adhesion between the nanofiller and matrix, while TEM allows direct visualization of the nanofiller size, shape, and distribution (Rhim et al., 2013). Spectroscopic techniques such as Fourier-transform infrared spectroscopy (FTIR) and X-ray diffraction (XRD) elucidate chemical interactions and crystalline structures. FTIR helps identify functional groups and intermolecular interactions between the biopolymer and nanofiller, while XRD provides information on the crystallinity and intercalation/exfoliation of layered nanofillers like clay minerals (Abraham et al., 2011).

Mechanical testing is crucial to evaluate the reinforcement effect of nanofillers on the biopolymer matrix. Tensile tests measure the strength, stiffness, and elongation at break of nanocomposite films or fibers. Nanoindentation techniques assess the local mechanical properties, such as hardness and elastic modulus, at the nanoscale (Mohanty et al., 2002). Dynamic mechanical analysis (DMA) provides insights into the viscoelastic behavior and thermal transitions of nanocomposites. Thermal analysis techniques, such as thermogravimetric analysis (TGA) and differential scanning calorimetry (DSC), assess the thermal stability, degradation behavior, and glass transition temperature of bio-based nanocomposites (Khalil et al., 2014).

Particular attention is given to the regenerability aspects of bio-based nanocomposites. The ability to maintain or restore material properties through sustainable processes is crucial for their long-term viability in agricultural applications. Strategies such as enzymatic degradation, composting, and recycling are explored to enable circular material flows. Enzymatic degradation involves the use of specific enzymes, such as cellulases, amylases, or proteases, to break down the biopolymer matrix into simpler components that can be assimilated by microorganisms in the environment (Mohanty et al., 2002). Composting studies assess the biodegradability of nanocomposites under controlled conditions, simulating soil or compost environments. The rate and extent of biodegradation are influenced by factors such as temperature, moisture, pH, and microbial activity (Abraham et al., 2011). Recycling strategies aim to recover and reuse the nanofillers or biopolymer components from end-of-life nanocomposites, reducing waste and conserving resources.

ENVIRONMENTAL AND HEALTH RISK ASSESSMENT OF BIO-BASED NANOCOMPOSITES

While bio-based nanocomposites offer significant sustainability advantages, it is crucial to address potential environmental and health risks associated with nanomaterials. The unique properties of nanoparticles that make them valuable for agricultural applications may also pose risks through mechanisms not observed with bulk materials. A comprehensive risk assessment framework is essential to ensure the safe deployment of these technologies.

Nanotoxicity and Bioaccumulation

The potential toxicity of nanomaterials to non-target organisms remains a critical concern. Studies have shown that certain nanoparticles can exhibit toxic effects on soil microorganisms, plants, and aquatic organisms at elevated concentrations. For instance, metal oxide nanoparticles such as ZnO and CuO have demonstrated phytotoxicity in some plant species, affecting germination rates, root elongation, and overall plant growth. The toxicity mechanisms may include oxidative stress generation, membrane damage, and interference with cellular processes. However, these effects are highly dependent on nanoparticle concentration, size, surface chemistry, and environmental conditions.

Bioaccumulation of nanoparticles in food chains is another important consideration. While bio-based nanocomposites are designed to biodegrade, the fate of released nanofillers in agricultural ecosystems requires careful study. Research has indicated that some nanoparticles can be taken up by plants and potentially transferred to higher trophic levels. Long-term studies are needed to assess whether bioaccumulation occurs at levels that could pose risks to ecosystem health or food safety.

Persistence and Environmental Fate

Understanding the environmental fate of bio-based nanocomposites is essential for predicting their long-term impacts. While the biopolymer matrix is biodegradable, the nanofillers may persist in the environment for extended periods. The persistence depends on the chemical nature of the nanomaterial, environmental conditions, and interactions with soil components. For example, clay nanoparticles may integrate into soil mineral fractions, while metal oxide nanoparticles may undergo transformation reactions such as dissolution, aggregation, or surface modifications.

The release kinetics of nanofillers from degrading nanocomposites also merit attention. As the biopolymer matrix breaks down, nanoparticles may be released into the soil or water environment. The rate and extent of release depend on degradation conditions and the strength of nanofiller-matrix interactions. Controlled degradation studies under various environmental scenarios are necessary to predict release patterns and potential accumulation hotspots.

Economic Feasibility and Scalability Challenges

The commercial viability of bio-based nanocomposites for agricultural applications depends on both economic feasibility and the ability to scale production from laboratory to industrial levels. While these materials show promising performance characteristics, several economic and technical challenges must be addressed to enable widespread adoption.

Production Costs and Cost-Benefit Analysis

Current production costs of bio-based nanocomposites are generally higher than conventional materials, primarily due to the costs of biopolymers, nanomaterial synthesis, and processing complexity. Starch, cellulose, and chitosan biopolymers can cost 2-5 times more than petroleum-based polymers, depending on purity requirements and source material. Nanofiller production adds additional costs, particularly for engineered nanomaterials like graphene oxide or functionalized nanoparticles.

Table 2. Cost Comparison of Bio-based Nanocomposites vs. Conventional Materials

Material Type	Approximate Cost (USD/kg)	Key Cost Drivers	Potential Cost Reduction Strategies
Conventional Plastic Mulch	$1.50-3.00	Petroleum prices, manufacturing	N/A
Starch-based Nanocomposite	$5.00-8.00	Biopolymer cost, nanofiller production	Use agricultural waste, optimize formulation
Chitosan-based Nanocomposite	$8.00-15.00	Chitosan extraction, nanoparticle synthesis	Improve extraction efficiency, scale-up
Conventional Fertilizer Coating	$0.50-1.00	Polymer coating materials	N/A
Nanocomposite Fertilizer Coating	$2.00-4.00	Biopolymer and nanofiller costs	Agricultural waste feedstocks, simplified processing

However, cost-benefit analysis should account for the broader value proposition of bio-based nanocomposites, including improved performance, environmental benefits, and potential savings from enhanced resource efficiency. For example, controlled release fertilizers may reduce overall fertilizer requirements by 20-30%, offsetting higher initial costs through reduced input expenses and improved crop yields. Similarly, biodegradable mulch films eliminate disposal costs associated with conventional plastic mulches.

Scalability and Manufacturing Challenges

Scaling production from laboratory to commercial scale presents several technical challenges. Maintaining consistent nanofiller dispersion in large batches is more difficult than in small-scale preparations, as mixing efficiency and shear forces differ significantly. Industrial-scale equipment for melt compounding or solution casting may require modifications to accommodate the unique properties of biopolymers and nanofillers.

Infrastructure limitations also affect scalability. Many biopolymer processing facilities are designed for food or pharmaceutical applications and may not have the capacity or equipment configuration needed for agricultural nanocomposite production. Establishing dedicated production lines requires substantial capital investment, which may be difficult to justify without demonstrated market demand.

Supply chain considerations add another layer of complexity. Consistent sourcing of high-quality biopolymers and nanofillers is essential for maintaining product quality. Agricultural waste-derived biopolymers may exhibit seasonal or regional variability, requiring robust quality control protocols. Establishing reliable supply networks for specialized nanomaterials can be challenging, particularly in regions with limited advanced materials infrastructure.

SUSTAINABILITY ASSESSMENT OF BIO-BASED NANOCOMPOSITES

A comprehensive sustainability assessment is vital to evaluate the true environmental credentials of bio-based nanocomposites. Life cycle assessment (LCA) is a powerful tool to quantify environmental impacts from cradle to grave, considering raw material extraction, production, use, and end-of-life stages (Priya et al., 2024). LCA follows a standardized methodology (ISO 14040/14044) that includes four main phases: goal and scope definition, life cycle inventory analysis, life cycle impact assessment, and interpretation (Salazar et al., 2014).

The goal and scope definition phase establishes the system boundaries, functional unit, and data requirements. For bio-based nanocomposites, the functional unit could be a specific amount of material (e.g., 1 kg) or a defined function (e.g., covering 1 hectare of agricultural land). The life cycle inventory analysis involves the collection and quantification of input and output data for each stage of the life cycle. This includes raw material consumption, energy use, emissions to air, water, and soil, and waste generation (Salazar et al., 2014). Data can be obtained from primary sources, such as direct measurements or surveys, or secondary sources, such as databases or literature.

Life cycle impact assessment translates the inventory data into potential environmental impacts using characterization factors. Common impact categories include global warming potential, acidification, eutrophication, ozone depletion, and ecotoxicity (Priya et al., 2024). The results are normalized and weighted to facilitate comparison and interpretation. The interpretation phase identifies significant issues, evaluates the completeness and consistency of the study, and draws conclusions and recommendations.

LCA studies on bio-based nanocomposites have shown promising results in terms of reduced environmental impacts compared to conventional materials. For example, Yates and Barlow (2013) conducted an LCA of polylactic acid (PLA)-clay nanocomposites and found that they had lower global warming potential, non-renewable energy use, and water consumption compared to neat PLA and polypropylene. The incorporation of biopolymers and renewable fillers reduces the reliance on fossil-based resources and associated greenhouse gas emissions. However, the environmental benefits of bio-based nanocomposites can be offset by factors such as land use change, fertilizer and pesticide use in biomass cultivation, and energy-intensive processing methods (Salazar et al., 2014).

The assessment also examines the effectiveness of regeneration processes in closing material loops and minimizing waste. It evaluates the biodegradability and compostability of bio-based nanocomposites under relevant environmental conditions. Biodegradation tests, such as soil burial or composting studies, provide insights into the rate and extent of material decomposition by microorganisms. The presence of nanofillers can influence biodegradation behavior, either by enhancing or hindering the microbial activity (Mohanty et al., 2002). Compostability assessments, following standards like ASTM D6400 or EN 13432, determine whether the nanocomposites can be safely composted in industrial or home composting facilities without leaving harmful residues.

The potential for recycling and upcycling is considered to extend material lifetimes and reduce virgin resource consumption. Mechanical recycling involves shredding, melting, and re-extrusion of nanocomposite waste into new products. However, the presence of nanofillers can pose challenges in terms of compatibility, dispersion, and property retention (Khalil et al., 2014). Chemical recycling, such as solvolysis or pyrolysis, can be used to recover the biopolymer and nanofiller components separately for reuse. Upcycling strategies aim to convert nanocomposite waste into higher-value products, such as reinforcing fillers or adsorbents for environmental remediation (Priya et al., 2024).

Economic feasibility is another important aspect of sustainability. The assessment analyzes production costs, market demand, and potential savings from reduced environmental externalities. The production of bio-based nanocomposites can be more expensive than conventional materials due to the higher cost of biopolymers and the specialized processing methods required (Khalil et al., 2014). However, the use of agricultural waste and by-products as low-cost feedstocks can improve the economic viability. Additionally, the potential savings from reduced environmental impacts, such as lower greenhouse gas emissions, water consumption, and waste management costs, can offset the higher production costs in the long term (Salazar et al., 2014).

Societal acceptance and stakeholder perceptions are also considered to ensure the adoption and long-term success of bio-based nanocomposite solutions. Public awareness and understanding of the benefits and risks associated with nanotechnology are crucial for informed decision-making. Engaging stakeholders, such as farmers, consumers, policymakers, and industry representatives, in the development and implementation of bio-based nanocomposites can help address concerns and build trust (Priya et al., 2024). Effective communication and transparency about the safety, performance, and sustainability aspects of nanocomposites are essential for fostering acceptance and driving widespread adoption.

AGRICULTURAL APPLICATIONS OF BIO-BASED NANOCOMPOSITES

Bio-based nanocomposites find diverse applications in sustainable agriculture, leveraging their unique properties to enhance crop productivity and resource efficiency. Some key application areas include soil conditioning, controlled release fertilizers, crop protection, and water management.

Soil Conditioning

Soil is a complex and dynamic system that plays a critical role in plant growth, nutrient cycling, and water retention. However, soil degradation due to erosion, compaction, salinization, and loss of organic matter poses significant challenges to agricultural productivity (Khalil et al., 2014). Bio-based nanocomposites can be used as soil conditioners to improve soil structure, water holding capacity, and nutrient availability. For example, starch-based nanocomposites with clay nanoparticles can form hydrogels that slowly release water and nutrients to plant roots (Wang et al., 2016). The incorporation of clay nanoparticles enhances the mechanical strength and stability of the hydrogel, preventing rapid degradation and leaching. The controlled release of water and nutrients reduces irrigation requirements and minimizes nutrient losses through runoff or leaching.

Nanocomposite hydrogels can also be designed to respond to external stimuli, such as pH, temperature, or light, enabling targeted and on-demand release of water and nutrients (Irewale et al., 2024). For instance, pH-responsive hydrogels can release nutrients in response to changes in soil pH, ensuring optimal nutrient availability for plant uptake. Temperature-responsive hydrogels can regulate water release based on soil temperature, conserving water during hot and dry periods. Light-responsive hydrogels can be triggered by sunlight to release nutrients during the day when plants are actively photosynthesizing (Mishra et al., 2024).

Another approach to soil conditioning is the use of nanocomposite mulches or films. Mulches are materials applied to the soil surface to conserve moisture, suppress weed growth, and regulate soil temperature (Khalil et al., 2014). Bio-based nanocomposite mulches, such as those made from starch, cellulose, or chitosan reinforced with nanofillers, offer biodegradability and enhanced performance compared to conventional plastic mulches. The nanofillers, such as clay minerals or nanocellulose, improve the mechanical strength, barrier properties, and thermal stability of the mulch films (Wang et al., 2016). The biodegradability of the nanocomposite mulches allows for their incorporation into the soil after use, reducing plastic waste accumulation and soil contamination.

Controlled Release Fertilizers

Fertilizers are essential inputs in agriculture, providing crops with the necessary nutrients for growth and development. However, the inefficient use of fertilizers can lead to environmental problems, such as eutrophication of water bodies, greenhouse gas emissions, and soil acidification (Savci, 2012). Controlled release fertilizers (CRFs) based on bio-based nanocomposites offer a promising solution to enhance nutrient use efficiency and minimize environmental losses. CRFs are designed to release nutrients slowly over an extended period, synchronizing nutrient supply with plant demand (Channab et al., 2024).

Nanocomposite coatings can be applied to conventional fertilizer granules to control nutrient release. Channab et al. (2024) developed a coating using starch, lignin, and graphene oxide nanoparticles for controlled phosphorus release from triple superphosphate fertilizers. The nanocomposite coating acts as a semi-permeable barrier, allowing water to penetrate and dissolve the fertilizer core while restricting the outward diffusion of nutrients. The graphene oxide nanosheets provide mechanical reinforcement and slow down the release rate by increasing the tortuosity of the diffusion path (Channab et al., 2024). The controlled release of phosphorus improved fertilizer use efficiency and reduced nutrient leaching, leading to better crop growth and yield.

Another approach to CRFs is the encapsulation of nutrients within nanocomposite matrices. Irewale et al. (2024) developed a bio-nanofertilizer by encapsulating nitrogen, phosphorus, and potassium (NPK) nutrients within a chitosan-alginate matrix reinforced with clay nanoparticles. The nanocomposite matrix protects the nutrients from rapid release and degradation, allowing for their slow and sustained release over time. The clay nanoparticles enhance the mechanical stability and water retention of the matrix, preventing premature disintegration and leaching (Irewale et al., 2024). The bio-nanofertilizer showed improved crop growth and nutrient uptake compared to conventional fertilizers, demonstrating the potential of nanocomposites in enhancing fertilizer efficiency.

Crop Protection

Crop protection against pests, diseases, and weeds is a major challenge in agriculture. The use of synthetic pesticides has raised concerns about their impacts on non-target organisms, pesticide resistance, and human health (Nicolopoulou-Stamati et al., 2016). Bio-based nanocomposites offer an alternative approach to crop protection by providing barrier properties, controlled release of active ingredients, and antimicrobial activity.

Nanocomposite films and coatings can be used as physical barriers to protect crops from pests and pathogens. Chitosan-based nanocomposites have shown promise in this regard due to their inherent antimicrobial properties and film-forming ability (Khalil et al., 2014). Mishra et al. (2024) developed a chitosan-based nanocomposite film reinforced with silver nanoparticles for protecting tomato plants against fungal diseases. The silver nanoparticles provide strong antimicrobial activity, while the chitosan matrix acts as a barrier to prevent the entry of pathogens. The nanocomposite film showed effective control of fungal diseases, reducing the need for chemical fungicides.

Nanocomposites can also be used for the controlled release of biopesticides, such as essential oils or plant extracts. Essential oils are volatile compounds derived from plants that have insecticidal, fungicidal, and herbicidal properties (Khalil et al., 2014). However, their high volatility and rapid evaporation limit their effectiveness in the field. Encapsulation of essential oils within nanocomposite matrices can prolong their release and improve their stability. Mohanty et al. (2002) developed a starch-based nanocomposite loaded with neem oil, a natural insecticide, for the controlled release of the oil. The nanocomposite showed sustained release of neem oil over several days, providing prolonged protection against insect pests.

Nanocellulose-reinforced biocomposites have also been explored for crop protection applications. Nanocellulose, derived from cellulosic biomass, has excellent mechanical properties, biodegradability, and compatibility with other biopolymers (Khalil et al., 2014). Abraham et al. (2011) developed a nanocellulose-reinforced chitosan film for the controlled release of a biopesticide, pyrethrins. The nanocellulose provides mechanical reinforcement to the chitosan matrix, improving its stability and durability. The controlled release of pyrethrins from the nanocomposite film showed effective insecticidal activity against agricultural pests.

Water Management

Water scarcity and quality are major challenges facing agriculture, particularly in arid and semi-arid regions. Efficient water management practices are essential to conserve water resources and ensure sustainable crop production. Bio-based nanocomposites can contribute to water management through their applications in water purification, desalination, and irrigation systems.

Nanocomposite membranes and adsorbents can be used for the removal of contaminants from agricultural water sources. Heavy metals, pesticides, and organic pollutants are common contaminants that can have detrimental effects on crop growth and food safety (Priya et al., 2024). Carbon-based nanomaterials derived from agricultural waste, such as biochar and activated carbon, have shown excellent adsorption capacities for various contaminants. Priya et al. (2024) developed a nanocomposite adsorbent by incorporating biochar derived from sugarcane bagasse into a chitosan matrix. The biochar-chitosan nanocomposite exhibited high adsorption efficiency for heavy metals like lead, cadmium, and chromium, making it suitable for water purification applications.

Membrane filtration is another approach to water purification and desalination. Nanocomposite membranes incorporating biopolymers and nanofillers can enhance the selectivity, permeability, and fouling resistance of the membranes (Mohanty et al., 2002). Khalil et al. (2014) developed a cellulose acetate-based nanocomposite membrane reinforced with nanocellulose for the removal of salt and organic contaminants from brackish water. The nanocellulose improves the mechanical strength and hydrophilicity of the membrane, reducing fouling and increasing water flux. The nanocomposite membrane showed high salt rejection and organic contaminant removal, demonstrating its potential for agricultural water desalination.

In irrigation systems, nanocomposites can be used for the controlled release of water and fertilizers. Hydrogel-based nanocomposites, as discussed in the soil conditioning section, can be applied as soil amendments to retain water and nutrients in the root zone. The slow release of water from hydrogel reduces evaporative losses and improves water use efficiency. Nanocomposite-based drip irrigation systems can also be developed for precise and targeted delivery of water and nutrients to crops. The incorporation of nanofillers, such as clay minerals or silica nanoparticles, can enhance the mechanical strength and durability of the drip lines, preventing clogging and leakage (Irewale et al., 2024).

CASE STUDIES

Several case studies demonstrate the successful implementation of bio-based nanocomposites in agricultural applications. These examples highlight the practical viability and benefits of nanocomposite materials in real-world settings.

Case Study 1: Starch-based Nanocomposite Coating for Controlled Release of Urea Fertilizer

Channab et al. (2024) developed a starch-based nanocomposite coating for the controlled release of urea fertilizer. The coating was prepared by blending starch with montmorillonite clay nanoparticles and applying it to urea granules. The nanocomposite coating reduced the dissolution rate of urea and prolonged its release over several weeks. Field trials conducted on maize crops showed that the nanocomposite-coated urea improved nitrogen use efficiency and crop yield compared to conventional urea fertilizer. The controlled release of nitrogen minimized losses through volatilization, leaching, and denitrification, leading to better nutrient uptake by the plants. This case study demonstrates the potential of bio-based nanocomposite coatings for enhancing fertilizer efficiency and reducing environmental impacts.

Case Study 2: Chitosan-based Nanocomposite Film for Post-harvest Preservation of Fruits

Mishra et al. (2024) developed a chitosan-based nanocomposite film for the post-harvest preservation of fruits. The nanocomposite film was prepared by incorporating zinc oxide nanoparticles into a chitosan matrix. The zinc oxide nanoparticles provide antimicrobial activity, while the chitosan matrix acts as a barrier to moisture and gas exchange. The nanocomposite film was applied as a coating on bananas to extend their shelf life. Bananas coated with the nanocomposite film showed reduced weight loss, delayed ripening, and inhibition of fungal growth compared to uncoated bananas. The nanocomposite coating maintained the quality and freshness of the bananas for up to 15 days at room temperature. This case study highlights the potential of bio-based nanocomposite films for post-harvest preservation of perishable crops, reducing food losses and ensuring food security.

Case Study 3: Nanocellulose-reinforced Biocomposite for Water Purification

Priya et al. (2024) developed a nanocellulose-reinforced biocomposite for the removal of heavy metals from agricultural wastewater. The biocomposite was prepared by blending chitosan with nanocellulose derived from rice straw. The nanocellulose provides mechanical reinforcement and increases the surface area of the biocomposite, enhancing its adsorption capacity. The biocomposite was used as an adsorbent for the removal of lead and cadmium from simulated wastewater. Batch adsorption experiments showed that the biocomposite had a high adsorption capacity and rapid adsorption kinetics for both heavy metals. The adsorption process was pH-dependent, with maximum removal efficiency observed at pH 5-6. The biocomposite could be regenerated using acid treatment and reused for multiple adsorption cycles. This case study demonstrates the potential of nanocellulose-reinforced biocomposites for the sustainable treatment of agricultural wastewater, preventing environmental contamination and protecting water resources.

These case studies provide real-world evidence of the benefits and applicability of bio-based nanocomposites in sustainable agriculture. They showcase the potential of nanocomposites to improve fertilizer efficiency, reduce post-harvest losses, and enhance water quality. The successful implementation of these nanocomposite materials highlights their promise as sustainable solutions for addressing agricultural challenges.

FUTURE TRENDS, RESEARCH DIRECTIONS, AND REGULATORY CONSIDERATIONS

The field of bio-based nanocomposites for agricultural applications is continuously evolving, with new materials, processing techniques, and functionalities being explored. Several emerging trends and research directions are shaping the future of this field.

One promising trend is the development of multi-functional nanocomposites that can perform multiple tasks simultaneously. For example, a nanocomposite fertilizer coating that provides controlled nutrient release, moisture retention, and pest repellence could be developed by combining different nanofillers and active ingredients (Irewale et al., 2024). Such multi-functional nanocomposites could offer a more holistic approach to crop management, addressing multiple challenges with a single material. Further research is needed to optimize the composition and synergistic effects of multi-functional nanocomposites for specific agricultural applications.

Another emerging trend is the integration of bio-based nanocomposites with precision agriculture technologies. Precision agriculture involves the use of sensors, data analytics, and automation to optimize crop management decisions based on site-specific conditions (Abraham et al., 2011). Bio-based nanocomposites could be designed to interact with sensors and respond to specific triggers, such as changes in soil moisture, pH, or pest infestation. For example, a nanocomposite hydrogel that releases water and nutrients in response to soil moisture levels detected by sensors could enable precise and automated irrigation. The integration of nanocomposites with precision agriculture could lead to more efficient and sustainable farming practices. Research efforts should focus on developing smart and responsive nanocomposites that can adapt to real-time crop and environmental conditions.

The development of stimuli-responsive nanocomposites is another area of active research. These materials can change their properties or behavior in response to external stimuli, such as temperature, light, pH, or magnetic fields (Mohanty et al., 2002). Stimuli-responsive nanocomposites could be used for targeted delivery of agrochemicals, controlled release of plant growth regulators, or self-healing of damaged crop tissues. For instance, a temperature-responsive nanocomposite could release pesticides only when the crop is under heat stress, minimizing the overall pesticide use (Khalil et al., 2014). The design and synthesis of stimuli-responsive nanocomposites with tunable and reversible properties are key research challenges that need to be addressed.

The valorisation of agricultural waste and by-products as feedstocks for bio-based nanocomposites is another promising trend. Agricultural residues, such as straw, husk, bagasse, and fruit peels, are abundant and renewable sources of cellulose, hemicellulose, and lignin (Priya et al., 2024). These biomass wastes can be processed into nanocellulose, nanofibers, or biochar and incorporated into nanocomposite materials. The utilization of agricultural waste not only reduces the environmental burden of waste disposal but also provides a low-cost and sustainable source of raw materials for nanocomposite production. Further research is needed to optimize the extraction, modification, and compatibility of biomass-derived nanofillers with different biopolymer matrices.

REGULATORY CONSIDERATIONS AND SAFETY ASSESSMENT

However, the development and application of bio-based nanocomposites in agriculture also raise important regulatory and safety considerations. The unique properties and behaviors of nanomaterials may pose potential risks to human health and the environment, which need to be carefully assessed and managed (Abraham et al., 2011). The lack of standardized testing protocols and risk assessment frameworks for nanomaterials in agriculture is a challenge that needs to be addressed. Collaborative efforts between researchers, regulatory agencies, and industry stakeholders are necessary to develop harmonized guidelines and best practices for the safe and responsible use of nanocomposites in agriculture.

Regulatory agencies, such as the U.S. Environmental Protection Agency (EPA) and the European Food Safety Authority (EFSA), have started to develop guidelines and regulations for the use of nanomaterials in agriculture and food production. These regulations aim to ensure the safety and effectiveness of nanomaterials while promoting innovation and sustainable practices (Irewale et al., 2024). However, the regulatory landscape for bio-based nanocomposites in agriculture is still evolving and varies across countries and regions. Continued research on the environmental fate, bioaccumulation, and toxicity of nanocomposites is necessary to inform evidence-based regulations and risk management strategies.

One key aspect of regulatory considerations is the assessment of the life cycle impacts of bio-based nanocomposites. While these materials are designed to be biodegradable and environmentally friendly, their production, use, and disposal may still have unintended consequences (Priya et al., 2024). Life cycle assessment (LCA) studies are needed to evaluate the cradle-to-grave environmental impacts of bio-based nanocomposites, including their energy and resource consumption, greenhouse gas emissions, and potential release of nanomaterials into the environment (Abraham et al., 2011). LCA can help identify hotspots and opportunities for improvement in the life cycle of nanocomposites, guiding the development of more sustainable and circular solutions.

Another important consideration is the safety and efficacy testing of bio-based nanocomposites in agricultural applications. The behavior and fate of nanomaterials in complex soil and plant systems are not fully understood and may vary depending on the specific composition and properties of the nanocomposite (Khalil et al., 2014). Adequate testing protocols and guidelines need to be developed to assess the potential effects of nanocomposites on soil health, plant growth, and food safety. Long-term field studies and monitoring programs are necessary to evaluate the performance and unintended consequences of nanocomposites under realistic agricultural conditions.

The communication and engagement of stakeholders, including farmers, consumers, and policymakers, are crucial for the responsible development and adoption of bio-based nanocomposites in agriculture. Transparency and public participation in the decision-making process can help build trust and ensure that the benefits and risks of nanocomposites are adequately considered (Abraham et al., 2011). Educational programs and outreach activities can raise awareness and promote informed choices about the use of nanocomposites in agriculture. Collaborative platforms and networks can facilitate knowledge sharing, technology transfer, and capacity building among researchers, industry, and end-users.

In conclusion, bio-based nanocomposites offer exciting opportunities for sustainable agriculture, but their development and application must be accompanied by responsible research, risk assessment, and regulation. Continued research efforts are needed to address knowledge gaps, optimize material properties, and scale up production processes. Collaborative and interdisciplinary approaches are essential to ensure the safe and effective integration of nanocomposites into agricultural systems. By proactively addressing regulatory and safety aspects, engaging stakeholders, and promoting sustainable practices, the full potential of bio-based nanocomposites can be realized for the benefit of agriculture, food security, and the environment.

CONCLUSION

Bio-based nanocomposites have emerged as a promising solution to address the challenges of sustainable agriculture and food security. By combining the unique properties of nanomaterials with the environmental benefits of biopolymers, these innovative materials offer a wide range of applications in crop production, water management, and post-harvest preservation.

This chapter has explored the synthesis, characterization, and sustainability assessment of bio-based nanocomposites, highlighting their potential to revolutionize agricultural practices. The use of green synthesis methods and the incorporation of renewable resources, such as agricultural waste and by-products, contribute to the environmental friendliness and circularity of these materials. Critical considerations of environmental and health risks, including nanotoxicity, bioaccumulation, and ecosystem persistence, have been addressed to ensure responsible development and deployment.

The diverse agricultural applications of bio-based nanocomposites, including soil conditioning, controlled release fertilizers, crop protection, and water purification, demonstrate their versatility and effectiveness in addressing critical challenges. Case studies of successful implementations provide real-world evidence of the benefits and practicality of nanocomposite solutions. Economic feasibility analysis and scalability assessments reveal both the challenges and opportunities for commercial viability, emphasizing the need for cost reduction strategies and infrastructure development.

Looking towards the future, several emerging trends and research directions are shaping the field of bio-based nanocomposites in agriculture. The development of multi-functional and stimuli-responsive nanocomposites, the integration with precision agriculture technologies, and the valorisation of agricultural waste as feedstocks are promising avenues for further research and innovation. These advancements can lead to more efficient, targeted, and sustainable agricultural practices.

However, the development and application of bio-based nanocomposites in agriculture also raise important regulatory and safety considerations. The assessment of potential risks to human health and the environment, as well as the development of standardized testing protocols and guidelines, are crucial for the responsible adoption of these materials. Collaborative efforts among researchers, policymakers, and stakeholders are essential to ensure the safe and effective use of nanocomposites in agriculture.

To fully harness the potential of bio-based nanocomposites for sustainable agriculture, several key actions are recommended:

1. Continued research and development: Further research is needed to optimize the synthesis, characterization, and performance of bio-based nanocomposites for specific agricultural applications. This includes exploring new biopolymers, nanofillers, and functionalization strategies, as well as investigating the long-term stability and environmental fate of nanocomposites.
2. Interdisciplinary collaboration: Fostering collaborations among materials scientists, agricultural researchers, environmental scientists, and social scientists is essential to address the complex challenges at the intersection of nanotechnology, agriculture, and sustainability. Interdisciplinary approaches can lead to more holistic and integrated solutions.
3. Stakeholder engagement and communication: Engaging stakeholders, including farmers, consumers, industry, and policymakers, in the development and implementation of bio-based nanocomposites is crucial for their successful adoption. Clear and transparent communication about the benefits, risks, and uncertainties associated with nanocomposites can help build trust and informed decision-making.
4. Capacity building and technology transfer: Providing training, education, and extension services to farmers and agricultural professionals is necessary to facilitate the uptake of bio-based nanocomposites in practice. Technology transfer mechanisms and partnerships between research institutions and industry can accelerate the commercialization and scale-up of nanocomposite solutions.
5. Policy support and regulation: Developing coherent and evidence-based policies and regulations for the use of bio-based nanocomposites in agriculture is essential to ensure their safety, effectiveness, and sustainability. This includes establishing standards, guidelines, and risk assessment frameworks that consider the unique properties and behaviors of nanomaterials in agricultural systems.

By addressing these key actions and considerations, the potential of bio-based nanocomposites can be unlocked to support sustainable agriculture, enhance food security, and contribute to the broader goals of sustainable development. As the world faces growing challenges in feeding a growing population while preserving the environment, bio-based nanocomposites offer a promising tool in the transition towards a more resilient and sustainable agricultural future.

The path forward requires a concerted effort from all stakeholders, including researchers, policymakers, farmers, and consumers. By working together, sharing knowledge, and promoting responsible innovation, we can harness the power of bio-based nanocomposites to transform agriculture and build a more sustainable and equitable world. The journey towards a sustainable agricultural future is complex and challenging, but with dedication, ingenuity, and a shared vision, we can create a future where food security, environmental stewardship, and social well-being are harmonized through the application of innovative materials and practices.

REFERENCES

Abraham, E., Deepa, B., Pothan, L. A., Jacob, M., Thomas, S., Cvelbar, U., & Anandjiwala, R. (2011). Extraction of nanocellulose fibrils from lignocellulosic fibres: A novel approach. *Carbohydrate Polymers*, *86*(4), 1468–1475. DOI: 10.1016/j.carbpol.2011.06.034

Channab, B.-E., Tayi, F., Aqlil, M., Akil, A., Essamlali, Y., Chakir, A., & Zahouily, M. (2024). Graphene oxide, starch, and kraft lignin bio-nanocomposite controlled-release phosphorus fertilizer: Effect on P management and maize growth. International Journal of Biological Macromolecules, 282(Part 5-137190). https://doi.org/ DOI: 10.1016/j.ijbiomac.2024.137190

Dasari, A., Yu, Z.-Z., & Mai, Y.-W. (2009). Fundamental aspects and recent progress on wear/scratch damage in polymer nanocomposites. *Materials Science and Engineering R Reports*, *63*(2), 31–80. DOI: 10.1016/j.mser.2008.10.001

Irewale, A. T., Dimkpa, C. O., Agunbiade, F. O., Oyetunde, O. A., Elemike, E. E., & Oguzie, E. E. (2024). Unlocking sustainable agricultural development in Africa via bio-nanofertilizer application - challenges, opportunities and prospects. *Scientific African*, *25*. Advance online publication. DOI: 10.1016/j.sciaf.2024.e02276

Khalil, H. P. S. A., Davoudpour, Y., Islam, M. N., Mustapha, A., Sudesh, K., Dungani, R., & Jawaid, M. (2014). Production and modification of nanofibrillated cellulose using various mechanical processes: A review. *Carbohydrate Polymers*, *99*, 649–665. DOI: 10.1016/j.carbpol.2013.08.069 PMID: 24274556

Mishra, B., Panda, J., Mishra, A. K., Nath, P. C., Nayak, P. K., Mahapatra, U., Sharma, M., Chopra, H., Mohanta, Y. K., & Sridhar, K. (2024). Recent advances in sustainable biopolymer-based nanocomposites for smart food packaging: A review. *International Journal of Biological Macromolecules*, *279*, 135583. DOI: 10.1016/j.ijbiomac.2024.135583 PMID: 39270899

Mohanty, A. K., Misra, M., & Drzal, L. T. (2002). Sustainable bio-composites from renewable resources: Opportunities and challenges in the green materials world. *Journal of Polymers and the Environment*, *10*(1), 19–26. DOI: 10.1023/A:1021013921916

Nicolopoulou-Stamati, P., Maipas, S., Kotampasi, C., Stamatis, P., & Hens, L. (2016). Chemical pesticides and human health: The urgent need for a new concept in agriculture. *Frontiers in Public Health*, *4*, 148. DOI: 10.3389/fpubh.2016.00148 PMID: 27486573

Priya, A. K., Muruganandam, M., & Suresh, S. (2024). Bio-derived carbon-based materials for sustainable environmental remediation and wastewater treatment. *Chemosphere, 362*, 142731. DOI: 10.1016/j.chemosphere.2024.142731 PMID: 38950744

Rhim, J.-W., Park, H.-M., & Ha, C.-S. (2013). Bio-nanocomposites for food packaging applications. *Progress in Polymer Science, 38*(10-11), 1629–1652. DOI: 10.1016/j.progpolymsci.2013.05.008

Salazar, R., Domenek, S., & Ducruet, V. (2014). Interactions of flavoured oil in-water emulsions with polylactide. *Food Chemistry, 148*, 138–146. DOI: 10.1016/j.foodchem.2013.10.018 PMID: 24262538

Savci, S. (2012). An agricultural pollutant: Chemical fertilizer. *International Journal of Environmental Sciences and Development, 3*(1), 73–80. DOI: 10.7763/IJESD.2012.V3.191

Vishnuvarthanan, M. (2024). Bio-nanocomposite and biofilm: An overview. In Elsevier Reference Module in Materials Science and Materials Engineering. https://doi.org/DOI: 10.1016/B978-0-443-15738-7.00051-9

Wang, W., Ma, X., Jiang, P., Hu, L., Zhi, Z., Chen, J., Ding, T., Ye, X., & Liu, D. (2016). Characterization of pectin from grapefruit peel: A comparison of ultrasound-assisted and conventional heating extractions. *Food Hydrocolloids, 61*, 730–739. DOI: 10.1016/j.foodhyd.2016.06.019

ADDITIONAL READING

1. FAO. (2017). The future of food and agriculture: Trends and challenges. Food and Agriculture Organization of the United Nations. https://www.fao.org/3/i6583e/i6583e.pdf

2. Mohanty, A. K., Vivekanandhan, S., Pin, J.-M., & Misra, M. (2018). Composites from renewable and sustainable resources: Challenges and innovations. *Science, 362*(6414), 536–542. DOI: 10.1126/science.aat9072 PMID: 30385569

3. Rhim, J.-W., Park, H.-M., & Ha, C.-S. (2013). Bio-nanocomposites for food packaging applications. *Progress in Polymer Science, 38*(10-11), 1629–1652. DOI: 10.1016/j.progpolymsci.2013.05.008

4. Vilarinho, F., Sanches Silva, A., Vaz, M. F., & Farinha, J. P. (2018). Nanocellulose in green food packaging. *Critical Reviews in Food Science and Nutrition, 58*(9), 1526–1537. DOI: 10.1080/10408398.2016.1270254 PMID: 28125279

5. Yates, M. R., & Barlow, C. Y. (2013). Life cycle assessments of biodegradable, commercial biopolymers---A critical review. *Resources, Conservation and Recycling*, *78*, 54–66. DOI: 10.1016/j.resconrec.2013.06.010

KEY TERMS AND DEFINITIONS

Agricultural waste valorisation: The process of converting agricultural waste and by-products into value-added products, such as bio-based materials or energy, reducing waste and promoting circular economy principles.

Bio-based nanocomposites: Hybrid materials composed of biopolymers derived from renewable resources and nanoscale fillers, exhibiting enhanced properties compared to conventional materials.

Biodegradability: The ability of a material to be decomposed by microorganisms into simpler substances, such as carbon dioxide, water, and biomass, under natural environmental conditions.

Controlled release fertilizers: Fertilizers designed to release nutrients slowly over an extended period, improving nutrient use efficiency and reducing environmental losses.

Green synthesis: An eco-friendly approach to synthesizing nanomaterials using natural reducing agents and stabilizers derived from plant extracts or agricultural waste, minimizing the use of hazardous chemicals.

Life cycle assessment (LCA): A methodological framework for assessing the environmental impacts of a product or process throughout its entire life cycle, from raw material extraction to end-of-life disposal.

Precision agriculture: An approach to farming that uses technology and data analytics to optimize crop management decisions, such as nutrient application, irrigation, and pest control, based on site-specific conditions.

Stimuli-responsive materials: Materials that can change their properties or behavior in response to external stimuli, such as temperature, light, pH, or magnetic fields.

Sustainable agriculture: Agricultural practices that aim to meet current food production needs without compromising the ability of future generations to meet their needs, while minimizing negative environmental impacts.

Sustainable development: Development that meets the needs of the present without compromising the ability of future generations to meet their own needs, balancing economic growth, social equity, and environmental protection.

Chapter 8
Bio-Nanocomposites as Functional Materials for the Future:
Integrating Green Chemistry and Nanotechnology for Global Sustainability Challenges

R. Venkatesh
https://orcid.org/0000-0002-0355-8857
PSNA College of Engineering and Technology, India

D. Gnanasangeetha
https://orcid.org/0000-0002-0219-4916
PSNA College of Engineering and Technology, India

Natarajan Sampathkumar
https://orcid.org/0000-0001-7329-7712
SSM College of Arts and Science, India

ABSTRACT

Bio-nanocomposites are leading materials that leverage the strengths of nanotechnology with sustainable, bio-based polymers and present a greener option for traditional composites. These are composites that comprise nanoscale fillers in the form of nanoparticles, nanofibers, and nanotubes, and bio-based matrices of starch, cellulose, and proteins. These are stronger in terms of mechanical, thermal, and barrier properties, and their green profile remains unscathed. Bio-nanocomposites present opportunities to answer global sustainability questions such as decreasing dependence on petroleum-based plastics, energy efficiency, and promoting the cir-

DOI: 10.4018/979-8-3373-2180-6.ch008

cular economy by combining green chemistry and nanotechnology. This chapter addresses synthesis routes, properties, characterization techniques, performance, and different applications in packaging, automotive, electronics, and agriculture. It also addresses future research directions to design the properties of bio-nanocomposites for large-scale commercialization.

1. INTRODUCTION

1.1 Definition of Bio-Nanocomposites

Bio-nanocomposites are multi-phase materials combining biobased polymers and nanometer-dimensioned fillers or nanoparticles, creating products that have higher mechanical, thermal, and functional properties with their eco-friendliness. Most of these products have a renewable or biodegradable matrix (e.g., cellulose, starch, chitosan, or proteins) reinforced with nanofillers (e.g., carbon nanotubes, nano clays, metal nanoparticles, or graphene) added for enhanced performance traits (Shibata, 2011). The nanoscale size of these fillers improves the matrix properties, providing substantial benefits over conventional composites, such as enhanced strength, flexibility, barrier properties, and lower environmental footprint. Bio-nanocomposites are a promising area in material science, combining the advantages of natural polymers with the special properties of nanomaterials for various applications (Thirugnanasambandan, 2022).

1.2 Historical Background

The development of bio-nanocomposites was spurred by the increasing need for green materials in the latter half of the 20th century due to environmental pressure on non-biodegradable plastics and hazardous industrial practices. Nanotechnology research, early on, was instrumental in revealing the distinct characteristics of nanomaterials, including high surface area, mechanical strength, and reactivity.

Table 1. Historical Development and Key Aspects of Bio-nanocomposites

Category	Description
Origin	The concept of bio-nanocomposites emerged in response to the increasing demand for sustainable materials, driven by environmental concerns regarding non-biodegradable plastics and toxic industrial processes.
Scientific Foundations	Early research in nanotechnology uncovered unique properties of nanomaterials such as increased surface area, mechanical strength, and reactivity. These properties laid the groundwork for integrating nanomaterials into renewable matrices.
Environmental Driving Force	A rising need to replace fossil fuel-based materials with renewable alternatives prompted the development of bio-nanocomposites as an eco-friendly solution.
Key Materials Involved	Bio-based Matrices: Starch, chitosan, cellulose, proteins. Nanofillers: Carbon nanotubes, nano clays, metal nanoparticles, graphene, etc.
Timeframe of Development	Late 20th Century: Emergence driven by environmental concerns. Early 2000s: Focus on integrating nanotechnology with bio-based materials. Ongoing: Advances in green chemistry have optimized synthesis processes.
Technological Advancements	Green chemistry techniques have made the synthesis of bio-nanocomposites more eco-friendly and economically viable, further improving their applicability across industries.
Global Challenges Addressed	Climate Change Resource Depletion Waste Management
Research and Development Focus	Bio-nanocomposites have become a focal point in material science research, driven by their potential to replace traditional materials and offer sustainable alternatives.

These discoveries, along with the growing necessity to substitute fossil fuel-derived materials with biomass-derived ones, culminated in the generation of bio-nanocomposites. In the early 2000s, researchers started investigating different matrices based on biota such as starch, chitosan, and cellulose, in conjunction with nanofillers such as carbon nanotubes, nano clays, and metal nanoparticles. Since then, green chemistry advancements have further optimized the synthesis pathways, rendering bio-nanocomposites more environmentally friendly, cost-effective, and industrially compatible. Their promise to solve world problems such as climate change, resource depletion, and waste disposal has rendered them a research and development focus area. This table (Table 1) successfully presents the historical background and evolution of bio-nanocomposites as well as their contribution to sustainability and solving global issues.

1.3 Literature Review

Nanotechnology has a central function in developing intelligent materials that contribute to human wellness and sustainable growth, consistent with the United Nations Sustainable Development Goals (UNSDGs). Nanotechnology demonstrates the ability of nanomaterials to provide groundbreaking solutions in various sectors such as smart food packaging, health sensing, waste reduction, food safety, and preservation of the environment. The research demonstrates how eco-friendly nanocomposites can develop a healthier and balanced future through connecting material science, technology, and society (Rajeev et al., 2024).

(Shchipunov, 2012) also emphasizes the bright prospects for bio-nanocomposites as alternatives to conventional synthetic polymers. It explains new approaches to the synthesis of these materials, with an emphasis on their compatibility and biocompatibility, which is a great benefit compared to conventional methods such as sol-gel processing. The research indicates the environmental value of bio-nanocomposites in decreasing dependence on fossil fuels, lowering greenhouse gas emissions, and being biodegradable. Overall, it proposes that biopolymers and their composites can be used to solve the problem of non-biodegradable plastics and pave the way towards more sustainable materials in different uses.

The article discusses biodegradable green polymers and natural resource-based nanocomposites and their advancements, and potential applications in packaging, biomedical engineering, and environmental sustainability. It summarizes the materials' mechanical and barrier properties, indicating that they are ideal for use in green applications, and observes limitations that need to be addressed to achieve widespread commercial applications. The article demands increased research and innovation in green nano polymer technology to solve environmental problems such as plastic pollution and ensure sustainable industrial practices (Ayesha Kausar & Ishaq Ahmad, 2023).

(Basavegowda & Baek, 2021) emphasizes the dramatic advancement in biopolymer nanocomposites for active food packaging, demonstrating their strengths over conventional packaging materials. The article addresses how the nanocomposites possess improved mechanical properties, barrier properties, and antimicrobial properties, allowing food to be preserved and made safe. It also emphasizes developing renewable and degradable materials to mitigate environmental issues due to traditional plastic packaging. In general, food packaging using nanotechnology is considered to be an encouraging means to enhance food quality and support sustainability.

The research investigates recent developments in metal oxide bio-nanocomposites for green food packaging. It discusses the fabrication process, applications, and future potential of these materials. The study emphasizes how these nanocomposites can render food packaging more sustainable, safer, and durable. The authors highlight their ability to tackle environmental concerns and respond to increasing demand for environmentally friendly packaging materials (Nandhini et al., 2024).

(Sahu et al., 2023) stresses the revolutionary prospects of nanotechnology in industries. It is centered on how materials change at the nanoscale (between 1-100 nanometers) and propel advances in electronics, medicine, materials science, and energy. The article explores how nanotechnology is reshaping technology, addressing global problems, and bringing new avenues for innovation and future discoveries.

1.4 Importance of Sustainable Materials in the Modern World

Traditionally, global material demand has been fulfilled using petrochemical-based synthetic polymers. With the problems facing the environment of fossil-based materials, such as resource depletion, pollution, and waste management, serious concerns have been raised. Consequently, sustainable materials are required that minimize the environmental footprint yet fulfill industrial performance needs. Sustainable materials are renewable, biodegradable, and less toxic from cradle to grave.

Bio-nanocomposites also have a very important role during this shift toward sustainability. When produced from biobased feedstocks, such materials provide renewable and biodegradable alternatives for traditional plastics to reduce the usage of petroleum products. In addition, bio-nanocomposites can be structured with particular end-of-life fate in mind such as compostability, recyclability, or biodegradability, thereby creating less waste and environmental contamination (Mondal, 2025). Their ability to displace conventional, non-biodegradable products in various sectors like packaging, automotive, electronics, and construction makes them significant in helping to build a more sustainable future.

1.5 Role of Green Chemistry and Nanotechnology in Addressing Global Sustainability Challenges

The meeting of green chemistry and nanotechnology has led to the emergence of a new paradigm for the synthesis of environmentally sustainable materials. Green chemistry focuses on designing products and processes that avoid or minimize the use and production of hazardous materials, promoting sustainable processes in chemical manufacturing and synthesis. In the case of bio-nanocomposites, green chemistry is used to direct the choice of renewable feedstocks, minimization of harmful solvents and reagents, and adoption of energy-conserving processes, which are of prime importance in minimizing environmental footprints and in making bio-nanocomposite technologies scalable (Sharma et al., 2024).

Nanotechnology, however, offers novel possibilities of material property enhancement at the nanoscale. Nanomaterials like nanoparticles, nanotubes, and nanofibers have extraordinary properties such as high surface area, mechanical strength, and electrical conductivity, which can be leveraged to increase the overall performance of bio-based matrices. The integration of these two disciplines allows for the production of bio-nanocomposites with enhanced properties that are suitable for the performance needs of various industrial applications (Mohan et al., 2024). Notably, nanotechnology can also make it possible to develop more energy-efficient and sustainable processes, such as low-temperature synthesis processes or non-toxic functionalization processes.

By combining green chemistry and nanotechnology, bio-nanocomposites are poised as an innovative solution to world sustainability problems. They provide a means to diminished reliance on fossil resources, lessened ecological footprint, and the solution to the increasing demand for materials conforming to circular economy principles and eco-efficiency (Goyal et al., 2024). As companies transition towards greener practices, bio-nanocomposites are leading the way, able to satisfy the sophisticated material needs of the future while helping to create a cleaner, more sustainable world.

2. METHODOLOGY

2.1 Synthesis Techniques: Top-Down vs. Bottom-Up Approaches

There are two principal synthesis methods of bio-nanocomposites: bottom-up and top-down (figure 1). They vary in terms of how the nanomaterials are introduced into the matrix, affecting properties and uses of the final bio-nanocomposites (Rawat, 2015).

Figure 1. Dual-Process Strategies in Nanofabrication

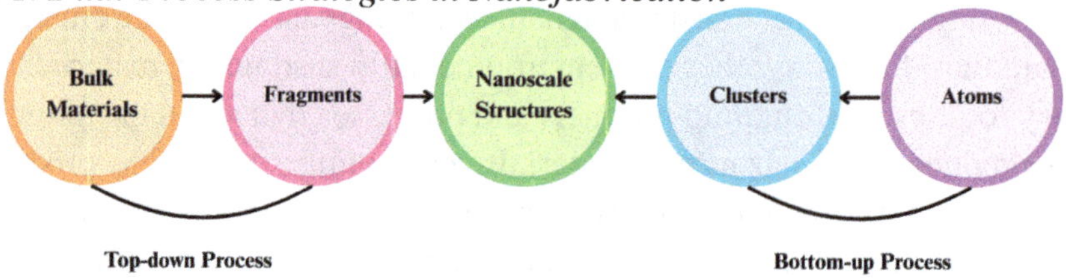

Top-Down Approach: In the top-down method, large materials or bulk materials are dissolved into small nanoscale parts. Mechanical processing, like grinding, milling, or exfoliation, is usually used in this method to decrease the dimensions of materials and form nanoparticles, nanofibers, or nanoplates. For instance, nano clays may be exfoliated to form single sheets that are afterward integrated into a biopolymer matrix. Concurrently, cellulose may be processed mechanically to obtain nanocellulose that is then compounded with other biologically based constituents (Tripathy, 2023). The top-down method tends to be less complex to implement in large volume but can offer controlled sizes and distributions of the nanomaterials, whereas high energy utilization is possible as a requirement along with the presence of surface faults or nanofiller irregularities.

Bottom-Up Approach: The bottom-up strategy is a method of creating nanomaterials from building blocks at the molecular or atomic level. The method usually utilizes chemical processes like sol-gel synthesis, chemical vapor deposition (CVD), or hydrothermal synthesis, in which small components are chemically attached to produce nanoscale structures. For instance, nanoparticles may be synthesized by a sol-gel procedure in which metallic or non-metallic precursors are subjected to a chemical reaction that results in nanoparticles, which are dispersed in the matrix (Grappa et al., 2024). The bottom-up process provides greater control over the nanomaterials' size, shape, and structure, hence improved-defined properties but at a greater complexity and expense than the top-down process.

Table 2. Advantages of Top-Down and Bottom-Up Synthesis Approaches for Bio-Nanocomposites

Criteria	Top-Down Approach	Bottom-Up Approach
Principle	Larger materials or bulk substances are broken down into smaller nanoscale components.	Nanomaterials are assembled from molecular or atomic building blocks.
Methods Used	Mechanical processing (e.g., grinding, milling, exfoliation).	Chemical processes (e.g., sol-gel synthesis, chemical vapor deposition, hydrothermal synthesis).
Scalability	Easier to scale up for industrial production.	More complex and challenging to scale up.
Energy Consumption	High energy input due to mechanical processing.	Generally, more energy-efficient for producing precise nanomaterials.
Control Over Nanomaterial Size	Less precise control over size and distribution.	Better control over size, shape, and structure of nanomaterials.
Surface Quality	May result in surface defects or irregularities in nanofillers.	High surface quality with minimal defects.
Cost	Generally lower cost for larger-scale production.	Can be more costly due to the complexity and precision involved.
Material Types	Suitable for a wide range of bulk materials, such as nano clays, cellulose.	Ideal for creating specific nanomaterials like metallic, ceramic, or polymer nanoparticles.
Applications	Typically used for bulk materials like nano clays or nanocellulose.	Ideal for applications requiring precise control over the nanostructure, such as functional coatings or high-performance composites.
Flexibility	More flexible for large-scale applications but less precise.	More precise and versatile for creating specialized nanomaterials but less adaptable to large-scale production.
Hybrid Approaches	Can be combined with bottom-up techniques for optimized performance.	Often paired with top-down techniques to improve scalability.

Both methods can be applied during the synthesis of bio-nanocomposites, frequently based on the desired properties of the final product, the nanofiller type, and matrix material. Hybrid methods combining both bottom-up and top-down methodologies are generally applied in most cases to optimize the performance of the nanocomposite (Rawas-Qalaji, 2016). This comparison table (Table 2) highlights the major advantages and features of the Top-Down and Bottom-Up methods in the synthesis of bio-nanocomposites, their strengths and weaknesses. The two methods have unparalleled strengths for different desired properties, cost, and application requirements of the bio-nanocomposites.

2.2 Selection of Matrix Materials and Nanofillers

Selection of matrix materials and nanofillers is the key parameter that influences the properties of bio-nanocomposites. The selection of both needs to be precise in order to attain the specific mechanical, thermal, and functional properties and sustainability and compatibility.

Matrix Materials: The matrix material in bio-nanocomposites is usually made of renewable, biobased polymers, which provide biodegradability and less environmental footprint. Some of the common matrix materials are:

a) Cellulose: A biodegradable and omnipresent biopolymer that gives it strength, rigidity, and biodegradability. It can be extracted from several sources of plants and can be further modified for its better compatibility with nanofillers.
b) Starch: An inexpensive, biodegradable biopolymer that can find future applications in food packaging, agriculture, and medical devices. Starch-based composites are easily processable but could need modification to improve their mechanical and thermal properties.
c) Chitosan: From chitin, this biopolymer has been found to be antimicrobial and is utilized in packaging and wound healing.
d) Proteins (e.g., soy or casein): Proteins are of interest because they are renewable and possess useful functions, like film-forming and antimicrobial properties, making them applicable in food packaging and biomedical fields.

The matrix material should be selected in terms of its compatibility with the chosen nanofillers and applicability for the desired application.

Nanofillers:

Nanofillers are those compounds that reinforce the properties of the biopolymer matrix at the nanoscale. The following are the frequently used nanofillers in bio-nanocomposites:

a) Nano clays: E.g., montmorillonite (MMT) or halloysite, that enhance mechanical properties, barrier character, and thermal stability of the matrix.
b) Nanocellulose: Obtained from natural cellulose fibers, it provides high strength, reinforcing capacity, and biodegradability. Cellulose nanocrystals (CNCs) or cellulose nanofibers (CNFs) are especially beneficial in enhancing the mechanical properties of bio-nanocomposites.
c) Carbon Nanotubes (CNTs) and Graphene: These nanomaterials based on carbon offer superior mechanical, electrical, and thermal characteristics. Yet, their biopolymer compatibility can be made necessary through functionalization or surface modification to improve dispersion in the matrix.
d) Metal Nanoparticles: E.g., silver, gold, or zinc oxide, that may endow antimicrobial or catalytic function in bio-nanocomposites, especially for food packaging or biomedical applications.

The choice of nanofillers is based on the intended properties of the final product and their uniform dispersion within the matrix, which is essential to achieve the optimum performance of the bio-nanocomposite (Anugraha et al., 2024).

2.3 Green Chemistry Principles in Bio-Nanocomposite Synthesis

Green chemistry principles (figure 2) control the design of eco-friendly synthesis processes for the bio-nanocomposites in terms of reducing environmental hazards, conserving energy, and avoiding waste. While synthesizing the bio-nanocomposites, green chemistry particularly aims to reduce the consumption of toxic chemicals, minimize the carbon footprint, and maximize resource usage efficiency (De Oliveira et al., 2024).

Figure 2. 12 Key Guidelines for Green Chemistry

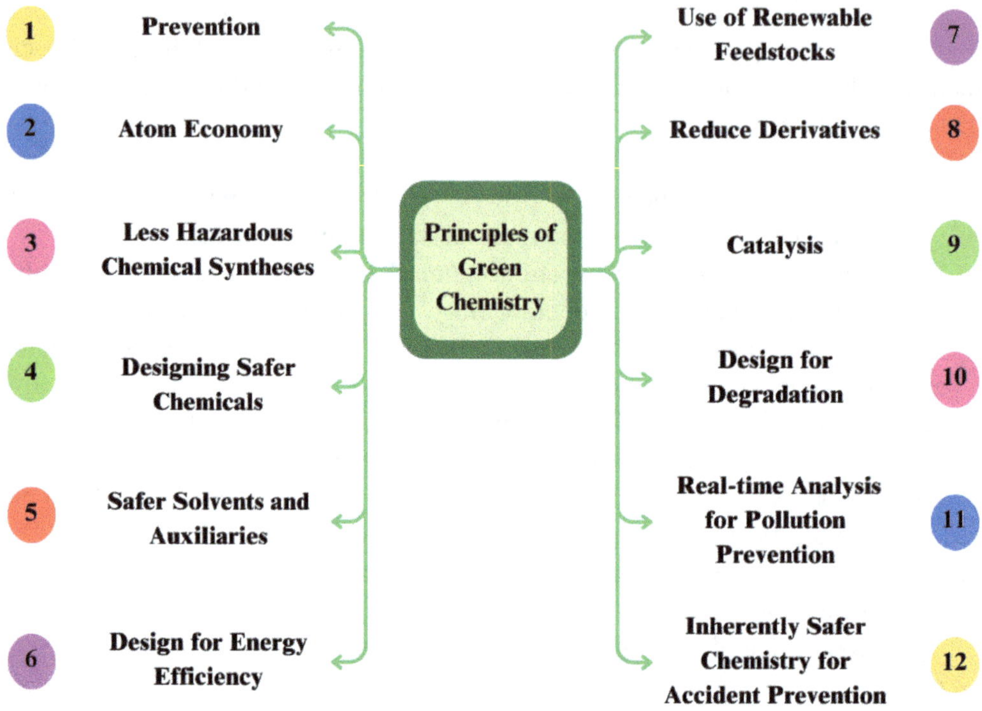

Major green chemistry principles used in the synthesis of bio-nanocomposites are:

a) Utilization of Renewable Resources: Polymers like cellulose, starch, and chitosan that are derived from biomass-based polymers utilize renewable natural resources, thus lessening the reliance on fossil fuels. Employment of these materials not only leads to the minimization of environmental effects but also contributes to the circular economy by facilitating biodegradability and compostability.
b) Solvent-Free and Low-Impact Synthesis: Most of the bio-nanocomposite synthesis processes have been moved towards solvent-free or water-based systems in order to remove toxic solvents and decrease energy usage. For instance, melt processing and extrusion methods are widely applied to prepare biopolymer-based nanocomposites in the absence of toxic solvents.
c) Energy-Efficient Processes: Green chemistry promotes the construction of energy-efficient processes that will reduce the amount of energy involved in synthesis. Low-temperature or ambient-pressure methods are mostly favored to cut down the carbon footprint of the manufacturing process. For instance, hydrothermal and solvent-free approaches can be applied to synthesize nanomaterials at comparably low temperatures.

d) Waste Reduction: Green chemistry promotes the reduction of waste formation in the synthesis of bio-nanocomposites. This may be done by using methods like closed-loop recycling, where unused material or waste products are utilized in the process, minimizing waste and resource utilization.
e) Safer Chemicals and Biocatalysts: The incorporation of safer, non-toxic chemicals or biocatalysts during the synthesis of bio-nanocomposites follows green chemistry concepts. Enzymatic systems, for example, can be used for functionalization of biopolymers or nanomaterial fabrication as an environmentally friendly approach compared to conventional chemical processing.

By integrating the principles of green chemistry in bio-nanocomposite synthesis, researchers can produce high-performance materials that also serve sustainability ends, including minimizing environmental footprint and enhancing resource efficiency. Synthesis aims to use environment-friendly techniques, select suitable nanofillers and matrices, and process optimization to produce functional sustainable bio-nanocomposites (Moosvi et al., 2021).

3. PROPERTIES OF BIO-NANOCOMPOSITES

Bio-nanocomposites have a distinct blend of improved properties over conventional materials, which makes them extremely promising for numerous applications in packaging, automobile, building, and electronics industries. These properties are largely due to synergistic interactions between nanofillers and biopolymer matrices at the nanometer level.

3.1 Mechanical Properties: Strength, Toughness, and Stiffness

Some of the key incentives for the innovation of bio-nanocomposites are to advance their mechanical property, particularly their strength, toughness, and rigidity. The nanofillers have very much improved mechanical functionality of the biopolymers, whose strength is commonly lower than plastic from petroleum (Greenfeld & Wagner, 2014).

a) Strength: Bio-nanocomposites show enhanced tensile and flexural strength due to reinforcement with nanofillers like nanocellulose, carbon nanotubes (CNTs), and nanoclays. Due to the small particle size and large surface area, such nanofillers have sufficient interactions with the biopolymer matrix to generate a strengthened material. Addition of cellulose nanocrystals (CNCs) or cellulose nanofibers (CNFs) has been observed to improve tensile strength in starch and chitosan-based composites.

b) Toughness: Toughness is the term used to define the ability of a material to withstand energy and deformation without failing. The presence of nanofillers will largely enhance the toughness of bio-nanocomposites by deterring crack progression and the enhancement of the global ductility of the material. Bio composites filled with nano clay, for instance, exhibit enhanced toughness due to the establishment of an efficient interfacial bond between the nanofiller and biopolymer.
c) Stiffness: Stiffness is connected to the material's ability to resist deformation under applied stress. CNTs and graphene nanofillers confer high stiffness to bio-nanocomposites by creating a stiff network in the biopolymer matrix. This renders the composite material suitable for applications demanding structural integrity, for example, in construction or automotive materials.

The enhanced mechanical properties render bio-nanocomposites ideal for replacing traditional materials in high-requirement applications, like packaging films, coatings, and structural elements.

3.2 Thermal Properties: Heat Resistance and Stability

The addition of nanofillers into biopolymer matrices has the ability to modify significantly the thermal behavior of bio-nanocomposites, enhancing heat resistance and stability over that of the pure biopolymer.

a) Heat Resistance: Most biopolymers undergo thermal degradation at comparatively low temperatures, which restricts their applications in high-temperature conditions. Incorporation of nanofillers like clay nanoparticles or CNTs may enhance the thermal stability of bio-nanocomposites by providing barriers that impede heat transfer and slow down the rate of thermal degradation. For instance, bio-nanocomposites filled with nano clay have better thermal stability than unfilled counterparts and can be applied in applications involving exposure to moderate heat.
b) Thermal Stability: Thermal stability is defined as the property of a material to retain its characteristics at higher temperatures. The thermal stability of bio-nanocomposites increases with less weight loss and minimum deformation upon heating at high temperatures. The inclusion of nanofillers like graphene or metal nanoparticles can enhance the thermal stability of biopolymer composites by strengthening the polymer chains and thereby decelerating the process of degradation.

Enhanced thermal characteristics allow bio-nanocomposites to be applied in a broad variety of applications that need high performance under thermal stress, such as packaging, automotive, and electronics sectors (GONG, 2004).

3.3 Barrier Properties: Moisture, Gas, and Light Permeability

Bio-nanocomposites are extensively studied for their capability to enhance the barrier quality of packaging materials, thus lowering the permeability of gases, moisture, and light. These functionalities are crucial for improving the shelf life of food, pharmaceuticals, and other sensitive products (Nah & Kader, 2010).

a) Moisture Permeability: Most biopolymers are in nature moisture-absorbing, which compromises their mechanical properties and diminishes their functionality. Moisture permeability of bio-nanocomposites is minimized with the incorporation of nanofillers like nano clays, CNCs, or CNTs. These nanofillers form a denser structure in the biopolymer matrix, restricting pathways for water vapor migration.
b) Gas Permeability: Equally, bio-nanocomposites can have lower permeability to gases like oxygen, carbon dioxide, and nitrogen. This is especially relevant in packaging, where lower gas exchange can aid in maintaining food products' freshness. The addition of nano clay or other nanofillers increases the barrier by creating a tortuous route that hinders gas diffusion, thereby increasing the material's protection capability against spoilage and oxidation.
c) Light Permeability: The incorporation of nanofillers can also enhance the light barrier function of bio-nanocomposites, especially in the protection of light-sensitive products like drugs or beverages. For example, the use of carbon-based nanomaterials such as CNTs or graphene can act as an ultraviolet (UV) light blocker, offering UV protection and inhibiting product degradation.

Their improved barrier property makes the bio-nanocomposites very useful for packaging materials, protective coatings, and other uses where protection of the sensitive products against environmental influences is needed.

3.4 Biodegradability and Environmental Impact

One of the major benefits of bio-nanocomposites compared to traditional synthetic materials is their biodegradable nature and lessened impact on the environment. With increasing environmental concerns related to plastic pollution and waste management across the globe, bio-nanocomposites are a greener alternative (Dean et al., 2012).

a) Biodegradability: Most biopolymers employed in bio-nanocomposites, including starch, cellulose, and chitosan, are biodegradable, i.e., they can degrade into natural elements when subjected to microorganisms in the environment. The addition of nanofillers may at times affect the biodegradability of the composite, depending on the nature of the filler and its compatibility with the biopolymer matrix. But most bio-nanocomposites are biodegradable and can degrade under controlled conditions, making them more eco-friendly compared to non-biodegradable man-made polymers.
b) Environmental Impact: The environmental footprint of bio-nanocomposites is usually less than that of conventional petroleum-based plastics. By using renewable materials for the matrix and minimizing the use of hazardous chemicals in processing, bio-nanocomposites provide a green alternative that meets international environmental standards. In addition, the use of renewable feedstocks like plant-based polymers minimizes carbon emissions related to raw material harvesting and processing.

The eco-friendliness of bio-nanocomposites, such as their biodegradability and low environmental footprint, makes them central materials in the creation of sustainable, circular economy-driven solutions for packaging, agriculture, and other sectors.

The bio-nanocomposites exhibit a variety of advanced properties which make them interesting for various industrial uses. Their strength, toughness, and stiffness mechanical properties are drastically enhanced upon introduction of the nanofillers, thus qualifying them for the replacement of the traditional materials in harsh applications. Their biodegradability, barrier characteristics, and thermal stability also play their part to qualify them as preferable for green growth (Mathew & Radhakrishnan, 2022). As the materials continue to develop, bio-nanocomposites will increasingly contribute toward solving global sustainability issues and fulfilling the performance demands of several industries.

4. CHARACTERIZATION TECHNIQUES

Characterization of bio-nanocomposites is essential to determine their structure, properties, and performance in different applications. Various sophisticated techniques are used to evaluate the structural, mechanical, thermal, and surface properties of bio-nanocomposites. The characterization techniques offer useful information regarding the efficiency of the dispersion of nanofillers, the interaction between the matrix and nanofillers, and the functionality of the composite material.

4.1 Structural and Morphological Analysis: SEM, TEM, XRD

Structural and morphological characterization is a dominant area in investigating the distribution and dispersion of the nanofillers in the biopolymer matrix. Optimization of the properties of bio-nanocomposites and confirmation that the nanofillers are suitably dispersed and in effect interacting with the matrix material calls for structural and morphological analysis (Solá, 2014). This table (Table 3) showcases the main contrast among SEM, TEM, and XRD on the basis of resolution, characteristics, findings, and precise utilization for characterizing the structural and morphological behavior of bio-nanocomposites.

Table 3. Comparison of Structural and Morphological Analysis Techniques for Bio-Nanocomposites: SEM, TEM, and XRD

Techniques	Resolution	Key Features	Insights Provided	Applications
Scanning Electron Microscopy (SEM)	Moderate	High-resolution surface images Surface morphology and filler distribution	Size, shape, and dispersion of nanofillers Surface features, aggregation, and filler-matrix interaction	Surface analysis, dispersion assessment, defect detection
Transmission Electron Microscopy (TEM)	Very High	Higher resolution than SEM Internal structure at the nanoscale	Detailed dispersion and alignment of nanoparticles Aggregation or defects of nanomaterials	Internal structure analysis, reinforcement mechanism
X-ray Diffraction (XRD)	Low to Moderate	Provides crystallinity and phase structure data Measures interlayer spacing	Degree of crystallinity Orientation and dispersion of nanofillers	Crystallinity assessment, phase structure, dispersion of nano clays

a) Scanning Electron Microscopy (SEM): One of the most employed methods for observing the surface topology of bio-nanocomposites is SEM. SEM gives good-resolution images which show the dimension, shape, and distribution of nanofillers in the biopolymer matrix. Through SEM, even the surface structures, including how well the fillers are distributed, aggregated, and any faults in the composite material, may be examined. Researchers can determine the interaction between matrix and nanofillers, along with measuring the homogeneity of the composite, through the utilization of SEM.

b) Transmission Electron Microscopy (TEM): TEM is more highly resolved than SEM and gives high-resolution images of the internal morphology of bio-nanocomposites at the nano level. TEM enables the visualization of the dispersion and orientation of individual nanofibers or nanoparticles in the matrix, which is essential in understanding the mechanism of reinforcement of the nanocomposites. TEM can also determine any defects or aggregation in the nanomaterials, which may affect the performance of the material.

c) X-ray Diffraction (XRD): XRD is a robust method applied to study the crystallinity and phase structure of bio-nanocomposites. It gives insight into the level of crystallinity in the biopolymer matrix and the nanofiller orientation. XRD patterns assist in the identification of changes in the crystal structure induced by the addition of nanofillers, which can result in enhanced mechanical properties or changed thermal stability.

4.2 Mechanical Testing: Tensile Strength, Flexural Strength

Mechanical testing is important to assess the performance of bio-nanocomposites based on their strength, flexibility, and overall load-carrying capacity. Some of the major mechanical properties are identified by standard testing techniques, which generate quantitative information about the composite's stress-resisting capacity (Nimbagal et al., 2021).

a) Tensile Strength: Tensile strength quantifies the resistance of a material to breakage or elongation under a tensile load. Tensile strength of bio-nanocomposites is generally greatly improved by the addition of nanofillers, which improve the mechanical properties of the biopolymer matrix. Typical tensile tests, e.g., as outlined by ASTM D638, are capable of defining the ultimate tensile strength (UTS), elongation at break, and Young's modulus, which are collectively useful for understanding the composite mechanical behavior.
b) Flexural Strength: Flexural strength is the capacity of a material to withstand bending or deformation caused by an applied load. Flexural tests (usually following ASTM D790 standards) yield the bending strength and modulus of bio-nanocomposites. This material property is of special interest in packaging, automotive, and building materials applications where the material needs to sustain structural integrity upon flexural loads.

Both tensile and flexural tests assist in determining the effect of nanofiller addition on the mechanical properties of bio-nanocomposites so that the material can be optimized for a given application.

4.3 Thermal Analysis: DSC, TGA

Thermal analysis becomes crucial for measuring the heat stability and resistance of bio-nanocomposites. The presence of nanofillers may impose a profound impact on the thermal properties of bio-polymer matrices, influencing their stability as well as behavior at elevated temperatures (Nurazzi et al., 2022).

a) Differential Scanning Calorimetry (DSC): DSC is employed for determining the flow of heat in or out of a material while it is going through temperature alteration. The technique offers information about the thermal transition of bio-nanocomposites such as glass transition temperature (Tg), melting temperature (Tm), and crystallization temperature (Tc). DSC can also be employed for assessing the influence of nanofillers on the thermal performance of the composite, such as changes in crystallinity, melting point, and thermal stability.
b) Thermogravimetric Analysis (TGA): TGA is a measure of the mass change of a material upon heating and gives information regarding the thermal stability and decomposition of the material. TGA can be used specifically to identify the thermal degradation temperature of bio-nanocomposites and to examine how the addition of nanofillers affects the thermal degradation resistance of the composite. TGA can also determine the nanocomposite composition by identifying the relative proportion of filler and matrix materials from their respective thermal degradation patterns.

4.4 Surface Properties: FTIR, Raman Spectroscopy

Surface characteristics of bio-nanocomposites are significant in comprehending the chemical interactions between nanofillers and the matrix, which affect their barrier, thermal, and mechanical properties.

a) Fourier Transform Infrared Spectroscopy (FTIR): FTIR is a versatile method of analyzing the chemical functional groups within a material. FTIR can be employed to examine the interaction between the biopolymer matrix and the nanofillers at the molecular level in bio-nanocomposites. By looking at the characteristic absorption peaks, FTIR can identify any chemical bonding between the fillers and matrix, e.g., hydrogen bonding, covalent bonding, or electrostatic interaction. FTIR also aids in detecting modifications in the chemical structure of the matrix as a result of the addition of nanofillers (Rath et al., 2024).
b) Raman Spectroscopy: Raman spectroscopy is employed to examine vibrational modes of molecules and obtain information regarding the molecular composition and chemical structure of bio-nanocomposites. It is especially helpful for

investigating carbonaceous nanomaterials, including CNTs and graphene, and their interactions with the biopolymer matrix (Rath et al., 2024). Raman spectra can indicate the quality and dispersion of nanofillers as well as any possible modification or functionalization during synthesis.

4.5 Rheological and Flow Behavior

Rheological measurements give important insight into the processability and flow behavior of bio-nanocomposites, as well as optimization of their processing and performance for practical applications.

a) Viscosity and Flow Behavior: Rheological testing quantifies bio-nanocomposite melt or suspension viscosity, which can inform whether the material can be processed with ease. Incorporation of nanofillers usually makes the biopolymer melts more viscous by causing interaction between the matrix and filler. Rheological measurements allow optimization of process parameters for bio-nanocomposites.
b) Storage and Loss Modulus: The storage modulus (G') and loss modulus (G”) derived from oscillatory rheological measurements give an idea about the elasticity and viscous nature of the bio-nanocomposites. Higher storage modulus means the material is more elastic, and higher loss modulus means the material is more viscous. It can be measured to identify the influence of nanofillers on the mechanical response of the composite during deformation.

A range of characterization methods is used to extensively evaluate the structural, mechanical, thermal, and surface characteristics of bio-nanocomposites. Methods like SEM, TEM, and XRD reveal information about the morphology and crystallinity of the material, whereas mechanical testing (tensile and flexural strength) determines its strength and flexibility. Thermal characterization (DSC and TGA) shows its heat stability and resistance, while surface characterization with FTIR and Raman spectroscopy facilitates chemical interactions between nanofillers and the biopolymer matrix (Mistretta et al., 2021). Rheological measurements are useful in comprehending the material's processability and flow characteristics.

5. PARAMETERS ANALYSIS

Properties and behavior of bio-nanocomposites are dependent on several factors such as the nature and content of nanofillers, the conditions of processing at the synthesis stage, and the intrinsic characteristics of the matrix material. Furthermore, environmental and economic factors also play a significant role in defining the viability and sustainability of bio-nanocomposites for practical uses.

5.1 Effect of Nanofiller Type and Loading on Properties

The composition of nanofiller and loading percentage in the biopolymer matrix are vital parameters determining the application performance of bio-nanocomposites. The parameters influence the material's processing characteristics, thermal, barrier, and mechanical properties.

Nanofiller Type: The type of nanofiller significantly influences the properties of bio-nanocomposites. Cellulose nanocrystals (CNCs), carbon nanotubes (CNTs), graphene, nano clays, and metal nanoparticles are some of the most common nanofillers. Each nanofiller type contributes unique properties to the composite material:

a) Cellulose Nanocrystals (CNCs): Owing to their high tensile strength and biodegradability, CNCs greatly improve the tensile strength, stiffness, and thermal stability of the matrix. They find special application in reinforcing biopolymers like starch or chitosan.
b) Carbon Nanotubes (CNTs): CNTs provide outstanding mechanical, electrical, and thermal characteristics and thus are suited for high-performance applications. But they are difficult to disperse in the matrix because of their tendency to agglomerate strongly.
c) Nano clays (e.g., Montmorillonite): These find broad applications in enhancing the barrier property of bio-nanocomposites, especially towards moisture and gas resistance. They can also be used to improve the thermal stability of the composite.
d) Graphene: Graphene-based fillers are very effective in improving the thermal stability, mechanical strength, and electrical conductivity of bio-nanocomposites.

Compatibility between the nanofiller and the biopolymer matrix is crucial for delivering requisite performance. It might be required to functionalize or surface modify nanofillers to enhance dispersion and interaction with the matrix (Njuguna et a., 2021).

5.2 Influence of Processing Conditions (Temperature, Time, Pressure)

Fabrication conditions in processing of bio-nanocomposites are critical to the resulting properties and performance. All three factors, temperature, time, and pressure, influence dispersion of nanofillers, crystallinity of the matrix, and interfacial adhesion between the matrix and filler (Nazir & Qayyum, 2023).

a) Temperature: The processing temperature affects the melt flow behavior, viscosity, and crystallization of the biopolymer. A high temperature could enhance the dispersion of some of the nanofillers (e.g., graphene or CNTs) and enhance interfacial contact between the matrix and filler. Excessive heating, however, can result in degradation of the biopolymer, which consequently results in weak mechanical properties.
b) Time: Processing time influences the extent of filler dispersion and matrix crystallization. Longer processing times can improve filler-matrix interaction, but can also result in thermal degradation or over-oxidation of the biopolymer. The ideal processing time is dependent on the particular biopolymer and nanofiller being used, as well as the processing technique (e.g., extrusion, molding, or electrospinning).
c) Pressure: Pressure contributes to uniform dispersion and enhancing interfacial adhesion between the matrix and nanofillers. In extrusion or injection molding processes, pressure application aids in attaining a homogeneous mixture, particularly in the case of high aspect ratio nanofillers such as CNTs or nano clays. Pressure also influences the crystallinity of the biopolymer, thereby affecting the mechanical properties of the composite.

The processing conditions should be precisely controlled to maximize the dispersion of nanofillers, avoid degradation, and produce the desired final bio-nanocomposite properties.

5.3 Relationship Between Matrix Properties and Bio-Nanocomposite Performance

The inherent characteristics of the biopolymer matrix greatly affect the behavior of bio-nanocomposites. The adhesion between the matrix and nanofillers controls how efficiently the nanofillers are able to strengthen the material's mechanical, thermal, and barrier properties (Farooq et al., 2024).

a) Polymer Matrix Composition: Various biopolymers, for instance, cellulose, starch, chitosan, or proteins, exhibit unique chemical and physical characteristics. For instance, matrices of cellulose are highly rigid and of high strength and thus are compatible with reinforcement with nanofillers to introduce mechanical property improvements. On the other hand, matrices of starch or proteins could be more compliant with more emphasis on barrier property improvement by introduction of nanofillers. The compatibility of the matrix and nanofillers influences the dispersion and the mechanical and thermal properties of the composite.

b) Crystallinity and Amorphousness: The crystallinity of the biopolymer matrix determines the strength, stiffness, and thermal stability of the bio-nanocomposite. A more crystalline matrix will generally result in enhanced mechanical properties, whereas an amorphous matrix can provide greater flexibility and ease of processing. The extent of crystallinity in the matrix can also impact the filler-matrix interaction and uniformity of filler dispersion.
c) Hydrophilicity and Hydrophobicity: The hydrophilic or hydrophobic character of the biopolymer matrix also influences the interaction with nanofillers. For example, hydrophilic matrices such as cellulose and chitosan have good interaction with hydrophilic nanofillers, enhancing dispersion and adhesion. In the case of hydrophobic nanofillers such as CNTs, the matrix or filler might need to be functionalized in order to enhance compatibility.

The functionality of bio-nanocomposites thus largely depends on the interaction between the matrix properties and the chosen nanofillers, with selection and optimization in consideration for particular applications.

5.4 Environmental and Economical Assessment of Bio-Nanocomposites

In assessing bio-nanocomposites for industrial use, their performance should not only be considered but also their environmental and economic viability. Bio-nanocomposites are usually assessed for their potential for sustainability, but resource availability, cost of production, and environmental impact should be evaluated.

a) Environmental Impact: The main environmental benefit of bio-nanocomposites is their renewable, biodegradable, and compostable nature. Through the use of biopolymers as matrix materials, bio-nanocomposites are able to provide significant decreases in plastic waste and pollution. The use of green chemistry principles for the synthesis of bio-nanocomposites also lowers the environmental impact. Nevertheless, the environmental influence of the nanofillers, especially carbon-based nanomaterials (e.g., CNTs or graphene), also has to be taken into account. These substances are usually based on non-renewable materials and can take energy-intensive production processes.
b) Economic Feasibility: Although bio-nanocomposites present substantial sustainability benefits, economic feasibility is a major factor for commercial production and market uptake. The price of biopolymer matrices (e.g., cellulose, starch, or chitosan) and nanofillers (e.g., CNTs, graphene, or nano clays) may range from inexpensive to very expensive depending on availability, scale of production, and requirements for functionalization. The preparation and dispersion costs of nanofillers further contribute to the overall production cost.

c) Life Cycle Assessment (LCA): LCA is an overarching approach applied in evaluating the life cycle environmental burdens of bio-nanocomposites from raw material harvesting to product manufacturing, utilization, and ultimate disposal. Through LCA, it is possible to determine principal phases where environmentally friendly impacts should be reduced while informing the formation of more environment-friendly bio-nanocomposite technologies.

The performance of bio-nanocomposites depends on various parameters such as the nature and loading of nanofillers, processing, and the nature of the matrix material. Further, a systemic evaluation of their environmental and economic importance is important in deciding their overall viability as sustainable materials for industrial applications. With optimized parameters, bio-nanocomposites can fulfill the increasing requirement for eco-friendly, high-performance materials (Riaz et al., 2024).

6. DISCUSSION

The fast-paced evolution of bio-nanocomposites has introduced tremendous opportunities for sustainable materials that can serve as alternatives to traditional, petroleum-derived polymers in many industrial applications. It addresses comparative performance of bio-nanocomposites to legacy materials, their promise in particular applications such as packaging, agriculture, and automotive, challenges of scaling up production, and the larger environmental and sustainability aspects.

6.1 Comparison of Bio-Nanocomposites with Conventional Materials

Bio-nanocomposites have numerous benefits compared to traditional materials, especially in aspects of sustainability, biodegradability, and performance. Due to these benefits, they are extremely desirable to replace synthetic polymers like polyethylene and polypropylene that are based on non-renewable petroleum-based resources (Mittal, 2011).

Table 4. Performance and Sustainability Comparison: Bio-Nanocomposites vs. Conventional Materials

Aspect	Bio-Nanocomposites	Conventional Materials
Source and Sustainability	Derived from renewable resources (biopolymers like starch, chitosan)	Derived from non-renewable petroleum-based resources (e.g., polyethylene, polypropylene)
Biodegradability	Biodegradable, breaks down naturally under microbial influence	Non-biodegradable, persists in the environment for centuries
Environmental Impact	Lower carbon footprint, reduces reliance on petroleum	High carbon footprint, contributes significantly to pollution
Mechanical Strength	Enhanced by nanofillers (e.g., CNCs, CNTs), higher strength and stiffness	High mechanical strength but lacks the reinforcement capabilities of nanofillers
Thermal Stability	Improved thermal stability with nanofillers (e.g., cellulose nanocrystals)	Good thermal stability but less advanced in some applications
Barrier Properties	Superior moisture, oxygen, and light resistance (e.g., nanoclays, graphene)	Limited barrier properties, often require additional coatings
Processing and Cost	Higher processing complexity and cost due to nanofiller incorporation	Easier processing, cost-effective at large scales
Application in High-Performance Fields	Suitable for advanced applications requiring enhanced strength and stability	Can be used in many fields but may not perform well under extreme conditions
Versatility in Use	Ideal for biodegradable packaging, medical, and high-performance applications	Widely used in everyday applications but lacks sustainability for long-term use
Consistency of Performance	Performance can be inconsistent due to processing challenges	Consistent performance, though not always suitable for high-performance or eco-friendly needs

Mechanical and Thermal Performance: Although traditional synthetic plastics have good mechanical properties and thermal stability, bio-nanocomposites can surpass them in certain applications because of the reinforcing action of nanofillers. For example, incorporating cellulose nanocrystals (CNCs) or carbon nanotubes (CNTs) into biopolymers like starch or chitosan leads to bio-nanocomposites with better mechanical strength, stiffness, and thermal stability. Such improved properties enable bio-nanocomposites to be utilized in high-performance applications where conventional biopolymers cannot meet the requirements.

Biodegradability and Environmental Footprint: One of the major benefits of bio-nanocomposites compared to traditional plastics is their biodegradability. Whereas traditional plastics last in the environment for centuries, bio-nanocomposites based on renewable feedstocks are intended to break down more rapidly under the influence of natural factors or microbes. This makes them a suitable substitute for use in plastic waste-generating applications like packaging and disposable products.

In addition, the utilization of biopolymers and renewable source-based nanofillers decreases the dependency on petroleum and the carbon footprint of the materials.

Barrier Properties: Bio-nanocomposites also possess better barrier properties than traditional materials. For instance, biopolymers loaded with nano clays or graphene can be more resistant to moisture, oxygen, and light, which are essential for packaging. This makes bio-nanocomposites a good substitute for food and beverage packaging, where shelf life is maximized and product quality is maintained.

Even with these benefits, bio-nanocomposites tend to be challenged by issues of cost, processing, and consistency of performance, which might restrict their direct application in some industries. This table (Table 4) points out the unique strengths and challenges of bio-nanocomposites over traditional synthetic materials.

6.2 Performance in Specific Applications (Packaging, Agriculture, Automotive, etc.)

Bio-nanocomposites are also being investigated for a broad variety of applications owing to their improved properties and eco-friendliness. Some of the major industries where bio-nanocomposites have tremendous potential are packaging, agriculture, and automotive.

Packaging: One of the most significant fields for the use of bio-nanocomposites is packaging. Conventional plastic packaging, including polyethylene or polypropylene, is a primary source of plastic waste. Bio-nanocomposites, especially those composed of biopolymers such as PLA (polylactic acid), cellulose, or starch, provide a more environmentally friendly option. With the addition of nanofillers like nano clays, cellulose nanofibers (CNFs), or graphene, bio-nanocomposites add enhanced mechanical strength, barrier qualities (resistance to moisture, gases, and light), and biodegradability. Thus, they find perfect applications in food packaging where performance and greenness are significant. But a large-scale production at low cost is still to be achieved.

Agriculture: Bio-nanocomposites also possess great potential in agricultural use, especially in the production of biodegradable films and coatings for soil conservation, mulch films, and seed coatings. The addition of nanomaterials like CNFs or biopolymer-based nano clays can enhance the mechanical properties and degradation rates of these materials to make them ideal for sustainable agricultural use. These composites can minimize the environmental impact of conventional synthetic polymers in agriculture. Also, bio-nanocomposites can be employed in controlled-release fertilizers and pesticides to offer sustainable options to traditional chemical-based products.

Automotive and Construction: In the automotive and construction industries, bio-nanocomposites can replace petroleum-based materials in structural components, interior parts, and insulation applications. The improved thermal stability and mechanical properties of bio-nanocomposites, along with their light weight, make them ideal for vehicular weight reduction and energy efficiency improvement (Susan et al., 2023). Bio-nanocomposites also find application in sustainable construction materials, for example, green composites for insulation and flooring. But scalability, cost, and durability issues over the long term need to be resolved for these materials to find large-scale application in these sectors.

6.3 Challenges in Scale-up and Commercialization

Though bio-nanocomposites are of much potential, some challenges exist for their large-scale production and full-scale commercialization.

Cost of Raw Materials: The raw materials used to produce bio-based products (such as biopolymers PLA, chitosan, and cellulose) are sometimes more costly than petro-based substitutes. Moreover, certain nanofillers (for example, carbon nanotubes (CNTs) or graphene) are presently still prohibitively expensive for bulk production. All these extra costs can lower the competitiveness of bio-nanocomposites, particularly in markets with price sensitives such as packaging.

Processing Challenges: Large-scale production of bio-nanocomposites calls for the development of cost-effective and efficient production methods. The dispersion of the nanofillers within the matrix is also important in determining the desired properties, but it is challenging to achieve uniform distribution on a large scale. Furthermore, the process conditions (temperature, pressure, time) should be optimized so that the degradation of the biopolymer does not occur and the filler dispersion is adequate. The requirement for high-tech equipment and sophisticated processing methods can contribute to the cost of production.

Performance Consistency: Consistent performance in bio-nanocomposites is another challenge since differences in nanofiller dispersion and matrix compatibility can cause inconsistencies in material properties. Scale-up also tends to add more complexities in maintaining uniformity, particularly when working with highly sensitive biopolymers and nanofillers.

6.4 Environmental Impact and Sustainability Considerations

One of the major driving factors in the evolution of bio-nanocomposites is their potential for environmental protection and sustainable development. Nevertheless, the environmental advantage has to be cautiously assessed considering the whole lifecycle of the material, from raw material harvesting to end-of-life waste disposal.

Renewable Resources: Bio-nanocomposites make use of renewable materials, including plant-based polymers (e.g., cellulose, starch, chitosan) and natural fibers, thus decreasing the dependency on fossil fuels. These materials are generally biodegradable, with the benefit of minimizing plastic waste in landfills and oceans. The environmental footprint of their production (e.g., energy use, water usage, and agricultural land use) needs to be considered in a life cycle assessment (LCA).

Biodegradability and End-of-Life Management: Bio-nanocomposites are generally more biodegradable than traditional plastics, minimizing the long-term environmental effect. Their biodegradability can, however, be subject to the particular formulation and nanofiller type. Certain nanofillers, especially those from non-renewable resources (e.g., CNTs, graphene), are not biodegradable and might become an environmental problem at the end of life. Moreover, the occurrence of nanomaterials could make waste disposal processes more challenging, as some fillers cannot be easily dissolved or may end up releasing toxic chemicals.

Carbon Footprint: While bio-nanocomposites can provide plastic pollution and biodegradability benefits, the net carbon footprint of these products should be assessed. The energy-demanding synthesis routes for the biopolymers and some nanofillers (such as CNTs, graphene) can eliminate some of the environmental advantages unless properly controlled. Green synthesis of nanofillers and green energy sources as production inputs are possible ways of mitigating the negative effects.

Bio-nanocomposites have great potential to replace traditional materials in different applications with great advantages of sustainability, performance, and biodegradability. Despite this, challenges persist in costs, scalability, processing, and the ecological footprint of some nanofillers (Vilaplana et al., 2010). Further research and development, as well as optimizing manufacturing methods and life cycle analyses, will be important to address these challenges and unlock the full potential of bio-nanocomposites to aid in a sustainable future.

7. FUTURE RESEARCH DIRECTIONS

7.1 Innovations in Synthesis and Functionalization of Bio-Nanocomposites

a) Nanomaterial Development: Future work can aim to investigate new nanomaterials, e.g., 2D nanomaterials (e.g., graphene, MXenes) or hybrid nanocomposites, to enhance further the mechanical, thermal, and barrier characteristics of bio-nanocomposites. Tailoring the properties of these nanomaterials is possible and can produce advanced bio-nanocomposites with optimized performance.
b) Sustainable Synthesis Routes: Innovations in greener synthesis methods, such as low-energy, solvent-free, and water-based methods, will be essential in min-

imizing the environmental footprint of the production of bio-nanocomposites. Furthermore, the production of nanofillers from renewable sources can further increase the sustainability of the materials.

c) Functionalization of Nanofillers: Surface treatment of nanofillers, e.g., functionalization with bioactive molecules or improved dispersion in the matrix, may enhance interfacial adhesion and, therefore, the mechanical properties and biodegradability of bio-nanocomposites. Functional group research, e.g., natural polymers or bioactive compounds, can result in the development of more specialized, multifunctional bio-nanocomposites for specific uses.

7.2 Enhancing Mechanical and Thermal Properties for Industrial Applications

a) Strength and Toughness Enhancement: One of the biggest challenges in bio-nanocomposite development is improving their mechanical properties to reach or surpass the levels of conventional materials such as plastics and metals. Studies could aim to achieve the optimal mixing of various nanofillers (e.g., carbon nanotubes, cellulose nanocrystals, clays) to achieve maximum structural integrity and resistance to stresses.

b) Thermal Stability and Processing: Enhancing the thermal stability of bio-nanocomposites will render them applicable for high-temperature usage, for example, in the automotive or electronics industry. Investigations on matrix choice and processing conditions may result in improved heat resistance and dimensional stability.

c) Multifunctional Materials: Future developments could also include the development of bio-nanocomposites that incorporate not just mechanical and thermal resistance but also other functionalities, including electromagnetic shielding, UV protection, or electrical conductivity. These multifunctional composites could find wide-ranging applications in industries such as electronics, automotive, and aerospace.

7.3 Addressing Recycling and Waste Management of Bio-Nanocomposites

a) Recycling Technologies: Establishing efficient recycling technologies for bio-nanocomposites will be crucial for loop closing in the circular economy. The studies may concentrate on chemical recycling technologies that degrade the bio-based matrices and reclaim high-value nanofillers to be recycled and reused in fresh composites.

b) End-of-Life Management: Knowledge of the compostability and biodegradability of bio-nanocomposites is central in resolving waste management issues. Exploring the formulation of bio-nanocomposites with completely biodegradable matrices and eco-compatible additives will reduce landfill waste as well as minimize environmental pollution.
c) Design for Disassembly: Future research would explore materials that are easily disassembled or dismantled into their component parts after application. This would enable recycling or re-use of bio-nanocomposites in accordance with circular economy practice, with minimal environmental footprint following product life cycles.

7.4 Interdisciplinary Approaches: Collaboration Between Nanotechnologists, Chemists, and Engineers

a) Cross-Disciplinary Synergy: Future development of bio-nanocomposites will necessitate tight cooperation between nanotechnologists, material scientists, chemists, and engineers. Nanotechnologists can investigate novel nanomaterials and their characteristics, chemists can develop optimized synthesis pathways and functionalization, whereas engineers can concentrate on scalable manufacturing techniques and integration into industrial processes.
b) Collaboration with Industry: Active engagement with industry players (e.g., producers, regulators, and consumers) will be key to confirming that bio-nanocomposites have the criteria for viable large-scale applications. Experiments and industries' experiences from real-life situations in fields such as packaging, automotive, and agriculture will inform the improvement of bio-nanocomposites to suit certain commercial purposes.
c) Advanced Computational Models: Material scientists and engineers can work together to create sophisticated computational models that mimic the nanoscale behavior of bio-nanocomposites, allowing for quicker optimization of materials and processes. Such models might forecast how variations in nanomaterial type, matrix, and processing conditions influence final properties, minimizing trial-and-error experimentation and speeding up product development.

The way forward for bio-nanocomposites is in an inter-disciplinary process where basic research as well as applications are being furthered. Progress in synthesis, functionalization, enhancement of properties, and recyclability, and excellent inter-disciplinary linkages, will be necessary for the large-scale development of these materials and bringing to their fullest utility against worldwide challenges to sustainability.

8. CONCLUSION

8.1 Summary of Key Findings and Insights

Bio-nanocomposites are new-generation materials that integrate nanotechnology with sustainable bio-based polymers, providing numerous advantages over conventional materials. They possess enhanced strength, thermal resistance, and barrier properties, which make them suitable for applications in packaging, automotive, electronics, and agriculture. Composed of green chemistry, they are eco-friendly and offer a sustainable option compared to petroleum-based plastics. Their characteristics can be designed by modifying the nanofillers and matrix material. Additional advances in surface modification and material dispersion also contribute to improved performance. Issues still need to be overcome in scaling up production and making bio-nanocomposites economically feasible for industrial-scale use.

8.2 The Role of Bio-Nanocomposites in Achieving Global Sustainability Goals

Bio-nanocomposites play a vital role in solving world sustainability issues. They have the potential to minimize the use of non-renewable plastics and enhance the use of renewable, biodegradable materials, advancing the fight against plastic pollution, carbon emissions reduction, and transitioning towards a circular economy. Bio-nanocomposites can minimize environmental footprint in sectors such as packaging, agriculture, and the automotive industry by substituting conventional plastics. Their recyclability and biodegradability also contribute to more sustainable product life cycles with less waste and closed-loop system support. As companies aim to be more sustainable, bio-nanocomposites will be crucial in helping reach objectives such as reducing resource exploitation, pollution reduction, and promotion of a circular economy.

8.3 Closing Remarks on Future Opportunities and Challenges

Bio-nanocomposites possess great potential for expansion, with continuous advancements in their manufacturing techniques, materials, and performance. With increasing affordability and scalability, they will likely substitute conventional materials across numerous applications. Challenges do persist, though, such as enhancing their mechanical and thermal properties for application in industry, recycling, and waste concerns, and optimizing manufacturing processes towards lower costs and sustainability. Interdisciplinary collaboration of specialists in fields such as nanotechnology, material science, and engineering will be central to future progress in bio-nanocomposites. More research is necessary to improve their functionality, refine their properties, and create environmentally sound disposal technologies. Ultimately, bio-nanocomposites can lead to tremendous advances toward more sustainable and green materials, revolutionizing industries and enabling a greener tomorrow.

REFERENCES

Anugraha, R. P., Tobing, S., Maniam, G. P., Zhang, D. H., & Chutima, P. (2024). Polymers, Composites, Nano- and Bio- Materials. In Materials Science Forum. Trans Tech Publications Ltd. https://doi.org/DOI: 10.4028/b-pm5ge6

Basavegowda, N., & Baek, K.-H. (2021). Advances in Functional Biopolymer-Based Nanocomposites for Active Food Packaging Applications. *Polymers*, *13*(23), 4198. DOI: 10.3390/polym13234198 PMID: 34883701

De Oliveira, M. L., Rodrigues, L. M., Da Veiga, M. M. S., & Souza, L. R. R. (2024). Advancing nanomaterial synthesis: Harnessing green chemistry for sustainable innovation. *Green Analytical Chemistry*, *11*, 100148. DOI: 10.1016/j.greeac.2024.100148

Dean, K. M., Sangwan, P., Way, C., & Nikolic, M. A. L. (2012). *Biodegradability Characterization of Polymer Nanocomposites. Characterization Techniques for Polymer Nanocomposites*. Portico., DOI: 10.1002/9783527654505.ch13

Farooq, K., Kumawat, Y. K., Sharma, K., & Kumar, V. (2024). Bio-nanocomposites: Fundamentals and Recent Advances. Novel Bio-Nanocomposites for Biomedical Applications, 1–35. https://doi.org/DOI: 10.1007/978-3-031-69654-1_1

GONG, F. (2004). Thermal properties of poly (vinyl chloride)/montmorillonite nanocomposites. *Polymer Degradation & Stability*, *84*(2), 289–294. DOI: 10.1016/j.polymdegradstab.2003.11.003

Goyal, R., Devi, M., Gautam, R. K., & Gupta, S. (2024). Insights into the prospects of green chemistry and nanotechnology in environmental sustainability. Role of Green Chemistry in Ecosystem Restoration to Achieve Environmental Sustainability, 3–9. https://doi.org/DOI: 10.1016/b978-0-443-15291-7.00030-4

Grappa, R., Venezia, V., Silvestri, B., Costantini, A., & Luciani, G. (2024). Synthesis of Lignin Nanoparticles: Top-Down and Bottom-Up Approaches. *One Health*, *2023*, 3. DOI: 10.3390/msf2024025003

Greenfeld, I., & Wagner, H. D. (2014). Nanocomposite toughness, strength and stiffness: Role of filler geometry. *Nanocomposites*, *1*(1), 3–17. DOI: 10.1179/2055033214y.0000000002

Kausar, A., & Ahmad, I. (2023). Degradable Green Polymers, Green Nano polymers and Green Nanocomposites Derived from Natural Systems: Statistics and Headways. *Nano-Horizons.*, *2*. Advance online publication. DOI: 10.25159/Nano-Horizons.49f1ef7cdb71

Mathew, S., & Radhakrishnan, E. K. (2022). Soil Burial and Biodegradability of Bio nanocomposites. Polymer Based Bio-Nanocomposites, 181–203. https://doi .org/DOI: 10.1007/978-981-16-8578-1_10

Mistretta, M. C., Botta, L., Arrigo, R., Leto, F., Malucelli, G., & La Mantia, F. P. (2021). Bio nanocomposite Blown Films: Insights on the Rheological and Mechanical Behavior. *Polymers*, *13*(7), 1167. DOI: 10.3390/polym13071167 PMID: 33916477

Mittal, V. (2011). Bio-nanocomposites: future high-value materials. Nanocomposites with Biodegradable Polymers, 1–27. https://doi.org/DOI: 10.1093/acprof:oso/9780199581924.003.0001

Mohan, C., Robinson, J., Vodwal, L., & Kumari, N. (2024). Sustainable Development Goals for addressing environmental challenges. Green Chemistry Approaches to Environmental Sustainability, 357–374. https://doi.org/DOI: 10.1016/b978-0-443-18959-3.00007-0

Mondal, S. (2025). Sustainable bio-polymeric materials. Advances in Sustainable Materials, 1–22. https://doi.org/DOI: 10.1016/b978-0-443-13849-2.00001-6

Moosvi, S. K., Gulzar Naqash, W., & Najar, Mohd. H. (2021). Green Chemistry. https://doi.org/DOI: 10.1515/9783110751895

Nah, C., & Kader, M. A. (2010). *Barrier Properties of Rubber Nanocomposites. Rubber Nanocomposites*. Portico., DOI: 10.1002/9780470823477.ch18

Nandhini, J., Bellarmin, M., Siva Prakash, S., Sowmya Sri, D., & Karthikeyan, E. (2024). Advancements in metal oxide bio-nanocomposites for sustainable food packaging: Fabrication, applications, and future prospectives. *Food Bioengineering*, *3*, 438–463. DOI: 10.1002/fbe2.12106

Nazir, R., & Qayyum, S. A. (2023). Influence of Nanofillers on Biodegradable Composites. Handbook of Nanofillers, 1–23. https://doi.org/DOI: 10.1007/978-981-99-3516-1_57-1

Nimbagal, V., Banapurmath, N. R., Sajjan, A. M., Patil, A. Y., & Ganachari, S. V. (2021). Studies on Hybrid Bio-Nanocomposites for Structural Applications. *Journal of Materials Engineering and Performance*, *30*(9), 6461–6480. DOI: 10.1007/s11665-021-05843-9

Njuguna, J., Ansari, F., Sachse, S., Rodriguez, V. M., Siqqique, S., & Zhu, H. (2021). Nanomaterials, nanofillers, and nanocomposites: types and properties. Health and Environmental Safety of Nanomaterials, 3–37. https://doi.org/DOI: 10.1016/b978-0-12-820505-1.00011-0

Nurazzi, N. M., Abdullah, N., Norrahim, M. N. F., Kamarudin, S. H., Ahmad, S., Shazleen, S. S., Rayung, M., Asyraf, M. R. M., Ilyas, R. A., & Kuzmin, M. (2022). Thermogravimetric Analysis (TGA) and Differential Scanning Calorimetry (DSC) of PLA/Cellulose Composites. Polylactic Acid-Based Nanocellulose and Cellulose Composites, 145–164. https://doi.org/DOI: 10.1201/9781003160458-7

Rajeev, A., Yin, L., Kalambate, P. K., Barjini Khabbaz, M., Trinh, B., Kamkar, M., Mekonnen, T. H., & Tang, S. (2024). Nano-enabled smart and functional materials toward human well-being and sustainable developments. *Nanotechnology*, *35*, 352003–352003. DOI: 10.1088/1361-6528/ad4dac PMID: 38768585

Rath, A. P., Krishnan, P. S. G., & Kanny, K. (2024). A comparison of physical, morphological, and mechanical properties of bio-polyester hybrid nanocomposites. Polymers for Advanced Technologies, 35(9). Portico. https://doi.org/DOI: 10.1002/pat.6566

Rawas-Qalaji, M. (2016). Combination of Bottom-Up and Top-Down Techniques for the Fabrication of Epinephrine Nanocrystals. https://doi.org/DOI: 10.26226/morressier.57d6b2b8d462b8028d88cfe0

Rawat, R. S. (2015). Dense plasma Focus - from alternative fusion source to versatile high energy density plasma source for plasma nanotechnology. *Journal of Physics: Conference Series*, *591*, 012021. DOI: 10.1088/1742-6596/591/1/012021

Riaz, F., Rasul, I., Azeem, F., Zubair, M., Nadeem, H., Imran, M., Muzammil, A., Afzal, M., & Siddique, M. H. (2024). Environmental impact, health implications, and life cycle assessment of bionanocomposites. Advances in Bionanocomposites, 387–406. https://doi.org/DOI: 10.1016/b978-0-323-91764-3.00009-7

Sahu MK, Yadav R, Tiwari SP (2023) Recent advances in nanotechnology. Int J Nanomater Nanotechnol Nanomed 9(2): 015-023. DOI: https://dx.doi.org/DOI: 10.17352/2455-3492.000053

Sharma, R. B., Kaur, A., Nautiyal, U., Kabra, A., & Arora, V. (2024). The role of green chemistry and nanotechnology in developing environmental sustainability. Role of Green Chemistry in Ecosystem Restoration to Achieve Environmental Sustainability, 33–40. https://doi.org/DOI: 10.1016/b978-0-443-15291-7.00009-2

Shchipunov, Y. (2012). Bio nanocomposites: Green sustainable materials for the near future. *Pure and Applied Chemistry*, *84*(12), 2579–2607. DOI: 10.1351/PAC-CON-12-05-04

Shibata, M. (2011). Bio-nanocomposites using bio-based epoxy resins. Nanocomposites with Biodegradable Polymers, 329–347. https://doi.org/DOI: 10.1093/acprof:oso/9780199581924.003.0013

Solá, F. (2014). *SEM and TEM Characterization of Polymer CNT Nanocomposites. Polymer Nanotube Nanocomposites*. Portico., DOI: 10.1002/9781118945964.ch4

Susan, Md. A. B. H., Harun-Ur-Rashid, M., & Imran, A. B. (2023). Green Polymer Nanocomposites in Automotive and Packaging Industries. *Current Pharmaceutical Biotechnology*, *24*(1), 145–163. DOI: 10.2174/1389201023666220506111027 PMID: 35524658

Thirugnanasambandan, T. (2022). Bio-nanocomposites in Biomedical Application. Polymer Based Bio-Nanocomposites, 275–291. https://doi.org/Bottom of FormDOI: 10.1007/978-981-16-8578-1_15

Tripathy, S. (2023). Top-down and Bottom-up Approaches for Synthesis of Nanoparticles. Nano biomaterials, 92–130. https://doi.org/DOI: 10.21741/9781644902370-4

Vilaplana, F., Strömberg, E., & Karlsson, S. (2010). Environmental and resource aspects of sustainable biocomposites. *Polymer Degradation & Stability*, *95*(11), 2147–2161. DOI: 10.1016/j.polymdegradstab.2010.07.016

KEY TERMS AND DEFINITIONS

Carbon Nanotubes (CNT): Cylindrical nanostructures made of carbon atoms with exceptional strength, conductivity, and unique properties, widely used in various applications.

Chemical Vapor Deposition (CVD): A process used to produce high-quality, high-performance materials by reacting gaseous precursors to form a solid material on a surface.

Differential Scanning Calorimetry (DSC): A technique that measures the heat flow into or out of a sample as it is heated or cooled, helping to analyze its thermal behavior, such as melting, crystallization, and phase transitions.

Fourier Transform Infrared Spectroscopy (FTIR): A technique used to identify the chemical composition and molecular structure of a material by measuring how it absorbs infrared light at different wavelengths.

Life Cycle Assessment (LCA): A technique used to evaluate the environmental impact of a product or process throughout its entire life cycle, from raw material extraction to production, use, and disposal.

MXenes: A class of two-dimensional materials made from transition metal carbides, nitrides, or carbonitrides, known for their excellent electrical conductivity, mechanical properties, and potential applications in energy storage, sensors, and electronics.

Scanning Electron Microscopy (SEM): A technique that uses focused electron beams to create high-resolution images of a sample's surface.

Thermogravimetric Analysis (TGA): A technique that measures the change in a material's weight as it is heated or cooled, helping to analyze its thermal stability and composition.

Transmission Electron Microscopy (TEM): A technique that transmits electrons through a thin sample to produce detailed images of its internal structure at a high resolution.

X-Ray Diffraction (XRD): A technique used to analyze the crystal structure, composition, and properties of materials by measuring the diffraction patterns produced when X-rays are directed at the sample.

Chapter 9
Innovations and Applications of Bio-Nanocomposites in Sustainable Food Packaging:
Biodegradable Materials With Nanotechnology for Enhanced Safety, Functionality, and Environment

Hari Shankar Biswas
https://orcid.org/0000-0003-1883-5244
Surendranath College, India

Antara Roy
University of Calcutta, India

Amit Kr. Kundu
https://orcid.org/0000-0002-7649-8154
Sripat Singh College, India

Dilip K. Maiti
https://orcid.org/0000-0002-8772-6685
University of Calcutta, India

DOI: 10.4018/979-8-3373-2180-6.ch009

ABSTRACT

The rising demand for sustainable food packaging has spurred advancements in bio-nanocomposites, which blend biodegradable biopolymers like starch, cellulose, and PLA with nanomaterials such as nanoclays, cellulose nanocrystals, carbon-based materials, and metal oxide nanoparticles. This combination enhances mechanical strength, thermal stability, and barrier properties, making them ideal for modern packaging. This chapter reviews synthesis methods—including solution casting, melt blending, and in situ polymerization—and explores their impact on material properties. Key functionalities such as antimicrobial activity, biodegradability, and shelf-life extension are discussed, along with emerging applications in active and intelligent packaging. Challenges like nanoparticle dispersion, polymer compatibility, and scalability are addressed. Future prospects include multifunctional systems, bio-derived nanomaterials, and advanced fabrication techniques. This chapter offers a comprehensive overview of bio-nanocomposites in sustainable food packaging.

INTRODUCTION

Food packaging is an essential component of the global food supply chain, playing a critical role in ensuring food safety, quality, and freshness throughout its storage and transportation (Aydın et al., 2024; Mahunu et al., 2024). As the world's population grows and consumer demand for convenience increases, packaging requirements have become more sophisticated, focusing not only on preserving food but also on maintaining its nutritional value and sensory attributes (Pascall et al., 2022; Ncube et al., 2021). Traditional packaging materials, primarily derived from petrochemical-based plastics, have dominated the market for decades due to their affordability, durability, and functional versatility (Hussain et al., 2024). However, these materials come with significant drawbacks, most notably their environmental impact. Non-biodegradability, limited recyclability, and the growing problem of plastic pollution have created an urgent need for alternative packaging materials that align with the principles of sustainability and environmental stewardship (Stanley et al., 2025).

In this context, bio-nanocomposites have emerged as a transformative innovation. These materials are engineered by integrating biopolymers, derived from renewable resources such as starch, cellulose, chitosan, and polylactic acid (PLA), with nanoscale fillers, including nanoclays, cellulose nanocrystals (CNCs), carbon-based nanomaterials, and metal oxide nanoparticles (Matin et al., 2024). The incorporation of nanomaterials addresses the inherent limitations of biopolymers, such as their relatively poor mechanical strength, thermal stability, and barrier properties, which have traditionally restricted their use in demanding packaging applications (Mishra et al., 2024). The unique properties of nanomaterials, including their high surface area, aspect ratio, and functionality, enable bio-nanocomposites to achieve exceptional performance that rivals or even surpasses conventional plastic packaging (del Rosario Herrera-Rivera et al., 2024).

The development of bio-nanocomposites aligns with the broader goals of sustainability and circular economy (Anwer et al., 2023). Unlike conventional plastics, bio-nanocomposites are often biodegradable and compostable, making them a promising solution for reducing the environmental footprint of food packaging (Dutta & Sit, 2024). Additionally, these materials can be tailored to provide advanced functionalities such as antimicrobial activity, oxygen and moisture barrier enhancement, and thermal resistance, addressing the diverse needs of the food industry. Recent advancements have also enabled the incorporation of intelligent and active packaging features, such as sensors and indicators, into bio-nanocomposites, offering real-time monitoring of food quality and spoilage (D'Almeida & de Albuquerque, 2024). This convergence of sustainability and functionality positions bio-nanocomposites as a cornerstone of next-generation food packaging technologies.

Despite their promising potential, the widespread adoption of bio-nanocomposites faces several challenges (Dalbanjan et al., 2024). Achieving uniform dispersion of nanomaterials within the polymer matrix remains a technical hurdle, as agglomeration can compromise the material's overall properties. Additionally, compatibility between nanomaterials and biopolymers must be carefully engineered to ensure optimal performance (Abady et al., 2024). The scalability of synthesis techniques and the economic feasibility of bio-nanocomposites are also critical considerations for their commercial viability. Furthermore, regulatory standards for food contact materials necessitate rigorous testing and validation, adding to the complexity of bringing bio-nanocomposite-based packaging to market (Chincholikar et al., 2023). Addressing these challenges requires a multidisciplinary approach, combining advances in materials science, nanotechnology, and manufacturing processes.

This chapter aims to provide a comprehensive overview of bio-nanocomposites for sustainable food packaging, focusing on their synthesis, properties, and applications. It explores the various types of biopolymers and nanomaterials used in these composites, detailing their unique contributions to the material's overall performance. Additionally, the chapter discusses the synthesis methods employed to create bio-nanocomposites, from solution casting to advanced techniques like electrospinning and in situ polymerization. Practical applications in active, intelligent, and edible packaging are highlighted, demonstrating the versatility and potential of these materials. Finally, the chapter addresses the environmental and economic implications of bio-nanocomposites, offering insights into their challenges and future directions. By bridging the gap between innovation and application, this chapter seeks to underscore the transformative potential of bio-nanocomposites in driving sustainable solutions for the food packaging industry.

2. BIOPOLYMERS FOR FOOD PACKAGING

2.1 Overview of Biopolymers

Biopolymers are naturally occurring polymers derived from renewable resources, such as plants, animals, and microorganisms (Figure 1) (Pinaeva & Noskov, 2024). Their production involves raw materials like starch, cellulose, chitin, proteins, and microbial fermentation products, positioning them as environmentally friendly alternatives to conventional petrochemical-based plastics (Getahun et al., 2024). Polysaccharides, such as starch, cellulose, and chitosan, represent a major category of biopolymers widely used in food packaging. Starch, a readily available and cost-effective material, can form films with acceptable barrier properties against oxygen but struggles with moisture sensitivity (Alibekov et al., 2024). Cellulose, derived from plant cell walls, exhibits high tensile strength and good film-forming capability, while chitosan, extracted from crustacean shells, offers antimicrobial properties, making it particularly suitable for active packaging applications (Yosri et al., 2024).

Figure 1. Types of Biopolymers

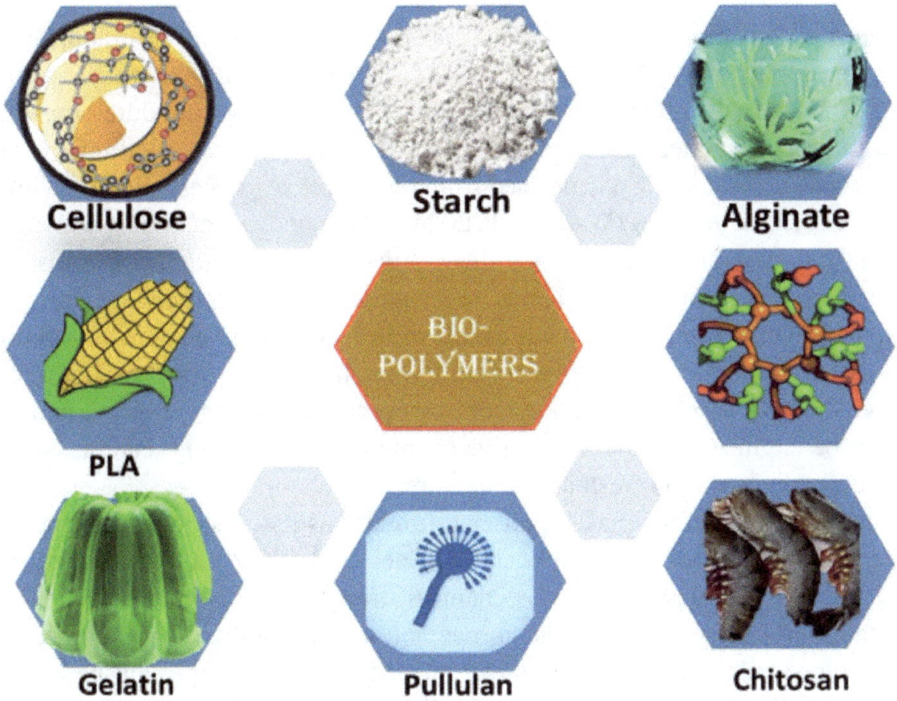

Proteins, such as whey protein and soy protein, are another prominent group of biopolymers. These materials are highly valued for their film-forming ability and biodegradability (Roy et al., 2024). Whey protein, a byproduct of the dairy industry, forms transparent and flexible films with excellent oxygen barrier properties, whereas soy protein-based films offer strong mechanical performance and compatibility with other additives (Mohammadi et al., 2025). Biodegradable polyesters, including polylactic acid (PLA) and polyhydroxyalkanoates (PHAs), constitute the third major group of biopolymers. PLA, derived from corn or sugarcane, is a widely used biopolymer due to its versatility and commercial availability. It offers good transparency and processability, making it ideal for applications such as food trays and containers (Khouri et al., 2024). PHAs, produced by bacterial fermentation, are compostable and demonstrate enhanced thermal and mechanical properties. These diverse biopolymers cater to a wide array of food packaging needs, contributing to the shift toward sustainable materials in the packaging industry (Yeo et al., 2024).

2.2 Advantages and Limitations

Biopolymers offer numerous advantages that make them highly desirable for food packaging applications (Dey et al., 2024). Their biodegradability and compostability significantly reduce the environmental impact of packaging waste, addressing one of the most pressing concerns associated with traditional plastics. Additionally, most biopolymers are non-toxic and derived from renewable resources, aligning with the principles of sustainability (Wang et al., 2024). Many of these materials also exhibit inherent functionalities that enhance their utility in food packaging. For instance, chitosan possesses antimicrobial properties, while PLA offers excellent clarity and stiffness. Proteins like whey and soy can form films with high oxygen barrier properties, which are essential for preserving the freshness of oxygen-sensitive foods (Pei et al., 2024). Moreover, biopolymers are compatible with food-grade additives and can be engineered to include active or intelligent packaging features, such as antimicrobial agents and freshness indicators.

However, biopolymers face significant limitations that restrict their standalone use in demanding packaging applications (Sani et al., 2024). One of the primary challenges is their relatively poor mechanical strength, which limits their ability to withstand physical stresses during storage and transportation. For example, films made from starch or cellulose often exhibit brittleness, while protein-based films are susceptible to moisture absorption, leading to compromised structural integrity (Thivya et al., 2024). Barrier properties, particularly against water vapor and gases, are another critical concern. Most biopolymers lack the ability to provide effective moisture barriers, making them unsuitable for packaging high-moisture or highly perishable foods. Thermal stability is also a limitation, as many biopolymers degrade at relatively low temperatures, complicating their processing and limiting their applicability in high-temperature environments (Wronka & Kowaluk, 2024).

To address these challenges, researchers have turned to the development of bio-nanocomposites, which integrate biopolymers with nanoscale fillers to enhance their properties (Sen et al., 2024). The incorporation of nanomaterials, such as nanoclays, cellulose nanocrystals, and carbon-based nanostructures, significantly improves the mechanical, thermal, and barrier properties of biopolymers (Priyadarshi et al., 2024). For instance, the high aspect ratio of nanoclays forms a tortuous path for gas molecules, enhancing oxygen and moisture barrier properties. Similarly, cellulose nanocrystals increase the tensile strength and modulus of biopolymer matrices, while metal oxide nanoparticles impart antimicrobial properties, extending the shelf life of packaged foods. By leveraging the synergies between biopolymers and nanomaterials, bio-nanocomposites overcome the inherent limitations of biopolymers, offering a sustainable yet high-performance solution for modern food packaging needs (Acharyya et al., 2024). This innovation not only addresses functional challenges but also aligns with global efforts to reduce environmental impact, making bio-nanocomposites a cornerstone of sustainable packaging technologies.

3. NANOMATERIALS IN BIONANOCOMPOSITE

3.1 Types of Nanomaterials

Nanomaterials play a pivotal role in enhancing the properties of bio-nanocomposites for food packaging. Their nanoscale dimensions and high surface area provide unique physical, chemical, and mechanical properties that significantly augment the performance of biopolymers (Rahman et al., 2024). Among the most widely used nanomaterials are nanoclays, cellulose nanocrystals, carbon-based nanomaterials, metal oxide nanoparticles, and natural nanoparticles (Karwatkar et al., 2025). Each of these categories contributes specific benefits, addressing the challenges associated with conventional biopolymers while maintaining environmental sustainability.

3.1.1 Nanoclays

Nanoclays are one of the earliest and most extensively studied nanomaterials for bio-nanocomposites (Korumilli et al., 2024). These layered silicates, primarily montmorillonite, possess a high aspect ratio and excellent dispersion characteristics. When incorporated into biopolymers, nanoclays create a tortuous path for gas molecules, significantly enhancing barrier properties against oxygen, carbon dioxide, and moisture (Sheibani et al., 2025). This feature is particularly critical for food packaging, where prolonged shelf life and protection against spoilage are essential. Moreover, nanoclays improve the mechanical properties of the composite, such as tensile strength and modulus, making the packaging more durable. The compatibility of nanoclays with various biopolymers, including PLA and chitosan, has made them a versatile choice in the development of advanced bio-nanocomposites.

3.1.2 Cellulose Nanocrystals (CNCs)

Cellulose nanocrystals (CNCs), derived from renewable sources such as wood, cotton, and agricultural residues, are a sustainable choice for bio-nanocomposites (Sinthiya et al., 2024). CNCs exhibit exceptional mechanical strength, stiffness, and thermal stability due to their highly crystalline structure. When added to biopolymers, they significantly enhance tensile strength and thermal resistance, addressing the inherent weaknesses of many biopolymers (Monfared et al., 2024). CNCs also contribute to improved barrier properties by creating a dense network within the polymer matrix, limiting the permeability of gases and moisture. Their biodegradable nature aligns with the overarching goal of sustainability, making CNCs an ideal reinforcement material for eco-friendly food packaging.

3.1.3 Carbon-Based Nanomaterials

Carbon-based nanomaterials, including graphene, graphene oxide, and carbon nanotubes (CNTs), offer advanced functionalities that extend the applications of bio-nanocomposites (Kumar et al., 2024). These nanomaterials are known for their exceptional electrical conductivity, mechanical strength, and thermal stability. Graphene and its derivatives, for instance, can enhance the structural integrity of biopolymer matrices and provide unique barrier properties (Pandey et al., 2025. CNTs, on the other hand, improve thermal conductivity and mechanical properties while imparting electrical conductivity, enabling the development of intelligent packaging systems. Although their application in food packaging is still in the early stages due to concerns about safety and scalability, carbon-based nanomaterials hold immense potential for creating multifunctional bio-nanocomposites.

3.1.4 Metal Oxide Nanoparticles

Metal oxide nanoparticles, such as zinc oxide (ZnO), titanium dioxide (TiO2), and silver oxide (Ag2O), are widely used in bio-nanocomposites for their antimicrobial properties (Nandhini et al., 2024). These nanoparticles inhibit the growth of bacteria, fungi, and other microorganisms, thereby extending the shelf life of packaged foods. ZnO and TiO2 are particularly popular due to their broad-spectrum antimicrobial activity, chemical stability, and low toxicity. Additionally, metal oxide nanoparticles enhance the mechanical and thermal properties of the composite. Their ability to provide UV protection is another advantage, as it prevents the degradation of food and packaging materials caused by ultraviolet radiation (Zhang & Jiang, 2023).

3.1.5 Natural Nanoparticles

Natural nanoparticles derived from bio-sources, such as chitin, starch, and alginate, represent a sustainable alternative to synthetic nanomaterials (Lizundia et al., 2022). Chitin nanofibers, extracted from crustacean shells, exhibit excellent mechanical properties, biodegradability, and antimicrobial activity, making them suitable for active food packaging. Starch nanoparticles, derived from sources like corn and potato, are inexpensive and biodegradable, offering good compatibility with biopolymers (Singh et al., 2024). These natural nanoparticles are often used to enhance the film-forming ability, strength, and barrier properties of biopolymer matrices. Additionally, their renewable origin and minimal environmental footprint align with the goals of green chemistry and sustainability. The diverse range of nanomaterials, from nanoclays to natural nanoparticles, enables the customization of bio-nanocomposites to meet specific requirements in food packaging (Khane et al., 2024). As research progresses, the focus will likely shift toward improving the safety, scalability, and cost-effectiveness of these nanomaterials, ensuring their broader adoption in commercial applications.

3.2 Properties of Nanomaterials

Nanomaterials have emerged as transformative elements in the development of bio-nanocomposites, imparting an array of enhanced properties that address the limitations of traditional biopolymers (Kumar et al., 2023). Their nanoscale dimensions, coupled with a high surface-area-to-volume ratio, allow them to interact extensively with the polymer matrix, resulting in composites with superior functional attributes. Key properties (Figure 2) conferred by nanomaterials include enhanced mechanical strength, improved barrier properties, antimicrobial activity, and thermal stability, each of which plays a crucial role in elevating the performance of food packaging materials.

Figure 2. Properties of Nanomaterials

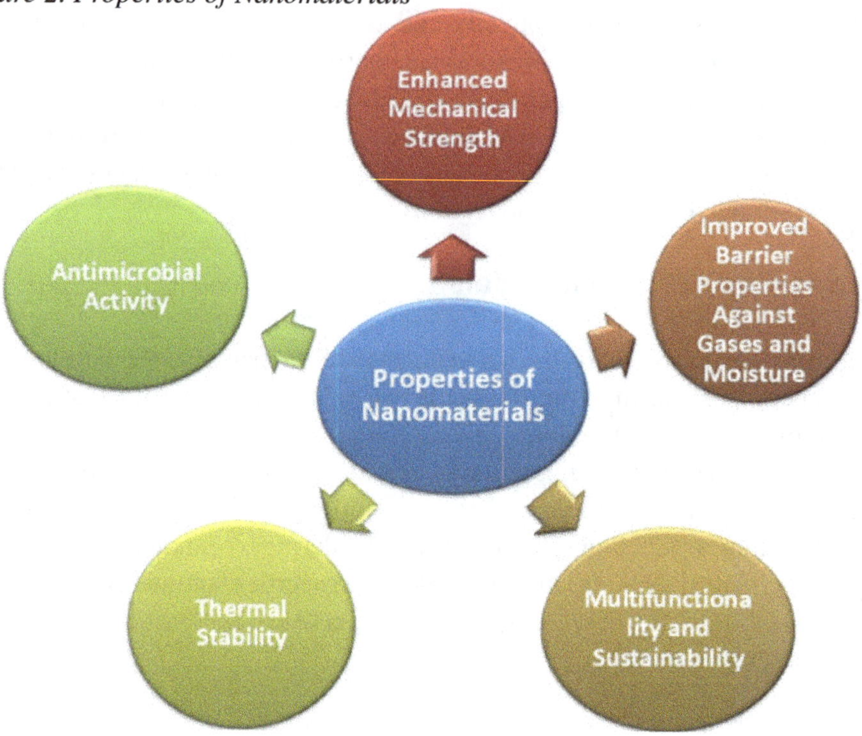

3.2.1 Enhanced Mechanical Strength

One of the most significant contributions of nanomaterials to bio-nanocomposites is the improvement in mechanical properties, including tensile strength, modulus, and flexibility (Olayil et al., 2022). Biopolymers like polylactic acid (PLA) and starch often lack the mechanical robustness required for demanding packaging applications. When nanomaterials such as cellulose nanocrystals (CNCs), nanoclays, or graphene oxide (GO) are incorporated, they reinforce the polymer matrix by creating a network of strong interactions (Wang et al., 2023). For instance, nanoclays exhibit a high aspect ratio and layer alignment, which distribute stress efficiently and prevent deformation under load. Similarly, CNCs and GO establish hydrogen bonds or covalent linkages with biopolymers, enhancing the stiffness and strength of the composite (Mavai et al., 2024). This mechanical enhancement ensures that bio-nanocomposites can withstand handling, transportation, and storage stresses without compromising food safety or integrity.

3.1.2 Improved Barrier Properties Against Gases and Moisture

Barrier properties are critical in food packaging to maintain freshness and prevent spoilage caused by exposure to oxygen, carbon dioxide, and moisture (Shah et al., 2024). Biopolymers, while biodegradable, often exhibit high permeability to gases and water vapor, limiting their effectiveness in preserving food quality. Nanomaterials such as nanoclays and CNCs significantly improve these barrier properties by forming a tortuous path within the polymer matrix, which slows the diffusion of molecules. Nanoclays, for example, create a layered structure that obstructs the passage of gas molecules, reducing permeability. CNCs, with their dense crystalline domains, restrict moisture transport, further enhancing the protective capabilities of the composite (de Lima et al., 2025). Additionally, graphene-based nanomaterials contribute to exceptional gas barrier properties due to their impermeable nature and high aspect ratio. These enhancements make bio-nanocomposites viable alternatives to conventional plastics in food packaging applications.

3.1.3 Antimicrobial Activity

The incorporation of antimicrobial nanomaterials into bio-nanocomposites adds a functional dimension to food packaging by preventing microbial contamination and spoilage (Jamwal & Mittal, 2024). Metal oxide nanoparticles such as zinc oxide (ZnO), titanium dioxide (TiO2), and silver nanoparticles are particularly effective in this regard. These nanomaterials exhibit broad-spectrum antimicrobial activity through mechanisms such as reactive oxygen species (ROS) generation, disruption of microbial membranes, and interaction with cellular components (Mondal et al., 2024). For example, silver nanoparticles are known for their potent antimicrobial efficacy against bacteria, fungi, and viruses, making them valuable in active packaging systems. Similarly, ZnO and TiO2 nanoparticles not only provide antimicrobial protection but also enhance UV resistance, preventing photodegradation of both the food and the packaging material. This dual functionality of antimicrobial activity and UV shielding ensures prolonged shelf life and food safety.

3.1.4 Thermal Stability

Thermal stability is another critical property that nanomaterials impart to bio-nanocomposites, enabling them to withstand processing and environmental conditions (Chatterjee & Dhar, 2025). Many biopolymers, such as polylactic acid and polyhydroxyalkanoates, exhibit limited thermal resistance, which can compromise their structural integrity during high-temperature processes like extrusion or thermal sealing (Lalonde et al., 2025). The addition of nanomaterials such as CNCs, nanoclays, and carbon-based nanomaterials enhances the thermal stability of these polymers by improving their crystallinity and reducing the mobility of polymer chains. CNCs, for instance, act as nucleating agents, promoting the formation of highly ordered crystalline regions within the polymer matrix (Ma et al., 2024). This increased crystallinity improves the thermal resistance and reduces the thermal expansion of the composite. Similarly, graphene-based nanomaterials, with their excellent thermal conductivity, dissipate heat efficiently, preventing thermal degradation. Enhanced thermal stability ensures that bio-nanocomposites remain functional and durable across a wide range of temperature conditions.

3.1.5 Multifunctionality and Sustainability

Beyond individual properties, nanomaterials contribute to the multifunctionality and sustainability of bio-nanocomposites, making them ideal for modern food packaging needs (Gorde et al., 2024). The combination of enhanced mechanical, barrier, antimicrobial, and thermal properties in a single material reduces the need for multi-layered or laminated packaging, simplifying recycling and reducing material consumption (Ahmed et al., 2024). Furthermore, the use of bio-based nanomaterials like CNCs and chitin nanofibers aligns with the principles of green chemistry, offering renewable and biodegradable solutions. These natural nanomaterials not only enhance the performance of bio-nanocomposites but also minimize their environmental footprint, addressing the growing demand for sustainable packaging alternatives (Nath et al., 2024). As research advances, the integration of bio-nanocomposites with smart technologies, such as sensors and indicators, will further expand their applications, creating intelligent packaging systems capable of monitoring food quality and safety in real time (Siciliano et al., 2024). Enhanced mechanical strength, improved barrier properties, antimicrobial activity, and thermal stability ensure that bio-nanocomposites meet the stringent requirements of the food packaging industry while adhering to sustainability goals (Khalid & Arif, 2022). Continued innovation in nanomaterial synthesis, functionalization, and incorporation techniques will further elevate the potential of bio-nanocomposites, paving the way for a new era of eco-friendly and efficient food packaging solutions.

4. SYNTHESIS OF BIO-NANOCOMPOSITES

The synthesis of bio-nanocomposites is a critical step that determines their structure, properties, and suitability for various applications (Karki et al., 2021). The choice of synthesis technique influences the dispersion of nanomaterials, interfacial interactions, and overall material performance. This section explores the most common synthesis methods and highlights the challenges associated with scaling and optimizing these processes.

4.1 Techniques

Bio-nanocomposites are synthesized using various techniques (Figure 3) that aim to integrate nanomaterials uniformly within a polymer matrix (Perera et al., 2022). The selection of a specific method depends on the desired properties, the compatibility between components, and the end-use applications. Below are the primary techniques employed in the synthesis of bio-nanocomposites:

Figure 3. Synthesis of Bio-nanocomposites

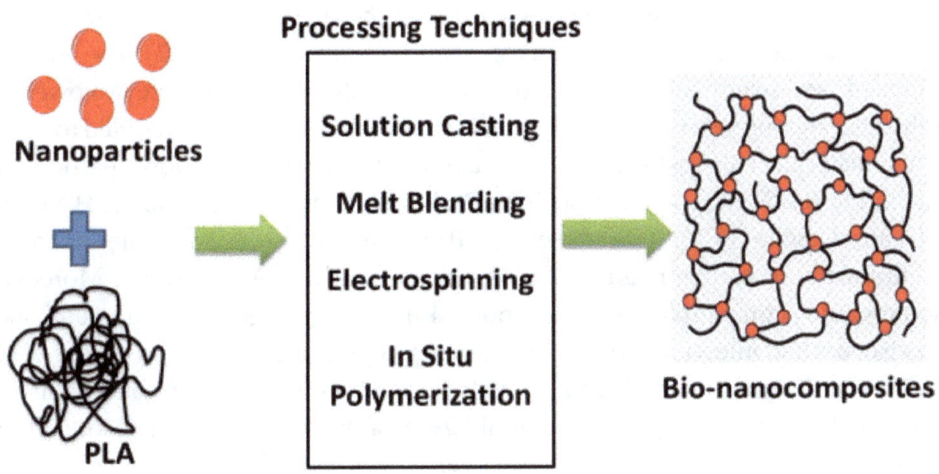

4.1.1 Solution Casting

Solution casting is a straightforward and widely adopted method for synthesizing bio-nanocomposites (Das et al., 2021). In this technique, the nanomaterials are first dispersed into a suitable solvent to form a stable suspension. The biopolymer is then dissolved in the same or a compatible solvent, and the two are mixed to create a homogeneous solution. The mixture is poured onto a flat surface, and the solvent is allowed to evaporate, leaving behind a thin film of bio-nanocomposite. This method is particularly advantageous for creating thin, flexible films with controlled nanomaterial dispersion (Assis et al., 2025). However, achieving a uniform dispersion of nanomaterials requires careful optimization of parameters such as solvent type, mixing time, and evaporation conditions. Common examples of solution-cast bio-nanocomposites include cellulose nanocrystals (CNC)-starch films and nanoclay-polylactic acid (PLA) composites. Despite its simplicity, solution casting is limited by its scalability and the use of organic solvents, which may pose environmental concerns (Norizan et al., 2022).

4.1.2 Melt Blending

Melt blending is a scalable and solvent-free technique that involves mixing nanomaterials and polymers at elevated temperatures (Ahmed et al., 2022). The polymer is melted, and nanomaterials are introduced into the molten matrix, followed by high-shear mixing to ensure even dispersion. The composite is then cooled to solidify the material. This method is well-suited for industrial applications, particularly for thermoplastic biopolymers such as PLA, polyhydroxyalkanoates (PHA), and polycaprolactone (PCL) (Lalonde et al., 2025). Melt blending offers the advantage of eliminating solvents, making it an environmentally friendly option. Moreover, the process is compatible with conventional polymer processing equipment, such as extruders and injection molders, facilitating large-scale production (Sadaf et al., 2024). Challenges in melt blending include maintaining the thermal stability of biopolymers and avoiding nanomaterial aggregation during mixing (Trotta et al., 2023). Additionally, achieving strong interfacial adhesion between hydrophobic polymers and hydrophilic nanomaterials often requires surface modification of the nanomaterials or the addition of compatibilizers (Dorigato & Fredi, 2024).

4.1.3 **Electrospinning**

Electrospinning is a versatile technique used to produce nanofiber-based bio-nanocomposites with exceptionally high surface area-to-volume ratios (Alosaimi et al., 2023). In this method, a polymer solution containing dispersed nanomaterials is subjected to a high-voltage electric field. The electrostatic forces stretch the polymer solution into fine fibers, which are collected on a grounded substrate to form a non-woven mat. Electrospinning is particularly effective for incorporating nanomaterials such as cellulose nanocrystals, carbon nanotubes, and metal oxide nanoparticles into biopolymer matrices (Kandhola et al., 2023). The resulting nanofiber composites exhibit enhanced mechanical, thermal, and barrier properties, making them ideal for food packaging applications (Min et al., 2022). Despite its advantages, electrospinning faces limitations in terms of scalability and production speed (Cimini et al., 2023). Furthermore, controlling the uniformity of fiber morphology and the dispersion of nanomaterials within the fibers requires precise optimization of solution properties and processing parameters (Ji et al., 2024).

4.1.4 In Situ Polymerization

In situ polymerization involves incorporating nanomaterials into the polymerization process, ensuring their uniform dispersion within the polymer matrix (Lages & Nicolas, 2023). In this method, nanomaterials are first dispersed in the monomer solution, and polymerization is initiated by adding a catalyst, heat, or light. As the polymer chains form and grow, they encapsulate the nanomaterials, creating a well-integrated bio-nanocomposite. This technique is particularly useful for achieving strong interfacial bonding between nanomaterials and the polymer matrix. For example, graphene oxide, nanoclays, or metal oxide nanoparticles can be functionalized to interact chemically with the monomers, enhancing the structural and functional properties of the composite (Das et al., 2025). While in situ polymerization offers excellent control over nanomaterial dispersion and polymer morphology, it is often more time-consuming and complex compared to other techniques (Shahid et al., 2025). The use of specific catalysts or initiators may also add to the cost and environmental impact of the process.

4.2 Challenges in Synthesis

While the synthesis of bio-nanocomposites holds immense promise, several challenges must be addressed to realize their full potential and enable large-scale production (Muhammad et al., 2025). Uniform dispersion of nanomaterials within the polymer matrix is essential for achieving optimal mechanical, thermal, and barrier properties (Manikandan et al., 2025). However, nanomaterials tend to aggregate due to their high surface energy and van der Waals forces, particularly at higher loadings. Overcoming this issue requires effective dispersion techniques, such as ultrasonication, high-shear mixing, or the use of surfactants and compatibilizers. The compatibility between nanomaterials and biopolymers is crucial for ensuring strong interfacial interactions. Many biopolymers are hydrophilic, while certain nanomaterials, such as graphene or carbon nanotubes, are hydrophobic (Thapliyal et al., 2025). Surface modification of nanomaterials, such as functionalization with hydroxyl or carboxyl groups, can improve compatibility and enhance composite performance. The scalability and cost-effectiveness of bio-nanocomposite synthesis remain significant hurdles (Hosseini et al., 2025). Techniques like electrospinning and in situ polymerization, while effective, may not be economically viable for large-scale applications due to their complexity and high energy requirements (Zhang et al., 2023). Developing simplified, low-cost methods that maintain material performance is essential for widespread adoption. Addressing these challenges requires interdisciplinary collaboration and innovative approaches to optimize synthesis techniques, enhance material properties, and reduce production costs. As research advances, the development of bio-nanocomposites will continue to pave the way for sustainable and high-performance food packaging solutions.

5. PROPERTIES OF BIO-NANOCOMPOSITES

Bio-nanocomposites are engineered to exhibit superior properties that address the limitations of conventional biopolymers while maintaining their sustainability (Farooq et al., 2024). These properties, enhanced by the integration of nanoscale fillers, make bio-nanocomposites ideal for food packaging applications. Below, the mechanical, barrier, antimicrobial, and biodegradability characteristics of bio-nanocomposites are discussed in detail.

5.1 Mechanical, Barrier and antimicrobial

Bio-nanocomposites significantly enhance mechanical, barrier, and antimicrobial properties, addressing limitations of native biopolymers in food packaging applications. Nanomaterials such as cellulose nanocrystals (CNCs), nanoclays, and graphene improve tensile strength, elasticity, and durability by reinforcing the polymer matrix and distributing stress uniformly. These materials ensure bio-nanocomposites can withstand transportation and storage while meeting specific packaging needs. Enhanced barrier properties, crucial for preserving food freshness, are achieved by incorporating nanoclays and graphene, which create tortuous paths within the matrix to hinder gas and moisture permeability. This not only extends shelf life but also reduces packaging weight and environmental impact. Additionally, integrating nanoparticles like silver (Ag), zinc oxide (ZnO), and titanium dioxide (TiO_2) imparts antimicrobial properties, effectively inhibiting microbial growth and ensuring food safety. These materials reduce reliance on preservatives and can be tailored for specific applications, offering a sustainable solution for packaging perishable items like fresh produce, dairy, and meat.

5.2 Biodegradability

One of the most significant advantages of bio-nanocomposites is their ability to retain the biodegradability of the base biopolymer (Mishra et al., 2024). This ensures that the materials degrade naturally in the environment, addressing the pressing issue of plastic pollution. The incorporation of nanomaterials, when derived from natural sources or in small concentrations, does not significantly compromise the compostability of the composite. Biodegradability is influenced by factors such as the type of biopolymer, the dispersion of nanomaterials, and the environmental conditions during degradation (Samir et al., 2022). Polymers like polylactic acid (PLA) and polyhydroxyalkanoates (PHA) are known for their rapid degradation under composting conditions, while the addition of naturally derived nanomaterials like chitin and cellulose nanocrystals further enhances the eco-friendliness of the composite (Arif et al., 2022). The biodegradability of bio-nanocomposites ensures that these materials leave a minimal environmental footprint, aligning with global sustainability goals. This property makes them a viable alternative to conventional plastic packaging, contributing to the circular economy and reducing the burden of non-biodegradable waste on ecosystems.

6. APPLICATIONS IN FOOD PACKAGING

Bio-nanocomposites have revolutionized food packaging by offering advanced functionalities beyond mere containment 9Figure 4) (Gorde et al., 2024). These materials enhance food preservation, ensure quality, and align with sustainability goals (Jamwal & Mittal, 2024). This section discusses the key applications of bio-nanocomposites in food packaging, focusing on active packaging, intelligent packaging, and edible packaging.

6.1 Active Packaging

Active packaging is an innovative approach that interacts with the food product or its surrounding environment to extend shelf life and enhance food safety (Firouz et al., 2021). Bio-nanocomposites play a pivotal role in active packaging by incorporating functional nanomaterials that offer antimicrobial, antioxidant, or oxygen-scavenging properties (Perera et al., 2022). Antimicrobial nanoparticles, such as silver (Ag), zinc oxide (ZnO), and titanium dioxide (TiO_2), integrated into bio-nanocomposites, actively inhibit microbial growth on food surfaces. This reduces the risk of spoilage and contamination, especially for perishable items like fresh produce, dairy, and meat. Additionally, the antioxidant properties of bio-nanocomposites, achieved by incorporating natural antioxidants like polyphenols or nanoparticles with catalytic activity, help delay oxidation processes, preserving food quality and flavor (Pires et al., 2023). Active packaging solutions using bio-nanocomposites not only improve food safety but also reduce reliance on chemical preservatives. For example, incorporating ZnO nanoparticles into a polylactic acid (PLA)-based bio-nanocomposite can actively protect packaged foods from microbial and oxidative deterioration. Such systems are particularly beneficial for organic food products where artificial additives are avoided.

Figure 4. Bio-nanocomposites Applications in Food Packaging

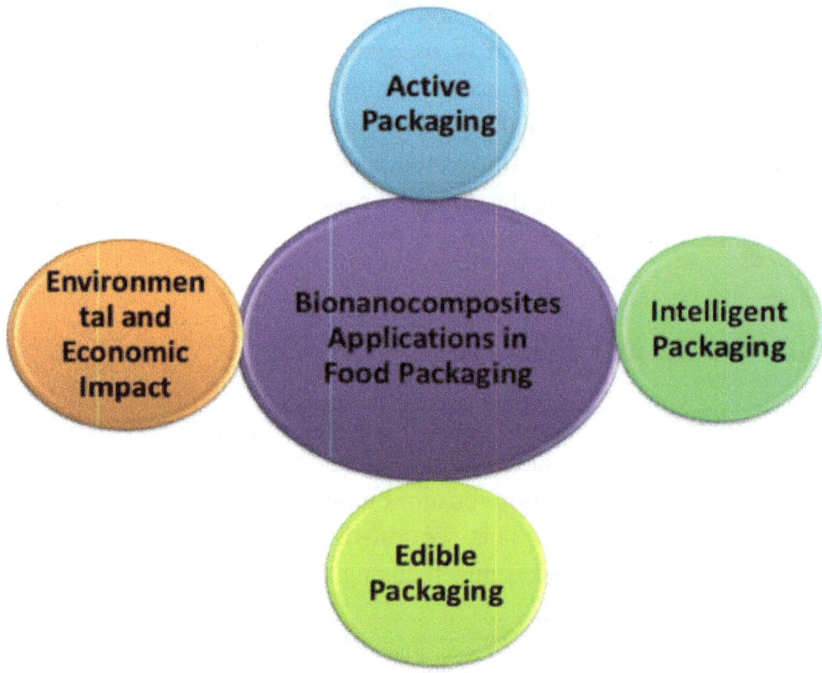

6.2 Intelligent Packaging

Intelligent packaging goes beyond protection by providing real-time monitoring of food quality and safety through embedded sensors or indicators (Dodero et al., 2021). Bio-nanocomposites serve as an excellent platform for developing intelligent packaging systems due to their ability to incorporate functional nanomaterials without compromising biodegradability (Kumar et al., 2023). Colorimetric indicators, often based on pH-sensitive materials, can be integrated into bio-nanocomposites to detect spoilage. For instance, bio-nanocomposites containing natural dyes or pH-responsive nanoparticles change color when exposed to gases like ammonia, which are produced during food spoilage (Raghuvanshi et al., 2023). Similarly, bio-nanocomposites embedded with oxygen-sensitive indicators can alert consumers to compromised packaging. Moreover, biosensors integrated into bio-nanocomposites can detect specific pathogens or toxins, providing an added layer of safety. Such systems are particularly valuable for monitoring high-risk foods like seafood or poultry (Kumar et al., 2022). Intelligent bio-nanocomposite packaging not only enhances consumer trust but also minimizes food waste by providing accurate information about food freshness.

6.3 Edible Packaging

Edible packaging is a unique application of bio-nanocomposites that addresses both food preservation and sustainability (Wani et al., 2023). Made from biopolymers like starch, chitosan, or gelatin, edible bio-nanocomposites can serve as an additional protective layer that is safe for consumption. Nanomaterials incorporated into edible films improve their mechanical strength, barrier properties, and functionality. For example, chitosan-based edible films enhanced with cellulose nanocrystals (CNCs) exhibit superior flexibility and resistance to moisture, making them ideal for wrapping fresh produce (Du et al., 2023). Additionally, edible films with embedded antimicrobial agents like silver nanoparticles can actively inhibit bacterial growth, ensuring food safety. The use of edible bio-nanocomposites is particularly relevant for single-use packaging applications, such as wrapping fruits or confectioneries (Verma et al., 2024). These materials reduce plastic waste and provide a biodegradable alternative that aligns with environmental sustainability goals. Moreover, edible packaging can be enriched with flavors, nutrients, or bioactive compounds, adding value to the packaged food (Figure 5).

Figure 5. Schematic representation of the packaging of food with UV protector, aroma and flavors protector

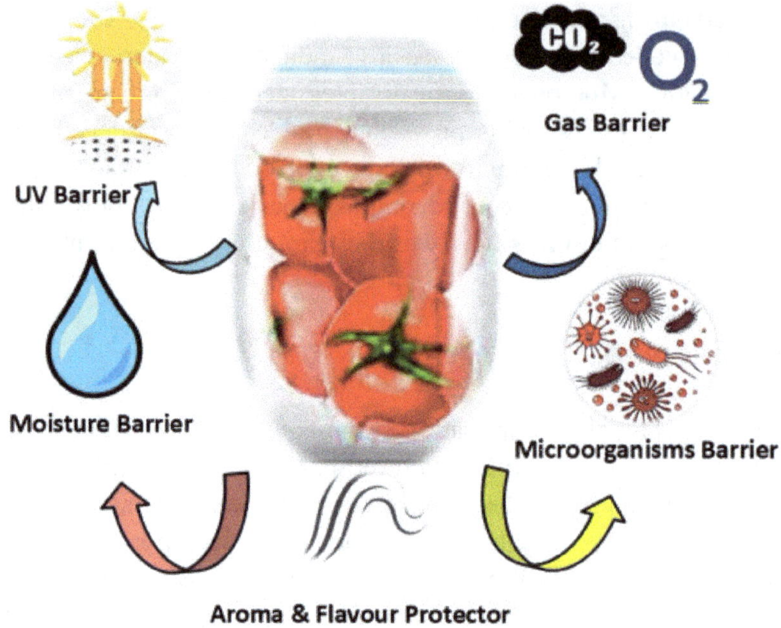

7. ENVIRONMENTAL AND ECONOMIC IMPACT

The widespread adoption of bio-nanocomposites in food packaging is driven by their dual promise of environmental sustainability and economic viability (Bharathi & Jayas, 2024). This section explores their environmental benefits and the economic considerations influencing their development and market adoption.

7.1 Environmental Benefits

The environmental benefits of bio-nanocomposites lie in their potential to mitigate the ecological issues associated with conventional plastic packaging (Buntinx et al., 2024). A significant advantage is their ability to reduce plastic waste, which accounts for a substantial portion of global pollution. Bio-nanocomposites, derived from renewable resources like plant-based polymers and natural nanomaterials, present a biodegradable alternative that breaks down into non-toxic byproducts under appropriate conditions, such as industrial composting (Gulati et al., 2024). These materials also contribute to lowering the carbon footprint of packaging. The production of biopolymers typically involves less energy and emits fewer greenhouse gases compared to petroleum-based plastics. Additionally, many bio-nanocomposites incorporate nanomaterials derived from agricultural or industrial waste, further promoting a circular economy by repurposing waste into high-value materials (Dadhwal et al., 2025). Compostability is another significant environmental advantage. Bio-nanocomposites that can degrade into nutrient-rich compost provide a sustainable waste management solution, especially for food packaging, which often becomes contaminated and is unsuitable for recycling (Sivakanthan & Kaumadi, 2022). For example, polylactic acid (PLA) films with embedded cellulose nanocrystals (CNCs) can decompose in industrial composting facilities, returning organic matter to the soil without leaving harmful residues (Naskar et al., 2023). Moreover, the reduced reliance on petroleum-based raw materials helps conserve non-renewable resources. By integrating natural nanomaterials such as chitosan or starch nanoparticles, bio-nanocomposites contribute to a more sustainable material economy, aligning with global efforts to achieve environmental goals such as reducing ocean pollution and carbon emissions.

7.2 Economic Considerations

While bio-nanocomposites offer clear environmental advantages, their economic feasibility plays a crucial role in determining their adoption in the market (Popescu et al., 2024). One of the primary challenges is the cost of raw materials and nanomaterials. Biopolymers, though renewable, often cost more to produce than conventional plastics due to limited economies of scale (Rech & Daugaard, 2025). Similarly, synthesizing nanomaterials, especially those requiring high purity or complex processing, can add to production costs. The scalability of synthesis techniques is another critical factor. Methods such as in situ polymerization or electrospinning, while effective in laboratory settings, may face challenges in scaling up to industrial levels. Developing cost-effective and efficient production methods that ensure uniform dispersion of nanomaterials while maintaining the desired properties of bio-nanocomposites is essential for economic viability (Ikram et al., 2021). Market adoption also depends on consumer acceptance and regulatory frameworks. While consumers are increasingly inclined toward sustainable products, higher costs can be a deterrent. Educating the public about the long-term benefits of bio-nanocomposites, such as reduced waste management expenses and environmental preservation, can drive demand (Buntinx et al., 2024). At the same time, governments and policymakers play a pivotal role by providing subsidies, tax incentives, and regulations that encourage the use of eco-friendly materials in packaging (Thapliyal et al., 2024). Another economic consideration is the potential for innovation-driven value addition. Bio-nanocomposites with advanced functionalities, such as antimicrobial properties or real-time food quality monitoring, can command a premium price in niche markets. Such innovations not only offset the initial production costs but also create opportunities for companies to differentiate themselves in competitive markets.

8. CHALLENGES AND FUTURE PERSPECTIVES

The development and implementation of bio-nanocomposites for sustainable food packaging have shown significant promise. However, several challenges must be addressed to fully realize their potential. Concurrently, emerging research and technological advancements pave the way for future innovations that can revolutionize the industry.

8.1 Challenges

One of the primary challenges in the adoption of bio-nanocomposites is ensuring food safety and regulatory compliance. For materials to be used in food packaging, they must meet stringent regulatory standards set by organizations like the FDA and EFSA. The incorporation of nanomaterials adds complexity, as their unique properties can sometimes raise concerns regarding migration into food products and potential health impacts. Establishing comprehensive safety assessments and standardized testing protocols is essential to address these concerns. Scaling up production processes poses another significant hurdle. While methods such as solution casting and electrospinning are effective at the laboratory scale, translating these processes to industrial levels requires overcoming technical and economic barriers. Achieving uniform dispersion of nanomaterials within biopolymer matrices on a large scale is challenging and often results in inconsistencies in product quality. Furthermore, the high costs of nanomaterial synthesis and the energy-intensive nature of some manufacturing processes can limit their feasibility for mass production. Balancing performance and cost-effectiveness is a critical challenge in the commercialization of bio-nanocomposites. Although these materials offer superior properties compared to traditional biopolymers, the added costs associated with nanomaterials and advanced processing methods can make them less competitive in price-sensitive markets. Developing cost-efficient synthesis techniques and optimizing material formulations to deliver high performance at a reasonable cost are essential for their widespread adoption.

8.2 Future Directions

The future of bio-nanocomposites lies in the development of multifunctional materials that go beyond conventional packaging roles. Researchers are exploring bio-nanocomposites that integrate advanced functionalities such as antimicrobial and antioxidant properties, real-time food quality monitoring, and active response to environmental changes. For instance, intelligent packaging systems embedded with sensors can provide consumers with real-time information about food freshness, thereby reducing food waste. Exploration of renewable and bio-derived nanomaterials is another promising avenue. Using nanomaterials derived from agricultural or industrial waste, such as cellulose nanocrystals from plant fibers or chitin nanoparticles from seafood shells, aligns with the principles of circular economy and sustainability. These renewable nanomaterials not only reduce reliance on fossil resources but also add value to waste streams, creating opportunities for more eco-friendly and cost-effective solutions. Advanced manufacturing technologies are expected to play a pivotal role in enabling large-scale production of bio-nanocomposites. Innovations such as additive manufacturing, roll-to-roll processing, and green chemistry approaches can enhance the scalability and sustainability of production processes. Moreover, advancements in dispersion techniques and surface modification of nanomaterials can improve their compatibility with biopolymers, ensuring consistent quality in industrial applications. Collaborative efforts among academia, industry, and policymakers will be crucial in driving progress in this field. Increased funding for research and development, supportive regulatory frameworks, and public awareness campaigns can accelerate the adoption of bio-nanocomposites. By addressing current challenges and leveraging future opportunities, bio-nanocomposites have the potential to transform the food packaging industry into a more sustainable and innovative sector.

9. CONCLUSION

Bio-nanocomposites have emerged as a transformative solution for sustainable food packaging, addressing the dual challenges of environmental sustainability and functional performance. By integrating biodegradable biopolymers with nanomaterials, these composites achieve a remarkable synergy that enhances mechanical strength, thermal stability, barrier properties, and even antimicrobial functionality. This unique combination positions bio-nanocomposites as a next-generation packaging material capable of preserving food quality, extending shelf life, and reducing the environmental footprint of packaging waste. Despite their immense potential, several challenges must be addressed for their widespread adoption. Regulatory compliance remains a critical area of focus, particularly with concerns about the safety of nanomaterials in food contact applications. Scaling up production to industrial levels while maintaining uniform quality and cost efficiency is another significant hurdle. Moreover, consumer acceptance and market competitiveness require balancing high-performance attributes with affordability. However, ongoing advancements in material science, manufacturing technologies, and regulatory frameworks hold promise in overcoming these barriers. The future of bio-nanocomposites lies in their ability to evolve into multifunctional materials, offering active and intelligent packaging solutions that meet the dynamic demands of modern food systems. The integration of renewable nanomaterials and innovations in green manufacturing will further enhance their sustainability and reduce reliance on non-renewable resources. Collaborative efforts among researchers, industry stakeholders, and policymakers will be instrumental in driving these developments forward.

ACKNOWLEDGMENTS

The author, Hari Shankar Biswas, sincerely acknowledges the financial support provided by the UGC Minor Project (Reference No. F.PSW-140/15-16). Heartfelt gratitude is extended to Tripurari Singha of the Department of Physics, Bose Institute, along with Dr. Suchandra Chatterjee and Dr. Lalita Das from the Department of Chemistry, Surendranath College, for their invaluable technical assistance and insightful guidance throughout the preparation of this manuscript.

REFERENCES

Abady, M. M., Shawky, A. M., Sakr, F. A., Mohammed, D. M., & Goda, E. S. (2024). Recent Advancements in Biosensors Using Biopolymers. In *Bio-Based Polymers: Farm to Industry. Volume 2: Current Trends and Applications* (pp. 81-112). American Chemical Society.

Acharyya, P. P., Sarma, M., & Kashyap, A. (2024). Recent advances in synthesis and bioengineering of bacterial nanocellulose composite films for green, active and intelligent food packaging. *Cellulose (London, England)*, *31*(12), 7163–7187.

Ahmed, M. J., Ashfaq, J., Sohail, Z., Channa, I. A., Sánchez-Ferrer, A., Ali, S. N., & Chandio, A. D. (2024). Lignocellulosic bioplastics in sustainable packaging–Recent developments in materials design and processing: A comprehensive review. *Sustainable. Materials Technology*, •••, 01077.

Ahmed, S. F., Mofijur, M., Rafa, N., Chowdhury, A. T., Chowdhury, S., Nahrin, M., & Ong, H. C. (2022). Green approaches in synthesising nanomaterials for environmental nanobioremediation: Technological advancements, applications, benefits and challenges. *Environmental Research*, *204*, 111967. PMID: 34450159

Alibekov, R. S., Urazbayeva, K. U., Azimov, A. M., Rozman, A. S., Hashim, N., & Maringgal, B. (2024). Advances in Biodegradable Food Packaging Using Wheat-Based Materials: Fabrications and Innovations, Applications, Potentials, and Challenges. *Foods*, *13*(18), 2964. PMID: 39335892

Alosaimi, A. M., Alorabi, R. O., Katowah, D. F., Al-Thagafi, Z. T., Alsolami, E. S., Hussein, M. A., & Rafatullah, M. (2023). Recent biomedical applications of coupling nanocomposite polymeric materials reinforced with variable carbon nanofillers. *Biomedicines*, *11*(3), 967. PMID: 36979948

Anwer, A. H., Ahtesham, A., Shoeb, M., Mashkoor, F., Ansari, M. Z., Zhu, S., & Jeong, C. (2023). State-of-the-art advances in nanocomposite and bio-nanocomposite polymeric materials: A comprehensive review. *Advances in Colloid and Interface Science*, •••, 102955. PMID: 37467558

Arif, Z. U., Khalid, M. Y., Sheikh, M. F., Zolfagharian, A., & Bodaghi, M. (2022). Biopolymeric sustainable materials and their emerging applications. *Journal of Environmental Chemical Engineering*, *10*(4), 108159.

Assis, D., Cardozo, O., Maia, R.Jr, Alves, S.Jr, de Melo, C. P., Justino, R., & Farias, P. M. (2025). Enhanced physical, electrical, and structural properties of chitosan thin films doped with Ag@ ZnO nanoparticles for energy storage applications. *Materials Chemistry and Physics*, *329*, 130003.

Aydın, A., Yüceer, M., Ulugergerli, E. U., & Caner, C. (2024). Improving food security as disaster relief using intermediate moisture foods and active packaging technologies. *Applied Food Research*, *4*(1), 100378.

Bharathi, V. S., & Jayas, D. S. (2024). Evolution of Bionanocomposites: Innovations and Applications in Food Packaging. *Foods*, *13*(23), 3787. PMID: 39682859

Buntinx, M., Vanheusden, C., & Hermans, D. (2024). Processing and Properties of Polyhydroxyalkanoate/ZnO Nanocomposites: A Review of Their Potential as Sustainable Packaging Materials. *Polymers*, *16*(21), 3061. PMID: 39518271

Buntinx, M., Vanheusden, C., & Hermans, D. (2024). Processing and Properties of Polyhydroxyalkanoate/ZnO Nanocomposites: A Review of Their Potential as Sustainable Packaging Materials. *Polymers*, *16*(21), 3061. PMID: 39518271

Chatterjee, N., & Dhar, P. (2025). Bio-nanocomposites: curse or miracle for the society? In *Characterization Techniques in Bionanocomposites* (pp. 3-39). Woodhead Publishing.

Chincholikar, P., Singh, K. R., Natarajan, A., Kerry, R. G., Singh, J., Malviya, J., & Singh, R. P. (2023). Green nanobiopolymers for ecological applications: A step towards a sustainable environment. *RSC Advances*, *13*(18), 12411–12429. PMID: 37091622

Cimini, A., Imperi, E., Picano, A., & Rossi, M. (2023). Electrospun nanofibers for medical face mask with protection capabilities against viruses: State of the art and perspective for industrial scale-up. *Applied Materials Today*, *32*, 101833. PMID: 37152683

D'Almeida, A. P., & de Albuquerque, T. L. (2024). Innovations in food packaging: From bio-based materials to smart packaging systems. *Processes (Basel, Switzerland), 12*(10), 2085.

Dadhwal, S., Ashton, L., Fraser, E. D., & Corradini, M. G. (2025). Life cycle assessment to evaluate bioplastics as alternatives to petroleum-based plastics in the food industry. In *Innovative Food Packaging and Processing Technologies* (pp. 449–473). Academic Press.

Dalbanjan, N. P., Eelager, M. P., Korgaonkar, K., Gonal, B. N., Kadapure, A. J., Arakera, S. B., & Kumar, S. P. (2024). Descriptive review on conversion of waste residues into valuable bionanocomposites for a circular bioeconomy. *Nano-Structures & Nano-Objects, 39*, 101265.

Das, N., Panda, S., Das, D. K., Nayak, S. K., Parthasarathy, S., & Patnaik, P. (2025). Polymer Nanocomposites and Nanostructured Polymers. In *Exploring Nanomaterial Synthesis, Characterization, and Applications* (pp. 213-240). IGI Global.

Das, S. K., Chakraborty, S., Naskar, S., & Rajabalaya, R. (2021). Techniques and methods used for the fabrication of bionanocomposites. In *Bionanocomposites in Tissue Engineering and Regenerative Medicine* (pp. 17–43). Woodhead Publishing.

de Lima, G. G., Zakaluk, I. C. B., Artner, M. A., Pedro, A. C., Gonzalez de Cademartori, P. H., Muniz, G. I. B. D., & Magalhães, W. L. E. (2025). Enhancing Barrier and Antioxidant Properties of Nanocellulose Films for Coatings and Active Packaging: A Review. *ACS Applied Nano Materials.*

del Rosario Herrera-Rivera, M., Torres-Arellanes, S. P., Cortés-Martínez, C. I., Navarro-Ibarra, D. C., Hernández-Sánchez, L., Solis-Pomar, F., & Román-Doval, R. (2024). Nanotechnology in food packaging materials: Role and application of nanoparticles. *RSC Advances, 14*(30), 21832–21858. PMID: 38984259

Dey, B., Prabhakar, M. R., Jayaraman, S., Gujjala, L. K. S., Venugopal, A. P., & Balasubramanian, P. (2024). Biopolymer-based solutions for enhanced safety and quality assurance: A review. *Food Research International*, •••, 114723. PMID: 39059918

Dodero, A., Escher, A., Bertucci, S., Castellano, M., & Lova, P. (2021). Intelligent packaging for real-time monitoring of food-quality: Current and future developments. *Applied Sciences (Basel, Switzerland), 11*(8), 3532.

Dorigato, A., & Fredi, G. (2024). Effect of nanofillers addition on the compatibilization of polymer blends. *Advanced Industrial and Engineering Polymer Research, 7*(4), 405–427.

Du, Y., Shi, B., Luan, X., Wang, Y., & Song, H. (2023). Chitosan/cellulose nanocrystal biocomposite coating for fruit postharvest preservation. *Industrial Crops and Products*, *205*, 117543.

Dutta, D., & Sit, N. (2024). *A comprehensive review on types and properties of biopolymers as sustainable bio-based alternatives for packaging*. Food Biomacromolecules.

. Farooq, K., Kumawat, Y. K., Sharma, K., & Kumar, V. (2024). Bio-nanocomposites: Fundamentals and Recent Advances. *Novel Bio-nanocomposites for Biomedical Applications*, 1-35.

Firouz, M. S., Mohi-Alden, K., & Omid, M. (2021). A critical review on intelligent and active packaging in the food industry: Research and development. *Food Research International*, *141*, 110113. PMID: 33641980

Getahun, M. J., Kassie, B. B., & Alemu, T. S. (2024). Recent advances in biopolymer synthesis, properties, & commercial applications: A review. *Process Biochemistry*.

Gorde, P. M., Dash, D. R., Singh, S. K., & Singha, P. (2024). Advancements in sustainable food packaging: A comprehensive review on utilization of nanomaterials, machine learning and deep learning. *Sustainable Chemistry and Pharmacy*, *39*, 101619.

Gulati, S., Amar, A., Chhabra, L., Katiyar, R., Sahu, T., & Varma, R. S. (2024). Greener nanobiopolymers and nanoencapsulation: Environmental implications and future prospects. *RSC Sustainability*, *2*(10), 2805–2832.

Hosseini, R., Rashidi, J., Mokhtariyan, M., & Landarani-Isfahani, A. (2025). Fabrication and Morphology of Biomaterials Based on the Used Synthesis Methods. In *Green Biomaterials in Tissue Engineering* (pp. 89–140). American Chemical Society.

Hussain, S., Akhter, R., & Maktedar, S. S. (2024). Advancements in sustainable food packaging: From eco-friendly materials to innovative technologies. *Sustainable Food Technology*, *2*(5), 1297–1364.

Ikram, R., Mohamed Jan, B., Abdul Qadir, M., Sidek, A., Stylianakis, M. M., & Kenanakis, G. (2021). Recent advances in chitin and chitosan/graphene-based bionanocomposites for energetic applications. *Polymers*, *13*(19), 3266. PMID: 34641082

Jamwal, V., & Mittal, A. (2024). Recent Progresses in Nanocomposite Films for Food-Packaging Applications: Synthesis Strategies, Technological Advancements, Potential Risks and Challenges. *Food Reviews International*, •••, 1–32.

Ji, D., Lin, Y., Guo, X., Ramasubramanian, B., Wang, R., Radacsi, N., & Ramakrishna, S. (2024). Electrospinning of nanofibres. *Nature Reviews. Methods Primers*, *4*(1), 1.

Kandhola, G., Park, S., Lim, J. W., Chivers, C., Song, Y. H., Chung, J. H., & Kim, J. W. (2023). Nanomaterial-based scaffolds for tissue engineering applications: A review on graphene, carbon nanotubes and nanocellulose. *Tissue Engineering and Regenerative Medicine*, *20*(3), 411–433. PMID: 37060487

Karki, S., Gohain, M. B., Yadav, D., & Ingole, P. G. (2021). Nanocomposite and bio-nanocomposite polymeric materials/membranes development in energy and medical sector: A review. *International Journal of Biological Macromolecules*, *193*, 2121–2139. PMID: 34780890

Karwatkar, P. A., Kulkarni, S. J., & Goswami, A. K. (2025). Bionanomaterials in Food Systems: Sources, Synthesis, Properties and Opportunities. *BioNanoScience*, *15*(1), 5.

Khalid, M. Y., & Arif, Z. U. (2022). Novel biopolymer-based sustainable composites for food packaging applications: A narrative review. *Food Packaging and Shelf Life*, *33*, 100892.

Khane, Y., Albukhaty, S., Sulaiman, G. M., Fennich, F., Bensalah, B., Hafsi, Z., & Al-aqbi, Z. T. (2024). Fabrication, characterization and application of biocompatible nanocomposites: A review. *European Polymer Journal*, •••, 113187.

Khouri, N. G., Bahú, J. O., Blanco-Llamero, C., Severino, P., Concha, V. O., & Souto, E. B. (2024). Polylactic Acid (PLA): Properties, Synthesis, and Biomedical Applications-A Review of the Literature. *Journal of Molecular Structure*, •••, 138243.

Korumilli, T., Abdullahi, A., Kumar, T. S., & Rao, K. J. (2024). Nanoclay-reinforced bio-composites and their packaging applications. In *Nanoclay-Based Sustainable Materials* (pp. 467–485). Elsevier.

Kumar, A., Gautam, Y. K., & Singh, N. (2024). State-of-the-art developments in surface functionalized carbon-based bio/nanocomposites for theranostic antibacterials, advanced bioimaging, and molecular bioelectronics inspired biosensing platforms. *Journal of Industrial and Engineering Chemistry*.

Kumar, A., Mishra, R. K., Verma, K., Aldosari, S. M., Maity, C. K., Verma, S., & Thakur, V. K. (2023). A comprehensive review of various biopolymer composites and their applications: From biocompatibility to self-healing. *Materials Today Sustainability*, *23*, 100431.

Kumar, R. S., Dhilipkumar, T., Jessie, J. A., Gaayathri, K. K., & Arumugam, S. (2023). Advancements in bio-polymeric composite materials for active and intelligent food packaging: A comprehensive review. *Materials Today: Proceedings.*

Kumar, V., Umrao, D., & Srivastava, A. (2022). Functional Nanomaterials for Food Packaging Applications. In *Food Packaging: The Smarter Way* (pp. 243–269). Springer Nature Singapore.

Lages, M., & Nicolas, J. (2023). In situ encapsulation of biologically active ingredients into polymer particles by polymerization in dispersed media. *Progress in Polymer Science, 137*, 101637.

Lalonde, J. N., Pilania, G., & Marrone, B. L. (2025). Materials designed to degrade: Structure, properties, processing, and performance relationships in polyhydroxyalkanoate biopolymers. *Polymer Chemistry, 16*(3), 235–265. PMID: 39464417

Lalonde, J. N., Pilania, G., & Marrone, B. L. (2025). Materials designed to degrade: Structure, properties, processing, and performance relationships in polyhydroxyalkanoate biopolymers. *Polymer Chemistry, 16*(3), 235–265. PMID: 39464417

Lizundia, E., Luzi, F., & Puglia, D. (2022). Organic waste valorisation towards circular and sustainable biocomposites. *Green Chemistry, 24*(14), 5429–5459.

Ma, F., Gao, Y., Xie, W., & Wu, D. (2024). Effect of hydrophobic modification of chitin nanocrystals on role as anti-nucleator in the crystallization of poly (ε-caprolactone)/polylactide blend. *International Journal of Biological Macromolecules, 269*, 132097. PMID: 38710249

Mahunu, G. K., Osei-Kwarteng, M., Ogwu, M. C., & Afoakwah, N. A. (2024). Safe food handling techniques to prevent microbial contamination. In *Food safety and quality in the global south* (pp. 427–461). Springer Nature Singapore.

Manikandan, D., Gandhi, V. C. S., Venkatesan, R., & Nellaiappan, T. A. (2025). Synthesis, Characterization, and Applications of Nanomaterials: Mechanical and Thermal Studies of Polymer Nanocomposites. In *Exploring Nanomaterial Synthesis, Characterization, and Applications* (pp. 197-212). IGI Global.

Matin, M. M., Amran, M., Islam, M. B., Hasnain, M., Begum, S. H., Rahman, M. R., & Bakri, M. K. B. (2024). Current and future development of nanocarbon and its biocomposites production. In *Advanced Nanocarbon Polymer Biocomposites* (pp. 49–103). Woodhead Publishing.

Mavai, S., Bains, A., Sridhar, K., Rashid, S., Elossaily, G. M., Ali, N., & Sharma, M. (2024). Formulation and application of poly lactic acid, gum, and cellulose-based ternary bioplastic for smart food packaging: A review. *International Journal of Biological Macromolecules*, •••, 131687. PMID: 38642692

Min, T., Zhou, L., Sun, X., Du, H., Zhu, Z., & Wen, Y. (2022). Electrospun functional polymeric nanofibers for active food packaging: A review. *Food Chemistry*, *391*, 133239. PMID: 35609463

Mishra, B., Panda, J., Mishra, A. K., Nath, P. C., Nayak, P. K., Mahapatra, U., & Sridhar, K. (2024). Recent advances in sustainable biopolymer-based nanocomposites for smart food packaging: A review. *International Journal of Biological Macromolecules*, •••, 135583. PMID: 39270899

Mohammadi, B., Pirsa, S., & Mahmudi, M. (2025). Animal and Vegetable Proteins: Applications as Film and Coating in Food Industry. *Journal of Polymers and the Environment*, •••, 1–21.

Mondal, S. K., Chakraborty, S., Manna, S., & Mandal, S. M. (2024). Antimicrobial nanoparticles: Current landscape and future challenges. *RSC Pharmaceutics*.

Monfared, A. R., Omranpour, H., Tuccitto, A. V., Zaoui, A., Rahman, S. S., Kheradmandkeysomi, M., & Park, C. B. (2024). Sustainable PLA Bio-Nanocomposites: Integration of TPU Nanofibrils and CNC for Enhanced Crystallization, Toughness, Stiffness, Transparency, and Oxygen Barrier Properties. *ACS Sustainable Chemistry & Engineering*, *12*(34), 13017–13029.

. Muhammad, M. A., Sulaiman, M., Raghav, S., & Raghav, J. (2025). Technologies for the Fabrication of Bio-nanomaterials. *Bio-Nanomaterials in Environmental Remediation: Industrial Applications*, 75-104.

Nandhini, J., Bellarmin, M., Siva Prakash, S., Sowmya Sri, D., & Karthikeyan, E. (2024). Advancements in metal oxide bio-nanocomposites for sustainable food packaging: Fabrication, applications, and future prospectives. *Food Bioengineering*.

Naskar, A., Sanyal, I., Nahar, N., Ghosh, D. D., & Chakraborty, S. (2023). Bionanocomposites films applied as active and smart food packaging: A review. *Polymer Engineering and Science*, *63*(9), 2675–2699.

Nath, P. C., Sharma, R., Mahapatra, U., Mohanta, Y. K., Rustagi, S., Sharma, M., & Sridhar, K. (2024). Sustainable production of cellulosic biopolymers for enhanced smart food packaging: An up-to-date review. *International Journal of Biological Macromolecules*, *273*, 133090. PMID: 38878920

Ncube, L. K., Ude, A. U., Ogunmuyiwa, E. N., Zulkifli, R., & Beas, I. N. (2021). An overview of plastic waste generation and management in food packaging industries. *Recycling*, *6*(1), 12.

Norizan, M. N., Shazleen, S. S., Alias, A. H., Sabaruddin, F. A., Asyraf, M. R. M., Zainudin, E. S., & Norrrahim, M. N. F. (2022). Nanocellulose-based nanocomposites for sustainable applications: A review. *Nanomaterials (Basel, Switzerland)*, *12*(19), 3483. PMID: 36234612

Olayil, R., & Prabu, V. A., DayaPrasad, S., Naresh, K., & Sreekanth, P. R. (2022). A review on the application of bio-nanocomposites for food packaging. *Materials Today: Proceedings*, *56*, 1302–1306.

Pandey, M., Nazar, R., Elella, M. H. A., Praharaj, S., Makhado, E., Rout, D., & Gilani, E. H. (2025). A Comprehensive Review of Recent Developments in Biomedical Materials Based on Graphene-Modified Bio-Nanocomposites. *BioNanoScience*, *15*(1), 125.

Pascall, M. A., DeAngelo, K., Richards, J., & Arensberg, M. B. (2022). Role and importance of functional food packaging in specialized products for vulnerable populations: Implications for innovation and policy development for sustainability. *Foods*, *11*(19), 3043. PMID: 36230119

Pei, J., Palanisamy, C. P., Srinivasan, G. P., Panagal, M., Kumar, S. S. D., & Mironescu, M. (2024). A comprehensive review on starch-based sustainable edible films loaded with bioactive components for food packaging. *International Journal of Biological Macromolecules*, •••, 133332. PMID: 38914408

Perera, K. Y., Jaiswal, S., & Jaiswal, A. K. (2022). A review on nanomaterials and nanohybrids based bio-nanocomposites for food packaging. *Food Chemistry*, *376*, 131912. PMID: 34971895

Pinaeva, L. G., & Noskov, A. S. (2024). Biodegradable biopolymers: Real impact to environment pollution. *The Science of the Total Environment*, •••, 174445. PMID: 38981547

Pires, J. R. A., Rodrigues, C., Coelhoso, I., Fernando, A. L., & Souza, V. G. L. (2023). Current applications of bionanocomposites in food processing and packaging. *Polymers*, *15*(10), 2336. PMID: 37242912

Popescu, C., Dissanayake, H., Mansi, E., & Stancu, A. (2024). Eco Breakthroughs: Sustainable Materials Transforming the Future of Our Planet. *Sustainability*, *16*(23), 10790.

Priyadarshi, R., Ghosh, T., Roy, S., & Rhim, J. W. (2024). Biopolymer-based antimicrobial nanocomposite materials for food packaging and preservation. In *Food Packaging and Preservation* (pp. 33–52). Academic Press.

Raghuvanshi, S., Khan, H., Saroha, V., Sharma, H., Gupta, H. S., Kadam, A., & Dutt, D. (2023). Recent advances in biomacromolecule-based nanocomposite films for intelligent food packaging-A review. *International Journal of Biological Macromolecules*, •••, 127420. PMID: 37852398

Rahman, A., Hasan, K., & Imran, A. B. (2024). Biopolymers for Supercapacitors. In *Bio-Based Polymers: Farm to Industry. Volume 3: Emerging Trends and Applications* (pp. 57-80). American Chemical Society.

Rech, A., & Daugaard, A. E. (2025). Thermoprocessing Biopolymers and Bio-Waste-Based Materials. *ACS Sustainable Resource Management*.

Roy, S., Malik, B., Chawla, R., Bora, S., Ghosh, T., Santhosh, R., & Sarkar, P. (2024). Biocompatible film based on protein/polysaccharides combination for food packaging applications: A comprehensive review. *International Journal of Biological Macromolecules*, •••, 134658. PMID: 39128751

Sadaf, M., Bragaglia, M., Slemenik Perše, L., & Nanni, F. (2024). Advancements in metal additive manufacturing: A comprehensive review of material extrusion with highly filled polymers. *Journal of Manufacturing and Materials Processing*, *8*(1), 14.

Samir, A., Ashour, F. H., Hakim, A. A., & Bassyouni, M. (2022). Recent advances in biodegradable polymers for sustainable applications. *NPJ Materials Degradation*, *6*(1), 68.

Sani, M. A., Priyadarshi, R., Zhang, W., Khezerlou, A., & Rhim, J. W. (2024). Innovative application of laccase enzyme in food packaging. *Trends in Food Science & Technology*, •••, 104623.

Sen, A., Kurian, M. S., Narayanan, D. P., Abraham, A., Jayalakshmi, P. S., Nair, S. G., & Simon, S. M. (2024). Bio-nanocomposites: Innovative solutions for addressing issues in health, agriculture, energy and environmental domains. *Nano-Structures & Nano-Objects*, *39*, 101270.

Shah, Y. A., Bhatia, S., Al-Harrasi, A., Tarahi, M., Almasi, H., Chawla, R., & Ali, A. M. M. (2024). Insights into recent innovations in barrier resistance of edible films for food packaging applications. *International Journal of Biological Macromolecules*, •••, 132354. PMID: 38750852

Shahid, M. A., Rahman, M. M., Hossain, M. T., Hossain, I., Sheikh, M. S., Rahman, M. S., & Hoque, M. I. U. (2025). Advances in Conductive Polymer-Based Flexible Electronics for Multifunctional Applications. *Journal of Composites Science*, *9*(1), 42.

Sheibani, S., Rozalli, N. H. M., & Nafchi, A. M. (2025). Barrier properties of bionanocomposites. In *Characterization Techniques in Bionanocomposites* (pp. 133–182). Woodhead Publishing.

Siciliano, S., Lopresto, C. G., & Lamonaca, F. (2024). From traditional packaging to smart bio-packaging for food safety: A review. *Euro-Mediterranean Journal for Environmental Integration*, •••, 1–16.

Singh, P., Pandey, V. K., Singh, R., Singh, K., Dash, K. K., & Malik, S. (2024). Unveiling the potential of starch-blended biodegradable polymers for substantializing the eco-friendly innovations. *Journal of Agriculture and Food Research*, *15*, 101065.

Sinthiya, P. A., Johnson, I., Prabu, H. J., Sahayaraj, A. F., & Selvan, M. T. (2024). Extraction of cellulose nanocrystals from Cissus quadrangularis for sustainable biocomposite production. *Biomass Conversion and Biorefinery*, •••, 1–13.

Sivakanthan, S., & Kaumadi, P. H. S. (2022). Environmentally Friendly Bionanocomposites in Food Industry. In *Biodegradable and Environmental Applications of Bionanocomposites* (pp. 237–263). Springer International Publishing.

Stanley, J., Culliton, D., Jovani-Sancho, A. J., & Neves, A. C. (2025). The Journey of Plastics: Historical Development, Environmental Challenges, and the Emergence of Bioplastics for Single-Use Products. *Eng*, *6*(1), 17.

Thapliyal, D., Karale, M., Diwan, V., Kumra, S., Arya, R. K., & Verros, G. D. (2024). Current status of sustainable food packaging regulations: Global perspective. *Sustainability*, *16*(13), 5554.

Thapliyal, D., Verros, G. D., & Arya, R. K. (2025). Nanoparticle-Doped Antibacterial and Antifungal Coatings. *Polymers*, *17*(2), 247. PMID: 39861318

Thivya, P., Gururaj, P. N., Reddy, N. B. P., & Rajam, R. (2024). Recent advances in protein-polysaccharide based biocomposites and their potential applications in food packaging: A review. *International Journal of Biological Macromolecules*, •••, 131757. PMID: 38657934

Trotta, F., Da Silva, S., Massironi, A., Mirpoor, S. F., Lignou, S., Ghawi, S. K., & Charalampopoulos, D. (2023). Silver bionanocomposites as active food packaging: Recent advances & future trends tackling the food waste crisis. *Polymers*, *15*(21), 4243. PMID: 37959923

. Verma, S. K., Prasad, A., & Katiyar, V. (2024). State of art review on sustainable biodegradable polymers with a market overview for sustainability packaging. *Materials Today Sustainability*, 100776.

Wang, L., Kelly, P. V., Ozveren, N., Zhang, X., Korey, M., Chen, C., & Gardner, D. J. (2023). Multifunctional polymer composite coatings and adhesives by incorporating cellulose nanomaterials. *Matter*, *6*(2), 344–372.

Wang, R., Feng, Y., Li, D., Li, K., & Yan, Y. (2024). Towards the sustainable production of biomass-derived materials with smart functionality: A tutorial review. *Green Chemistry*.

Wani, N. R., Dar, A. H., Dash, K. K., Pandey, V. K., Srivastava, S., Jan, S. Y., & Sabahi, N. (2023). Recent advances in the production of bionanomaterials for development of sustainable food packaging: A comprehensive review. *Environmental Research*, •••, 116948. PMID: 37611789

Wronka, A., & Kowaluk, G. (2024). Multiphase Biopolymers Enriched with Suberin Extraction Waste: Impact on Properties and Sustainable Development. *Materials (Basel)*, *17*(22), 5472. PMID: 39597296

Yeo, J. C. C., Muiruri, J. K., Fei, X., Wang, T., Zhang, X., Xiao, Y., & Li, Z. (2024). Innovative biomaterials for food packaging: Unlocking the potential of polyhydroxyalkanoate (PHA) biopolymers. *Biomaterials Advances*, •••, 213929. PMID: 39024863

Yosri, N., Khalifa, S. A., Attia, N. F., Du, M., Yin, L., Abolibda, T. Z., & El-Seedi, H. R. (2024). Advancing sustainability in the green engineering of nanocomposites based on marine-derived polymers and their applications: A comprehensive review. *International Journal of Biological Macromolecules*, •••, 133249. PMID: 38906361

Zhang, F., Si, Y., Yu, J., & Ding, B. (2023). Electrospun porous engineered nanofiber materials: A versatile medium for energy and environmental applications. *Chemical Engineering Journal*, *456*, 140989.

Zhang, Y., & Jiang, W. (2023). Effective strategies to enhance ultraviolet barrier ability in biodegradable polymer-based films/coatings for fruit and vegetable packaging. *Trends in Food Science & Technology*, •••, 104139.

Chapter 10
AI-Enabled Bio-Nanocomposite Materials for Soil Health and Environmental Sustainability

Mrutyunjay Padhiary
https://orcid.org/0000-0002-2236-568X
Assam University, Silchar, India

ABSTRACT

Bio-nanostructured composite materials, composed of biological and nanostructured elements, have added a new dimension to the latest agricultural developments. This chapter explores how these materials can be optimized using artificial intelligence to enhance soil fertility, structure, and ecosystem resilience. Precision design, performance prediction, and real-time monitoring of bio-nano composites, ensuring their targeted functionality in soil remediation, nutrient delivery, and pollution mitigation, are being enabled by AI algorithms, including machine learning and deep learning. AI-integrated systems might support frameworks for sustainable land management, forecast environmental effects, and replicate soil-microbe-nano synthesis. Smart nano-fertilizers, AI-assisted clean-up methods, and sensors in materials for checking soil quality demonstrate a shift toward smarter, more sustainable farming and caring for the environment. It also explores the ethical and hazardous aspects of AI-enabled bio-nano technologies and their potential impact on future-proof agri-environmental systems.

DOI: 10.4018/979-8-3373-2180-6.ch010

1. INTRODUCTION

1.1 Global Soil Health Crisis and Environmental Degradation

Soil degradation is now one of the most pressing environmental concerns globally, threatening food security, biodiversity, and ecosystem balance (Gomiero, 2016). The primary drivers of soil degradation are multifaceted, comprising physical, chemical, and biological processes (Lehman et al., 2015). Salinization, or an excess of soluble salts in the soil profile, is primarily caused by poor irrigation management, inadequate drainage, and rising water tables, which are more common in arid and semi-arid areas (Padhiary & Kumar, 2024). Salinization causes osmotic stress in plants, reduces microbial activity in soil, and lowers crop yields. Similarly, soil acidification, in some instances caused by overuse of ammonium fertilizers, acid rain, and mining, impinges on the pH value of the soil, which indirectly influences the nutrient content, microbial population structure, and soil fertility (Huang et al., 2023). Water and wind erosion remain trendy topsoil-loss reasons, with deforestation, intensive tillage, and overgrazing standing out. These processes of degradation are subsequently accelerated by urbanization, industrialization, and global warming, leading to a vast loss of fertile land and soil fertility on continents (Quinton & Fiener, 2024).

Soil degradation is a self-reinforcing process that cascades into impacts in agricultural ecosystems, climate, and ecosystem processes. In farming, the soils have reduced fertility, compromised water-holding capacity, and organic matter loss, causing lower crop yield and vulnerability to pests and diseases. Further, degraded soils emit large quantities of greenhouse gases like nitrous oxide and carbon dioxide, which drive feedback processes related to climate change (Lal, 2012). Soil organic carbon stock loss deteriorates the soil as a carbon sink, accelerating global warming. Beyond agriculture, soil erosion diminishes important ecosystem services like water filtration, nutrient cycling, and the provision of habitat for the biota inhabiting the soil. The loss of these services reduces biodiversity, makes ecosystems more vulnerable, and harms the livelihoods of millions of people who depend on productive soils for farming, forestry, and pastoralism. Problems such as these require transformative solutions with an intersection of technological, environmental, and policy interventions (Dhakal & Kattel, 2019).

Ecosystem stability is threatened by fundamental sustainability issues, including decreased biodiversity and falling soil fertility. While bio-nano composites and AI tools present potential solutions, barriers such as regulatory gaps, high costs, and low farmer awareness hinder their implementation (Samson Prince et al., 2025). It is crucial to address these limitations to ensure that emerging technologies promote long-term soil health and sustainable land management.

1.2 Role of Emerging Technologies

Materials science research has developed new possibilities for soil rehabilitation and environmental sustainability. Bio-nano composite materials, biochar, zeolites, and engineered materials have shown capability in improving the quality of the soil by increasing nutrient retention, water-holding capability, and pollutant adsorption. These materials have multifunctionality, which makes them both slow-release nutrient vectors, pH buffers, and soil conditioners (Kukowska & Szewczuk-Karpisz, 2024). Their nanoscale nature, with high surface area and reactivity, allows them to target interaction with soil components and thus become effective tools in remediating contaminated soil and rehabilitating degraded land. The materials will also be able to encourage beneficial interactions between microbes, which will make the soil even more resilient and fertile. However, their performance and safety in advanced soil conditions need intelligent design as well as monitoring methods to maximize their advantages while ensuring minimum potential risks (Raj et al., 2024).

At the same time as there has been innovation in materials, digital technologies, particularly artificial intelligence, remote sensing, and the Internet of Things, have revolutionized environmental monitoring and precision soil management (Sharma & Shivandu, 2024). Satellite imaging at high spatial resolution, proximal soil sensing, and drone-based aerial surveys now enable spatially explicit, real-time monitoring of soil moisture, nutrient concentration, pH, and contamination levels (Inoue, 2020). AI algorithms further enhance these capabilities by processing large, complex datasets to generate predictive models, risk assessments, and decision support tools. These digital systems enable proactive and site-specific soil management practices, optimizing the use of inputs, reducing environmental footprints, and improving the sustainability of agricultural and land management systems (Padhiary, Hoque, et al., 2025). The convergence of digital and materials technologies presents unprecedented opportunities to develop smart, responsive, and sustainable soil management frameworks (Rani et al., 2024).

1.3 The Synergistic Role of AI and Bio-Nano Composite Materials

Traditionally, soil amendments have been applied uniformly across landscapes using both organic and inorganic materials, leading to inefficiencies and unintended environmental impacts. The advent of nanotechnology and biotechnology has brought about new developments of bio-nano composite materials that changed this by providing specific solutions to restore soil health (Liu & Lal, 2012). At the same time, AI technologies have revolutionized data analysis, modeling, and decision-making processes in agriculture and environmental science. The convergence of

AI with bio-nano composites is an emerging innovation that uses the forecasting and optimization power of AI to design, implement, and regulate nano-enabled soil amendments (Ashique et al., 2024). Trends now are shifting to higher adoption of using machine learning algorithms for optimizing composite formulations, forecasting environmental interactions, and evaluating long-term risks. In addition, AI-based models are also being used to engineer composites with programmable release profiles, biocompatibility, and low ecotoxicity to hasten the transition towards sustainable and precision soil management practices (Uddin et al., 2025). Smart nanomaterials, AI-enhanced nanosensors, and machine-learning systems for soil restoration are examples of recent developments in AI-integrated nanotechnology for soil and environmental management. Important research reports on sustainable agricultural nanotechnology (Pagano et al., 2025), AI-IoT frameworks for soil health (Pandey et al., 2025), and environmental sensing (Pawar et al., 2025). Also, studies on nano-enabled remediation methods (Mishra et al., 2025) and nanoscale delivery systems for climate-resilient crops (Singh et al., 2025) are investigated. With a focus on real-time monitoring and predictive analytics, these results show a substantial move toward integrated AI–bio-nano ecosystems for sustainable management.

This chapter aims to discuss the role of AI-powered bio-nano composite materials in improving soil health and promoting environmental sustainability. It highlights the principles, mechanisms, and applications of these materials, focusing on AI's contributions to their design, deployment, and monitoring. Important topics include the synthesis and properties of bio-nano composites, their impacts on soil fertility, remediation of contaminants, and water resource management, as well as AI-driven optimization and risk assessments. Case studies showcase the implementation of AI-bio-nano systems in various agroecosystems, addressing their advantages, limitations, and scalability. Also, the chapter examines challenges such as environmental risks, ethical issues, and regulatory considerations, proposing future research and innovation directions for sustainable soil management. It criticizes the different academic views on how to combine nanomaterials and AI, stressing the need for an evidence-based understanding of the long-term effects on the environment and the need to address biases. The overall goal connects fundamental soil science principles with material characteristics and AI enhancement strategies, advocating for the responsible incorporation of bio-nano composites into sustainable practices.

2. FUNDAMENTALS OF BIO-NANO COMPOSITE MATERIALS

2.1 Classification of Bio-Nano Composites

Bio-nano composite materials are a new class of functional materials in which biological components, such as biopolymers, microorganisms, or natural fibers, are combined with nanomaterials, such as metal nanoparticles, carbon nanostructures, or nano-clays.

These organic nanocomposites are grouped according to their function, structure, and composition (Shchipunov, 2012). Composed mainly of bio-based matrices like chitosan, cellulose, starch, alginate, or lignin, organic nanocomposites are loaded with nanoparticles like nanocellulose, carbon nanotubes (CNTs), or graphene oxide (GO). These biodegradable and biocompatible polymers can be used in sustainable soil systems. Their enhanced mechanical strength, regulated breakdown, and capacity to discharge nutrients or bioactive compounds play a role in improving the soil structure and fertility through which they impact mechanical strength (Wang et al., 2023).

Inorganic nanocomposites are usually composed of mineral-based matrices, such as zeolites, clays, or silica, which are incorporated with metallic or metal oxide nanoparticles like ZnO, TiO_2, or Fe_3O_4. In contaminant immobilization and soil remediation, such composites play an important role because of their photocatalytic activity, high adsorption capacity, and effectiveness in pollutant removal. However, the potential longevity of these materials in the environment requires careful risk assessment and adherence to regulated modes of application (Mustapha et al., 2020). Hybrid nanocomposites combine organic and inorganic components and are often functionalized with certain ligands or biomolecules to improve their reactivity, stability, and selectivity under conditions of soil (Rando et al., 2022). A hybrid of chitosan-iron oxide nanoparticles is able to immobilize heavy metals and increase the availability of soil nutrients. Also increasing their use in precision soil management, functionalization with peptides, amino acids, or enzymes enables targeted interactions with roots, soil bacteria, or contaminants (Yasmin et al., 2024).

2.2 Synthesis Methods and Material Engineering

Green synthesis methods focus on making bio-nano composites using plant extracts, microorganisms, or natural reducing agents. Such techniques guarantee the sustainability and safety of the produced composites by avoiding the use of harmful chemicals and high-energy processes. The green chemistry paradigm in soil amendment materials is exemplified by methods that include biosynthesis of nanoparticles using bacterial cultures or the use of agricultural waste as raw materials for nanomaterial manufacture (Gunasena et al., 2025; Padhiary, Roy, & Dey, 2024).

Surface engineering of bio-nano composites controls their release profiles and modifies their interactions with soil components. Functionalizing methods like conjugation with functional groups (e.g., carboxyl, amine, or thiol), grafting, or coating with biodegradable polymers let the composites interact especially with soil contaminants, nutrients, or microbial populations. These alterations have the potential to improve the materials' dispersibility, stability, and reactivity under a wide range of soil conditions, hence enhancing their effectiveness and lowering the risks they pose to the environment (Kumar et al., 2024).

Scaling up the production of bio-nano composites for agricultural applications presents significant engineering challenges. Industrial approaches optimize raw material inputs, energy utilization, and cost-effectiveness by aligning laboratory-scale synthesis techniques with continuous flow reactors. To create formulations that are user-friendly and suitable for field application, techniques such as spray drying, extrusion, and pelletization are utilized. In the commercialization of these materials for environmental and ecological sustainability, it is essential to ensure consistency, quality control, and regulatory compliance (Mishra et al., 2023).

2.3 Mechanisms of Interaction with Soil Components

Various physico-chemical interactions between soil constituents and bio-nano composites such as minerals, organic matter, and pore water occur. These include redox reactions, ion exchange, and adsorption-desorption processes. Iron oxide nanoparticles can, for example, adsorb phosphate ions and, therefore, regulate their availability in the soil. Through ion-exchange processes, nano-clays can also sequester heavy metals. Soil ionic strength, pH, and the availability of competing ions influence such interactions and require precise characterization and modelling (Yang et al., 2024). Of utmost importance is the biological interactions of bio-nanocomposites with soil biota. Composts may carry beneficial bacteria, thus promoting rhizosphere colonization and plant growth. Also of influence are microbial community structure, enzyme activity, and nutrient cycling processes. Additionally, plant root interactions may trigger variations in the pattern of root exudation, thereby promoting symbiotic relationships or enhancing plant resistance to abiotic stress. More focus is directed towards developing composites that enhance beneficial microbial interactions and avoid toxicity (Shi et al., 2022). Applications in soil conditions are safe depending on awareness of the environmental fate, mobility, transformation, and persistence of bio-nano composites. Their conglomeration, migration in soil pores, and potential leaching into groundwater are subject to factors such as particle size, surface charge, and functionalization (Wahab et al., 2024). Degradation processes, including chemical alteration, photodegradation, and microbiological degradation, shift the characteristics and impacts of nanomaterials over the long term. We need to conduct long-term field studies and comprehensive life cycle assessments to evaluate their behavior and potential risks under various environmental conditions (Zaaba & Jaafar, 2020).

Nanomaterials in soil undergo environmental processes like aggregation, dissolution, and interaction with organic matter, which affect their mobility and persistence (Miklasová et al., 2025). They may accumulate in specific soil layers or migrate to groundwater, posing risks such as toxicity to microorganisms, disruption of nutrient cycling, and altered soil enzymatic activity. Understanding these processes is crucial for assessing the environmental safety of nano-enabled soil amendments and evaluating the design and functionality of advanced bio-nano composites in soil systems. As shown in Figure 1, bio-nano composites engage with soil components through several channels, such as adsorption on mineral surfaces, interaction with root exudates, and the provocation of microbial colonization, thus enhancing nutrient cycling and pollution immobilizing effect. Depending on their composition, application, and ecological impact, Table 1 compares soil amendment products. Due to sluggish nutrient release and restricted biodegradability, conventional additions with excellent adsorption capacity and persistence are biochar and zeolite. By contrast, smart nano-fertilizers and chitosan nanocomposites provide controlled release systems, targeted delivery, and improved biodegradability. While microbial carrier composites promote positive biological interactions in the rhizosphere, hybrid organic-inorganic composites become flexible solutions combining nutrient delivery with remedial purposes (Behl et al., 2024). It is revealed that, while persistence and potential toxicity incline inorganic composites toward causing environmental issues, bio-based and hybrid composites strike an ecological safety versus usefulness balance. This comparison study informs the rationale for selecting suitable bio-nanocomposites for specific soil fertility and environmental applications.

Figure 1. Interaction pathways of bio-nano composites in soil

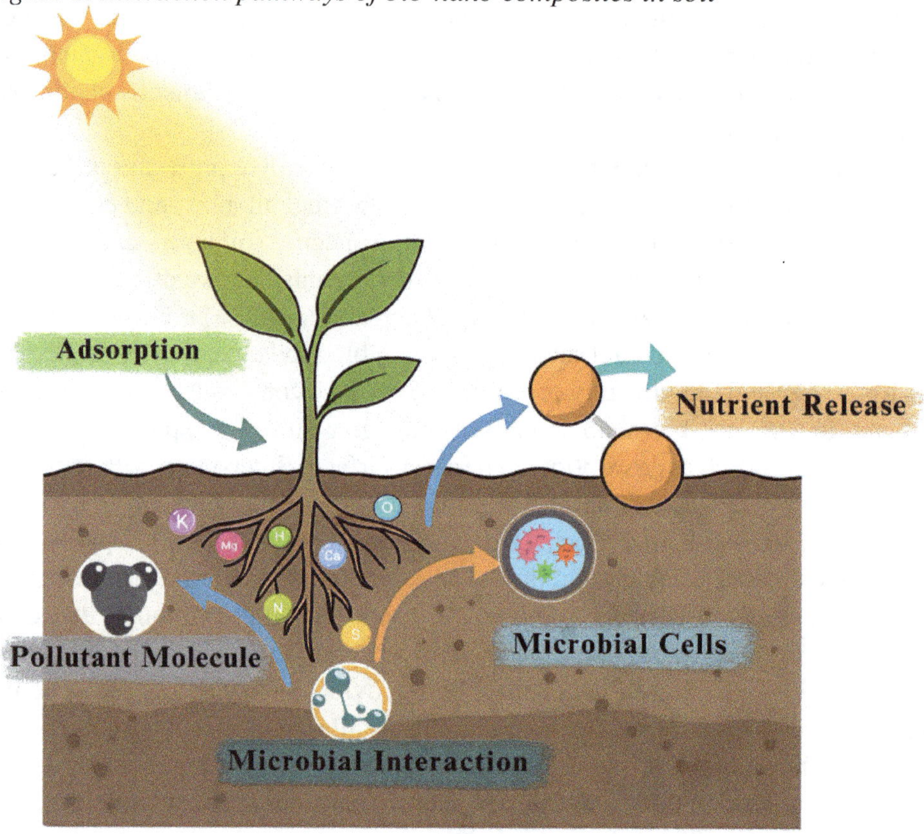

Table 1. Comparative properties of soil amendments

Soil Amendment Type	Composition	Key Functions	Mechanism of Action	Nutrient Release	Adsorption Capacity	Environmental Persistence	Biodegradability	Risk Assessment	References
Biochar	Pyrolyzed biomass	Soil conditioning	Adsorption, pH buffering	Slow	High	High	Low	Low	(Arwenyo et al., 2023)
Zeolite	Aluminosilicate minerals	Nutrient retention	Ion exchange	Slow	High	High	Negligible	Low	(Nakhli et al., 2017)
Chitosan nanocomposite	Chitosan + nanoparticles	Fertility enhancement	Biodegradation, release	Moderate	Moderate	Low	High	Minimal	(Othman et al., 2023)
Graphene oxide composite	GO + biopolymer	Soil structure improvement	Adsorption, interaction with microbes	Moderate	High	Medium	Medium	Moderate	(Kovtun et al., 2020)
Nano-clay composite	Montmorillonite + nanomaterials	Pollutant immobilization	Ion exchange, sorption	Slow	High	High	Low	Low	(Ayalew, 2022)

continued on following page

Table 1. Continued

Soil Amendment Type	Composition	Key Functions	Mechanism of Action	Nutrient Release	Adsorption Capacity	Environmental Persistence	Biodegradability	Risk Assessment	References
Metal oxide composite	Fe_3O_4, ZnO nanoparticles	Contaminant removal	Redox reaction, sorption	Minimal	High	Medium	Low	High	(Naseem & Durrani, 2021)
Hybrid organic-inorganic	Chitosan + Fe_3O_4 + bioactive	Multifunctional (fertility + remediation)	Controlled release, microbial interaction	Controlled	High	Medium	Medium	Moderate	(Liu et al., 2021)
Microbial carrier composite	Biochar + microbes	Biofertilizer delivery	Microbial colonization	Slow	Moderate	Medium	Medium	Low	(Saravanan et al., 2023)
Enzyme-functionalized composite	Biopolymer + enzymes	Soil enzyme enhancement	Enzymatic reactions	Fast	Low	Low	High	Minimal	(Efremenko et al., 2023)
Smart nano-fertilizer	Nanoencapsulated nutrients	Precision nutrient delivery	Triggered release mechanisms	Controlled	Moderate	Low	High	Moderate	(Abd-Elsalam & Alghuthaymi, 2024)

2.4 Theoretical Foundations

The fusion of theoretical paradigms, such as systems thinking, sustainability science, and material informatics, is the foundation for the development of AI-enabled bio-nano composite materials. Systems thinking depends critically on awareness of the interdependencies and feedback loops seen in complicated systems. One instance of this is soil ecosystems, dynamic environments influenced by many factors. This method lets scientists replicate how bio-nanostructures would interact with soil components, react to climatic stimuli, and influence more general agro-ecological results. Sustainability science promotes renewable materials, low environmental impact, circular design, and socially inclusive technologies. It calls for fair technology deployment, non-toxic nanoparticles, and biodegradable polymers (Zhang et al., 2021). Material informatics identifies potential material combinations and iteratively refines them using artificial intelligence and big data; hence it accelerates material discovery and optimization. These theoretical foundations ensure that the advancement of AI-enabled bio-nano composites is technically sound, ethically accountable, environmentally sustainable, and systemically cohesive. This methodology is structured to create next-generation materials that address the complexity of real-world soil and environmental challenges (Batra et al., 2020).

3. AI IN BIO-NANO COMPOSITE MATERIAL DESIGN AND OPTIMIZATION

AI combined with material science has brought a revolutionary method to the design, synthesis, and assessment of bio-nano composite materials. Predictive modeling, optimization, and risk assessment in the complexity of interactions among bio-based matrices, nanomaterials, soil environments, and biological entities require data-intensive methodologies. AI algorithms, with their ability to analyze large, multivariate datasets, provide a structured framework for accelerating the innovation pipeline of bio-nano composites. This framework allows for precise design, effective synthesis, and thorough evaluations of environmental safety (Liu et al., 2013).

3.1 Material components: Biopolymers, enzymes, microbes, nanoparticles

Biopolymers, enzymes, microbes, and nanoparticles are incorporated into bio-nano composite materials. Biopolymers such as chitosan, cellulose, alginate, pectin, and starch are biodegradable and non-toxic to living organisms. They can be chemically altered to enhance their mechanical strength, water retention, and controlled-release characteristics (Saraswat et al., 2023). Urease and phosphatase are some of the enzymes that assist in causing biochemical alterations of the soil, thus improving the availability of nutrients and minimizing losses. Microbes, specifically useful bacteria and fungi, enable soil recovery through nitrogen fixation, phosphorus solubilization, and organic matter decomposition. A lot of their capability to thrive and perform under poor soil conditions could be enhanced through immobilizing or encapsulating them inside nanostructured carriers (Gong et al., 2022). Researchers use nano-clays, nano-silica, ZnO, and FeO nanoparticles because of their high surface area, catalytic capacity, and ability to be functionalized. These improve nutrient use efficiency, purify toxins, and enhance plant growth. By bringing together many components in a single platform that can perform multiple things, bio-nano composites can help with specific soil problems. They may, for example, enhance fertility, purge pollutants, or conserve water. It is crucial to choose and ensure material compatibility to ascertain the composites' ecological efficiency and sustainability.

3.2 Green Synthesis Methods and Eco-friendly Nanofabrication

The production of bio-nano composites is increasingly making use of environmentally friendly nanofabrication methods and green synthesis to achieve sustainability goals. Green synthesis is the reduction of metal salts to nanoparticles or the polymerization and crosslinking of biopolymers through the use of biological agents such as plants, bacteria, enzymes, or biomolecules. As stabilizing and reducing agents, plant extracts make the nanoparticles more stable and biocompatible compared to those manufactured by more conventional methods (Nandhini et al., 2024). Microbial synthesis is used for the production of nanoparticles, providing a scalable and low-energy method that utilizes organisms such as Aspergillus, Bacillus, or Lactobacillus. Mixing man-made nanoparticles with biopolymer blends is referred to as nanofabrication (Chauhan & Punia, 2023). This process may be achieved by solvent casting, electrospinning, emulsion polymerization, or spray drying. Heat-free and toxic chemical-free biodegradable capsules can be achieved using methods like ionic gelation and ultrasonic-assisted synthesis. Formulation of bio-nanocomposites that improve soil health without compromising ecosystem integrity is the ultimate goal of green synthesis and production.

3.3 Functionalization Techniques for Specific Soil Applications

Bio-nano composites can selectively couple with soil elements through the functionalization process, which is a major part of their formulation (Pulido-Reyes et al., 2017). It involves altering the surface properties, reactivity, or load of the composite to meet the desired soil challenge: nutrient deficit, salinity, contamination, or poor water retention. For leaching reduction, nanoparticles or biopolymers are functionalized to couple and release key macro- and micronutrients, like nitrogen. The composites are functionalized in polluted soils with active groups to adsorb or detoxify organic pollutants and heavy metals. Hydrophilic functional groups are responsible for the retention of moisture and lowering of pH, while microbial functionalization allows bio-nano composites to transfer enzymes or living organisms across the material. Enzyme-functionalized nanoparticles can make soil more productive by breaking down organic matter or cycling nutrients (Aliyari Rad et al., 2023). Another aspect of advanced functionalization is the presence of stimuli-responsive systems. The release of substances in response to environmental cues or artificial intelligence distinguishes these systems, enhancing accuracy and reducing waste. These approaches make it possible to tailor bio-nano composites to match specific soils, crops, climatic regions, and degradation patterns, thus offering sustainable, efficient, and targeted soil management options.

3.4 AI Techniques for Predictive Modeling of Composite Behavior

Predictive modeling using AI enables the prediction of physicochemical properties, environmental fate, and activity of bio-nano composites under different soil and environmental conditions (Sharmile et al., 2025). Trial and error methods are being replaced by more and more machine learning and deep learning techniques that study past experimental data and generate predictive models that are efficient at modeling the complex behavior of composites.

Different machine learning algorithms, such as support vector machines, random forests, and artificial neural networks, have been extensively used to predict key properties of bio-nano composites, i.e., adsorption capacity, release kinetics, surface area, and mechanical strength (Kibrete et al., 2023). SVM models are highly effective in handling small data sets that possess non-linear characteristics, making them suitable for initial studies. RF models exhibit robust strengths in predicting the efficacy of soil amendments by well capturing the interactions between many factors, like particle size, surface functionalization, and environmental pH. ANN models, with their multi-layered structure, are effective at revealing hidden patterns in large datasets to predict complex interactions like nutrient release profiles and pollutant degradation efficiencies.

Deep learning models, especially CNN and RNN, are employed in complex predictions, such as modeling dynamic interactions between bio-nano composites and soil microbiomes and predicting real-time changes in composite performance under altering environmental circumstances (Khaki et al., 2020; Padhiary, Powar, et al., 2025). The models can incorporate data from different sources like spectroscopy, electron microscopy, and environmental sensors, allowing for comprehensive understanding of composite behavior, degradation processes, and potential interactions between soil, plants, and microbes.

3.5 AI in Material Synthesis Optimization

AI is essential for the optimization of the synthesis parameters of bio-nano composites, which guarantees environmental safety, scalability, and cost-efficiency (Darie-Ni ă &Frąckowiak, 2025). By integrating AI with experimental design methodologies, it is possible to streamline synthesis processes, minimizing energy consumption, chemical usage, and batch variability.

Algorithms like genetic algorithms (GA), particle swarm optimization (PSO), and Bayesian optimization are utilized to enhance synthesis conditions, focusing on factors such as temperature, pH, reaction time, and precursor concentrations (Acosta-Angulo et al., 2022). The models forecast the ideal combination of parameters that enhance the desired characteristics, such as adsorption capacity or release control, while reducing unwanted traits like toxicity or instability. The integration of conventional design of experiments methodologies, including response surface methodology and Taguchi designs, with AI algorithms significantly improves the efficiency and predictive capabilities of the optimization process (Sibalija & Majstorovic, 2012).

3.6 AI in Environmental Risk Assessment of Bio-Nano Composites

AI methodologies are also becoming essential instruments for the evaluation of environmental risks associated with bio-nano composites (Samson Prince et al., 2025). It is necessary to analyze multidimensional datasets that include material qualities, environmental parameters, and biological endpoints to predict possible toxicity, bioaccumulation, and environmental destiny. To predict the potential toxicological effects of bio-nano composites on soil microbiomes, plant systems, and aquatic organisms, AI-based toxicity models employ data from in vitro and in silico studies (Singh et al., 2020). By comparing the structural characteristics of composites with existing toxicity data, quantitative structure-activity relationship (QSAR) models combined with machine learning techniques aid in the prediction of ecotoxicity.

Lifecycle analysis (LCA) models integrated with AI facilitate the simulation of degradation pathways, allowing for the estimation of the environmental footprint of bio-nano composites throughout their production to end-of-life stages (Jesus et al., 2022). Predictive models can anticipate degradation rates, by-product formation, and environmental hotspots, thereby facilitating the development of safer and more sustainable material designs. Figure 2 visually represents the AI-driven approach to bio-nano composite investigation, showcasing the interconnected workflow of data-driven design, predictive modeling, and validation. Table 2 summarizes the different AI models utilized at each research stage, highlighting the diverse applications of AI in improving the efficiency, safety, and sustainability of bio-nano composite materials.

Table 2. AI models and their application

AI Model/ Algorithm	Application Area	Composite Property or Process Targeted	Advantages	References
Support vector machine (SVM)	Prediction of adsorption capacity	Heavy metal removal efficiency	High accuracy with limited data	(Hlihor, 2015)
Random forest (RF)	Nutrient release kinetics prediction	Controlled-release fertilizers	Robust to outliers and overfitting	(Irfan et al., 2020)
Artificial neural network (ANN)	Pollutant degradation efficiency	Organic pollutant breakdown pathways	Captures complex non-linear relationships	(Ayodele et al., 2021)
Convolutional neural network (CNN)	Image-based structural analysis	Nanocomposite morphology prediction	High-throughput data processing	(Kadulkar et al., 2021)

continued on following page

Table 2. Continued

AI Model/ Algorithm	Application Area	Composite Property or Process Targeted	Advantages	References
Recurrent neural network (RNN)	Real-time soil interaction modeling	Dynamic nutrient release & soil conditions	Time-series prediction capabilities	(Wang et al., 2023)
Genetic algorithm (GA)	Synthesis parameter optimization	Reaction conditions and yields	Multi-objective optimization	(Na et al., 2017)
Particle swarm optimization (PSO)	Process efficiency improvement	Energy-efficient composite synthesis	Fast convergence on optimal solutions	(Zhang et al., 2015)
Bayesian optimization	Multi-parameter synthesis optimization	Fine-tuning synthesis process parameters	Effective in handling uncertainty	(Kundacina et al., 2025)
QSAR modeling + ML	Ecotoxicity prediction	Soil and aquatic organism toxicity	Predictive with limited experimental data	(Lavado et al., 2022)
AI-based lifecycle modeling	Environmental degradation forecasting	Lifecycle impact and by-product formation	Full lifecycle environmental assessment	(Dinesh & Rahul Prasad, 2024)

Figure 2. AI workflow for designing and validating bio-nano composites

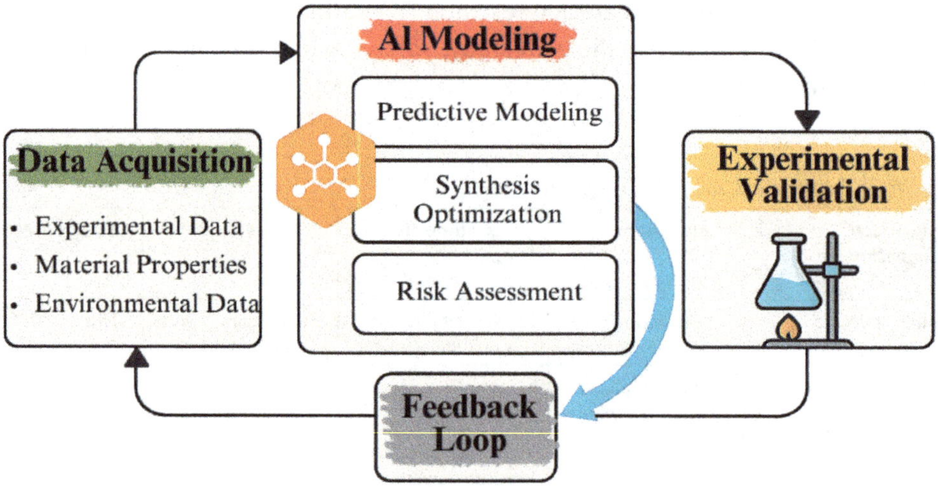

4. APPLICATIONS IN SOIL HEALTH ENHANCEMENT

In this chapter, emerging technologies like bio-nano composites and artificial intelligence are linked to common soil issues such as nutrient loss, poor water retention, salinity, erosion, and contamination. The focus shifts from merely technical mechanisms to the practical solutions these innovations can provide, enhancing crop productivity and supporting sustainable land use. These applications also reinforce the broader objectives of the edited volume by demonstrating how AI-enabled bio-nano solutions can enhance sustainable soil management and support future-ready agricultural systems (Kumar et al., 2025). AI and bio-nano composites together create advanced applications for soil improvement, offering precise interventions including optimized nutrient delivery, increased water retention, better soil structure, and effective remediation of contamination (Pathan et al., 2020). AI-driven models ensure efficiency and environmental sustainability, reducing input overuse and minimizing negative ecological impacts.

4.1 AI-Optimized Nano-Fertilizers

AI-driven innovations in nano-fertilizer formulation facilitate the precise delivery of nutrients to plants, thereby improving soil health and reducing environmental runoff and nutrient leaching (Munir et al., 2024). The fertilizers utilize nanoscale carriers, including nanoclays, carbon nanotubes, or biochar nanoparticles, specifically engineered to release nutrients in a controlled fashion.

AI algorithms, especially genetic algorithms and machine learning models, play a crucial role in optimizing nano-fertilizer formulations by examining extensive datasets related to soil characteristics, plant nutrient needs, and environmental factors (Danai-Varsou et al., 2023). The design of nano-carriers with customized release profiles is facilitated by AI, which guarantees that nutrients are released in accordance with the absorption patterns of plants. Furthermore, artificial intelligence has the capability to enhance the coating thickness, porosity, and surface functionalization of nanoparticles, allowing for precise adjustments in nutrient release kinetics tailored to particular soil conditions.

AI-driven modeling enables researchers to predict and engineer the release kinetics of nutrients from bio-nano composites, thereby guaranteeing that plants have access to essential elements like nitrogen, phosphorus, and potassium at the precise moment they are required (Amin et al., 2025). These systems utilize soil data inputs, such as moisture levels, pH, and microbial activity, to activate nutrient release through stimuli-responsive mechanisms. This precision approach prevents the accumulation of superfluous fertilizers in soil and water bodies, enhances soil nutrient balance, and reduces nutrient wastage.

4.2 Precision agriculture and intelligent nutrient management

Precision agriculture employs advanced AI-driven bio-nano composite materials to enhance crop yields while reducing environmental impacts. The incorporation of nanocarriers and biopolymers in these materials facilitates a precise, localized, and timely release of nutrients, tailored to the immediate requirements of soil and crops (Hoque & Padhiary, 2024; Machado et al., 2022). Artificial intelligence systems evaluate information gathered from Internet of Things sensors, unmanned aerial vehicles, and satellite imagery to generate nutrient maps, pinpointing areas that need targeted interventions (Padhiary, Kumar, et al., 2024). These maps facilitate the strategic application of bio-nano composites, guaranteeing that fertilizers are distributed precisely in the right locations and at the appropriate times. The intelligent composites are capable of modulating nutrient release in response to soil temperature, moisture levels, pH, and root exudates, thereby enhancing nutrient utilization efficiency (Yin et al., 2021). Algorithms in machine learning analyze patterns of plant nutrient uptake and enhance the composition and dosing schedule of composites. This combination enhances agricultural practices by optimizing yields, lowering fertilizer expenses, minimizing environmental impact, and safeguarding soil microbiomes.

4.3 Enhancing microbial diversity and promoting soil biodiversity

The diversity of life in the soil can be enhanced by bio-nano composites, which provide favorable environments for microorganisms such as biocontrol agents, bacteria that accumulate nitrogen, and fungi that dissolve phosphorus (Sheershwal et al., 2025). To create these composites, biocompatible and bioactive components, including polysaccharides, lignin, cellulose, and enzymes, can be utilized in the design process. Artificial intelligence has the potential to determine the best material combinations by analyzing factors such as soil type, climate conditions, and crop species (Zhang et al., 2021). Microbial inoculants, activated in response to specific soil signals, have the potential to enhance smart composites. Predictive models ensure that the release timing and dosage align with the microbial colonization periods, thereby enhancing the establishment's success. Certain composites can sequester harmful substances, thereby fostering a more conducive environment for a variety of life forms. Remote sensing and AI-powered analysis of soil metagenomes enable the monitoring of microbial community structure and facilitate dynamic adjustments to management practices (Yang et al., 2024). Not only does the promotion of resilient, diverse microbial ecosystems increase soil fertility and disease resistance, but it also improves ecosystem services such as carbon sequestration, water regulation, and crop productivity (Roy et al., 2025).

4.4 Carbon sequestration and climate-resilient soil systems

Soils are the major carbon sinks on Earth, and it is imperative to optimize their carbon sequestration capabilities to mitigate climate change (Rodrigues et al., 2023). The application of artificial intelligence in bio-nano composites, including nano-biochar carriers and polysaccharide-rich matrices, has the potential to enhance the stability of organic matter, improve soil cohesion, and increase the activity of carbon-storing microbes. These composites have the capacity to bind organic carbon molecules and safeguard them from microbial degradation, thereby enabling the development of more durable carbon-storing materials. Improving the delivery of nutrients and water can greatly boost plant productivity and root biomass, which in turn leads to an indirect rise in carbon inputs from below ground. The integration of remote sensing technologies facilitates ongoing observation of soil organic carbon dynamics, thereby guiding adaptive management strategies (Li et al., 2024). These methods not only reduce greenhouse gas emissions but also enhance soil structure, increase water retention, and bolster resilience against extreme weather events.

4.5 Water Retention and Soil Structure Enhancement

Bio-nano composites contribute to the development of soil structure and the enhancement of water retention capacity, both of which are essential in sustainable agriculture, particularly in conditions that are frequently characterized by drought (Menossi et al., 2022). The production of stable soil aggregates is facilitated by nanocomposites, which include hybrids of nano-clay and biochar (Abd-Elsalam et al., 2024). These nanocomposites operate as binding agents in the soil. These aggregates improve porosity, aeration, and the quality of the habitat for microorganisms. A soil modeling process that is helped by artificial intelligence can assist in determining the ideal particle size, surface functionalization, and dose of nanocomposites that are necessary to enhance aggregation and improve the water-holding capacity of different types of soil.

For the purpose of predicting the water retention behavior of soils that have been altered with nanocomposites, artificial intelligence models such as ANN and SVM are utilized (Lowe et al., 2022). To develop predictive models of soil moisture retention, these models combine data on soil texture, organic matter content, and nanomaterial properties. This provides farmers and land managers with the ability to make informed decisions regarding soil amendment techniques.

4.6 Remediation of Contaminated Soils

AI-guided bio-nano composites offer innovative approaches for the remediation of contaminated soils, especially those impacted by heavy metals, pesticides, and industrial pollutants (M. Santos et al., 2025). Bio-nano composites that are functionalized, including amino-functionalized carbon nanotubes and magnetic biochar nanoparticles, demonstrate the ability to selectively adsorb heavy metals from soils (Yang et al., 2019). Algorithms driven by artificial intelligence enhance the design of these composites through the prediction of adsorption capacities, kinetic behavior, and selectivity across different soil pH levels, ionic strengths, and types of contaminants (Akhtar et al., 2024). This enhances the advancement of effective, targeted remediation technologies.

The merging of AI is crucial for enhancing bioremediation protocols that utilize bio-nano composites in conjunction with microbial consortia (Blessing & Olateru, 2025). By integrating models such as RF and DL, it is possible to predict the synergistic effects of nanomaterials and microbes, optimizing conditions such as nutrient supplementation, soil aeration, and pH adjustment to maximize contaminant degradation and soil detoxification. Figure 3 shows that AI-optimized nano-fertilizers have many benefits for soil health, such as releasing nutrients in a precise way, improving soil structure, and being able to adsorb pollutants. All of these benefits are managed by AI-driven decision-making systems. The quantitative comparison of efficiency gains between these AI-integrated fertilizers and conventional approaches is presented in Table 3, highlighting their superior performance in nutrient use efficiency, soil health metrics, and environmental sustainability. The presence of heavy metals, organic pollutants, and emerging threats such as microplastics in soil represents a considerable risk to ecosystem health and food safety.

Bio-nano composites present cutting-edge solutions for soil remediation, functioning as selective, efficient, and eco-friendly agents for the removal of pollutants. Biopolymeric matrices incorporating nanoparticles can bind, degrade, or immobilize detrimental substances (Annu et al., 2024). One example is magnetic nanoparticles that are coated with biodegradable polymers. Composites that are functionalized with enzymes can catalyze the degradation of organic pollutants into non-toxic forms. AI assists in tailoring these composites to specific pollutants by predicting their adsorption capabilities, binding affinities, and degradation pathways, all informed by the characteristics of the soil and the chemical composition of the pollutants (TavallaieNejad et al., 2025). Researchers are currently developing innovative composites that can identify microplastic particles and break them down through enzymatic processes.

Farmers are facing significant soil-related challenges such as nutrient depletion, salinity, erosion, contamination, low organic matter, and varying water availability, all of which threaten crop yields and economic stability. Although solutions like bio-nano composites and AI technologies exist, their adoption is limited by barriers including high costs, lack of training, usability issues, and challenges in scaling to different farm sizes and soil types (Sheershwal et al., 2025). Smallholder farmers encounter even greater obstacles due to restricted access to resources, insufficient extension services, and market volatility, which complicates technology implementation. Observational data indicate that the financial uncertainty discourages smallholders from utilizing soil amendments and precision farming tools, emphasizing the critical need to link scientific progress with practical, inclusive strategies for sustainable soil management.

Table 3. Comparison of conventional and AI-optimized fertilizers

Parameter	Conventional Fertilizers	AI-Optimized Nano-Fertilizers	Efficiency Gain (%)	References
Nutrient use efficiency (%)	30-50	70-90	+70%	(Thirumalaisamy et al., 2025)
Nutrient leaching reduction (%)	Low	High	+60%	(Thirumalaisamy et al., 2025)
Crop yield increase (%)	20-30	+35	+40%	(Tiwari, 2022)
Soil organic matter improvement	Negligible	Significant (via carrier support)	+40%	(Xueping, 2015)
Water retention capacity (%)	Unaffected	Enhanced	+35%	(Elsabagh et al., 2024)
Heavy metal adsorption efficiency (%)	Low	High	67-163%	(Qian et al., 2023)
Microbial diversity impact	Less	Supportive (via smart release)	+60%	(Dang & Morrissey, 2024)
Environmental footprint (kg CO2 eq.)	High	Low	+39%	(Yu et al., 2022)
Cost efficiency in long term	Low (due to overuse, runoff)	High (due to precise application)	+65%	(Jayara et al., 2023)

Figure 3. Mechanisms of AI-optimized nano-fertilizers in soil

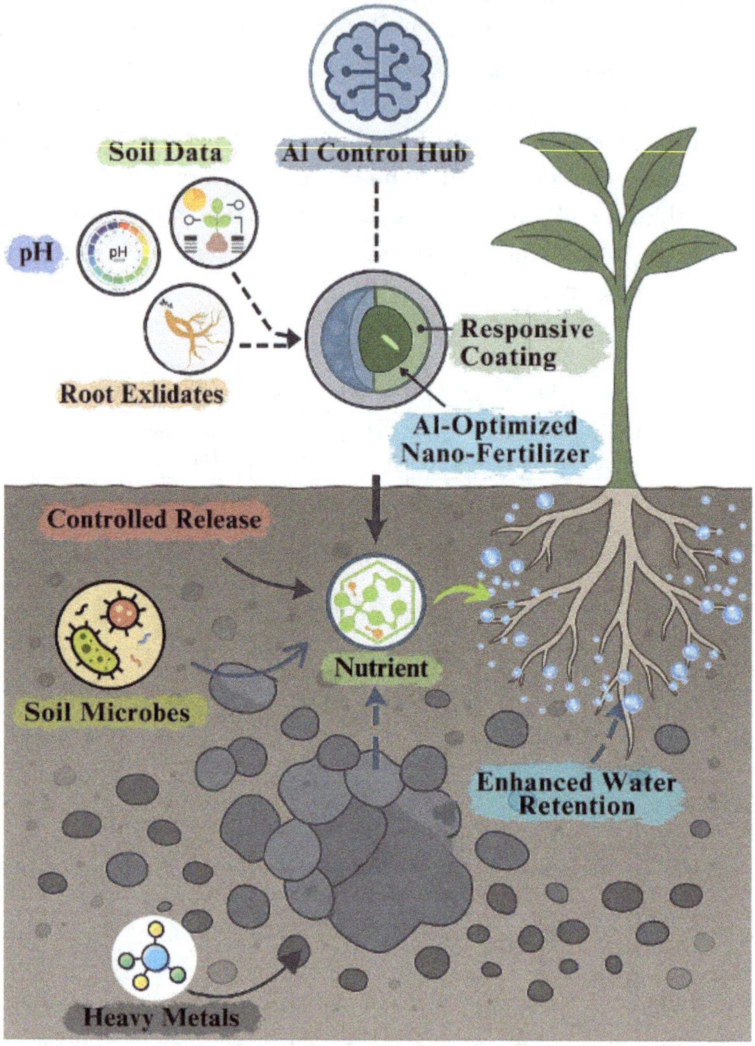

5. ENVIRONMENTAL MONITORING AND SMART SOIL MANAGEMENT SYSTEMS

By combining bio-nano composites with AI and IoT, we are introducing intelligent soil management systems. These modern devices, which provide real-time, accurate monitoring of soil health indices, enable proactive decision-making and sustainable exploitation of soil resources. Optimization of soil management procedures can be achieved by farmers and land managers through the use of nanocomposites embedded in sensor networks coupled with analytics and decision support systems powered by artificial intelligence. This enhances production, conserves resources, and mitigates environmental concerns (Padhiary, Roy, Dey, et al., 2024).

5.1 Machine learning algorithms for composite optimization

Machine learning algorithms are significant in improving bio-nano composites by optimizing the most important properties, including nutrient release, mechanical stability, environmental safety, and production cost. Supervised learning algorithms, such as decision trees and gradient boosting, analyze historical experimental data to identify critical material parameters that influence performance. Unsupervised learning techniques, such as clustering algorithms, systematically categorize material formulations according to behavioral similarities, allowing for identification of new composite designs free from human bias. Autonomous experimentation with synthetic routes or functionalization routes accelerates discovery processes (Volk et al., 2023). Genetic algorithms and PSO enable the exploration of multiple potential solutions, facilitating the identification of optimal combinations of biopolymers, nanoparticles, and functional agents (Rostamzadeh-Renani et al., 2024). Bayesian optimization methods enable effective optimization subject to limitations of experimental runs, which becomes vital in situations where material synthesis is expensive or time-consuming. Application of machine learning for optimization can enhance bio-nano composites developed for particular soil issues to attain performance and sustainability objectives simultaneously.

5.2 Sensor-Embedded Bio-Nano Composites

The development of intelligent soil sensors based on bio-nano composites represents a groundbreaking achievement in environmental monitoring. The sensors take advantage of the unique properties of nanomaterials, such as high surface area, reactivity, and tunable selectivity, to analyze numerous soil parameters with outstanding sensitivity and accuracy. The sensor matrices are loaded with nanomaterials like graphene oxide, metal oxide nanoparticles, and carbon nanotubes to recognize crucial indicators such as pH, nutrient levels, soil moisture, temperature, and contaminant levels (Thakur & Kumar, 2022). Additionally, the selectivity of these composites is considerably enhanced by the bio-functionalization of them using enzymes, microbial proteins, or aptamers, which enables the detection of specific soil-borne pathogens or pollutants. The sensors are built right into the soil profile, so they can collect data in real time without disturbing the soil ecology too much.

The large datasets that these bio-nanocomposite sensors make require advanced AI algorithms to be useful. Various types of data streams are analyzed using machine learning models like CNN, SVM, and decision trees to remove noise and uncover useful insights (Bansal et al., 2022). The algorithms use past soil data, weather data, and plant development patterns to make models that can anticipate soil health trends. This makes it easier to detect problems with soil health, nutrient imbalances, or contamination occurrences early on. AI algorithms make it easier to personalize recommendations by customizing models to fit the individual characteristics of a site, the kind of soil, and the cropping system. This makes sure that soil management tactics are effective and specific to each area.

5.3 AI Systems for Real-Time Soil Health Monitoring

AI systems enhanced by IoT technology are essential for real-time soil health monitoring, since they connect diverse sensor networks to cloud-based analytical platforms. Bio-nano composites containing embedded sensors send soil data wirelessly via low-power wide-area networks (LPWAN) or 5G systems to centralized artificial intelligence servers. To generate thorough assessments of soil health, these cloud-based systems combine many soil metrics and perform continuous data processing. With the combination of IoT, AI, and nanotechnology, data collection and analysis can be done automatically, and the technology can be used on a large scale in many ecological systems (Hoque, Padhiary, Prasad, et al., 2025; Taha et al., 2025). This real-time feedback loop enables farmers to implement data-informed strategies like precision fertilization, irrigation scheduling, or soil amendment applications.

To improve user engagement and support informed choices, AI-powered soil health monitoring systems feature user-friendly dashboards and visualization tools. These platforms utilize geospatial mapping, heat maps, and temporal trend graphs to effectively convey soil quality parameters in a format that is easy to interpret. Algorithms in artificial intelligence identify irregularities, recommend remedial measures, and issue notifications for essential limits (Larson et al., 2021). The incorporation of mobile applications and web portals enables farmers, extension agents, and policymakers to remotely access real-time soil health data, thereby enhancing the timeliness and accuracy of interventions. Through the spatial and temporal visualization of soil health dynamics, these dashboards connect data collection with practical soil management strategies.

5.4 AI Decision Support Systems for Sustainable Soil Management

Advanced AI decision support systems integrate multi-parameter soil health decision models that take into account the intricate interactions of physical, chemical, and biological soil characteristics (Zhao & Wan, 2023). The models utilize ensemble learning techniques like Random Forests and Gradient Boosting, combining real-time data from bio-nano sensors with historical land-use records, weather forecasts, and crop growth models. This allows for a thorough evaluation of soil health conditions, recognition of constraining elements, and modeling of different soil management approaches. These platforms facilitate sustainable soil management by providing customized recommendations, forecasting the long-term effects of various practices, and enhancing the efficiency of resource utilization.

AI-driven soil health decision support systems are seamlessly incorporated with precision agriculture platforms, thereby creating a cohesive framework for customized crop and soil management (Tsakiridis et al., 2023). This interface makes it possible to use real-time soil data to guide variable rate applications of fertilizers, lime, irrigation, and soil conditioners. This ensures the correct use of inputs and minimizes environmental effects. In addition, by combining data from autonomous farm machinery, drone imaging, and remote sensing with soil health monitoring, completely automated cycles of crop management and soil health may be set up. This will pave the way for regenerative agriculture and climate-smart farming (James et al., 2024). Figure 4 shows the AI-integrated real-time soil health monitoring system. This system uses bio-nano sensor networks, AI analytics, and user interfaces to give you complete, data-driven solutions for managing your soil. The system constantly checks important soil quality measures, as shown in Table 4. This makes it possible to use adaptive, location-specific tactics that improve sustainable land management and precision farming.

Figure 4. AI-integrated real-time soil health monitoring system

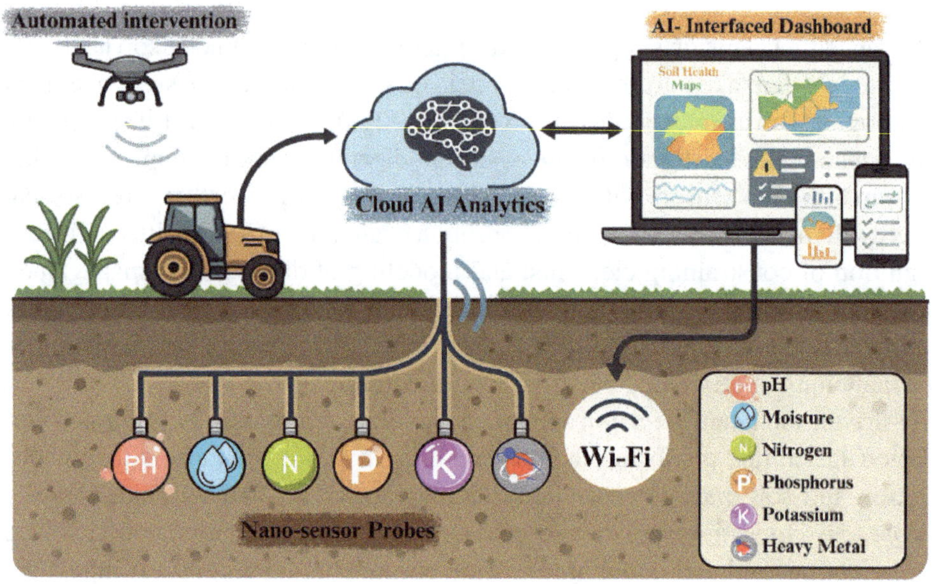

Table 4. Soil quality parameters monitored by AI-integrated nanocomposites

Parameter	Sensor Technology	AI Analytical Method	Environmental Significance	References
Soil moisture (%)	Graphene oxide-based capacitive sensors	ANN, Regression models	Water conservation and irrigation scheduling	(Han & Yun, 2024)
Soil pH	Carbon nanotube-based potentiometric sensors	SVM, Fuzzy Logic	Soil acidity/alkalinity management	(Srivastava et al., 2018)
Nitrogen (N) content (ppm)	Nanoparticle-enhanced colorimetric biosensors	Decision Trees, KNN	Precision fertilization	(Carrillo-Lopez et al., 2024)
Phosphorus (P) content (ppm)	Nano-clay embedded optical sensors	Random Forests	Optimizing P availability	(Vidal & Govan, 2024)
Potassium (K) content (ppm)	Ion-selective electrode sensors with nanofilms	Ensemble Learning	Balanced fertilization strategies	(Zhai et al., 2024)
Soil organic carbon (%)	Surface-enhanced raman spectroscopy (sers)	Deep learning cnn	Soil health and carbon sequestration	(Xiuquan et al., 2025)

continued on following page

Table 4. Continued

Parameter	Sensor Technology	AI Analytical Method	Environmental Significance	References
Heavy Metals (Pb, Cd, Hg) (ppm)	Magnetic biochar nanosensors	Predictive Modeling, Clustering	Soil contamination monitoring	(Shi et al., 2025)
Soil temperature (°c)	Nanocomposite thermoresistive sensors	Regression Analysis	Crop stress mitigation	(Phadkule & Sarma, 2023)
Microbial activity (atp levels)	Enzyme-functionalized nanobiosensors	Neural Networks	Biological soil health assessment	(Sharma et al., 2022)
Soil salinity (ds/m)	Nano-enabled EC sensors	Classification Algorithms	Salinity management	(El-Ramady et al., 2024)
Pesticide residues (mg/kg)	Aptamer-nano biosensors	Deep Learning (LSTM, CNN)	Food safety and soil detoxification	(Y. Chen et al., 2024)

6. CASE STUDIES AND FUTURE SCENARIOS

The application of AI-driven bio-nano composites for soil management is advancing to become a vital component of sustainable agriculture and environmental restoration. This section presents two pragmatic case studies and a prospective foresight scenario that demonstrate how AI systems enhance the application, optimization, and evaluation of bio-nano composites in challenging soil conditions, including those impacted by salinity and heavy metal pollution.

6.1 Case Study 1: AI-Guided Application of Nano-Composites in Saline Soils

Saline soil degradation presents considerable challenges to crop productivity, particularly in arid and semi-arid areas. This case study involved the integration of AI algorithms with soil salinity sensors and nanocomposite application systems to create an adaptive treatment strategy. The deployment strategy entailed the spatial mapping of soil electrical conductivity (EC) through the utilization of sensor arrays integrated with AI-driven geostatistical interpolation models (Barathkumar et al., 2025). The models pinpointed areas of significant impact and directed the targeted use of calcium carbonate and biochar-infused nanocomposites, intending to displace sodium ions and enhance soil structure.

The AI system used deep learning methods, especially convolutional neural networks, to analyze past salinity and crop yield data to predict how long recovery would take and to adjust treatment amounts as needed (Wai et al., 2022). This adaptive management technique optimized resources, reduced leaching risks, and accelerated soil health restoration. Following the adoption of the AI-driven methodology, critical soil recovery parameters were recorded throughout two cropping cycles. Metrics included lower EC levels, better soil porosity, more microbial biomass, and higher crop yields. The findings revealed a 48% reduction in EC, a 35% enhancement in water infiltration rate, and a 2.3-fold rise in microbial activity, demonstrating successful soil recovery (Wang et al., 2022). The results illustrate the importance of AI in improving the accuracy and effectiveness of nanocomposite applications in saline soils.

6.2 Case Study 2: AI-Driven Bioremediation of Heavy Metal Contaminated Soils

AI-enabled nano-bioremediation was implemented to detoxify soils that were contaminated with lead (Pb) and cadmium (Cd). Integrated sensing platforms characterized the site, an abandoned industrial area, by measuring heavy metal concentrations, pH levels, and organic matter content. Algorithms utilizing artificial intelligence for clustering effectively categorized soil zones based on the severity of contamination, facilitating the customized application of bio-nano composites, including iron oxide nanoparticles paired with organic ligands (Youden et al., 2022). AI optimization protocols employed genetic algorithms to model treatment scenarios, incorporating variables such as composite dosage, soil pH adjustments, and rainfall predictions (Alprol et al., 2024). The protocols enhanced treatment schedules, reduced environmental risks, and guaranteed adherence to environmental thresholds (Padhiary, 2025).

The evaluation conducted after treatment indicated a significant reduction in heavy metal bioavailability, with a decrease of 63% for cadmium and 58% for lead over a period of four months (Zhou et al., 2020). The application of AI modeling effectively maintained the treatment within ecotoxicological safety limits, simultaneously achieving a 22% reduction in operational expenses. This experience illustrates the ability of AI to effectively handle intricate remediation situations and modify protocols dynamically, guaranteeing sustainable results.

6.3 Foresight Scenario: AI-Enabled Closed-Loop Soil Management Systems

AI-enabled closed-loop soil management systems are anticipated to transform soil sustainability in the years leading up to 2035. The integration of predictive analytics, real-time sensors, bio-nano composite delivery platforms, and autonomous decision support systems is essential for advancing these systems. Models developed using extensive soil, climate, and crop datasets will forecast declines in soil health, proactively implement nanocomposite applications, and guarantee minimal ecological disturbances (Rabehi et al., 2024).

These systems will integrate advanced metrics focused on sustainability, including assessments of soil organic carbon levels, biodiversity indices, and models of ecosystem resilience (Davis et al., 2023). Based on reinforcement learning frameworks, these models will align agricultural productivity goals with ecosystem services, promoting resilient, adaptive, and sustainable soil management globally (Table 5). Figure 5 illustrates a model for AI-enhanced decision-making in nano-bioremediation, showcasing the evolution of data from site sensing to AI-guided risk assessment and optimized application procedures.

Table 5. Global AI-enabled bio-nano composite soil applications

Region	Soil Issue	Bio-Nano Composite Used	AI Approach Applied	Key Sensors Used	Application Method	Outcomes	References
India (Gujarat)	Saline soil	Biochar-caco3 nanocomposite	CNN geospatial models	EC sensors, moisture probes	Site-specific application	48% EC reduction	(Gojiya et al., 2024)
Egypt (Nile Delta)	Salinity stress	Zeolite-nano biochar	Lstm yield forecasting	Ec, soil moisture	Automated drip integration	35% yield increase	(Haroon, 2021)
Brazil (Amazonas)	Acidic soil degradation	Nano-lime with bio-inoculants	Decision tree models	Ph sensors, orp sensors	Direct soil injection	Ph increase from 4.2 to 6.0	(Teixeira et al., 2020)
Bangladesh (Sundarbans)	Salt intrusion	Nano-encapsulated polymers	Svm soil risk mapping	Ec, soil salinity	Precision spraying via uav	40% salinity decrease	(Sahana et al., 2020)
Spain (Murcia)	Organic matter depletion	Nano-biochar with fungal inoculants	Ai-powered carbon models	Soc sensors, moisture	Incorporation via tillage	33% soc increase	(Chausali et al., 2021)
South Korea (Incheon)	Urban soil pollution	Carbon nanotube adsorbents	Ann pollution mapping	Voc sensors, heavy metals	Robotic soil remediation units	60% pollution load reduction	(Sharma & Kumar, 2024)
USA (Mississippi)	Heavy metal contamination	Biochar-graphene encapsulated nzvi (bc-g@ fe^0)	Predictive modeling	Icp-ms, soil ph sensors	Fixed-bed continuous flow	Up to 92.2% as^{3+} immobilization	(Karunaratne et al., 2023)
China (Jiangsu)	Chromium (cr(vi)) pollution	Cmc-stabilized nzvi on biochar (cmc-nzvi@bc)	Decision tree analysis	Cr(vi) sensors, soil ph meters	Soil amendment (3 g/kg)	97.96% cr(vi) immobilization	(Huang et al., 2024)
South Africa	Organic pollutant (methylparaben)	$Cofe_2o_4$-biochar nanocomposite	Ai-powered adsorption modeling	Organic pollutant sensors	Adsorption in constructed wetlands	85.6% methylparaben removal	(Bhandari et al., 2023)

Figure 5. AI risk assessment framework for bio-nano composites in soil ecosystems

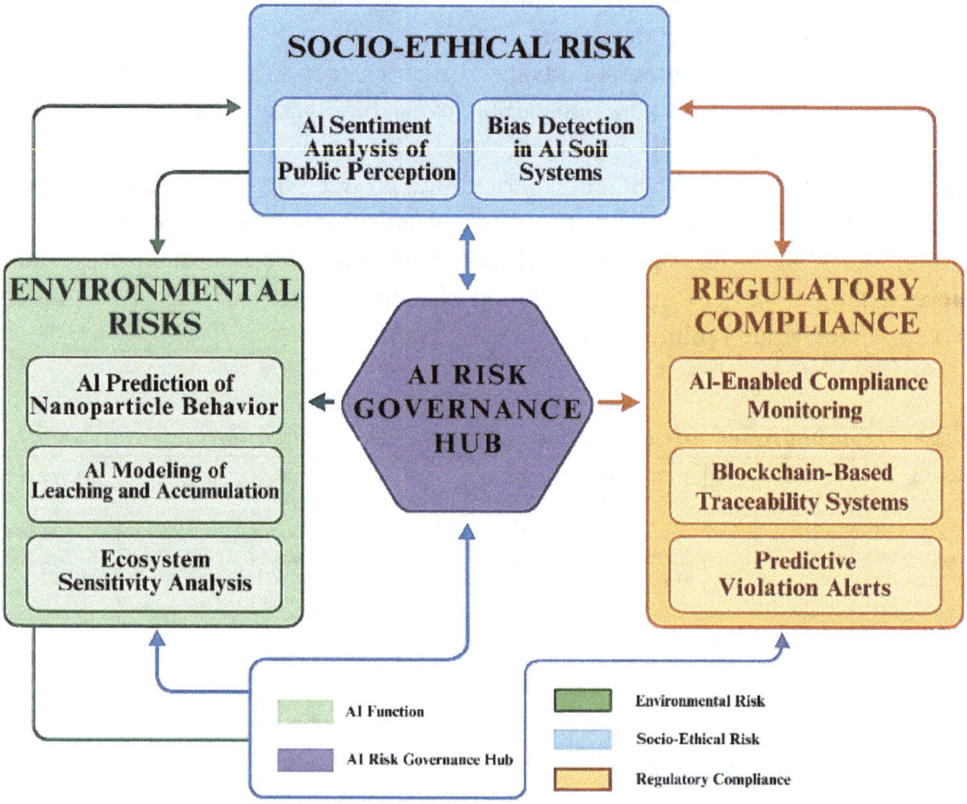

7. CHALLENGES, RISKS, AND ETHICAL CONSIDERATIONS

The use of bio-nano composites in soil management has enormous transformational potential, but it also poses a number of environmental, health, socio-ethical, and governance concerns that must be thoroughly addressed. Incorporating AI technologies into deployment and monitoring operations intensifies the complexity of these difficulties (Esmaeilzadeh, 2024). This section examines these dimensions by emphasizing the urgent hazards linked to bio-nano composites and the intricate concerns posed by AI-driven decision-making. Recent studies on nanomaterial environmental risk assessment highlight persistent gaps in long-term ecotoxicity, soil persistence, and trophic transfer, pointing to the importance of more rigorous, multi-season field evaluations. Recent reviews have shown that many engineered nanomaterials exhibit complex behaviors in soil systems, including slow degradation, surface modifications, and potential microbial toxicity, which complicate predictions of environmental fate (Kaur & Bhushanwar, 2025; Santos et al., 2025). Updated assessments further emphasize the importance of evaluating bioaccumulation pathways, groundwater mobility, and real-field exposure scenarios, as these factors remain insufficiently characterized for many agricultural applications (Lallawmkimi et al., 2025; Rahim et al., 2021). Emerging policy-oriented research calls for standardized soil-testing protocols, harmonized regulatory benchmarks, and lifecycle-based evaluation frameworks to ensure responsible development and deployment of nano-enabled soil amendments (Chaudhary et al., 2025; Pagano et al., 2025).

7.1 Environmental and Health Risks of Bio-Nano Composites

The environmental hazards associated with bio-nano composites mostly stem from their interactions with soil organisms, potential build-up within ecosystems, and unintended consequences for non-target species (Wahab et al., 2024). Although these composites have the potential to improve nutrient use efficiency and address pollution issues, research has raised important concerns regarding the persistence of nanoparticles, their transfer through trophic levels, and the risks of bioaccumulation. These factors may pose threats to soil microbial communities and disrupt food webs. Furthermore, the leaching of leftover nanoparticles into groundwater systems poses additional health risks, impacting the quality of drinking water and the integrity of aquatic ecosystems. Artificial intelligence possesses significant potential to evaluate and mitigate these environmental issues. Machine learning models, particularly those employing ensemble learning and deep learning neural networks, can predict regions with a heightened probability of nanoparticle buildup or unintentional exposure (Singh et al., 2020). These models examine past soil chemistry, hydrology, and ecological datasets to identify areas at risk, allowing for precise and careful composite applications. Combining environmental sensitivity indices into AI risk models allows for the alignment of composite deployment with ecological thresholds, thereby ensuring minimal collateral damage (Parul & Das, 2024).

AI-driven predictive models utilize Monte Carlo simulations and agent-based modeling to analyze the mobility, deterioration, and leaching of nanocomposites across different temporal and spatial scales. These tools offer insights into optimal dosages and application timings while accounting for cumulative environmental impacts, essential for proactive management. Understanding nanomaterials' interactions with variables like soil pH, organic carbon, clay minerals, and microbial communities is crucial for safe usage, as these factors affect nanoparticle mobility, persistence, and toxicity. Safety evaluations adhere to international standards set by the International Organization for Standardization (ISO) and the Organization for Economic Co-operation and Development (OECD), which outline methods for assessing physicochemical properties, conducting particle-size analyses, and evaluating ecotoxicity in soil organisms, including nematodes and earthworms (Pandey & Pandey, 2025). OECD Test Guidelines ensure validated assessments of dispersion stability, soil adsorption, leaching potential, and overall toxicity, thus addressing environmental impact and ecological risk comprehensively (Caorsi et al., 2025; Chen et al., 2025). However, the effectiveness of AI models is limited by sparse field data and the complexity of soil biological and chemical interactions, leading to a need for continuous calibration and validation. Implementing a life-cycle approach is vital since nanoparticles' production, application, and degradation can result in emissions or transformation products in soils. Additional challenges include economic factors, insufficient awareness among farmers, and inconsistent regulatory frameworks, highlighting the necessity for cost-effective manufacturing, thorough farmer training, and standardized certification systems.

7.2 Socio-Ethical Considerations

The socio-ethical challenges associated with the use of AI-guided bio-nano composites in agriculture are intricately linked to public perception, cultural acceptability, and the nuances of risk communication. Investigations indicate that the general populace continues to harbor doubts regarding the safety of nanotechnology in agriculture, concerned about unseen hazards and uncertain long-term consequences (Singh & Gurjar, 2022). The fact that AI isn't completely clear makes it harder for people to accept, because they want AI-driven decisions to be open and accountable. Without broad stakeholder participation and debate between science and society, the application of these technologies may face significant resistance, especially among traditional farming communities (Cheyns& Riisgaard, 2014).

Systems that utilize artificial intelligence for soil management gather extensive and sensitive information, such as detailed soil profiles at the farm level, the health of crops, and patterns of land use (Kalantzopoulos et al., 2024). This brings to light significant ethical issues regarding data ownership, privacy, and consent. Furthermore, bias in AI algorithms, often caused by the inadequate representation of certain agro-ecological zones or socio-economic groups in training datasets, can lead to inequitable recommendations, thus marginalizing smallholder farmers and leading to regional disparities in technological access.

7.3 Regulatory and Governance Frameworks

Worldwide, regulatory frameworks for nanomaterials in agriculture are still under development and often lack precision about bio-nano composites and their environmental routes (Saraswat et al., 2023). The advancement of AI regulatory frameworks is ongoing; however, they usually concentrate on agro-environmental applications specifically. The absence of standardized protocols for nano-bio materials and AI models engenders substantial governance issues, leading to uncertainties around accountability, liability, and the demands for continuous governance.

AI can be effectively utilized as a tool within regulatory frameworks to facilitate the monitoring of compliance. The integration of AI-driven remote sensing with blockchain traceability systems can guarantee that bio-nanocomposite materials are utilized within regulatory limits while also ensuring that environmental monitoring data remains transparent and secure from tampering (Rana et al., 2025). Predictive compliance models can foresee regulatory infractions or ecological threshold breaches, facilitating preemptive enforcement actions.

8. CONCLUSION

AI-driven sustainable soil management is a future that combines predictive, adaptable, and climate-resilient strategies. This approach combines AI, IoT, bio-nano composites, and regenerative agriculture practices to create proactive, tailored, and climate-resilient soil management systems. The combination of AI-driven bio-nano composites with precision irrigation, cover cropping, and agroforestry practices will enhance systemic resilience, minimizing susceptibility to droughts, floods, and various climatic disturbances. AI-driven closed-loop decision support systems will use data about the soil, weather forecasts, market trends, and ecological restrictions to offer you all the knowledge you need to manage your soil. These technologies will enable farmers to comprehend risks, opportunities, and methods to enhance the sustainability of their farms, watersheds, and regions in real time. However, significant research gaps persist, hindering the comprehensive use of AI-enabled bio-nano composite technologies for environmentally sustainable soil management. A major challenge is the lack of field-scale validation and calibration of AI models used in soil systems. To make models more accurate, relevant to the context, and scalable, we need many field trials, collaborative validation frameworks with farmers, and large datasets from multiple seasons. Incorporating uncertainty quantification modules into AI models will improve transparency and build stakeholder trust in AI-generated recommendations. Collaborative research frameworks that integrate soil science, artificial intelligence, nanotechnology, ethical considerations, and policy aspects are necessary for the future of AI-enabled bio-nano composite soil management. Innovation roadmaps should emphasize inclusive pathways, encouraging collaboration among AI developers, soil ecologists, agronomists, social scientists, regulators, and end-users. Creating living labs and open innovation platforms for co-development, testing, and iterative enhancement of AI models and bio-nano composites will expedite the responsible adoption process.

REFERENCES

Abd-Elsalam, K. A., & Alghuthaymi, M. A. (Eds.). (2024). *Nanofertilizers for sustainable agroecosystems: Recent advances and future trends*. Springer.

Abd-Elsalam, K. A., Mehmood, M. A., Ashfaq, M., Abdelkhalek, T. E., Hassan, R. K., & Ravichandran, M. (2024). Liquid Nanoclay: Synthesis and Applications to Transform an Arid Desert into Fertile Land. *Soil Systems*, *8*(3), 3. Advance online publication. DOI: 10.3390/soilsystems8030073

Acosta-Angulo, B., Diaz-Angulo, J., Lara-Ramos, J., Torres-Palma, R., Martínez-Pachón, D., Moncayo-Lasso, A., & Machuca-Martínez, F. (2022). Analysis of the Applications of Particle Swarm Optimization and Genetic Algorithms on Reaction Kinetics: A Prospective Study for Advanced Oxidation Processes. *ChemElectroChem*, *9*(13), e202200229. DOI: 10.1002/celc.202200229

Akhtar, M. S., Ali, S., & Zaman, W. (2024). Innovative Adsorbents for Pollutant Removal: Exploring the Latest Research and Applications. *Molecules (Basel, Switzerland)*, *29*(18), 18. Advance online publication. DOI: 10.3390/molecules29184317 PMID: 39339312

Aliyari Rad, S., Nobaharan, K., Pashapoor, N., Pandey, J., Dehghanian, Z., Senapathi, V., Minkina, T., Ren, W., Rajput, V. D., & Asgari Lajayer, B. (2023). Nano-Microbial Remediation of Polluted Soil: A Brief Insight. *Sustainability*, *15*(1), 1. Advance online publication. DOI: 10.3390/su15010876

Alprol, A. E., Mansour, A. T., Ibrahim, M. E. E.-D., & Ashour, M. (2024). Artificial Intelligence Technologies Revolutionizing Wastewater Treatment: Current Trends and Future Prospective. *Water (Basel)*, *16*(2), 314. DOI: 10.3390/w16020314

Amin, H., Tayyab, A., Umair, M., Naveed, M. R., Yaseen, H. R., Qasim, M., Muzammal, M., Saif, S., Ali, M. A., & Sultan, M. A. (2025). Recent Progress in Wastewater Treatment: Exploring the Roles of Zero-Valent Iron and Titanium Dioxide Nanoparticles. *Indus Journal of Bioscience Research*, *3*(4), 4. Advance online publication. DOI: 10.70749/ijbr.v3i4.1141

Annu, M., Tripathi, S., & Shin, D. K. (2024). Biopolymeric Nanocomposites for Wastewater Remediation: An Overview on Recent Progress and Challenges. *Polymers*, *16*(2), 2. Advance online publication. DOI: 10.3390/polym16020294 PMID: 38276702

Arwenyo, B., Varco, J. J., Dygert, A., Brown, S., Pittman, C. U., & Mlsna, T. (2023). Contribution of modified P-enriched biochar on pH buffering capacity of acidic soil. *Journal of Environmental Management, 339*, 117863. DOI: 10.1016/j.jenvman.2023.117863 PMID: 37080104

Ashique, S., Raikar, A., Jamil, S., Lakshminarayana, L., Gajbhiye, S. A., De, S., & Kumar, S. (2024). Artificial Intelligence Integration with Nanotechnology: A New Frontier for Sustainable and Precision Agriculture. *Current Nanoscience, 21*(2), 242–273. DOI: 10.2174/0115734137275111231206072049

Ayalew, A. A. (2022). A critical review on clay-based nanocomposite particles for application of wastewater treatment. *Water Science and Technology, 85*(10), 3002–3022. DOI: 10.2166/wst.2022.150 PMID: 35638802

Ayodele, B. V., Alsaffar, M. A., Mustapa, S. I., Cheng, C. K., & Witoon, T. (2021). Modeling the effect of process parameters on the photocatalytic degradation of organic pollutants using artificial neural networks. *Process Safety and Environmental Protection, 145*, 120–132. DOI: 10.1016/j.psep.2020.07.053

Bansal, M., Goyal, A., & Choudhary, A. (2022). A comparative analysis of K-Nearest Neighbor, Genetic, Support Vector Machine, Decision Tree, and Long Short Term Memory algorithms in machine learning. *Decision Analytics Journal, 3*, 100071. DOI: 10.1016/j.dajour.2022.100071

Barathkumar, S., Sellamuthu, K. M., Sathyabama, K., Malathi, P., Kumaraperumal, R., & Devagi, P. (2025). Advancements in Soil Quality Assessment: A Comprehensive Review of Machine Learning and AI-Driven Approaches for Nutrient Deficiency Analysis. *Communications in Soil Science and Plant Analysis, 56*(2), 251–276. DOI: 10.1080/00103624.2024.2406484

Batra, R., Song, L., & Ramprasad, R. (2020). Emerging materials intelligence ecosystems propelled by machine learning. *Nature Reviews. Materials, 6*(8), 655–678. DOI: 10.1038/s41578-020-00255-y

Behl, K., Jaiswal, P., & Pabbi, S. (2024). Recent advances in microbial and nanoformulations for effective delivery and agriculture sustainability. *Biocatalysis and Agricultural Biotechnology, 58*, 103180. DOI: 10.1016/j.bcab.2024.103180

Bhandari, G., Gangola, S., Dhasmana, A., Rajput, V., Gupta, S., Malik, S., & Slama, P. (2023). Nano-biochar: Recent progress, challenges, and opportunities for sustainable environmental remediation. *Frontiers in Microbiology, 14*, 1214870. DOI: 10.3389/fmicb.2023.1214870 PMID: 37547682

Blessing, A.-A., & Olateru, K. (2025). AI-driven optimization of bioremediation strategies for river pollution: A comprehensive review and future directions. *Frontiers in Microbiology, 16*, 1504254. DOI: 10.3389/fmicb.2025.1504254 PMID: 40371099

Caorsi, G., Candotto Carniel, F., Légnani, M., Flahaut, E., González, V. J., Vázquez, E., Prato, M., & Tretiach, M. (2025). Applicability of the OECD Test Guideline 201 to graphene-related materials: Dispersion stability matters. *Ecotoxicology and Environmental Safety, 292*, 117888. DOI: 10.1016/j.ecoenv.2025.117888 PMID: 39987683

Carrillo-Lopez, L. M., Villanueva-Verduzco, C., Villanueva-Sánchez, E., Fajardo-Franco, M. L., Aguilar-Tlatelpa, M., Ventura-Aguilar, R. I., & Soto-Hernández, R. M. (2024). Nanomaterials for Plant Disease Diagnosis and Treatment: A Review. *Plants, 13*(18), 2634. DOI: 10.3390/plants13182634 PMID: 39339607

Chaudhary, S., Prasanna, D., Trivedi, P., Nandy, R., Prasad, S. S., Sunitha, S., & Panigrahi, C. K. (2025). Nanotechnology Applications in Agriculture for Crop Protection and Nutrient Delivery: A Review. *International Journal of Plant and Soil Science, 37*(9), 106–119. DOI: 10.9734/ijpss/2025/v37i95690

Chauhan, N. S., & Punia, A. (2023). Heavy Metal Removal and Recovery: Sustainable and Efficient Approaches. In Singh, P., Verma, P., & Singh, R. P. (Eds.), *Wastewater Resource Recovery and Biological Methods* (pp. 87–124). Springer International Publishing., DOI: 10.1007/978-3-031-40198-5_5

Chausali, N., Saxena, J., & Prasad, R. (2021). Nanobiochar and biochar based nanocomposites: Advances and applications. *Journal of Agriculture and Food Research, 5*, 100191. DOI: 10.1016/j.jafr.2021.100191

Chen, X., Wang, S., Lai, H., Deng, L., Zhong, Q., Okoye, C. O., Niu, Q., Jing, Y., Wang, J., & Jiang, J. (2025). Nano-Biotechnology in Soil Remediation: Use of Nanomaterials to Promote Plant Growth and Stress Tolerance. *Nanomaterials (Basel, Switzerland), 15*(22), 1743. DOI: 10.3390/nano15221743 PMID: 41295648

Chen, Y., Li, Kejun, Yang, Yuanjiao, Yuan, Xin, Zhang, Mei, & and Huang, K. (2024). Recent developments in optical aptamer sensors for pesticide residues detectio. *Applied Spectroscopy Reviews, 59*(9), 1147–1182. DOI: 10.1080/05704928.2024.2308132

Cheyns, E., & Riisgaard, L. (2014). Introduction to the symposium: The exercise of power through multi-stakeholder initiatives for sustainable agriculture and its inclusion and exclusion outcomes. *Agriculture and Human Values, 31*(3), 409–423. DOI: 10.1007/s10460-014-9508-4

Danai-Varsou, D., Zhang, P., Afantitis, A., Guo, Z., Lynch, I., & Melagraki, G. (2023). Nanoinformatics and artificial intelligence for nano-enabled sustainable agriculture. In *Nano-Enabled Sustainable and Precision Agriculture* (pp. 503–531). Elsevier., DOI: 10.1016/B978-0-323-91233-4.00015-6

Dang, C., & Morrissey, E. M. (2024). The size and diversity of microbes determine carbon use efficiency in soil. *Environmental Microbiology*, *26*(5). Advance online publication. DOI: 10.1111/1462-2920.16633 PMID: 38733078

Darie-Ni ă, R. N., & Frąckowiak, S. (2025). An Overview of Potential Applications of Environmentally Friendly Hybrid Polymeric Materials. *Polymers*, *17*(2), 252. DOI: 10.3390/polym17020252 PMID: 39861324

Davis, A. G., Huggins, D. R., & Reganold, J. P. (2023). Linking soil health and ecological resilience to achieve agricultural sustainability. *Frontiers in Ecology and the Environment*, *21*(3), 131–139. DOI: 10.1002/fee.2594

Dhakal, B., & Kattel, R. R. (2019). Effects of global changes on ecosystems services of multiple natural resources in mountain agricultural landscapes. *The Science of the Total Environment*, *676*, 665–682. DOI: 10.1016/j.scitotenv.2019.04.276 PMID: 31051371

Dinesh, A., & Rahul Prasad, B. (2024). Predictive models in machine learning for strength and life cycle assessment of concrete structures. *Automation in Construction*, *162*, 105412. DOI: 10.1016/j.autcon.2024.105412

Efremenko, E., Stepanov, N., Aslanli, A., Lyagin, I., Senko, O., & Maslova, O. (2023). Combination of Enzymes with Materials to Give Them Antimicrobial Features: Modern Trends and Perspectives. *Journal of Functional Biomaterials*, *14*(2), 64. DOI: 10.3390/jfb14020064 PMID: 36826863

El-Ramady, H., Prokisch, J., Mansour, H., Bayoumi, Y. A., Shalaby, T. A., Veres, S., & Brevik, E. C. (2024). Review of Crop Response to Soil Salinity Stress: Possible Approaches from Leaching to Nano-Management. *Soil Systems*, *8*(1), 1. Advance online publication. DOI: 10.3390/soilsystems8010011

Elsabagh, S. S., Elkhatib, E. A., & Rashad, M. (2024). Novel nano-fertilizers derived from drinking water industry waste for sustained release of macronutrients: Performance, kinetics and sorption mechanisms. *Scientific Reports*, *14*(1). Advance online publication. DOI: 10.1038/s41598-024-56274-0 PMID: 38454001

Esmaeilzadeh, P. (2024). Challenges and strategies for wide-scale artificial intelligence (AI) deployment in healthcare practices: A perspective for healthcare organizations. *Artificial Intelligence in Medicine, 151*, 102861. DOI: 10.1016/j.artmed.2024.102861 PMID: 38555850

Gojiya, K. M., Karangiya, B. A., Chavda, S. K., Gaadhe, S. K., Gojiya, D. K., & Rank, H. D. (2024). Soil Moisture Modelling Using Remote Sensing and Artificial Neural Networks: A Study of Devbhumi Dwarka Region, Gujarat, India. *International Journal of Environment and Climate Change, 14*(1), 1–13. DOI: 10.9734/ijecc/2024/v14i13788

Gomiero, T. (2016). Soil Degradation, Land Scarcity and Food Security: Reviewing a Complex Challenge. *Sustainability, 8*(3), 281. DOI: 10.3390/su8030281

Gong, Y.-Z., Niu, Q.-Y., Liu, Y.-G., Dong, J., & Xia, M.-M. (2022). Development of multifarious carrier materials and impact conditions of immobilised microbial technology for environmental remediation: A review. *Environmental Pollution, 314*, 120232. DOI: 10.1016/j.envpol.2022.120232 PMID: 36155222

Gunasena, M. D. K. M., Galpaya, G. D. C. P., Abeygunawardena, C. J., Induranga, D. K. A., Priyadarshana, H. V. V., Millavithanachchi, S. S., Bandara, P. K. G. S. S., & Koswattage, K. R. (2025). Advancements in Bio-Nanotechnology: Green Synthesis and Emerging Applications of Bio-Nanoparticles. *Nanomaterials (Basel, Switzerland), 15*(7), 7. Advance online publication. DOI: 10.3390/nano15070528 PMID: 40214574

Han, J.-H., & Yun, K.-S. (2024). Real-Time Soil Moisture Monitoring Using a Graphene Oxide Sensor Powered by a Solar Cell-Based Energy Harvester. *Journal of Electrical Engineering & Technology, 19*(5), 3331–3337. DOI: 10.1007/s42835-024-01790-2

Haroon, S. (2021). Manipulation of Bio-nanotechnology in Agri-Sector for Pest Control with Special Emphasis to Nematode Problems in Egypt. *Annals of Agricultural Sciences, Moshtohor, 59*(2), 33–36. DOI: 10.21608/assjm.2021.183354

Hlihor, R. M. (2015). *Experimental analysis and mathematical prediction of Cd(II) removal by biosorption using support vector machines and genetic algorithms. 32*(3).

Hoque, A., & Padhiary, M. (2024). Automation and AI in Precision Agriculture: Innovations for Enhanced Crop Management and Sustainability. *Asian Journal of Research in Computer Science, 17*(10), 95–109. DOI: 10.9734/ajrcos/2024/v17i10512

Hoque, A., Padhiary, M., Kumar, K., & Barbhuiya, J. A. (2025). The Role of Robotics in Sustainable Agriculture and Waste Management. In Rajest, S. S., & Baron, M. (Eds.), *A. J. Obaid, Muthmainnah* (pp. 265–302). Multidisciplinary Advancements in Human-AI Augmentation. IGI Global Scientific Publishing., DOI: 10.4018/979-8-3373-1987-2.ch012

Hoque, A., Padhiary, M., Prasad, G., & Kumar, K. (2025). Real-Time Data Processing in Agricultural Robotics. In Bora, D. J., & Bania, R. K. (Eds.), *Computer Vision Techniques for Agricultural Advancements* (pp. 431–468). IGI Global., DOI: 10.4018/979-8-3693-8019-2.ch014

Huang, K., Li, M., Li, R., Rasul, F., Shahzad, S., Wu, C., Shao, J., Huang, G., Li, R., Almari, S., Hashem, M., & Aamer, M. (2023). Soil acidification and salinity: The importance of biochar application to agricultural soils. *Frontiers in Plant Science*, *14*, 1206820. DOI: 10.3389/fpls.2023.1206820 PMID: 37780526

Huang, T., Li, Y., Qian, J., Liu, S., & Wang, J. (2024). Remediation of Cr(VI)-contaminated soil by double-modified nanoscale zero-valent iron: Performance and mechanism. *Journal of Soils and Sediments*, *24*(4), 1724–1738. DOI: 10.1007/s11368-024-03768-z

Inoue, Y. (2020). Satellite- and drone-based remote sensing of crops and soils for smart farming – a review. *Soil Science and Plant Nutrition*, *66*(6), 798–810. DOI: 10.1080/00380768.2020.1738899

Irfan, S. A., Azeem, B., Irshad, K., & Algarni, S., KuShaari, K., Islam, S., & Abdelmohimen, M. A. H. (2020). Machine Learning Model for Nutrient Release from Biopolymers Coated Controlled-Release Fertilizer. *Agriculture*, *10*(11). Advance online publication. DOI: 10.3390/agriculture10110538

James, J., Choudhary, P., & Singh, S. Archana, & Sharma, N. (2024). Regenerative Agriculture: Potential, Progress, Opportunities, and Challenges. In S. Kumar, R. S. Meena, P. Sheoran, & M. K. Jhariya (Eds.), *Regenerative Agriculture for Sustainable Food Systems* (pp. 49–82). Springer Nature Singapore. https://doi.org/DOI: 10.1007/978-981-97-6691-8_3

Jayara, A. S., Kumar, R., Pandey, P., Singh, S., Shukla, A., Singh, A. P., Pandey, S., Meena, R. L., & Reddy, K. I. (2023). BOOSTING NUTRIENT USE EFFICIENCY THROUGH FERTILIZER USE MANAGEMENT. *Applied Ecology and Environmental Research*, *21*(4), 2931–2952. DOI: 10.15666/aeer/2104_29312952

Jesus, J. D., Oliveira-Esquerre, K., & Medeiros, D. L. (2022). Environmental model using life cycle assessment and artificial intelligence techniques to predict impacts on industrial water treatment. *IOP Conference Series. Materials Science and Engineering, 1250*(1), 012002. DOI: 10.1088/1757-899X/1250/1/012002

Kadulkar, S., Howard, M. P., Truskett, T. M., & Ganesan, V. (2021). Prediction and Optimization of Ion Transport Characteristics in Nanoparticle-Based Electrolytes Using Convolutional Neural Networks. *The Journal of Physical Chemistry B, 125*(18), 4838–4849. DOI: 10.1021/acs.jpcb.1c02004 PMID: 33914555

Kalantzopoulos, G., Paraskevopoulos, P., Domalis, G., Liopa-Tsakalidi, A., Tsesmelis, D. E., & Barouchas, P. E. (2024). The Western Greece Soil Information System (WESIS)—A Soil Health Design Supported by the Internet of Things, Soil Databases, and Artificial Intelligence Technologies in Western Greece. *Sustainability, 16*(8), 3478. DOI: 10.3390/su16083478

Karunaratne, T. N., Rodrigo, P. M., Oguntuyi, D. O., Mlsna, T. E., Zhang, J., & Zhang, X. (2023). Unraveling biochar surface area on structure and heavy metal removal performances of carbothermal reduced nanoscale zero-valent iron. *Journal of Bioresources and Bioproducts, 8*(4), 388–398. DOI: 10.1016/j.jobab.2023.06.003

Kaur, R., & Bhushanwar, K. (2025). Enhancing Sustainability, Climate Resilience, and Resource Efficiency with IoT-Based Precision Agriculture. *International Journal of Research and Review in Applied Science, Humanities, and Technology*, 172–179. https://doi.org/DOI: 10.71143/7db36796

Khaki, S., Wang, L., & Archontoulis, S. V. (2020). A CNN-RNN Framework for Crop Yield Prediction. *Frontiers in Plant Science, 10*, 1750. DOI: 10.3389/fpls.2019.01750 PMID: 32038699

Kibrete, F., Trzepieciński, T., Gebremedhen, H. S., & Woldemichael, D. E. (2023). Artificial Intelligence in Predicting Mechanical Properties of Composite Materials. *Journal of Composites Science, 7*(9), 9. Advance online publication. DOI: 10.3390/jcs7090364

Kovtun, A., Campodoni, E., Favaretto, L., Zambianchi, M., Salatino, A., Amalfitano, S., Navacchia, M. L., Casentini, B., Palermo, V., Sandri, M., & Melucci, M. (2020). Multifunctional graphene oxide/biopolymer composite aerogels for microcontaminants removal from drinking water. *Chemosphere, 259*, 127501. DOI: 10.1016/j.chemosphere.2020.127501

Kukowska, S., & Szewczuk-Karpisz, K. (2024). Biochar and Zeolite Uses in Improving Immobilization of Nutrients and Pollutants in Soils. *Separation and Purification Reviews*, •••, 1–24. DOI: 10.1080/15422119.2024.2426157

Kumar, A. K., Gautam, Y., & Singh, N. (2024). State-of-the-art developments in surface functionalized carbon-based bio/nanocomposites for theranostic antibacterials, advanced bioimaging, and molecular bioelectronics inspired biosensing platforms. *Journal of Industrial and Engineering Chemistry*, *140*, 103–145. DOI: 10.1016/j.jiec.2024.05.065

Kumar, K., Hoque, A., Padhiary, M., & Saikia, P. (2025). Emerging Trends in Nanotechnology, Autonomous Farms, and Collaborative Agri-Industry Marketing: The Future of AI in Food Production. In Whig, P., & Elngar, A. (Eds.), (pp. 239–278). AI Innovations for Transforming Food Production. IGI Global Scientific Publishing., DOI: 10.4018/979-8-3373-0842-5.ch010

Kundacina, I., Kundacina, O., Miskovic, D., & Radonic, V. (2025). Advancing microfluidic design with machine learning: A Bayesian optimization approach. *Lab on a Chip*, *25*(4), 657–672. DOI: 10.1039/D4LC00872C PMID: 39887214

Lal, R. (2012). Climate Change and Soil Degradation Mitigation by Sustainable Management of Soils and Other Natural Resources. *Agricultural Research*, *1*(3), 199–212. DOI: 10.1007/s40003-012-0031-9

Lallawmkimi, M. C., Patil, S., Upadhyay, D. K., Majumdar, N. B. A., Kumar, G. S., & Panigrahi, C. K. (2025). Application of Nanotechnology in Agriculture: Opportunities and Challenges in the Context of Environmental Sustainability. *Archives of Current Research International*, *25*(1), 37–53. DOI: 10.9734/acri/2025/v25i11035

Larson, D. B., Harvey, H., Rubin, D. L., Irani, N., Tse, J. R., & Langlotz, C. P. (2021). Regulatory Frameworks for Development and Evaluation of Artificial Intelligence–Based Diagnostic Imaging Algorithms: Summary and Recommendations. *Journal of the American College of Radiology*, *18*(3), 413–424. DOI: 10.1016/j.jacr.2020.09.060 PMID: 33096088

Lavado, G. J., Baderna, D., Carnesecchi, E., Toropova, A. P., Toropov, A. A., Dorne, J. L. C. M., & Benfenati, E. (2022). QSAR models for soil ecotoxicity: Development and validation of models to predict reproductive toxicity of organic chemicals in the collembola *Folsomia candida*. *Journal of Hazardous Materials*, *423*, 127236. DOI: 10.1016/j.jhazmat.2021.127236 PMID: 34844354

Lehman, R., Cambardella, C., Stott, D., Acosta-Martinez, V., Manter, D., Buyer, J., Maul, J., Smith, J., Collins, H., Halvorson, J., Kremer, R., Lundgren, J., Ducey, T., Jin, V., & Karlen, D. (2015). Understanding and Enhancing Soil Biological Health: The Solution for Reversing Soil Degradation. *Sustainability*, *7*(1), 988–1027. DOI: 10.3390/su7010988

Li, T., Cui, L., Kuhnert, M., McLaren, T. I., Pandey, R., Liu, H., Wang, W., Xu, Z., Xia, A., Dalal, R. C., & Dang, Y. P. (2024). A comprehensive review of soil organic carbon estimates: Integrating remote sensing and machine learning technologies. *Journal of Soils and Sediments*, *24*(11), 3556–3571. DOI: 10.1007/s11368-024-03913-8

Liu, R., & Lal, R. (2012). Nanoenhanced Materials for Reclamation of Mine Lands and Other Degraded Soils: A Review. *Journal of Nanotechnology*, *2012*, 1–18. DOI: 10.1155/2012/461468

Liu, X., Tang, Harper, Steevens, J., Xu, R., & Harper. (2013). Predictive modeling of nanomaterial exposure effects in biological systems. *International Journal of Nanomedicine*, *31*. Advance online publication. DOI: 10.2147/IJN.S40742 PMID: 24098077

Liu, X., Wu, Y., Zhao, X., & Wang, Z. (2021). Fabrication and applications of bioactive chitosan-based organic-inorganic hybrid materials: A review. *Carbohydrate Polymers*, *267*, 118179. DOI: 10.1016/j.carbpol.2021.118179 PMID: 34119147

Lowe, M., Qin, R., & Mao, X. (2022). A Review on Machine Learning, Artificial Intelligence, and Smart Technology in Water Treatment and Monitoring. *Water (Basel)*, *14*(9), 9. Advance online publication. DOI: 10.3390/w14091384

Machado, T. O., Grabow, J., Sayer, C., de Araújo, P. H. H., Ehrenhard, M. L., & Wurm, F. R. (2022). Biopolymer-based nanocarriers for sustained release of agrochemicals: A review on materials and social science perspectives for a sustainable future of agri- and horticulture. *Advances in Colloid and Interface Science*, *303*, 102645. DOI: 10.1016/j.cis.2022.102645 PMID: 35358807

Menossi, M., Casalongué, C., & Alvarez, V. A. (2022). Bio-nanocomposites for Modern Agricultural Applications. In Mallakpour, S., & Hussain, C. M. (Eds.), *Handbook of Consumer Nanoproducts* (pp. 1201–1237). Springer Nature., DOI: 10.1007/978-981-16-8698-6_68

Miklasová, M., Komendová, R., Brtnický, M., Čelan Korošin, N., & Kučerík, J. (2025). Short- and long-term influences of rutile and anatase nanoparticles on soil organic matter structure, ageing and moisture. *Journal of Thermal Analysis and Calorimetry*, *150*(19), 15103–15116. DOI: 10.1007/s10973-025-14741-4

Mishra, A. K., Paul, R. A. I., Grover, D., Das, R., Mishra, S., Akshu, A., & Pavithra, P. Roohi, Soni, P. G., & Sharma, S. (2025). Nanotechnology for Soil and Water Management. In K. A. Abd-Elsalam (Ed.), *Agricultural Nanotechnology* (pp. 371–418). Royal Society of Chemistry. https://doi.org/DOI: 10.1039/9781837677313-00371

Mishra, B., Mohanta, Y. K., Reddy, C. N., Reddy, S. D. M., Mandal, S. K., Yadavalli, R., & Sarma, H. (2023). Valorization of agro-industrial biowaste to biomaterials: An innovative circular bioeconomy approach. *Circular Economy*, *2*(3), 100050. DOI: 10.1016/j.cec.2023.100050

Munir, A., Salah, M. A., Ali, M., Ali, B., Saleem, M. H., Samarasinghe, K. G. B. A., De Silva, S. I. S., Ercisli, S., Iqbal, N., & Anas, M. (2024). Advancing Agriculture: Harnessing Smart Nanoparticles for Precision Fertilization. *BioNanoScience*, *14*(4), 3846–3863. DOI: 10.1007/s12668-024-01597-5

Mustapha, S., Ndamitso, M. M., Abdulkareem, A. S., Tijani, J. O., Shuaib, D. T., Ajala, A. O., & Mohammed, A. K. (2020). Application of TiO2 and ZnO nanoparticles immobilized on clay in wastewater treatment: A review. *Applied Water Science*, *10*(1), 49. DOI: 10.1007/s13201-019-1138-y

Na, J., Kshetrimayum, K. S., Lee, U., & Han, C. (2017). Multi-objective optimization of microchannel reactor for Fischer-Tropsch synthesis using computational fluid dynamics and genetic algorithm. *Chemical Engineering Journal*, *313*, 1521–1534. DOI: 10.1016/j.cej.2016.11.040

Nakhli, S. A. A., Delkash, M., Bakhshayesh, B. E., & Kazemian, H. (2017). Application of Zeolites for Sustainable Agriculture: A Review on Water and Nutrient Retention. *Water, Air, and Soil Pollution*, *228*(12), 464. DOI: 10.1007/s11270-017-3649-1

Nandhini, J., Karthikeyan, E., & Rajeshkumar, S. (2024). Eco-friendly bionanocomposites: Pioneering sustainable biomedical advancements in engineering. *Discover Nano*, *19*(1), 86. DOI: 10.1186/s11671-024-04007-7 PMID: 38724698

Naseem, T., & Durrani, T. (2021). The role of some important metal oxide nanoparticles for wastewater and antibacterial applications: A review. *Environmental Chemistry and Ecotoxicology*, *3*, 59–75. DOI: 10.1016/j.enceco.2020.12.001

Othman, S. H., Ronzi, N. D. A., Shapi'i, R. A., Dun, M., Ariffin, S. H., & Mohammed, M. A. P. (2023). Biodegradability of Starch Nanocomposite Films Containing Different Concentrations of Chitosan Nanoparticles in Compost and Planting Soils. *Coatings*, *13*(4), 777. DOI: 10.3390/coatings13040777

Padhiary, M. (2025). Benefits, Risks, and Responsible Use of Nanomaterials in Wastewater Remediation. In Agarwal, N., Shah, M. P., Solanki, V. S., & Singh, N. (Eds.), *Nanomaterials in Wastewater Research: Progress and Challenges* (pp. 443–464). Springer Nature., DOI: 10.1007/978-981-96-3908-3_19

Padhiary, M., Hoque, A., Prasad, G., Kumar, K., & Sahu, B. (2025). Precision Agriculture and AI-Driven Resource Optimization for Sustainable Land and Resource Management. In Ruiz-Vanoye, J. A., & Díaz-Parra, O. (Eds.), *Smart Water Technology for Sustainable Management in Modern Cities* (pp. 197–232). IGI Global., DOI: 10.4018/979-8-3693-8074-1.ch009

Padhiary, M., & Kumar, R. (2024). Assessing the Environmental Impacts of Agriculture, Industrial Operations, and Mining on Agro-Ecosystems. In Azrour, M., Mabrouki, J., Alabdulatif, A., Guezzaz, A., & Amounas, F. (Eds.), *Smart Internet of Things for Environment and Healthcare* (Vol. 1165, pp. 107–126). Springer Nature Switzerland., DOI: 10.1007/978-3-031-70102-3_8

Padhiary, M., Kumar, R., & Sethi, L. N. (2024). Navigating the Future of Agriculture: A Comprehensive Review of Automatic All-Terrain Vehicles in Precision Farming. *Journal of The Institution of Engineers (India): Series A, 105*(3), 767–782. https://doi.org/DOI: 10.1007/s40030-024-00816-2

Padhiary, M., Powar, S., Asokan, A., Sundar Rajan, M., Al-Said, K., & Al Said, N. (2025). Combining Convolutional Neural Networks with Reinforcement Learning for Autonomous Robotics. *2025 International Conference on Pervasive Computational Technologies (ICPCT)*, 421–425. https://doi.org/DOI: 10.1109/ICPCT64145.2025.10940484

Padhiary, M., Roy, D., & Dey, P. (2024). Mapping the Landscape of Biogenic Nanoparticles in Bioinformatics and Nanobiotechnology: AI-Driven Insights. In Das, S., Khade, S. M., Roy, D. B., & Trivedi, K. (Eds.), *Synthesizing and Characterizing Plant-Mediated Biocompatible Metal Nanoparticles* (pp. 337–376). IGI Global., DOI: 10.4018/979-8-3693-6240-2.ch014

Padhiary, M., Roy, P., Dey, P., & Sahu, B. (2024). Harnessing AI for Automated Decision-Making in Farm Machinery and Operations: Optimizing Agriculture. In Hai-Jew, S. (Ed.), *Enhancing Automated Decision-Making Through AI* (pp. 249–282). IGI Global., DOI: 10.4018/979-8-3693-6230-3.ch008

Pagano, M., Lunetta, E., Belli, F., Mocarli, G., Cocozza, C., & Cacciotti, I. (2025). Advancements in Agricultural Nanotechnology: An Updated Review. *Plants, 14*(18), 2939. DOI: 10.3390/plants14182939 PMID: 41012090

Pandey, H., & Pandey, R. K. (2025). Ensuring Safe Nanotechnology: A Biosafety and Regulatory Perspective. In Esquivel Escalante, K., Feregrino Pérez, A. A., Rosales Pérez, A., & Prasad, R. (Eds.), *Green Nanomaterials* (pp. 257–272). Springer Nature Switzerland., DOI: 10.1007/978-3-032-04492-1_12

Pandey, U., Vashisth, A., Mishra, A., & Kumari, M. (2025). *AI and IoT-Driven Soil Health Restoration: A Machine Learning Approach for Sustainable Agriculture*. In Review. https://doi.org/DOI: 10.21203/rs.3.rs-6490610/v1

Parul, & Das, V. (2024). Environmental Sustainability, Compliance and Life Cycle Assessment of AI-Integrated Systems. In S. Gulati (Ed.), *Application of Artificial Intelligence in Wastewater Treatment* (pp. 225–243). Springer Nature Switzerland. https://doi.org/DOI: 10.1007/978-3-031-69433-2_9

Pathan, M., Patel, N., Yagnik, H., & Shah, M. (2020). Artificial cognition for applications in smart agriculture: A comprehensive review. *Artificial Intelligence in Agriculture*, *4*, 81–95. DOI: 10.1016/j.aiia.2020.06.001

Pawar, Dr. P., Goyal, Dr. J., Gupta, A. K., Y, S. K., Kuttimarks, Dr. M. S., & Desale, Dr. S. V. (2025). Smart Nanomaterials And AI-Enhanced Technologies For Environmental Separation And Sensing. *International Journal of Environmental Sciences*, *11*(8s), 856–877. DOI: 10.64252/hr71wx22

Phadkule, S. S., & Sarma, S. (2023). Progress in nanocomposite based flexible temperature sensors: A review. *Measurement. Sensors*, *27*, 100692. DOI: 10.1016/j.measen.2023.100692

Pulido-Reyes, G., Leganes, F., Fernández-Piñas, F., & Rosal, R. (2017). Bio-nano interface and environment: A critical review. *Environmental Toxicology and Chemistry*, *36*(12), 3181–3193. DOI: 10.1002/etc.3924 PMID: 28731222

Qian, L., Mei, C., Luo, W., Liu, W., Chen, M., Zhang, X., Li, X., Cheng, B., & Ma, H. (2023). *A Versatile Biochar Fertilizer: Excellent Adsorption Performance for Multi-Heavy Metals and Enhanced Effect on Plant Growth in Metal Contaminated Soil*. Elsevier BV. https://doi.org/DOI: 10.2139/ssrn.4535379

Quinton, J. N., & Fiener, P. (2024). Soil erosion on arable land: An unresolved global environmental threat. *Progress in Physical Geography*, *48*(1), 136–161. DOI: 10.1177/03091333231216595

Rabehi, A., Helal, H., Zappa, D., & Comini, E. (2024). Advancements and Prospects of Electronic Nose in Various Applications: A Comprehensive Review. *Applied Sciences (Basel, Switzerland)*, *14*(11), 4506. DOI: 10.3390/app14114506

Rahim, H. U., Qaswar, M., Uddin, M., Giannini, C., Herrera, M. L., & Rea, G. (2021). Nano-Enable Materials Promoting Sustainability and Resilience in Modern Agriculture. *Nanomaterials (Basel, Switzerland)*, *11*(8), 2068. DOI: 10.3390/nano11082068 PMID: 34443899

Raj, P., & Ghosh, A., Ravi, Jadav, S. K., Patra, H., Shehnaz, Panotra, N., Sharma, V., Sharma, R., Pandey, S. K., & Singh, B. V. (2024). Nanomaterials for Sustainable Soil Remediation and Contaminant Immobilization. *Asian Journal of Soil Science and Plant Nutrition*, *10*(4), 684–697. DOI: 10.9734/ajsspn/2024/v10i4439

Rana, A., Gautam, D., Kumar, P., & Kumar Das, A. (2025). Architectures, Benefits, Security, and Privacy Issues of Internet of Nano Things: A Comprehensive Survey, Opportunities, and Research Challenges. *IEEE Communications Surveys and Tutorials*, *27*(2), 1152–1190. DOI: 10.1109/COMST.2024.3423477

Rando, G., Sfameni, S., Galletta, M., Drommi, D., Cappello, S., & Plutino, M. R. (2022). Functional Nanohybrids and Nanocomposites Development for the Removal of Environmental Pollutants and Bioremediation. *Molecules (Basel, Switzerland)*, *27*(15), 15. Advance online publication. DOI: 10.3390/molecules27154856 PMID: 35956804

Rani, S., & Baloda, S., Dinesh, & Mehta, A. (2024). AI-Driven Insights into Soil Physio-Chemical Properties. *International Journal of Environment and Climate Change*, *14*(12), 834–845. DOI: 10.9734/ijecc/2024/v14i124666

Rodrigues, C. I. D., Brito, L. M., & Nunes, L. J. R. (2023). Soil Carbon Sequestration in the Context of Climate Change Mitigation: A Review. *Soil Systems*, *7*(3), 3. Advance online publication. DOI: 10.3390/soilsystems7030064

Rostamzadeh-Renani, R., Jasim, D. J., Baghoolizadeh, M., Rostamzadeh-Renani, M., Andani, H. T., Salahshour, S., & Baghaei, Sh. (2024). Multi-objective optimization of rheological behavior of nanofluids containing CuO nanoparticles by NSGA II, MOPSO, and MOGWO evolutionary algorithms and group method of data handling artificial neural networks. *Materials Today. Communications*, *38*, 107709. DOI: 10.1016/j.mtcomm.2023.107709

Roy, P., Padhiary, M., Hoque, A., Sahu, B., Roy, D., & Kumar, K. (2025). Machine Learning for Precision Agriculture and Crop Yield Optimization: Techniques and Applications. In Emara, T., Hassan, E., Trinh, T., Li, G., & Saber, A. (Eds.), *The Role of Artificial Intelligence in Advancing Applied Life Sciences* (pp. 189–234). IGI Global., DOI: 10.4018/979-8-3693-9208-9.ch007

Sahana, M., Rehman, S., Patel, P. P., Dou, J., Hong, H., & Sajjad, H. (2020). Assessing the degree of soil salinity in the Indian Sundarban Biosphere Reserve using measured soil electrical conductivity and remote sensing data–derived salinity indices. *Arabian Journal of Geosciences*, *13*(24). Advance online publication. DOI: 10.1007/s12517-020-06310-w

Samson Prince, H., Macarius, N. M. N., Chandrashekhar Walkikar, S., Das, S., Rajan, A. P., & Deepankumar, K. (2025). Harmonizing Nature and Innovation: Sustainable Environmental Solutions Through Nano-biotech Synergy. In Jeyaseelan, A., & Murugasen, K. (Eds.), *Sustainable Environmental Remediation: Avenues in Nano and Biotechnology* (pp. 35–83). Springer Nature Switzerland., DOI: 10.1007/978-3-031-78483-5_2

Santos, M., Rebola, S., & Evtuguin, D. V. (2025). Soil Remediation: Current Approaches and Emerging Bio-Based Trends. *Soil Systems*, *9*(2), 2. Advance online publication. DOI: 10.3390/soilsystems9020035

Santos, P. A., Biraku, X., Nielsen, E., Ozketen, A. C., Ozketen, A. A., & Hakki, E. E. (2025). Agricultural nanotechnology for a safe and sustainable future: Current status, challenges, and beyond. *Journal of the Science of Food and Agriculture*, *105*(6), 3159–3169. DOI: 10.1002/jsfa.13922 PMID: 39319478

Saraswat, P., Singh, S., Prasad, M., Misra, R., Rajput, V. D., & Ranjan, R. (2023). Applications of bio-based nanomaterials in environment and agriculture: A review on recent progresses. *Hybrid Advances*, *4*, 100097. DOI: 10.1016/j.hybadv.2023.100097

Saravanan, A., Swaminaathan, P., Kumar, P. S., Yaashikaa, P. R., Kamalesh, R., & Rangasamy, G. (2023). A comprehensive review on immobilized microbes - biochar and their environmental remediation: Mechanism, challenges and future perspectives. *Environmental Research*, *236*, 116723. DOI: 10.1016/j.envres.2023.116723 PMID: 37487925

Sharma, D., Teli, G., Gupta, K., Bansal, G., Gupta, G. D., & Chawla, P. A. (2022). Nano-biosensors from Agriculture to Nextgen Diagnostic Tools. *Current Nanomaterials*, *7*(2), 110–138. DOI: 10.2174/2405461507666220131104843

Sharma, K., & Shivandu, S. K. (2024). Integrating artificial intelligence and Internet of Things (IoT) for enhanced crop monitoring and management in precision agriculture. *Sensors International*, *5*, 100292. DOI: 10.1016/j.sintl.2024.100292

Sharma, R., & Kumar, V. (2024). Nano enabled agriculture for sustainable soil. *Waste Management Bulletin*, *2*(1), 152–161. DOI: 10.1016/j.wmb.2024.01.002

Sharmile, N., Chowdhury, R. R., & Desai, S. (2025). A Comprehensive Review of Quality Control and Reliability Research in Micro–Nano Technology. *Technologies*, *13*(3), 94. DOI: 10.3390/technologies13030094

Shchipunov, Y. (2012). Bionanocomposites: Green sustainable materials for the near future. *Pure and Applied Chemistry*, *84*(12), 2579–2607. DOI: 10.1351/PAC-CON-12-05-04

Sheershwal, A., Singh, A., Sharma, V., & Trivedi, B. (2025). Microbial synthesis of nanoparticles for sustainable agricultural advancements: A comprehensive review. *Nanotechnology for Environmental Engineering*, *10*(1), 16. DOI: 10.1007/s41204-025-00405-z

Shi, Y., Zhang, S., Zhou, H., Dong, Y., Liu, G., Ye, W., He, R., & Zhao, G. (2025). Recent Developments in Heavy Metals Detection: Modified Electrodes, Pretreatment Methods, Prediction Models and Algorithms. *Metals*, *15*(1), 1. Advance online publication. DOI: 10.3390/met15010080

Shi, Y., Zhong, S., Wang, X., & Feng, C. (2022). A review of the removal of heavy metal ions in wastewater by modified montmorillonite. *Water Policy*, *24*(10), 1590–1609. DOI: 10.2166/wp.2022.033

Sibalija, T. V., & Majstorovic, V. D. (2012). An integrated approach to optimise parameter design of multi-response processes based on Taguchi method and artificial intelligence. *Journal of Intelligent Manufacturing*, *23*(5), 1511–1528. DOI: 10.1007/s10845-010-0451-y

Singh, A. V., Ansari, M. H. D., Rosenkranz, D., Maharjan, R. S., Kriegel, F. L., Gandhi, K., Kanase, A., Singh, R., Laux, P., & Luch, A. (2020). Artificial Intelligence and Machine Learning in Computational Nanotoxicology: Unlocking and Empowering Nanomedicine. *Advanced Healthcare Materials*, *9*(17), 1901862. DOI: 10.1002/adhm.201901862 PMID: 32627972

Singh, A. V., Rosenkranz, D., Ansari, M. H. D., Singh, R., Kanase, A., Singh, S. P., Johnston, B., Tentschert, J., Laux, P., & Luch, A. (2020). Artificial Intelligence and Machine Learning Empower Advanced Biomedical Material Design to Toxicity Prediction. *Advanced Intelligent Systems*, *2*(12), 2000084. DOI: 10.1002/aisy.202000084

Singh, D., & Gurjar, B. R. (2022). Nanotechnology for agricultural applications: Facts, issues, knowledge gaps, and challenges in environmental risk assessment. *Journal of Environmental Management*, *322*, 116033. DOI: 10.1016/j.jenvman.2022.116033

Singh, S. P., Paneru, P., & Deepak, D. (2025). Nanotechnological approaches for enhancing climate resilience and sustainable adaptation in agriculture. In S. P. Singh (Ed), P. Paneru (Ed), & K. K. Singh (Ed), *Deep Science Publishing*. Deep Science Publishing. https://doi.org/DOI: 10.70593/978-93-49307-12-4_14

Srivastava, A. K., Dev, A., & Karmakar, S. (2018). Nanosensors and nanobiosensors in food and agriculture. *Environmental Chemistry Letters*, *16*(1), 161–182. DOI: 10.1007/s10311-017-0674-7

Taha, M. F., Mao, H., Zhang, Z., Elmasry, G., Awad, M. A., Abdalla, A., Mousa, S., Elwakeel, A. E., & Elsherbiny, O. (2025). Emerging Technologies for Precision Crop Management Towards Agriculture 5.0: A Comprehensive Overview. *Agriculture, 15*(6), 582. DOI: 10.3390/agriculture15060582

TavallaieNejad, A., Vila, M. C., Paneiro, G., & Baptista, J. S. (2025). A Systematic Review of Machine Learning Algorithms for Soil Pollutant Detection Using Satellite Imagery. *Remote Sensing, 17*(7), 1207. DOI: 10.3390/rs17071207

Teixeira, W. G., Víctor Hugo Alvarez, V., & Neves, J. C. L. (2020). New methods for estimating lime requirement to attain desirable pH values in Brazilian soils. *Revista Brasileira de Ciência do Solo, 44*. Advance online publication. DOI: 10.36783/18069657rbcs20200008

Thakur, A., & Kumar, A. (2022). Recent advances on rapid detection and remediation of environmental pollutants utilizing nanomaterials-based (bio)sensors. *The Science of the Total Environment, 834*, 155219. DOI: 10.1016/j.scitotenv.2022.155219 PMID: 35421493

Thirumalaisamy, R., Nimithap, S., Harini, S., Rajiniganth, R., Haritha, S., Ameer, K., Vadivel, P., & Selvankumar, T. (2025). Potential Application of Nanoemulsion in Agriculture: A Current Perspective. *International Journal of Advanced Science and Engineering, 11*(3), 4328–4350. DOI: 10.29294/IJASE.11.3.2025.4328-4350

Tiwari, K. N. (2022). Nano technology based P fertilizers for higher efficiency and agriculture sustainability. *ANNALS OF PLANT AND SOIL RESEARCH, 24*(2), 198–207. DOI: 10.47815/apsr.2022.10149

Tsakiridis, N. L., Samarinas, N., Kalopesa, E., & Zalidis, G. C. (2023). Cognitive Soil Digital Twin for Monitoring the Soil Ecosystem: A Conceptual Framework. *Soil Systems, 7*(4), 88. DOI: 10.3390/soilsystems7040088

Uddin, Md. H., Mulla, M. H., Abedin, T., Manap, A., Yap, B. K., Rajamony, R. K., Shahapurkar, K., Khan, T. M. Y., Soudagar, M. E. M., & Nur-E-Alam, M. (2025). Advances in natural fiber polymer and PLA composites through artificial intelligence and machine learning integration. *Journal of Polymer Research, 32*(3), 76. DOI: 10.1007/s10965-025-04282-7

Vidal, C. L. A., & Govan, J. E. (2024). Machine Learning Techniques for Improving Nanosensors in Agroenvironmental Applications. *Agronomy (Basel), 14*(2), 2. Advance online publication. DOI: 10.3390/agronomy14020341

Volk, A. A., Epps, R. W., Yonemoto, D. T., Masters, B. S., Castellano, F. N., Reyes, K. G., & Abolhasani, M. (2023). AlphaFlow: Autonomous discovery and optimization of multi-step chemistry using a self-driven fluidic lab guided by reinforcement learning. *Nature Communications*, *14*(1), 1403. DOI: 10.1038/s41467-023-37139-y PMID: 36918561

Wahab, A., Muhammad, M., Ullah, S., Abdi, G., Shah, G. M., Zaman, W., & Ayaz, A. (2024). Agriculture and environmental management through nanotechnology: Eco-friendly nanomaterial synthesis for soil-plant systems, food safety, and sustainability. *The Science of the Total Environment*, *926*, 171862. DOI: 10.1016/j.scitotenv.2024.171862 PMID: 38527538

Wai, K. P., Chia, M. Y., Koo, C. H., Huang, Y. F., & Chong, W. C. (2022). Applications of deep learning in water quality management: A state-of-the-art review. *Journal of Hydrology (Amsterdam)*, *613*, 128332. DOI: 10.1016/j.jhydrol.2022.128332

Wang, C., Ji, Y., Cao, X., Yue, L., Chen, F., Li, J., Yang, H., Wang, Z., & Xing, B. (2022). Carbon Dots Improve Nitrogen Bioavailability to Promote the Growth and Nutritional Quality of Soybeans under Drought Stress. *ACS Nano*, *16*(8), 12415–12424. DOI: 10.1021/acsnano.2c03591 PMID: 35946591

Wang, H., Zhang, L., & Zhao, J. (2023). Application of a Fusion Attention Mechanism-Based Model Combining Bidirectional Gated Recurrent Units and Recurrent Neural Networks in Soil Nutrient Content Estimation. *Agronomy (Basel)*, *13*(11), 2724. DOI: 10.3390/agronomy13112724

Wang, X., Tarahomi, M., Sheibani, R., Xia, C., & Wang, W. (2023). Progresses in lignin, cellulose, starch, chitosan, chitin, alginate, and gum/carbon nanotube (nano) composites for environmental applications: A review. *International Journal of Biological Macromolecules*, *241*, 124472. DOI: 10.1016/j.ijbiomac.2023.124472 PMID: 37076069

Xiuquan, Z., Wang, G., Li, Z., Zheng, D., & Song, H. (2025). *Soil Organic Matter Estimation Using Fusion Raman Spectroscopy Based on Multimodal Deep Learning* (SSRN Scholarly Paper No. 5244519). Social Science Research Network. https://doi.org/DOI: 10.2139/ssrn.5244519

Xueping, X. (2015). *Organic fertilizer capable of improving fertilizer efficiency of soil and preparation method thereof.*

Yang, J., Hou, B., Wang, J., Tian, B., Bi, J., Wang, N., Li, X., & Huang, X. (2019). Nanomaterials for the Removal of Heavy Metals from Wastewater. *Nanomaterials (Basel, Switzerland)*, *9*(3), 3. Advance online publication. DOI: 10.3390/nano9030424 PMID: 30871096

Yang, J., Li, C., Lo, L. S. H., Zhang, X., Chen, Z., Gao, J. U. C., Dai, Z., Nakaoka, M., Yang, H., & Cheng, J. (2024). Artificial Intelligence-Assisted Environmental DNA Metabarcoding and High-Resolution Underwater Optical Imaging for Noninvasive and Innovative Marine Environmental Monitoring. *Journal of Marine Science and Engineering, 12*(10), 1729. DOI: 10.3390/jmse12101729

Yang, X., Zhou, Y., Hu, J., Zheng, Q., Zhao, Y., Lv, G., & Liao, L. (2024). Clay minerals and clay-based materials for heavy metals pollution control. *The Science of the Total Environment, 954*, 176193. DOI: 10.1016/j.scitotenv.2024.176193 PMID: 39278488

Yasmin, K., & Hossain, Md. S., & Li, W. C. (2024). Simultaneous immobilization strategy of anionic metalloids and cationic metals in agricultural systems: A review. *Chemosphere, 364*, 143106. DOI: 10.1016/j.chemosphere.2024.143106 PMID: 39153530

Yin, H., Cao, Y., Marelli, B., Zeng, X., Mason, A. J., & Cao, C. (2021). Soil Sensors and Plant Wearables for Smart and Precision Agriculture. *Advanced Materials, 33*(20), 2007764. DOI: 10.1002/adma.202007764 PMID: 33829545

Youden, B., Jiang, R., Carrier, A. J., Servos, M. R., & Zhang, X. (2022). A Nanomedicine Structure–Activity Framework for Research, Development, and Regulation of Future Cancer Therapies. *ACS Nano, 16*(11), 17497–17551. DOI: 10.1021/acsnano.2c06337 PMID: 36322785

Yu, X., Keitel, C., Zhang, Y., Wangeci, A. N., & Dijkstra, F. A. (2022). Global meta-analysis of nitrogen fertilizer use efficiency in rice, wheat and maize. *Agriculture, Ecosystems & Environment, 338*, 108089. DOI: 10.1016/j.agee.2022.108089

Zaaba, N. F., & Jaafar, M. (2020). *A review on degradation mechanisms of polylactic acid: Hydrolytic, photodegradative, microbial, and enzymatic degradation*. https://doi.org/DOI: 10.1002/pen.25511

Zhai, J., Duan, S., Luo, B., Jin, X., Dong, H., & Wang, X. (2024). Classification techniques of ion selective electrode arrays in agriculture: A review. *Analytical Methods, 16*(47), 8068–8079. DOI: 10.1039/D4AY01346H PMID: 39543972

Zhang, P., Guo, Z., Ullah, S., Melagraki, G., Afantitis, A., & Lynch, I. (2021). Nanotechnology and artificial intelligence to enable sustainable and precision agriculture. *Nature Plants, 7*(7), 864–876. DOI: 10.1038/s41477-021-00946-6 PMID: 34168318

Zhang, Y., Wang, S., & Ji, G. (2015). A Comprehensive Survey on Particle Swarm Optimization Algorithm and Its Applications. *Mathematical Problems in Engineering, 2015*, 1–38. DOI: 10.1155/2015/931256

Zhao, J., & Wan, S. (2023). Artificial Intelligence and Hyperspectral Modeling for Soil Management. In Clarke, N., Peng, D., & Clarke, J. L. (Eds.), *Innovation for Environmentally-friendly Food Production and Food Safety in China* (pp. 67–91). Springer Nature Singapore., DOI: 10.1007/978-981-99-2828-6_4

Zhou, J., Du, B., Liu, H., Cui, H., Zhang, W., Fan, X., Cui, J., & Zhou, J. (2020). The bioavailability and contribution of the newly deposited heavy metals (copper and lead) from atmosphere to rice (Oryza sativa L.). *Journal of Hazardous Materials*, *384*, 121285. DOI: 10.1016/j.jhazmat.2019.121285 PMID: 31577969

Chapter 11
Synthesis, Properties, and Application of Bio–Nanocomposites:
Application of Nanobiochar and Biochar Nanocomposites in Environment

Jeremaih David Bala
Federal University of Technology, Minna, Nigeria

Oloruntoba Samuel Job
https://orcid.org/0009-0005-8571-695X
Federal University of Technology, Minna, Nigeria

Innocent Ojeba Musa
https://orcid.org/0000-0002-8843-6911
Skyline University, Nigeria

Al Hassan Abdullahi Abdul-Rahman
Federal University, Lokoja, Nigeria

Wama Binga Emmanuel
Taraba State University, Nigeria

Adamu Mustapha
Federal University of Technology, Minna, Nigeria

Nathaniel Adunbarin Abiodun
https://orcid.org/0009-0007-7637-9183
National Institute for Freshwater Fisheries Research, Nigeria

Nathaniel Nwachukwu Friday
https://orcid.org/0009-0007-3805-0571
Federal University of Technology, Minna, Nigeria

ABSTRACT

Modern society relies on nanobiochar and biochar nanocomposites as innovative materials to resolve major environmental issues alongside agricultural difficulties. Engineered biochar-based materials boost adsorption performance which makes

DOI: 10.4018/979-8-3373-2180-6.ch011

them efficient tools for wastewater bioremediation mainly focused on heavy metal and organic substance elimination. Advanced surface properties coupled with improved porosity and nanoparticle functionalization allows them to efficiently adsorb pollutants which include arsenic and chromium along with lead, cadmium and copper. Research findings show nanobiochar exhibits better capabilities for extracting pollutants from water sources which renders them critical components in environmental sustainability operations. Nanobiochar proves crucial for agriculture by performing various functions outside wastewater treatment. Nanobiochar supports plant development through better soil nutritive value and enhanced nutrient absorption while maintaining soil acidity at stable levels. Nanobiochar functions as a soil additive that improves organic matter composition while preserving liquid retention while preventing harmful metal accumulation. Nanobiochar has demonstrated chemical residue removal ability by absorbing dangerous pesticide residues that target environmental toxicity. The antimicrobial capabilities of nanobiochar enable disease suppression in soil whereas chemical pesticides have less environmental benefits. kullanım of nanobiochar promotes beneficial microbial growth that upgrades fertilizer cycling operations while strengthening soil well-being. Through fertilizer application nanobiochar supports effective nutrient dispersal that prevents waste and decreases pollution impacts on the environment. Both nanobiochar and biochar nanocomposites serve as decisive elements for implementing sustainable environmental measures and agricultural practices. These materials show remarkable value by improving all elements of water purification and soil quality and crop productivity and microbial balance to resolve international concerns about environmental pollution and food shortages. Research development needs focused material optimization to establish these materials as crucial elements for ecological protection and resource management systems.

1.0 INTRODUCTION

The worsening environmental problems which include soil degradation alongside water pollution and climate change escalate because of increasing population numbers coupled with urban development and industrial activities. The current situation requires innovative strategies which deliver both sustainability and performance for solving these problems (Dijoo and Khurshid, 2022). Biochar has become a promising solution because it serves various purposes in environmental management through its ability to sequester carbon and enhance soil quality as well as clean up pollution sources. However, conventional biochar often faces limitations in terms of performance, especially in complex environmental matrices. The regulated thermal decomposition of biomass feedstock, such as agricultural waste and other lignocellulosic materials, in the absence (pyrolysis) or restricted presence (gasification) of oxygen yields biochar, a solid carbonised byproduct. The production of biochar exists through several thermochemical processes which operate with solid waste together with animal manure. Biochar technology attracts intense scientific research because it lowers greenhouse gas emissions and strengthens plant growth and enhances soil conditions while eliminating pollutants according to Alimohammadi et al. (2011) and Ashfaq et al. (2017). Studies at numerous levels have evaluated the adsorption capacity of materials derived from biochar. The use of biochar-based materials plays a pivotal role in carbon sequestration reduction which helps plants grow better and fights against climate change (Saxena et al., 2013; Chen et al., 2017). The sorption ability of biochar-based materials supports the adsorption of multiple pollutants from water solutions according to Naghdi et al. (2018) because of their physical along with chemical surface characteristics. These encompass an extensive surface area, environmental abundance, significant microporosity, exceptional sorption capabilities, and notable structural features. The thermo-chemical treatment creates biofuels and syngas alongside biochar which makes this process financially superior to biomass combustion utilization (Naghdi et al., 2018). Biochar production through nanotechnology integration leads to the development of nanobiochar and biochar nanocomposites which bring advanced functionalities that create extensive opportunities for their applications (Zhou et al., 2024). The process of limiting oxygen during organic biomass pyrolysis produces biochar as a carbon-abundant substance. Researchers value biochar because it features both extensive pores and significant surface area and pollutant interception properties which make it suitable for numerous applications such as soil improvement and carbon capture and water purification. The traditional utilization of biochar for enhancing soil performance and minimizing greenhouse gas release expands into multiple additional valuable applications beyond basic usage. Recent advancements have sought to improve the physicochemical properties of biochar to make it more effective in tackling a variety of environmental challenges. This has led to the exploration of biochar modifications, including the incorporation of nanomaterials (Uday et al., 2022). Nanotechnology, which involves the manipulation of materials at the nanoscale (1-100 nanometers), has revolutionized various scientific fields, including environmental science. Materials at the nanoscale exhibit unique physical, chemical, and biological properties that differ significantly from their bulk counterparts. These properties include increased surface area, enhanced reactivity, and superior mechanical strength, making nanomaterials ideal for environmental applications (Khan et al., 2022). By integrating nanotechnology with biochar, researchers have created nanobiochar, a hybrid material that combines the benefits of biochar and nanomaterials. Nanobiochar exhibits superior adsorption capacity, catalytic activity, and reusability compared to traditional biochar. Nanobiochar refers to biochar materials that are modified or enhanced with nanostructures. This can involve reducing biochar to the nanoscale or incorporating nanoparticles into its structure. On the other hand, biochar nanocomposites are hybrid materials in which biochar serves as a matrix or support for nanomaterials. These composites leverage the synergistic properties of biochar and nanomaterials to achieve improved performance. For instance, the addition of metal oxides, carbon nanotubes, or graphene to biochar can significantly enhance its adsorption capacity and stability, making it more effective for applications such as water purification and soil remediation (Chausali et al., 2021).

Nanobiochar was developed to address the limits of biochar by improving its physical, chemical, and structural properties. According to Noreen and Abd-Elsalam (2021), one of its distinguishing and favourable characteristics is its increased surface area, which is due to its nanosize. It also features a large number of functional groups, which contribute to its increased mechanical and thermal properties. It has a better adsorption capacity than macrobiochar due to its smaller hydrodynamic radius, greater number of functional groups with oxygen and carbon defects, and more negative zeta potentials (Ramanayaka et al., 2020). Topological or edge defects in carbon nanomaterials frequently affect the material's physiochemical properties. Creating material edges with a high number of unpaired electrons can minimise formation energy and accelerate electron transmission. Carbon flaws also have a significant impact on catalytic activity during oxygen reduction (Ramanayaka et al., 2020).

Because of its porous shape, advantageous physical and chemical properties, and utility as an enzyme support and soil amendment, nanobiochar is currently being explored for these tasks. Gámiz et al. (2017) found that nanobiochar particles have a larger specific surface area, micro-porosity, and surface hydrophobicity. These qualities make them effective at absorbing a wide range of pollutants, including herbicides, heavy metals, PCBs, and PAHs. Furthermore, their greater alkalinity and pH make them a possible solution to soil acidity. As a result, nanocomposites containing nanobiochar have been produced for a variety of applications, including wastewater treatment, energy production, and pollution removal such as organic dyes (Yu et al., 2021). Furthermore, its porous structure, enhanced surface functional group density, and high surface-to-volume ratio made it a useful enzyme immobilisation support material (Perveen et al., 2022). Nanocomposites' pore characteristics, functional groups, and active sites frequently demonstrated significant increase in a variety of areas, including catalytic destruction capability, active sites, and separation ease and speed (Goswami et al., 2022). According to Goswami et al. (2022), a hybrid nanocomposite composed of carbon nanotubes and biochar showed tremendous potential as an inexpensive sorbent material for cleaning up organic contaminants and dyes. According to Ramadan and Abd-Elsalam (2020), these nanocomposites have the potential to be a new adsorbent for the cost-effective and environmentally friendly removal of heavy metals such as Cr(VI) from water.

Despite its numerous benefits, nanobiochar has prompted concerns regarding the possibility of groundwater contamination due to biochar nanoparticle migration via drainage, surface runoff, and irrigation. The development of nanobiochar-heavy-metal composites by adsorbing heavy metals increases the danger of soil and water contamination because the heavy metals can desorb from the composites and travel to other regions (Liu et al., 2021). Nanobiochar and biochar nanocomposites hold immense promise for addressing complex environmental challenges. Their unique properties and multifaceted applications make them valuable tools in the quest for sustainable solutions. As research progresses, these materials are poised to play an increasingly important role in pollution remediation, soil improvement, and climate change mitigation, paving the way for a more sustainable future. Therefore, this study aimed to look at their application in the environment.

2.0 SYNTHESIS AND CHARACTERISATION

2.1 Synthesis of Nanobiochar and Nanocomposites

According to Xu et al., (2024), the synthesis of nanobiochar and its composites involves several critical steps and processes that significantly influence their final properties and applications. The primary method for biochar production is pyrolysis, which involves the thermal decomposition of biomass under oxygen-limited conditions. This process typically occurs at temperatures ranging from 300°C to 900°C, with the specific temperature significantly affecting the resulting biochar's properties. Lower temperature pyrolysis (300-500°C) generally produces biochar with more functional groups and higher cation exchange capacity, while higher temperatures (600-900°C) result in materials with greater surface area and carbon content (Amusat et al., 2021). During the pyrolysis process, the biomass undergoes several transformations, including dehydration, decomposition of cellulose and hemicellulose, and lignin degradation. The residence time in the pyrolysis reactor also plays a crucial role, typically ranging from 30 minutes to several hours depending on the desired properties. Fast pyrolysis, with shorter residence times, produces different characteristics compared to slow pyrolysis, which allows for more complete carbonization (Xu et al., 2024). The incorporation of nanomaterials into biochar represents a significant advancement in creating enhanced composite materials. This process can be accomplished through various methods, including in-situ synthesis and post-pyrolysis modification. In-situ synthesis involves adding nanomaterial precursors to the biomass before pyrolysis, allowing for the simultaneous formation of biochar and nanoparticles. Common nanomaterials incorporated include metal oxides (such as Fe_3O_4, TiO_2, and ZnO), carbon nanotubes, and graphene oxide. These additions can significantly enhance the biochar's functionality, including its adsorption capacity, catalytic activity, and magnetic properties (Zhang et al., 2024). Functionalisation and modification methods further expand the possibilities for tailoring nanobiochar properties to specific applications. Surface modification techniques include acid or base treatment, oxidation, and the grafting of specific functional groups. Chemical activation using agents such as KOH, H3PO4, or ZnCl2 can significantly increase surface area and pore development. Physical activation using steam or CO2 at high temperatures also enhances porosity and surface characteristics. These modifications can introduce specific functional groups, alter surface charge, and improve the material's interaction with target compounds (Zhao et al., 2021; Gęca et al., 2023; Zhang et al., 2024).

2.2 Characterisation Techniques of nanobiochar and biochar nanoccomposites

Xia et al. (2023), reported that comprehensive characterisation of nanobiochar and its composites is essential for understanding their properties and potential applications. Microscopic techniques play a fundamental role in this analysis. Scanning Electron Microscopy (SEM) provides detailed information about surface morphology and pore structure at the microscale, revealing important features such as pore distribution, surface roughness, and particle aggregation. The addition of Energy Dispersive X-ray Spectroscopy (EDS) capability allows for elemental mapping and composition analysis of the surface (Xia et al., 2023). Transmission Electron Microscopy (TEM) offers even higher resolution imaging, capable of visualizing individual nanoparticles and their distribution within the biochar matrix. This technique is particularly valuable for examining the integration of nanomaterials, their size distribution, and the nature of interfaces between different phases in the composite. High-resolution TEM can also provide information about crystal structure and lattice arrangements at the atomic scale (Xia et al., 2023). Spectroscopic methods provide crucial information about the chemical composition and structure of nanobiochar materials. Fourier Transform Infrared Spectroscopy (FTIR) identifies functional groups present on the surface, offering insights into the chemical modifications and surface chemistry of the material. Common features observed include O-H, C=O, C-O, and various aromatic ring vibrations, which can change significantly with different synthesis conditions or modifications (Xia et al., 2023). X-ray Diffraction (XRD) analysis reveals the crystalline structure of the material and any incorporated nanophases. This technique is particularly useful for identifying the presence and nature of mineral components, crystalline nanomaterials, and determining their phase purity. The broadening of XRD peaks can also provide information about crystallite size using the Scherrer equation (Xia et al., 2023).

Raman spectroscopy complements FTIR by providing detailed information about carbon structures and graphitization degree. The characteristic D and G bands in Raman spectra indicate the presence of disordered and graphitic carbon structures, respectively. The intensity ratio of these bands (ID/IG) serves as an indicator of the degree of graphitization and structural order in the carbon matrix (Ahmad et al., 2023). Surface area and porosity analysis using the Brunauer-Emmett-Teller (BET) method is crucial for understanding the material's physical structure and potential applications. This technique provides information about specific surface area, pore size distribution, and pore volume. The analysis typically involves nitrogen adsorption-desorption isotherms at 77 K, although other gases may be used for specific applications. The shape of the isotherm and the presence of hysteresis loops provide information about pore types (micropores, mesopores, or macropores) and their interconnectivity. Advanced analysis using methods such as the Barrett-Joyner-Halenda (BJH) algorithm can provide detailed pore size distributions, while t-plot analysis can distinguish between micropore and external surface areas (Das aand Panda, 2022; Ahmad et al., 2023). These characterization techniques, when used in combination, provide a comprehensive understanding of the nanobiochar and nanocomposite materials' physical and chemical properties. This information is essential for optimizing synthesis conditions, validating modification procedures, and determining the suitability of materials for specific applications. The data obtained from these analyses also helps in understanding structure-property relationships and guides the development of improved materials for various environmental and industrial applications Das and Panda, 2022; Ahmad et al., 2023).

3.1 Nanobiochar production

Raczkiewicz et al. (2024) reported that the production of nanobiochar represents a significant advancement in biomass conversion technology, combining traditional biochar production methods with modern nanomaterial science. According to Ramanayaka et al., (2020), The process of nanobiochar involves several crucial stages and considerations to ensure the production of high-quality nanoscale biochar materials with desired properties and functionalities. The following are the stages od nanaobiochar below:

Raw Material Selection and Preparation

The production of nanobiochar begins with careful selection of appropriate biomass feedstock. Common sources include agricultural residues (rice husks, corn stalks, wheat straw), woody biomass (pine, oak, bamboo), and organic wastes (sewage sludge, food waste). The selection of feedstock significantly influences the final properties of the nanobiochar, including its porosity, surface chemistry, and potential applications. Prior to processing, the raw materials undergo thorough cleaning to remove contaminants and are typically dried to achieve a moisture content below 10% to optimize the pyrolysis process (Ramanayaka et al., 2020). The biomass is then subjected to size reduction through mechanical processing, which may include grinding, milling, or crushing. This step is crucial as it determines the initial particle size distribution and affects the uniformity of the final product. Advanced grinding techniques, such as ball milling or jet milling, are often employed to achieve preliminary size reduction to the microscale level (Abed Hussein et al., 2022).

Pyrolysis Process and Parameters

The core of nanobiochar production lies in the pyrolysis process, which can be categorized into several types based on operating conditions. Slow pyrolysis, conducted at heating rates of 5-30°C/min with residence times of several hours, promotes the formation of a more stable carbon structure. Fast pyrolysis, with heating rates of hundreds of degrees per minute and shorter residence times, can lead to different physical and chemical properties in the final product (Ramanayaka et al., 2020; Abed Hussein et al., 2022; Raczkiewicz et al., 2024). Temperature control during pyrolysis is critical, typically ranging from 300°C to 900°C. Lower temperatures (300-500°C) tend to preserve more functional groups and result in biochar with higher cation exchange capacity, while higher temperatures (600-900°C) promote greater carbonization and surface area development. The atmosphere during pyrolysis must be carefully controlled, usually maintaining inert conditions through the continuous flow of nitrogen or argon gas to prevent oxidation (Ramanayaka et al., 2020; Abed Hussein et al., 2022; Raczkiewicz et al., 2024).

Nanostructure Development

The development of nanostructured features in biochar can be achieved through several approaches. One method involves the incorporation of templating agents during the pre-pyrolysis stage. These templates, such as silica nanoparticles or metal-organic frameworks (MOFs), create nano-sized pores and channels within the biochar structure. After pyrolysis, the templates can be removed through chemical etching, leaving behind a network of nanopores.

Another approach involves the use of activation processes to develop nanoscale features. Physical activation using steam or CO_2 at elevated temperatures (700-900°C) creates additional porosity through controlled oxidation of carbon atoms. Chemical activation using agents such as KOH, H_3PO_4, or $ZnCl_2$ promotes the formation of nanoscale pores and increases surface area through chemical reactions with the carbon matrix (Ramanayaka et al., 2020; Abed Hussein et al., 2022; Raczkiewicz et al., 2024).

Surface Modification and Functionalization

The surface chemistry of nanobiochar can be modified through various treatment methods to enhance its functionality. Acid treatment using HNO3, H2SO4, or H3PO4 introduces oxygen-containing functional groups and can help in developing nanoscale surface features. Basic treatment with NaOH or KOH can increase the negative surface charge and create additional porosity. Advanced functionalization techniques include plasma treatment, which can introduce specific functional groups and create nanoscale surface roughness. Surface grafting with organic molecules or polymers can provide specific binding sites for target compounds. These modifications significantly influence the material's performance in various applications, such as environmental remediation or catalyst support (Ramanayaka et al., 2020; Abed Hussein et al., 2022; Raczkiewicz et al., 2024).

Nanoparticle Incorporation

The incorporation of nanoparticles into biochar can be achieved through several methods. In-situ synthesis involves adding metal precursors to the biomass before pyrolysis, leading to the simultaneous formation of biochar and metal nanoparticles. Post-synthesis deposition methods include impregnation, precipitation, and sol-gel processes, where nanoparticles are deposited onto the pre-formed biochar surface. Common nanoparticles incorporated include metal oxides (Fe3O4, TiO2, ZnO), noble metals (Ag, Au), and carbon-based nanomaterials (carbon nanotubes, graphene oxide). The selection of nanoparticles depends on the intended application and desired properties of the final composite material (Ramanayaka et al., 2020; Abed Hussein et al., 2022; Raczkiewicz et al., 2024).

Quality Control and Characterization

Throughout the production process, various quality control measures are implemented to ensure consistency and desired properties. This includes monitoring of process parameters such as temperature, pressure, and gas flow rates during pyrolysis. Regular sampling and analysis help maintain product quality and optimize production conditions. Key parameters monitored during production include particle size distribution, surface area, pore size distribution, and chemical composition. Advanced analytical techniques such as dynamic light scattering (DLS) and nanoparticle tracking analysis (NTA) are employed to verify the nanoscale dimensions of the produced materials (Ramanayaka et al., 2020; Abed Hussein et al., 2022; Raczkiewicz et al., 2024).

Scale-up and Industrial Considerations

The scale-up of nanobiochar production from laboratory to industrial scale presents several challenges that must be addressed. These include maintaining uniform heating and temperature distribution in larger reactors, ensuring consistent quality across batches, and implementing effective heat recovery systems to improve energy efficiency. Safety considerations are particularly important in industrial-scale production, including dust control measures, proper handling of nanomaterials, and implementation of appropriate personal protective equipment protocols. Environmental considerations include the treatment of pyrolysis gases and the management of any waste streams generated during the production process. The successful production of nanobiochar requires careful attention to each stage of the process, from raw material selection through final quality control. The combination of traditional biochar production methods with modern nanotechnology approaches offers numerous opportunities for developing advanced materials with enhanced properties and performance characteristics (Ramanayaka et al., 2020; Abed Hussein et al., 2022; Raczkiewicz et al., 2024).

Mechanisms of Action by Nanobiochar and Biochar Naanocomposites

Ramanayaka et al. (2020) reported that the mechanisms of action in nanobiochar systems involve complex interactions between the material's surface chemistry, structural properties, and various environmental constituents. Understanding these mechanisms is crucial for optimising nanobiochar performance in environmental applications and developing more effective materials for specific purposes.

Surface Chemistry and Adsorption Mechanisms

The surface chemistry of nanobiochar plays a fundamental role in its interaction with various compounds and pollutants. The surface contains numerous functional groups, including carboxyl (-COOH), hydroxyl (-OH), carbonyl (C=O), and phenolic groups, which are formed during the pyrolysis process and subsequent modifications. These functional groups create active sites for various interaction mechanisms, including electrostatic attraction, hydrogen bonding, π-π interactions, and surface complexation (Zhang et al., 2022). Electrostatic interactions occur between charged surface groups and ionic species in solution. The point of zero charge

(PZC) of nanobiochar, which typically ranges from pH 4 to 10 depending on the production conditions and modifications, determines the surface charge at different pH values. When the solution pH is below the PZC, the surface becomes positively charged, favoring the adsorption of anions. Conversely, above the PZC, the negative surface charge promotes cation adsorption. This mechanism is particularly important for the removal of heavy metal ions and ionic organic pollutants (Hamadeen and Elkhatib, 2022). Hydrogen bonding occurs between surface functional groups and molecules containing electronegative atoms such as oxygen, nitrogen, or fluorine. This mechanism is significant in the adsorption of polar organic compounds, including pharmaceuticals and personal care products. The strength and number of hydrogen bonds depend on the type and density of surface functional groups, which can be modified through various treatment methods (Jena et al., 2022). π-π interactions are prominent in nanobiochar systems due to the presence of aromatic carbon structures. These interactions are particularly important for the adsorption of aromatic compounds, including polycyclic aromatic hydrocarbons (PAHs), phenolic compounds, and aromatic dyes. The degree of graphitization and the arrangement of aromatic sheets in the biochar structure influence the strength of these interactions (Jena et al., 2022).

Role of Nanomaterials in Enhancing Biochar Properties

Zhao et al. (2021) reported that the incorporation of nanomaterials into biochar significantly enhances its properties and expands its functional capabilities. These enhancements occur through several mechanisms, including increased surface area, introduction of new functional groups, and creation of additional active sites for specific interactions. Magnetic nanoparticles, such as Fe_3O_4, provide magnetic separation capabilities and can also participate in redox reactions. The magnetic properties facilitate easy recovery of the material from aqueous solutions, while the iron oxide surfaces can participate in both adsorption and catalytic degradation of pollutants. The presence of iron oxides can also promote Fenton-like reactions, generating reactive oxygen species that can degrade organic contaminants (Venkateswarlu et al., 2021). Metal oxide nanoparticles (TiO_2, ZnO) enhance photocatalytic activity and provide additional surface sites for pollutant binding. Under UV or visible light irradiation, these materials generate electron-hole pairs that can initiate the degradation of organic pollutants through advanced oxidation processes. The combination of photocatalytic activity with the adsorptive properties of biochar creates a synergistic effect, improving overall pollutant removal efficiency (Ramesh et al., 2021). Metal oxide nanoparticles significantly enhance photocatalytic activity, making them effective for pollutant degradation. Their ability to improve charge transfer and increase surface area, along with modifications like doping, expands their applicability in environmental remediation. Despite challenges like UV light dependency, ongoing research and development continue to optimize their performance for sustainable applications as reported by Danish et al. (2021). In a similar form carbon-based nanomaterial, such as graphene oxide or carbon nanotubes, increase electrical conductivity and provide additional surface area for adsorption. These materials can enhance electron transfer processes and create new pathways for pollutant degradation. The high surface area and unique electronic properties of carbon nanomaterials also contribute to improved adsorption capacity and selectivity (Maiti et al., 2019).

Interaction with Pollutants and Environmental Matrices

The interaction between nanobiochar and environmental pollutants occurs through multiple mechanisms that often operate simultaneously. For heavy metal removal, surface complexation and ion exchange are dominant mechanisms. Metal ions can form inner-sphere complexes with surface functional groups or undergo ion exchange with existing cations on the biochar surface. The presence of nanomaterials can provide additional binding sites and enhance the strength of metal-surface interactions (Pathak et al., 2024). Organic pollutant removal involves both physical and chemical mechanisms. Physical adsorption occurs through van der Waals forces and hydrophobic interactions, particularly important for non-polar organic compounds. The nanoporous structure provides size-selective adsorption sites, while the presence of nanomaterials can enhance both adsorption capacity and selectivity (Pathak et al., 2024). The interaction with environmental matrices significantly influences nanobiochar performance. In soil systems, nanobiochar can interact with soil organic matter, minerals, and microorganisms. These interactions can affect both the stability of the material and its effectiveness in pollutant removal. The presence of competing ions and natural organic matter in water systems can impact adsorption efficiency through competitive binding and pore blocking mechanisms (Ramanayaka et al., 2021). The pH of the environmental matrix plays a crucial role in determining the effectiveness of various mechanisms. pH affects surface charge, functional group protonation/deprotonation, and the speciation of pollutants. Understanding these pH-dependent interactions is essential for optimizing nanobiochar performance in different environmental conditions (Ahmaruzzaman, 2021). Pathak et al. (2024) reported that Temperature influences the kinetics and thermodynamics of adsorption processes. Higher temperatures typically increase adsorption rates but may decrease overall adsorption capacity for exothermic processes. The presence of nanomaterials can affect the temperature dependence of various mechanisms, potentially providing more stable performance across different temperature ranges. Also, the aging of nanobiochar in environmental matrices can lead to changes in surface chemistry and physical properties. Oxidation of surface groups, accumulation of mineral phases, and biological interactions can alter the material's effectiveness over time. The incorporation of nanomaterials may help maintain long-term stability and performance by providing additional functional sites and protecting the biochar structure. Understanding these mechanisms of action is essential for designing more effective nanobiochar materials and optimizing their application in environmental remediation. The complex interplay between surface chemistry, nanomaterial properties, and environmental conditions requires careful consideration in both material design and application strategies. Continued research into these mechanisms will lead to improved materials and more efficient environmental treatment technologies (Pathak et al., 2024).

Environmental Applications of Nanobiochar and Biochar Nanocomposites

The environmental potential of macro-biochar has been extensively explored and documented, but recent advancements have shifted attention toward nanobiochar and its enhanced properties for diverse environmental applications. Nanobiochar has emerged as a multifunctional material for applications such as carbon sequestration, renewable energy production, pollutant removal, wastewater treatment, and phytoremediation. Unlike conventional biochar, nanobiochar exhibits improved surface characteristics, including increased surface area, porosity, and functional groups, which make it highly effective in addressing various environmental challenges (Chausali et al., 2021). One notable environmental benefit of nanobiochar lies in its role in carbon sequestration. Biomass combustion, when integrated with nanobiochar systems, is considered a carbon-neutral process, as the carbon dioxide released is eventually reabsorbed by plants during photosynthesis. Moreover, biochar's stable carbon structure ensures its resistance to degradation, enabling it to sequester carbon for extended periods (ranging from centuries to millennia) when applied to agricultural soils. This dual capability of energy production and carbon storage has positioned biochar and nanobiochar as carbon-negative technologies that contribute to climate change mitigation (Winters et al., 2022). Additionally, nanobiochar is gaining recognition for its advanced adsorption capabilities, particularly for the removal of emerging contaminants such as pharmaceuticals, pesticides, heavy metals, and other micropollutants. These contaminants often enter the environment through improper waste management, industrial discharge, and agricultural runoff, posing significant risks to ecosystems and human health. Nanobiochar, with its superior adsorption capacity and surface functionality, has been proven effective in trapping and immobilizing these pollutants. Biochar composites enhanced with nanomaterials, such as nanoparticles or nanoclays, have demonstrated remarkable efficiency in remediating contaminated soils and water systems, outperforming conventional biochar (Chausali et al., 2021). Moreover, biochar's ability to adsorb harmful chemicals and immobilize toxic metals has made it a promising solution for bioremediation. Its application in pollution control not only mitigates environmental risks but also enhances soil quality, thereby supporting sustainable agricultural practices. As a result, nanobiochar is increasingly being recognized as a versatile agent capable of addressing a wide range of environmental challenges (Chausali et al., 2021).

Nanobiochar and Nanocomposites: Bridging Energy, Catalysis, and Materials Research

Nanobiochar has recently gained attention as a sustainable material with numerous potential uses in fields such as energy generation, fuel additives, catalysis, biomedicine, and material science (Chausali et al., 2021). Fuel additives are one of its most prominent uses in fuel technology, which aims to decrease harmful vehicle emissions. Because of their low toxicity and lack of negative effects on the environment, bio-nano additives have recently become more popular than their metal-based counterparts in this setting. The use of nanobiochar particles in diesel-biodiesel-ethanol mixes improved engine performance and significantly decreased exhaust emissions, according to Song et al. (2022). Nanobiochar made from sugarcane performed better than metal-based fuel additives, according to research by Moulick et al. (2025), which also used bagasse as a source. Nanobiochar and its composites have shown promise as acid catalysts and enzyme supports, in addition to fuel applications. Biocatalytic systems built on nanobiochar, as shown by Ahmad et al. (2023), can increase enzyme loading capacity, surface area, and mass transfer resistance, all of which contribute to enhanced performance. A perfect immobilization material for enzyme systems, nanobiochar has a porous structure, a high surface-to-volume ratio, and several functional groups. Enzyme activity and stability can be significantly enhanced by utilizing these features. Nanocomposites made of biochar exhibited enhanced catalytic capabilities, including an increase in the number of active sites, an improvement in pore characteristics, and an abundance of functional groups. Plus, compared to previous approaches, this mineral separation procedure is both faster and easier.

Additionally, Ahmad et al. (2023) looked into the possibility of heterogeneous acid catalysis using nanobiochar. Magnetic nanobiochar made from biomass waste, including oil palm empty fruit bunches, could be a good heterogeneous acid catalyst, according to Jenie et al. (2017). During esterification processes, the researchers found that sulfonated magnetic nanobiochar functioned better than commercial solid and homogenous acid catalysts. These enhancements highlight the versatility and use of nanobiochar in industrial catalytic procedures. For a number of electrochemical systems, nanobiochar has also served as a sensor. To identify microcystin-LR (MCLR) poisons in water, for instance, Ambaye et al. (2023) developed a paper immunosensor based on biochar. The anti-MCLR antibody is deposited onto filter paper by dipping-drying nanobiochar particles, which are highly conductive and dispersible, in this innovative method. Nanobiochar has promising uses in environmental monitoring and safety applications, since this approach demonstrated great sensitivity and efficiency in toxin detection. Research has increased the variety of applications for nanobiochar. A renewable substitute for carbon black in reinforcing styrene-butadiene rubber, the process of converting waste lignin into nanobiochar was detailed by Chausali et al. (2021). This enhancement highlights the role of nanobiochar as a renewable resource and broadens the possibilities for sustainable material development. Wastewater treatment is another area where nanobiochar's environmental benefits are readily apparent. Remediating biomass gasification effluent was demonstrated in Chausali et al. (2021) using biochar-based bio-supports in an up-flow packed bed bioreactor (UFPBBR). This method proved that nanobiochar is effective in cleaning water and repairing damaged ecosystems, with a high detoxification rate.

Advanced Bioremediation: Nanobiochar and Biochar Nanocomposites Solutions for Wastewater Treatment

Heavy metal contamination in water is a significant environmental issue due to its accumulation and toxicity to aquatic ecosystems and human health. Unlike organic and inorganic contaminants that can degrade over time, heavy metals persist in the environment, making their removal a major challenge. Various methods, such as ion exchange, chemical precipitation, and adsorption, have been employed for heavy metal removal. Among these, adsorption has emerged as the most widely accepted approach due to its cost-effectiveness and efficiency. However, conventional adsorbents like activated carbon, silica gel, zeolite, and carbon-based nanofibers often exhibit limitations, including modest adsorbent capacity, poor oxidation resistance, and lower selectivity, making the development of advanced alternatives crucial (Musa et al., 2024a). Biochar, derived from the thermal decomposition of organic materials in the absence of oxygen, has emerged as a promising adsorbent for removing contaminants from water. Its porous structure and surface chemistry allow it to adsorb a wide range of pollutants, providing a sustainable solution for water pollution control. Recent advancements have focused on enhancing the functionality of biochar by incorporating nanotechnology, resulting in nanobiochar and biochar nanocomposites with improved adsorptive and catalytic properties. These advanced materials are particularly effective in the bioremediation of wastewater, as they combine the adsorption capabilities of biochar with the enhanced reactivity of nanoparticles (He et al., 2017; Alagha et al., 2020). Nanobiochar and biochar nanocomposites have demonstrated exceptional efficiency in removing various heavy metals, including arsenic, cadmium, chromium, lead, copper, and mercury, from wastewater. For instance, layered double hydroxides, magnetic biochar, graphitic nanobiochar, and zinc oxide-coated biochar have been employed for the removal of multiple contaminants. These materials exhibit enhanced adsorption capacities compared to conventional biochar due to their increased surface area, improved pore structure, and modified surface chemistry. In addition to adsorption, some nanobiochar-based composites also enable the catalytic degradation of organic pollutants, further enhancing their utility in wastewater remediation (Ahuja et al., 2022). Modified biochar with specific nanomaterials significantly improves its performance. For example, photosynthetic bacteria loaded onto biochar nanocomposites have demonstrated the ability to remove phosphate, ammonia, and organic matter from wastewater with remarkable efficiency. Similarly, the use of alkaline-treated biochar, $KMnO4$-impregnated biochar, and iron oxide-coated biochar has proven effective for addressing cadmium contamination. These modifications enhance the material's sorption capacity and adsorption rate, enabling faster and more effective removal of pollutants (Noreen and Abd-Elsalam, 2021). Also, notable advancement involves the use of magnetic biochar, which allows for easy separation and recovery after the adsorption process. Magnetic nanobiochar composites, such as those derived from wheat straw and rice husk, have shown high removal efficiencies for contaminants like mercury, tetracycline, and arsenic. The application of these composites in real-world scenarios has highlighted their potential for large-scale wastewater treatment (Chausali et al., 2021). Incorporating nanotechnology into biochar production has revolutionized its use in bioremediation. These materials not only address the limitations of traditional adsorbents but also provide sustainable solutions for environmental management. Their ability to simultaneously adsorb and degrade heavy metals and organic pollutants makes them a valuable tool for wastewater treatment. As research in this field continues to evolve, the development of novel nanobiochar composites and their application in diverse environmental settings is expected to further enhance the effectiveness of bioremediation strategies (Alagha et al., 2020).

Adsorption-Driven Phytoremediation: The Role of Nanobiochar and Biochar Nanocomposites in Contaminant Removal

Biochar has long been recognized for its exceptional adsorption capacity, making it a widely studied material for pollutant removal. More recently, nanocarbons, including engineered nanomaterials such as activated carbon and nanobiochar, have shown great promise in adsorbing natural organic matter and contaminants from the environment (Awad et al., 2020; Chandel et al., 2022). These materials are particularly effective in addressing pollutants like heavy metals, pharmaceuticals, dyes, and pesticides, which pose significant environmental and health challenges (Chandel et al., 2022). The advanced sorption properties of nanobiochar and biochar-based nanocomposites, as compared to conventional biochar, have spurred considerable interest among researchers. Nanobiochar, due to its smaller particle size, increased surface area, and enhanced functional groups, demonstrates superior adsorption efficiency. For instance, pinewood-derived nanobiochar exhibited a 95% removal efficiency for carbamazepine (CBZ), a pharmaceutical micropollutant, from aqueous media (Naghdi et al., 2019). Similarly, oxygen-functionalized nanobiochar derived from pinewood achieved removal efficiencies of 83% and 86% for carbamazepine in spiked water and secondary effluents, respectively, showcasing its versatility in diverse water matrices (Naghdi et al., 2017). Functionalized nanocomposites further enhance the adsorption potential of biochar. For example, sugarcane bagasse and hickory chip-derived carbon nanotube-biochar nanocomposites displayed a methylene blue (MB) adsorption capacity of 6.2 mg/g, highlighting their efficacy in removing organic dyes from wastewater. Likewise, magnetic corn stover biochar (MCSBC) showed significant fluoride adsorption, with capacities of 4.11 mg/g at 25°C and 3.41 mg/g at 45°C, illustrating its potential in inorganic contaminant removal as reported in Chausali et al. (2021). The integration of nanominerals into biochar matrices has also proven to be a game-changer for pollutant remediation. Graphitic nanobiochar derived from waste and dendro biochar achieved high adsorption capacities for glyphosate (83 mg/g) and oxytetracycline (520 mg/g), both of which are common agricultural and pharmaceutical contaminants. Similarly, hematite-modified magnetic biochar (HPB) derived from pinewood demonstrated enhanced arsenic (As V) adsorption, with capacities increasing from 265 mg/kg for unmodified biochar to 429 mg/kg for HPB. These findings underscore the effectiveness of nanomineral-modified biochar in tackling complex environmental contaminants (Chausali et al., 2021). Among other noteworthy examples, magnesium oxide (MgO) nanoflakes and $Mg(OH)_2$ particles have shown remarkable phosphate removal capabilities, with adsorption capacities of up to 835 mg/g and correlation to biochar magnesium content. These nanometal oxide-modified biochars have proven particularly effective in nutrient recovery and mitigating agricultural runoff. Similarly, clay-biochar composites derived from hickory chips and bamboo bagasse exhibited methylene blue adsorption capacities, showcasing their potential for organic dye removal. The role of nanobiochar in heavy metal remediation is equally significant. For instance, $KMnO_4$-treated biochar demonstrated adsorption capacities for copper (Cu II), cadmium (Cd II), and lead (Pb II), emphasizing its utility in immobilizing toxic metals (Wang et al., 2015). Furthermore, manganese oxide-modified biochar achieved nearly 99% lead

(Pb II) removal efficiency, a substantial improvement from 6.4% for unmodified biochar. These results highlight the potential of metal-oxide-modified nanobiochar in addressing metal contamination in various environmental settings (Wang et al., 2018). Biochar composites have also been employed for advanced wastewater treatment. For example, sodium hydroxide-coated nanobiochar derived from artichoke leaves achieved 97% removal of metformin hydrochloride (MFH), a pharmaceutical contaminant, from wastewater, outperforming its performance in tap and sea water. Similarly, magnetic nanobiochar derived from wheat straw showed remarkable adsorption capacities for tetracycline and mercury (Hg II), demonstrating its application in both organic and inorganic contaminant remediation. Another promising application of nanobiochar is its role in nutrient recovery from agricultural runoff. For example, vermiculite-treated algal biomass biochar (VBC) achieved phosphate adsorption capacities, which is higher than pristine biochar. This demonstrates its potential to mitigate eutrophication while recovering valuable nutrients for reuse in agriculture (Chausali et al., 2021).

Transforming Agriculture with the Application and Use of Nanobiochar and Biochar Nanocomposites

According to Musa et al., (2024b), agriculture faces significant challenges, including soil degradation, crop productivity decline, and chemical residues from pesticides and fertilizers. Addressing these concerns, nanobiochar and biochar nanocomposites have emerged as revolutionary materials with the potential to transform agricultural practices. These materials, created by incorporating nanoparticles into biochar, possess enhanced chemical, physical, and biological properties that benefit various aspects of farming, including plant growth, pesticide remediation, soil health, disease control, fertilizer delivery, and microbial activity (Musa et al., 2024c). Nanobiochar significantly enhances plant growth by improving nutrient availability and water retention in the soil. Its porous structure facilitates the storage and gradual release of essential nutrients, ensuring plants receive a consistent supply for optimal growth. This helps reduce nutrient deficiencies and promotes healthier roots, which translates to improved crop yields. Moreover, nanobiochar's capacity to regulate soil pH enables better nutrient absorption by plants, making it especially beneficial for crops grown in marginal or nutrient-poor soils (Ahmad et al., 2024). The issue of pesticide residues in agricultural soils and water sources is another area where nanobiochar demonstrates its transformative potential. With its large surface area and functionalized surface properties, nanobiochar adsorbs and immobilizes pesticide residues, preventing their movement into water systems and protecting surrounding ecosystems. This ability not only mitigates environmental pollution but also reduces the harmful effects of pesticide toxicity on crops and soil.

By addressing these issues, nanobiochar contributes to safer and more sustainable farming practices (Chausali et al., 2021). In addition to its role in nutrient delivery and pesticide remediation, nanobiochar is highly effective as a soil amendment. Its application improves soil structure, porosity, and water retention capacity, creating an environment conducive to plant growth. Nanobiochar also enhances the organic carbon content in the soil, improving its fertility and making it more resilient to environmental stressors. Furthermore, it binds with heavy metals and contaminants, detoxifying the soil and ensuring its suitability for agricultural activities. This makes nanobiochar particularly valuable for rehabilitating degraded soils or areas impacted by industrial or agricultural pollution (Ramadan and Abd-Elsalam, 2020). The management of plant diseases is another critical agricultural challenge addressed by nanobiochar and biochar nanocomposites. These materials possess antimicrobial properties that inhibit the growth of soilborne pathogens, thereby reducing the incidence of crop diseases. Additionally, nanobiochar can act as a carrier for biocontrol agents and beneficial microbes, delivering them directly to the plant root zone where they help combat harmful pathogens. This reduces reliance on synthetic chemical treatments and promotes a healthier, more balanced agricultural ecosystem (Ramadan and Abd-Elsalam, 2020). Nanobiochar has also revolutionized the delivery of fertilizers by serving as a nutrient carrier. Its porous structure allows it to adsorb fertilizers and release them gradually, minimizing losses due to leaching or volatilization. This ensures that crops receive a steady supply of nutrients throughout their growth cycle, improving fertilizer use efficiency and lowering input costs. The integration of nanoparticles further enhances the interaction between fertilizers and plant roots, enabling more effective nutrient uptake. This targeted approach reduces the environmental impact of fertilizer use while maximizing agricultural productivity (Chausali et al., 2021). The promotion of microbial growth in the soil is another significant advantage of nanobiochar and biochar nanocomposites. These materials provide a supportive habitat for beneficial microorganisms, such as nitrogen-fixing bacteria and phosphate-solubilizing microbes, which play an essential role in soil nutrient cycling. By fostering microbial activity, nanobiochar enhances soil fertility and helps create a more balanced soil ecosystem. The synergy between nanobiochar and these microorganisms improves nutrient availability and plant health, further contributing to sustainable agricultural practices (Udomkun et al., 2024).

CONCLUSION

Nanobiochar and biochar nanocomposites have revolutionised the fields of wastewater bioremediation and sustainable agriculture by offering effective and eco-friendly solutions to pressing environmental issues. Their exceptional adsorption capacity has been instrumental in removing heavy metals and organic contaminants from polluted water sources. Functionalized biochar nanocomposites, including magnetic biochar and chitosan-coated biochar, have demonstrated remarkable efficiency in capturing toxic substances, mitigating water pollution, and improving the quality of aquatic environments. The ability of nanobiochar to adsorb multiple contaminants highlights its significance in advancing clean water technologies. In agriculture, nanobiochar has shown tremendous potential in enhancing soil health, optimizing plant growth, and improving resource utilization. As a soil amendment, it enriches soil structure, increases moisture retention, and minimizes heavy metal toxicity. Its ability to adsorb pesticide residues makes it a sustainable alternative for reducing agrochemical pollution. Moreover, nanobiochar plays a crucial role in disease management by suppressing soilborne pathogens and serving as a carrier for beneficial microbes and biocontrol agents. This not only reduces dependence on chemical pesticides but also fosters a more balanced and resilient agricultural ecosystem. Fertilizer efficiency is another critical area where nanobiochar provides significant benefits. By enabling controlled nutrient release, it minimizes losses due to leaching and volatilization, thereby reducing environmental pollution while maximizing crop productivity. Additionally, nanobiochar promotes microbial growth, supporting beneficial soil microorganisms that contribute to nutrient cycling and overall soil fertility. As agriculture and environmental management face increasing challenges due to climate change and population growth, the adoption of nanobiochar and biochar nanocomposites represents a sustainable pathway toward resilience and efficiency. Their multifunctionality in wastewater treatment and agricultural enhancement makes them indispensable tools for achieving long-term ecological balance. Moving forward, continued research and innovation in biochar nanotechnology will be essential in optimizing these materials for broader applications, ensuring their role in creating a cleaner and more sustainable future.

REFERENCES

Abed Hussein, B., Mahdi, A. B., Emad Izzat, S., Acwin Dwijendra, N. K., Romero Parra, R. M., Barboza Arenas, L. A., & Thaeer Hammid, A. (2022). Production, structural properties nano biochar and effects nano biochar in soil: A review. *Egyptian Journal of Chemistry*, *65*(12), 607–618.

Ahmad, A., Zahra, M., e Alam, F., Ali, S., Pervaiz, M., Saeed, Z., & Luque, R. (2023). RETRACTED: A sustainable approach for the multi-dimensional exploitation of mixed biochar based nano-composites.

Ahmad, A., Zahra, M., e Alam, F., Ali, S., Pervaiz, M., Saeed, Z., & Luque, R. (2023). RETRACTED: A sustainable approach for the multi-dimensional exploitation of mixed biochar based nano-composites.

Ahmaruzzaman, M. (2021). Biochar based nanocomposites for photocatalytic degradation of emerging organic pollutants from water and wastewater. *Materials Research Bulletin*, *140*, 111262. DOI: 10.1016/J.MATERRESBULL.2021.111262

Alagha, O., Manzar, M., Zubair, M., Anıl, I., Mu'azu, N., & Qureshi, A. (2020). Comparative Adsorptive Removal of Phosphate and Nitrate from Wastewater Using Biochar-MgAl LDH Nanocomposites: Coexisting Anions Effect and Mechanistic Studies. *Nanomaterials (Basel, Switzerland)*, *10*. Advance online publication. DOI: 10.3390/nano10020336 PMID: 32079126

Alimohammadi, M., Xu, Y., Wang, D., Biris, A. S., & Khodakovskaya, M. V. (2011). Physiological responses induced in tomato plants by a two-component nanostructural system composed of carbon nanotubes conjugated with quantum dots and its in vivo multimodal detection. *Nanotechnology*, *22*(29), 295101. PMID: 21673379

Ambaye, A. D., Fito, J., Nxumalo, E., & Nkambule, T. T. I. (2023). The challenges of micropollutants and their environmental detection through biochar-based nanostructured electrochemical sensors: A review. *International Journal of Environmental Science and Technology*, *20*(10), 11741–11762.

Amusat, S. O., Kebede, T. G., Dube, S., & Nindi, M. M. (2021). Ball-milling synthesis of biochar and biochar–based nanocomposites and prospects for removal of emerging contaminants: A review. *Journal of Water Process Engineering*, *41*, 101993.

Ashfaq, M., Verma, N., & Khan, S. (2017). Carbon nanofibers as a micronutrient carrier in plants: Efficient translocation and controlled release of Cu nanoparticles. *Environmental Science. Nano*, *4*(1), 138–148.

Awad, A., Jalab, R., Benamor, A., Nasser, M., Ba-Abbad, M., El-Naas, M., & Mohammad, A. (2020). Adsorption of organic pollutants by nanomaterial-based adsorbents: An overview. *Journal of Molecular Liquids, 301*, 112335. DOI: 10.1016/j.molliq.2019.112335

Chandel, M., Kaur, K., Sahu, B. K., Sharma, S., Panneerselvam, R., & Shanmugam, V. (2022). Promise of nano-carbon to the next generation sustainable agriculture. *Carbon, 188*, 461–481.

Chausali, N., Saxena, J., & Prasad, R. (2021). Nanobiochar and biochar based nanocomposites: Advances and applications. *Journal of Agriculture and Food Research, 5*, 100191.

Chen, L., Chen, X. L., Zhou, C. H., Yang, H. M., Ji, S. F., Tong, D. S., & Chu, M. Q. (2017). Environmental-friendly montmorillonite-biochar composites: Facile production and tunable adsorption-release of ammonium and phosphate. *Journal of Cleaner Production, 156*, 648–659.

Danish, M., Estrella, L., Alemaida, I., Lisin, A., Moiseev, N., Ahmadi, M., Nazari, M., Wali, M., Zaheb, H., & Senjyu, T. (2021). Photocatalytic Applications of Metal Oxides for Sustainable Environmental Remediation. *Metals*. Advance online publication. DOI: 10.3390/met11010080

Das, R., & Panda, S. N. (2022). Preparation and applications of biochar based nanocomposite: A review. *Journal of Analytical and Applied Pyrolysis, 167*, 105691.

Dijoo, Z. K., & Khurshid, R. (2022). Environmental degradation as a multifaceted consequence of human development. In *Environmental Biotechnology* (pp. 39–56). Apple Academic Press.

Gámiz, B., Cox, L., Hermosín, M. C., Spokas, K., & Celis, R. (2017). Assessing the effect of organoclays and biochar on the fate of abscisic acid in soil. *Journal of Agricultural and Food Chemistry, 65*(1), 29–38. PMID: 27959547

Gęca, M., Khalil, A. M., Tang, M., Bhakta, A. K., Snoussi, Y., Nowicki, P., & Chehimi, M. M. (2023). Surface treatment of biochar—methods, surface analysis and potential applications: A comprehensive review. *Surfaces, 6*(2), 179–213.

Goswami, L., Kushwaha, A., Kafle, S., & Kim, B. (2022). Surface Modification of Biochar for Dye Removal from Wastewater. *Catalysts*. Advance online publication. DOI: 10.3390/catal12080817

Hamadeen, H. M., & Elkhatib, E. A. (2022). New nanostructured activated biochar for effective removal of antibiotic ciprofloxacin from wastewater: Adsorption dynamics and mechanisms. *Environmental Research, 210*, 112929. PMID: 35167852

He, S., Zhong, L., Duan, J., Feng, Y., Yang, B., & Yang, L. (2017). Bioremediation of Wastewater by Iron Oxide-Biochar Nanocomposites Loaded with Photosynthetic Bacteria. *Frontiers in Microbiology, 8*. Advance online publication. DOI: 10.3389/fmicb.2017.00823 PMID: 28588556

Jena, S., Dutta, J., Tulsiyan, K. D., Sahu, A. K., Choudhury, S. S., & Biswal, H. S. (2022). Noncovalent interactions in proteins and nucleic acids: Beyond hydrogen bonding and π-stacking. *Chemical Society Reviews, 51*(11), 4261–4286. PMID: 35560317

Jenie, S. N., Kristiani, A., Kustomo, K., Simanungkalit, S., & Mansur, D. (2017, November). Preparation of nanobiochar as magnetic solid acid catalyst by pyrolysis-carbonization from oil palm empty fruit bunches. In AIP Conference Proceedings (Vol. 1904, No. 1). AIP Publishing.

Khan, Y., Sadia, H., Ali Shah, S. Z., Khan, M. N., Shah, A. A., Ullah, N., & Khan, M. I. (2022). Classification, synthetic, and characterization approaches to nanoparticles, and their applications in various fields of nanotechnology: A review. *Catalysts, 12*(11), 1386.

Liu, Y., Sun, H., Sun, J., Meng, X., Jiang, Y., & Wang, N. (2021). Transport of micro/nano biochar in quartz sand modified by three different clay minerals. *Environmental Pollutants and Bioavailability, 33*(1), 113–121.

Maiti, D., Tong, X., Mou, X., & Yang, K. (2019). Carbon-Based Nanomaterials for Biomedical Applications: A Recent Study. *Frontiers in Pharmacology, 9*. Advance online publication. DOI: 10.3389/fphar.2018.01401 PMID: 30914959

Moulick, D., Mukherjee, A., Mahanta, S., Chowdhara, B., Pattnaik, B. K., Mukhopadhyay, A. K., & Hazra, S. (2025). Nano-composites and nano-biochar: a novel solution to alleviate heavy metal stress in crops. In *Nanomaterials and Nano-Biochar in Reducing Soil Stress* (pp. 183–224). Apple Academic Press.

Musa, I. O., Isibor, P. O., Job, O. S., Mustapha, A., Mustapha, A., Akande, S., & Adeniji, H. (2024c). Introduction to Nanotoxicology. In *Environmental Nanotoxicology: Combatting the Minute Contaminants* (pp. 1–22). Springer Nature Switzerland.

Musa, I. O., Job, O. S., Adams, M., Abdulsalam, M., Nathaniel, V., Maude, A. M., Adedayo, O. A., & Tiamiyu, A. G. T. (2024b). Soil Erosion, Mineral Depletion and Regeneration. In *Prospects for Soil Regeneration and Its Impact on Environmental Protection* (pp. 159–172). Springer Nature Switzerland.

Musa, I. O., Mustapha, A., Abdulsalam, M., Livinus, M. U., Job, S. O., Elelu, S. A., Salami, O. L., & Muhammad, A. S. (2024). Soil Microbes and Soil Contamination. In *Soil Microbiome in Green Technology Sustainability* (pp. 3–35). Springer Nature Switzerland.

Naghdi, M., Taheran, M., Brar, S. K., Kermanshahi-Pour, A., Verma, M., & Surampalli, R. Y. (2017). Immobilized laccase on oxygen functionalized nanobiochars through mineral acids treatment for removal of carbamazepine. *The Science of the Total Environment*, *584*, 393–401. PMID: 28117156

Naghdi, M., Taheran, M., Brar, S. K., Kermanshahi-Pour, A., Verma, M., & Surampalli, R. Y. (2018). Pinewood nanobiochar: A unique carrier for the immobilization of crude laccase by covalent bonding. *International Journal of Biological Macromolecules*, *115*, 563–571. PMID: 29689286

Naghdi, M., Taheran, M., Pulicharla, R., Rouissi, T., Brar, S. K., Verma, M., & Surampalli, R. Y. (2019). Pine-wood derived nanobiochar for removal of carbamazepine from aqueous media: Adsorption behavior and influential parameters. *Arabian Journal of Chemistry*, *12*(8), 5292–5301.

Noreen, S., & Abd-Elsalam, K. (2021). Biochar-based nanocomposites: A sustainable tool in wastewater bioremediation., 185-200. .DOI: 10.1016/B978-0-12-821141-0.00023-9

Noreen, S., & Abd-Elsalam, K. A. (2021). Biochar-based nanocomposites: A sustainable tool in wastewater bioremediation. In *Aquananotechnology* (pp. 185–200). Elsevier.

Pathak, H. K., Seth, C. S., Chauhan, P. K., Dubey, G., Singh, G., Jain, D., & Khoo, K. S. (2024). Recent advancement of nano-biochar for the remediation of heavy metals and emerging contaminants: Mechanism, adsorption kinetic model, plant growth and development. *Environmental Research*, 119136. PMID: 38740295

Perveen, S., Nadeem, R., Nosheen, F., Asjad, M., Awrejcewicz, J., & Anwar, T. (2022). Biochar-Mediated Zirconium Ferrite Nanocomposites for Tartrazine Dye Removal from Textile Wastewater. *Nanomaterials (Basel, Switzerland)*, *12*. Advance online publication. DOI: 10.3390/nano12162828 PMID: 36014692

Raczkiewicz, M., Ostolska, I., Mašek, O., & Oleszczuk, P. (2024). Effect of the pyrolysis conditions and type of feedstock on nanobiochars obtained as a result of ball milling. *Journal of Cleaner Production*, *458*, 142456.

Ramadan, M. M., & Abd-Elsalam, K. A. (2020). Micro/nano biochar for sustainable plant health: Present status and future prospects. Carbon nanomaterials for agri-food and environmental applications, 323-357.

Ramanayaka, S., Vithanage, M., Alessi, D. S., Liu, W. J., Jayasundera, A. C., & Ok, Y. S. (2020). Nanobiochar: Production, properties, and multifunctional applications. *Environmental Science. Nano*, *7*(11), 3279–3302.

Ramesh, K., Gnanavel, B., & Shkir, M. (2021). Enhanced visible light photocatalytic degradation of bisphenol A (BPA) by reduced graphene oxide (RGO)–metal oxide (TiO2, ZnO and WO3) based nanocomposites. *Diamond and Related Materials*, *118*, 108514. DOI: 10.1016/J.DIAMOND.2021.108514

Saxena, J., Rana, G., & Pandey, M. (2013). Impact of addition of biochar along with Bacillus sp. on growth and yield of French beans. *Scientia Horticulturae*, *162*, 351–356.

Song, B., Cao, X., Gao, W., Aziz, S., Gao, S., Lam, C. H., & Lin, R. (2022). Preparation of nano-biochar from conventional biorefineries for high-value applications. *Renewable & Sustainable Energy Reviews*, *157*, 112057.

Uday, V., Harikrishnan, P. S., Deoli, K., Zitouni, F., Mahlknecht, J., & Kumar, M. (2022). Current trends in production, morphology, and real-world environmental applications of biochar for the promotion of sustainability. *Bioresource Technology*, *359*, 127467. PMID: 35710048

Udomkun, P., Chandi, K., Boonupara, T., & Kaewlom, P. (2024). Innovative approaches: Exploring nano-biochar technology's impact on soil properties, alachlor retention, and microbial populations. *Environmental Technology & Innovation*, *35*, 103659.

Venkateswarlu, S., Yoon, M., & Kim, M. (2021). An environmentally benign synthesis of Fe3O4 nanoparticles to Fe3O4 nanoclusters: Rapid separation and removal of Hg(II) from an aqueous medium. *Chemosphere*, *286*(Pt 2), 131673. DOI: 10.1016/j.chemosphere.2021.131673 PMID: 34358889

Wang, H., Gao, B., Wang, S., Fang, J., Xue, Y., & Yang, K. (2015). Removal of Pb (II), Cu (II), and Cd (II) from aqueous solutions by biochar derived from KMnO4 treated hickory wood. *Bioresource Technology*, *197*, 356–362. PMID: 26344243

Wang, Y. Y., Liu, Y. X., Lu, H. H., Yang, R. Q., & Yang, S. M. (2018). Competitive adsorption of Pb (II), Cu (II), and Zn (II) ions onto hydroxyapatite-biochar nanocomposite in aqueous solutions. *Journal of Solid State Chemistry*, *261*, 53–61.

Winters, D., Boakye, K., & Simske, S. (2022). Toward carbon-neutral concrete through biochar–cement–calcium carbonate composites: A critical review. *Sustainability, 14*(8), 4633.

Xia, C., Liang, Y., Li, X., Al Garalleh, H., Garaleh, M., Hill, J. M., & Pugazhendhi, A. (2023). Remediation competence of nanoparticles amalgamated biochar (nanobiochar/nanocomposite) on pollutants: A review. *Environmental Research, 218*, 114947. PMID: 36462692

Xu, J., Xie, Y., Yao, Q., Lv, L., & Chu, H. (2024). Advances in sustainable nanobiochar: Precursors, synthesis methods and applications. *Nanoscale, 16*(32), 15009–15032. PMID: 39041285

Yu, F., Tian, F., Zou, H., Ye, Z., Peng, C., Huang, J., Zheng, Y., Zhang, Y., Yang, Y., Wei, X., & Gao, B. (2021). ZnO/biochar nanocomposites via solvent free ball milling for enhanced adsorption and photocatalytic degradation of methylene blue. *Journal of Hazardous Materials, 415*, 125511. DOI: 10.1016/j.jhazmat.2021.125511 PMID: 33740715

Zhang, X., Wells, M., Niazi, N. K., Bolan, N., Shaheen, S., Hou, D., & Wang, Z. (2022). Nanobiochar-rhizosphere interactions: Implications for the remediation of heavy-metal contaminated soils. *Environmental Pollution, 299*, 118810. PMID: 35007673

Zhao, Y., Qamar, S., Qamar, M., Bilal, M., & Iqbal, H. (2021). Sustainable remediation of hazardous environmental pollutants using biochar-based nanohybrid materials. *Journal of Environmental Management, 300*, 113762. DOI: 10.1016/j.jenvman.2021.113762 PMID: 34543967

Zhou, T., Deng, J., Zeng, Y., Liu, X., Song, B., Ye, S., Li, M., Yang, Y., Wang, Z., & Zhou, C. (2024). Biochar Meets Single-Atom: A Catalyst for Efficient Utilization in Environmental Protection Applications and Energy Conversion. *Small, 2404254*. Advance online publication. DOI: 10.1002/smll.202404254 PMID: 38984755

Compilation of References

Abady, M. M., Shawky, A. M., Sakr, F. A., Mohammed, D. M., & Goda, E. S. (2024). Recent Advancements in Biosensors Using Biopolymers. In *Bio-Based Polymers: Farm to Industry. Volume 2: Current Trends and Applications* (pp. 81-112). American Chemical Society.

Abd-Elsalam, K. A., & Alghuthaymi, M. A. (Eds.). (2024). *Nanofertilizers for sustainable agroecosystems: Recent advances and future trends*. Springer.

Abd-Elsalam, K. A., Mehmood, M. A., Ashfaq, M., Abdelkhalek, T. E., Hassan, R. K., & Ravichandran, M. (2024). Liquid Nanoclay: Synthesis and Applications to Transform an Arid Desert into Fertile Land. *Soil Systems*, *8*(3), 3. Advance online publication. DOI: 10.3390/soilsystems8030073

Abebe, B., Zereffa, E. A., Tadesse, A., & Murthy, H. C. A. (2020). A Review on Enhancing the Antibacterial Activity of ZnO: Mechanisms and Microscopic Investigation. *Nanoscale Research Letters*, *15*(1), 1–19. DOI: 10.1186/S11671-020-03418-6/TABLES/2 PMID: 33001404

Abed Hussein, B., Mahdi, A. B., Emad Izzat, S., Acwin Dwijendra, N. K., Romero Parra, R. M., Barboza Arenas, L. A., & Thaeer Hammid, A. (2022). Production, structural properties nano biochar and effects nano biochar in soil: A review. *Egyptian Journal of Chemistry*, *65*(12), 607–618.

Abraham, E., Deepa, B., Pothan, L. A., Jacob, M., Thomas, S., Cvelbar, U., & Anandjiwala, R. (2011). Extraction of nanocellulose fibrils from lignocellulosic fibres: A novel approach. *Carbohydrate Polymers*, *86*(4), 1468–1475. DOI: 10.1016/j.carbpol.2011.06.034

Acharyya, P. P., Sarma, M., & Kashyap, A. (2024). Recent advances in synthesis and bioengineering of bacterial nanocellulose composite films for green, active and intelligent food packaging. *Cellulose (London, England)*, *31*(12), 7163–7187.

Acosta-Angulo, B., Diaz-Angulo, J., Lara-Ramos, J., Torres-Palma, R., Martínez-Pachón, D., Moncayo-Lasso, A., & Machuca-Martínez, F. (2022). Analysis of the Applications of Particle Swarm Optimization and Genetic Algorithms on Reaction Kinetics: A Prospective Study for Advanced Oxidation Processes. *ChemElectroChem*, *9*(13), e202200229. DOI: 10.1002/celc.202200229

Adams, L. K., Lyon, D. Y., & Alvarez, P. J. J. (2006). Comparative eco-toxicity of nanoscale TiO2, SiO2, and ZnO water suspensions. *Water Research*, *40*(19), 3527–3532. DOI: 10.1016/J.WATRES.2006.08.004 PMID: 17011015

Adeosun, S. O., Lawal, G. I., Balogun, S. A., & Akpan, E. I. (2012). Review of green polymer nanocomposites. *Journal of Minerals & Materials Characterization & Engineering*, *11*(4), 385–416.

Adrar, S., & Ajji, A. (2024). Effect of blending sequence and epoxy functionalized compatibilizer on barrier and mechanical properties of PBS and PBS/PLA nanocomposite blown films. *Journal of Vinyl and Additive Technology*, *30*(4), 883–894.

Afif, K., Rebolledo, C., & Roy, J. (2021). Drivers, barriers and performance outcomes of sustainable packaging: A systematic literature review. *British Food Journal*. Advance online publication. DOI: 10.1108/BFJ-02-2021-0150

Agwuncha, S. C., Sadiku, E. R., Ibrahim, I. D., Aderibigbe, B. A., Owonubi, S. J., Agboola, O., Reddy, A. B., Bandla, M., Varaprasad, K., & Bayode, B. L. (2017). Poly (lactic acid) biopolymer composites and nanocomposites for biomedicals and biopackaging applications. *Handb Compos Renew Mater*, *8*, 135–170.

Ahmad, A., Zahra, M., e Alam, F., Ali, S., Pervaiz, M., Saeed, Z., & Luque, R. (2023). RETRACTED: A sustainable approach for the multi-dimensional exploitation of mixed biochar based nano-composites.

Ahmaruzzaman, M. (2021). Biochar based nanocomposites for photocatalytic degradation of emerging organic pollutants from water and wastewater. *Materials Research Bulletin*, *140*, 111262. DOI: 10.1016/J.MATERRESBULL.2021.111262

Ahmed, M. J., Ashfaq, J., Sohail, Z., Channa, I. A., Sánchez-Ferrer, A., Ali, S. N., & Chandio, A. D. (2024). Lignocellulosic bioplastics in sustainable packaging–Recent developments in materials design and processing: A comprehensive review. *Sustainable. Materials Technology*, •••, 01077.

Ahmed, S. F., Mofijur, M., Rafa, N., Chowdhury, A. T., Chowdhury, S., Nahrin, M., & Ong, H. C. (2022). Green approaches in synthesising nanomaterials for environmental nanobioremediation: Technological advancements, applications, benefits and challenges. *Environmental Research*, *204*, 111967. PMID: 34450159

Aida, H. J., Nadlene, R., Mastura, M. T., Yusriah, L., Sivakumar, D., & Ilyas, R. A. (2021). Natural fibre filament for fused deposition modelling (FDM): A review. *International Journal of Sustainable Engineering*, *14*(6), 1988–2008. DOI: 10.1080/19397038.2021.1962426

Aisida, S. O., Madubuonu, N., Alnasir, M. H., Ahmad, I., Botha, S., Maaza, M., & Ezema, F. I. (2019). Biogenic synthesis of iron oxide nanorods using Moringa oleifera leaf extract for antibacterial applications. *Applied Nanoscience*, *10*(1), 305–315. DOI: 10.1007/s13204-019-01099-x

Akaraonye, E., Keshavarz, T., & Roy, I. (2010). Production of polyhydroxyalkanoates: The future green materials of choice. *Journal of Chemical Technology and Biotechnology*, *85*, 732–743.

Akhtar, M. S., Ali, S., & Zaman, W. (2024). Innovative Adsorbents for Pollutant Removal: Exploring the Latest Research and Applications. *Molecules (Basel, Switzerland)*, *29*(18), 18. Advance online publication. DOI: 10.3390/molecules29184317 PMID: 39339312

Akinnawo, S. O. (2024). Advance Nanocomposites from Biopolymers and Fillers: Sources, Characterization, and End-Use Applications. *Polymer-Plastics Technology and Materials*, *63*(5), 570–604.

Alagha, O., Manzar, M., Zubair, M., Anıl, I., Mu'azu, N., & Qureshi, A. (2020). Comparative Adsorptive Removal of Phosphate and Nitrate from Wastewater Using Biochar-MgAl LDH Nanocomposites: Coexisting Anions Effect and Mechanistic Studies. *Nanomaterials (Basel, Switzerland)*, *10*. Advance online publication. DOI: 10.3390/nano10020336 PMID: 32079126

Alakrach, A. M., Al-Rashdi, A. A., Alqadi, T., Saadi, M. A. A., Ting, S. S., Dahham, O. S., & Zulkepli, N. N. (2021). Polylactic Acid/Halloysite Nanotubes-Titanium Dioxide Nanocomposites: Antibacterial and Mechanical Properties. *Materials Science Forum*, *1021*, 270–279.

Alam, T., Khan, R. A., Ali, A., Sher, H., Ullah, Z., & Ali, M. (2018). Biogenic synthesis of iron oxide nanoparticles via Skimmia laureola and their antibacterial efficacy against bacterial wilt pathogen Ralstonia solanacearum. *Materials Science and Engineering C*, *98*, 101–108. DOI: 10.1016/j.msec.2018.12.117 PMID: 30812984

Alexis, F. (2005).. . *Polymer International*, *54*, 36–46.

Alibekov, R. S., Urazbayeva, K. U., Azimov, A. M., Rozman, A. S., Hashim, N., & Maringgal, B. (2024). Advances in Biodegradable Food Packaging Using Wheat-Based Materials: Fabrications and Innovations, Applications, Potentials, and Challenges. *Foods, 13*(18), 2964. PMID: 39335892

Alimohammadi, M., Xu, Y., Wang, D., Biris, A. S., & Khodakovskaya, M. V. (2011). Physiological responses induced in tomato plants by a two-component nanostructural system composed of carbon nanotubes conjugated with quantum dots and its in vivo multimodal detection. *Nanotechnology, 22*(29), 295101. PMID: 21673379

Aliyari Rad, S., Nobaharan, K., Pashapoor, N., Pandey, J., Dehghanian, Z., Senapathi, V., Minkina, T., Ren, W., Rajput, V. D., & Asgari Lajayer, B. (2023). Nano-Microbial Remediation of Polluted Soil: A Brief Insight. *Sustainability, 15*(1), 1. Advance online publication. DOI: 10.3390/su15010876

Allaker, R. P. (2010). The use of nanoparticles to control oral biofilm formation. *Journal of Dental Research, 89*(11), 1175–1186. DOI: 10.1177/0022034510377794 PMID: 20739694

Alosaimi, A. M., Alorabi, R. O., Katowah, D. F., Al-Thagafi, Z. T., Alsolami, E. S., Hussein, M. A., & Rafatullah, M. (2023). Recent biomedical applications of coupling nanocomposite polymeric materials reinforced with variable carbon nanofillers. *Biomedicines, 11*(3), 967. PMID: 36979948

Alphandéry, E. (2020). Bio-synthesized iron oxide nanoparticles for cancer treatment. *International Journal of Pharmaceutics, 586*, 119472. DOI: 10.1016/j.ijpharm.2020.119472 PMID: 32590095

Alprol, A. E., Mansour, A. T., Ibrahim, M. E. E.-D., & Ashour, M. (2024). Artificial Intelligence Technologies Revolutionizing Wastewater Treatment: Current Trends and Future Prospective. *Water (Basel), 16*(2), 314. DOI: 10.3390/w16020314

Ambaye, A. D., Fito, J., Nxumalo, E., & Nkambule, T. T. I. (2023). The challenges of micropollutants and their environmental detection through biochar-based nanostructured electrochemical sensors: A review. *International Journal of Environmental Science and Technology, 20*(10), 11741–11762.

Amin, H., Tayyab, A., Umair, M., Naveed, M. R., Yaseen, H. R., Qasim, M., Muzammal, M., Saif, S., Ali, M. A., & Sultan, M. A. (2025). Recent Progress in Wastewater Treatment: Exploring the Roles of Zero-Valent Iron and Titanium Dioxide Nanoparticles. *Indus Journal of Bioscience Research, 3*(4), 4. Advance online publication. DOI: 10.70749/ijbr.v3i4.1141

Amorim, W. L., Costa, H. O., Souza, F. C., Castro, M. G., & Silva, L. (2009). Experimental study of the tissue reaction caused by the presence of cellulose produced. *Brazilian Journal of Otorhinolaryngology*, *75*(2), 200–207. PMID: 19575105

Amusat, S. O., Kebede, T. G., Dube, S., & Nindi, M. M. (2021). Ball-milling synthesis of biochar and biochar–based nanocomposites and prospects for removal of emerging contaminants: A review. *Journal of Water Process Engineering*, *41*, 101993.

Anderson, J. M. (2001)... *Annual Review of Materials Research*, *31*(1), 1–401.

Annu, M., Tripathi, S., & Shin, D. K. (2024). Biopolymeric Nanocomposites for Wastewater Remediation: An Overview on Recent Progress and Challenges. *Polymers*, *16*(2), 2. Advance online publication. DOI: 10.3390/polym16020294 PMID: 38276702

Anugraha, R. P., Tobing, S., Maniam, G. P., Zhang, D. H., & Chutima, P. (2024). Polymers, Composites, Nano- and Bio- Materials. In Materials Science Forum. Trans Tech Publications Ltd. https://doi.org/DOI: 10.4028/b-pm5ge6

Anwer, A. H., Ahtesham, A., Shoeb, M., Mashkoor, F., Ansari, M. Z., Zhu, S., & Jeong, C. (2023). State-of-the-art advances in nanocomposite and bio-nanocomposite polymeric materials: A comprehensive review. *Advances in Colloid and Interface Science*, *318*, 102955. PMID: 37467558

Aoki, N., Nishikawa, M., & Hattori, K. (2003). Synthesis of chitosan derivatives bearing cyclodextrin and adsorption of p-nonylphenol and bisphenol A. *Carbohydrate Polymers*, *52*(3), 219–223.

Arif, Z. U., Khalid, M. Y., Sheikh, M. F., Zolfagharian, A., & Bodaghi, M. (2022). Biopolymeric sustainable materials and their emerging applications. *Journal of Environmental Chemical Engineering*, *10*(4), 108159.

Arkles, B. (1997). *Kirk-Othmer encyclopedia of chemical technology*. John Wiley.

Arvanitoyannis, I. S., & Tserkezou, P. (2009). In Long, Y. (Ed.), *Biodegradable polymer blends and composites from renewable resources* (pp. 55–86). John Wiley.

Arwenyo, B., Varco, J. J., Dygert, A., Brown, S., Pittman, C. U., & Mlsna, T. (2023). Contribution of modified P-enriched biochar on pH buffering capacity of acidic soil. *Journal of Environmental Management*, *339*, 117863. DOI: 10.1016/j.jenvman.2023.117863 PMID: 37080104

Ashfaq, M., Verma, N., & Khan, S. (2017). Carbon nanofibers as a micronutrient carrier in plants: Efficient translocation and controlled release of Cu nanoparticles. *Environmental Science. Nano*, *4*(1), 138–148.

Ashique, S., Raikar, A., Jamil, S., Lakshminarayana, L., Gajbhiye, S. A., De, S., & Kumar, S. (2024). Artificial Intelligence Integration with Nanotechnology: A New Frontier for Sustainable and Precision Agriculture. *Current Nanoscience*, *21*(2), 242–273. DOI: 10.2174/0115734137275111231206072049

Ashrafi-Saiedlou, S., Rasouli-Sadaghiani, M., Fattahi, M., & Ghosta, Y. (2025). Biosynthesis and characterization of iron oxide nanoparticles fabricated using cell-free supernatant of Pseudomonas fluorescens for antibacterial, antifungal, antioxidant, and photocatalytic applications. *Scientific Reports*, *15*(1). Advance online publication. DOI: 10.1038/s41598-024-84974-0 PMID: 39762412

Assis, D., Cardozo, O., Maia, R.Jr, Alves, S.Jr, de Melo, C. P., Justino, R., & Farias, P. M. (2025). Enhanced physical, electrical, and structural properties of chitosan thin films doped with Ag@ ZnO nanoparticles for energy storage applications. *Materials Chemistry and Physics*, *329*, 130003.

Ates, B., Koytepe, S., Ulu, A., Gurses, C., & Thakur, V. K. (2020). Chemistry, Structures, and Advanced Applications of Nanocomposites from Biorenewable Resources. *Chemical Reviews*, *120*(17), 9304–9362. PMID: 32786427

Averous, L., & Boquillon, N. (2004). Biocomposites based on plasticized starch: Thermal and mechanical behaviours. *Carbohydrate Polymers*, *56*, 111–122.

Avérous, L., & Pollet, E. (Eds.). (2023). *Biodegradable and biobased polymers for environmental and biomedical applications* (2nd ed.). Wiley.

Awad, A., Jalab, R., Benamor, A., Nasser, M., Ba-Abbad, M., El-Naas, M., & Mohammad, A. (2020). Adsorption of organic pollutants by nanomaterial-based adsorbents: An overview. *Journal of Molecular Liquids*, *301*, 112335. DOI: 10.1016/j.molliq.2019.112335

Ayalew, A. A. (2022). A critical review on clay-based nanocomposite particles for application of wastewater treatment. *Water Science and Technology*, *85*(10), 3002–3022. DOI: 10.2166/wst.2022.150 PMID: 35638802

Aydın, A., Yüceer, M., Ulugergerli, E. U., & Caner, C. (2024). Improving food security as disaster relief using intermediate moisture foods and active packaging technologies. *Applied Food Research*, *4*(1), 100378.

Ayodele, B. V., Alsaffar, M. A., Mustapa, S. I., Cheng, C. K., & Witoon, T. (2021). Modeling the effect of process parameters on the photocatalytic degradation of organic pollutants using artificial neural networks. *Process Safety and Environmental Protection*, *145*, 120–132. DOI: 10.1016/j.psep.2020.07.053

Azadi, E., Dinari, M., Derakhshani, M., Reid, K. R., & Karimi, B. (2024). Sources and Extraction of Biopolymers and Manufacturing of Bio-Based Nanocomposites for Different Applications. *Molecules (Basel, Switzerland), 29*(18), 4406. PMID: 39339400

Azevedo, V. M., Dias, M. V., Borges, S. V., Costa, A. L. R., Silva, E. K., Medeiros, É. A. A., & Nilda de Fátima, F. S. (2015). Development of whey protein isolate bio-nanocomposites: Effect of montmorillonite and citric acid on structural, thermal, morphological and mechanical properties. *Food Hydrocolloids, 48*, 179–188.

Bansal, M., Goyal, A., & Choudhary, A. (2022). A comparative analysis of K-Nearest Neighbor, Genetic, Support Vector Machine, Decision Tree, and Long Short Term Memory algorithms in machine learning. *Decision Analytics Journal, 3*, 100071. DOI: 10.1016/j.dajour.2022.100071

Baradaran, S., Moghaddam, E., Basirun, W. J., Mehrali, M., Sookhakian, M., Hamdi, M., Moghaddam, M. N., & Alias, Y. (2014). Mechanical properties and biomedical applications of a nanotube hydroxyapatite-reduced graphene oxide composite. *Carbon, 69*, 32–45.

Barathkumar, S., Sellamuthu, K. M., Sathyabama, K., Malathi, P., Kumaraperumal, R., & Devagi, P. (2025). Advancements in Soil Quality Assessment: A Comprehensive Review of Machine Learning and AI-Driven Approaches for Nutrient Deficiency Analysis. *Communications in Soil Science and Plant Analysis, 56*(2), 251–276. DOI: 10.1080/00103624.2024.2406484

Basavegowda, N., & Baek, K.-H. (2021). Advances in Functional Biopolymer-Based Nanocomposites for Active Food Packaging Applications. *Polymers, 13*(23), 4198. DOI: 10.3390/polym13234198 PMID: 34883701

Bascones, A., Vega, J. M., Olmo, N., Turnay, J. G., Gavilanes, G., & Lizarbe, M. A. (2002). Dental and maxillofacial surgery applications of polymers. In *Polymeric biomaterials* (pp. 426–454). Marcel Dekker.

Batley, G. E., Kirby, J. K., & McLaughlin, M. J. (2013). Fate and risks of nanomaterials in aquatic and terrestrial environments. *Accounts of Chemical Research, 46*, 854–862. PMID: 22759090

Batool, F., Iqbal, M. S., Khan, S., Khan, J., Ahmed, B., & Qadir, M. I. (2021). Biologically synthesized iron nanoparticles (FeNPs) from Phoenix dactylifera have anti-bacterial activities. *Scientific Reports, 11*(1). Advance online publication. DOI: 10.1038/s41598-021-01374-4 PMID: 34764312

Batra, R., Song, L., & Ramprasad, R. (2020). Emerging materials intelligence ecosystems propelled by machine learning. *Nature Reviews. Materials*, *6*(8), 655–678. DOI: 10.1038/s41578-020-00255-y

Baumann, M. D., Kang, C. E., & Stanwick, J. C.. (2009). An injectable drug delivery platform for sustained combination therapy. *Journal of Controlled Release*, *138*(3), 205–213. PMID: 19442692

Bayne, S. C., Thompson, J. Y., Swift, E. J., Stamatiades, P., & Wilkerson, M. (1998). A CHARACTERIZATION OF FIRST-GENERATION FLOWABLE COMPOSITES. *The Journal of the American Dental Association*, *129*(5), 567–577. DOI: 10.14219/JADA.ARCHIVE.1998.0274 PMID: 9601169

Behera, A. K., Srivastava, R., & Das, A. B. (2022). Mechanical and degradation properties of thermoplastic starch reinforced nanocomposites. *Starch*, *74*(3–4). Advance online publication. DOI: 10.1002/star.202100270

Beheshtkhoo, N., Kouhbanani, M. J., Savardashtaki, A., Amani, A. M., & Taghizadeh, S. (2018). Green synthesis of iron oxide nanoparticles by aqueous leaf extract of Daphne mezereum as a novel dye removing material. *Applied Physics. A, Materials Science & Processing*, *124*(5). Advance online publication. DOI: 10.1007/s00339-018-1782-3

Behl, K., Jaiswal, P., & Pabbi, S. (2024). Recent advances in microbial and nanoformulations for effective delivery and agriculture sustainability. *Biocatalysis and Agricultural Biotechnology*, *58*, 103180. DOI: 10.1016/j.bcab.2024.103180

Bensy, A. D., Christobel, G. J., Muthusamy, K., Alfarhan, A., & Anantharaman, P. (2022). Green synthesis of iron nanoparticles from Ulva lactuca and bactericidal activity against enteropathogens. *Journal of King Saud University. Science*, *34*(3), 101888. DOI: 10.1016/j.jksus.2022.101888

Besun, N., Peker, S., Kokturk, U., & Yilmaz, H. (1997).. . *Colloid & Polymer Science*, *275*, 378–389.

Bhandari, G., Gangola, S., Dhasmana, A., Rajput, V., Gupta, S., Malik, S., & Slama, P. (2023). Nano-biochar: Recent progress, challenges, and opportunities for sustainable environmental remediation. *Frontiers in Microbiology*, *14*, 1214870. DOI: 10.3389/fmicb.2023.1214870 PMID: 37547682

Bharathi, V. S., & Jayas, D. S. (2024). Evolution of Bionanocomposites: Innovations and Applications in Food Packaging. *Foods*, *13*(23), 3787. PMID: 39682859

Bhattarai, N., Edmondson, D., Veiseh, O., Matsen, F. A., & Zhang, M. Q. (2005). Electrospun chitosan-based nanofibers and their cellular compatibility. *Biomaterials, 26*(31), 6176–6184. PMID: 15885770

Biswas, A., Vanlalveni, C., Lalfakzuala, R., Nath, S., & Rokhum, S. L. (2021). Mikania mikrantha leaf extract mediated biogenic synthesis of magnetic iron oxide nanoparticles: Characterization and its antimicrobial activity study. *Materials Today: Proceedings, 42*, 1366–1373. DOI: 10.1016/j.matpr.2021.01.108

Bledzki, A. K., & Jaszkiewicz, A. (2020). Mechanical performance of biocomposites based on PLA and PHBV reinforced with natural fibres — A comparative study to PP. *Composites Science and Technology, 200*(12), 1687–1696.

Blessing, A.-A., & Olateru, K. (2025). AI-driven optimization of bioremediation strategies for river pollution: A comprehensive review and future directions. *Frontiers in Microbiology, 16*, 1504254. DOI: 10.3389/fmicb.2025.1504254 PMID: 40371099

Brinker, C. J., & Scherer, G. W. (1990). Elasticity of DLCA model gels with loops. In *Sol–gel science: The physics and chemistry of sol–gel processing*. Academic Press.

Brown, R. M. (1998). Microbial cellulose: A new resource for wood, paper, textiles, food and specialty products [Position paper]. http://www.botany.utexas.edu/facstaff/facpages/mbrown/position1.html

Brydson, J. A. (1999). *Plastics materials*. Butterworth-Heinemann.

Buarki, F., AbuHassan, H., Hannan, F. A., & Henari, F. Z. (2022). Green Synthesis of Iron Oxide Nanoparticles Using Hibiscus rosa sinensis Flowers and Their Antibacterial Activity. *Journal of Nanotechnology, 2022*, 1–6. DOI: 10.1155/2022/5474645

Buntinx, M., Vanheusden, C., & Hermans, D. (2024). Processing and Properties of Polyhydroxyalkanoate/ZnO Nanocomposites: A Review of Their Potential as Sustainable Packaging Materials. *Polymers, 16*(21), 3061. PMID: 39518271

Butt, B. Z., Sumera Javad, S. J., & Saba Zia, S. Z. (2018). Biogenic synthesis of iron nanoparticles from Catharanthus roseus. *Asian Journal of Agriculture and Biology, 6*(1), 83–89.

Caf, F. (2022). Biogenic Synthesis of Iron Oxide Nanoparticle Using Padina Pavonica Extract: Application for Photocatalytic Degradation of Congo Red Dye, Neurotoxicity and Antioxidant Activity. *Turkish Journal of Fisheries and Aquatic Sciences, 23*(2). Advance online publication. DOI: 10.4194/trjfas21398

Camargo, P. H. C., Satyanarayana, K. G., & Wypych, F. (2009). Nanocomposites: Synthesis, structure, properties and new application opportunities. *Materials Research, 12*, 1–39.

Cano, L., Pollet, E., Avérous, L., & Tercjak, A. (2017). Effect of TiO2 nanoparticles on the properties of thermoplastic chitosan-based nano-biocomposites obtained by mechanical kneading. *Composites. Part A, Applied Science and Manufacturing, 93*, 33–40. DOI: 10.1016/J.COMPOSITESA.2016.11.012

Caorsi, G., Candotto Carniel, F., Légnani, M., Flahaut, E., González, V. J., Vázquez, E., Prato, M., & Tretiach, M. (2025). Applicability of the OECD Test Guideline 201 to graphene-related materials: Dispersion stability matters. *Ecotoxicology and Environmental Safety, 292*, 117888. DOI: 10.1016/j.ecoenv.2025.117888 PMID: 39987683

Carrillo-Lopez, L. M., Villanueva-Verduzco, C., Villanueva-Sánchez, E., Fajardo-Franco, M. L., Aguilar-Tlatelpa, M., Ventura-Aguilar, R. I., & Soto-Hernández, R. M. (2024). Nanomaterials for Plant Disease Diagnosis and Treatment: A Review. *Plants, 13*(18), 2634. DOI: 10.3390/plants13182634 PMID: 39339607

Castillo, L. A., López, O. V., Ninago, M. D., Versino, F., Barbosa, S. E., García, M. A., & Villar, M. A. (2017). Effect of mineral and organic fillers on processing, structure, and final properties of starch. In *Composites and nanocomposites based on starches*. Elsevier.

Catriona, M., & Jonathan, W. (2014). The potential of 3D printing to reduce the environmental impacts of production. In *ECEEE Industrial Summer Study Proceedings*. Sea Green Tree.

Cavassin, E. D., de Figueiredo, L. F. P., Otoch, J. P., Seckler, M. M., de Oliveira, R. A., Franco, F. F., Marangoni, V. S., Zucolotto, V., Levin, A. S. S., & Costa, S. F. (2015). Comparison of methods to detect the in vitro activity of silver nanoparticles (AgNP) against multidrug resistant bacteria. *Journal of Nanobiotechnology, 13*(1). Advance online publication. DOI: 10.1186/S12951-015-0120-6 PMID: 26438142

Chandel, M., Kaur, K., Sahu, B. K., Sharma, S., Panneerselvam, R., & Shanmugam, V. (2022). Promise of nano-carbon to the next generation sustainable agriculture. *Carbon, 188*, 461–481.

Chandu, B., Kurmarayuni, C. M., Kurapati, S., & Bollikolla, H. B. (2020). Green and economical synthesis of graphene–silver nanocomposite exhibiting excellent photocatalytic efficiency. *Carbon Letters, 30*(2), 225–233.

Channab, B.-E., Tayi, F., Aqlil, M., Akil, A., Essamlali, Y., Chakir, A., & Zahouily, M. (2024). Graphene oxide, starch, and kraft lignin bio-nanocomposite controlled-release phosphorus fertilizer: Effect on P management and maize growth. International Journal of Biological Macromolecules, 282(Part 5-137190). https://doi.org/ DOI: 10.1016/j.ijbiomac.2024.137190

Chatterjee, N., & Dhar, P. (2025). Bio-nanocomposites: curse or miracle for the society? In *Characterization Techniques in Bionanocomposites* (pp. 3-39). Woodhead Publishing.

Chaudhary, S., Prasanna, D., Trivedi, P., Nandy, R., Prasad, S. S., Sunitha, S., & Panigrahi, C. K. (2025). Nanotechnology Applications in Agriculture for Crop Protection and Nutrient Delivery: A Review. *International Journal of Plant and Soil Science, 37*(9), 106–119. DOI: 10.9734/ijpss/2025/v37i95690

Chauhan, N. S., & Punia, A. (2023). Heavy Metal Removal and Recovery: Sustainable and Efficient Approaches. In Singh, P., Verma, P., & Singh, R. P. (Eds.), *Wastewater Resource Recovery and Biological Methods* (pp. 87–124). Springer International Publishing., DOI: 10.1007/978-3-031-40198-5_5

Chausali, N., Saxena, J., & Prasad, R. (2021). Nanobiochar and biochar based nanocomposites: Advances and applications. *Journal of Agriculture and Food Research, 5*, 100191. DOI: 10.1016/j.jafr.2021.100191

Cheng, L., Zhang, K., Weir, M. D., Melo, M. A. S., Zhou, X., & Xu, H. H. K. (2015). Nanotechnology strategies for antibacterial and remineralizing composites and adhesives to tackle dental caries. *Nanomedicine (London, England), 10*(4), 627–641. DOI: 10.2217/NNM.14.191 PMID: 25723095

Chen, H., Wang, J., Cheng, Y., Wang, C., Liu, H., Bian, H., Pan, Y., Sun, J., & Han, W. (2019). Application of Protein-Based Films and Coatings for Food Packaging: A Review. *Polymers, 11*(12), 2039. PMID: 31835317

Chen, J., Huang, W., Chen, Y., Zhou, Z., Liu, H., Zhang, W., & Huang, J. (2022). Facile Preparation of Chitosan-Based Composite Film with Good Mechanical Strength and Flame Retardancy. *Polymers, 14*(7), 1337. PMID: 35406210

Chen, J., Xu, J., Wang, K., Cao, X., & Sun, R. (2023). Cellulose acetate fibers prepared from different raw materials with rapid synthesis method. *Carbohydrate Polymers, 301*, 115545. PMID: 26686180

Chen, L., Chen, X. L., Zhou, C. H., Yang, H. M., Ji, S. F., Tong, D. S., & Chu, M. Q. (2017). Environmental-friendly montmorillonite-biochar composites: Facile production and tunable adsorption-release of ammonium and phosphate. *Journal of Cleaner Production*, *156*, 648–659.

Chen, S., & Wang, Y. (2001). Study on β-cyclodextrin grafting with chitosan and slow release of its inclusion complex with radioactive iodine. *Journal of Applied Polymer Science*, *82*(10), 2414–2421.

Chen, X., Wang, S., Lai, H., Deng, L., Zhong, Q., Okoye, C. O., Niu, Q., Jing, Y., Wang, J., & Jiang, J. (2025). Nano-Biotechnology in Soil Remediation: Use of Nanomaterials to Promote Plant Growth and Stress Tolerance. *Nanomaterials (Basel, Switzerland)*, *15*(22), 1743. DOI: 10.3390/nano15221743 PMID: 41295648

Chen, Y., Li, Kejun, Yang, Yuanjiao, Yuan, Xin, Zhang, Mei, & and Huang, K. (2024). Recent developments in optical aptamer sensors for pesticide residues detectio. *Applied Spectroscopy Reviews*, *59*(9), 1147–1182. DOI: 10.1080/05704928.2024.2308132

Chen, Z., Mo, X., & Qing, F. (2007). Electrospinning of collagen–chitosan complex. *Materials Letters*, *61*(16), 3490–3494.

Cheung, H. Y., Lau, K. T., Tao, X. M., & Hui, D. (2008). A potential material for tissue engineering: Silkworm silk/PLA biocomposite. *Composites. Part B, Engineering*, *39*(6), 1026–1033. DOI: 10.1016/J.COMPOSITESB.2007.11.009

Cheyns, E., & Riisgaard, L. (2014). Introduction to the symposium: The exercise of power through multi-stakeholder initiatives for sustainable agriculture and its inclusion and exclusion outcomes. *Agriculture and Human Values*, *31*(3), 409–423. DOI: 10.1007/s10460-014-9508-4

Chincholikar, P., Singh, K. R., Natarajan, A., Kerry, R. G., Singh, J., Malviya, J., & Singh, R. P. (2023). Green nanobiopolymers for ecological applications: A step towards a sustainable environment. *RSC Advances*, *13*(18), 12411–12429. PMID: 37091622

Choy, J. H., Oh, J., & Choi, S. J. (2008). *Bio-inorganic hybrid nanomaterials*. Wiley-VCH.

Chung, D. D. L. (2010). *Composite Materials*. https://doi.org/DOI: 10.1007/978-1-84882-831-5

Cimini, A., Imperi, E., Picano, A., & Rossi, M. (2023). Electrospun nanofibers for medical face mask with protection capabilities against viruses: State of the art and perspective for industrial scale-up. *Applied Materials Today*, *32*, 101833. PMID: 37152683

Csiszár, E., Kun, D., & Fekete, E. (2021a). The Role of Structure and Interactions in Thermoplastic Starch–Nanocellulose Composites. *Polymers*, *13*(18), 18. PMID: 34578087

Cui, C., Shao, C., Meng, L., & Yang, J. (2019). High-Strength, Self-Adhesive, and Strain-Sensitive Chitosan/Poly(acrylic acid) Double-Network Nanocomposite Hydrogels Fabricated by Salt-Soaking Strategy for Flexible Sensors. *ACS Applied Materials & Interfaces*, *11*(42), 39228–39237. PMID: 31550132

Czaja, W. K., Young, D. J., Kawecki, M., & Brown, R. M.Jr. (2007). The future prospects of microbial cellulose in biomedical applications. *Biomacromolecules*, *8*(1), 1–12. PMID: 17206781

D'Almeida, A. P., & de Albuquerque, T. L. (2024). Innovations in food packaging: From bio-based materials to smart packaging systems. *Processes (Basel, Switzerland)*, *12*(10), 2085.

Da'na, E., Taha, A., & Afkar, E. (2018). Green Synthesis of Iron Nanoparticles by Acacia nilotica Pods Extract and Its Catalytic, Adsorption, and Antibacterial Activities. *Applied Sciences (Basel, Switzerland)*, *8*(10), 1922. DOI: 10.3390/app8101922

Dadhwal, S., Ashton, L., Fraser, E. D., & Corradini, M. G. (2025). Life cycle assessment to evaluate bioplastics as alternatives to petroleum-based plastics in the food industry. In *Innovative Food Packaging and Processing Technologies* (pp. 449–473). Academic Press.

Dalbanjan, N. P., Eelager, M. P., Korgaonkar, K., Gonal, B. N., Kadapure, A. J., Arakera, S. B., & Kumar, S. P. (2024). Descriptive review on conversion of waste residues into valuable bionanocomposites for a circular bioeconomy. *Nano-Structures & Nano-Objects*, *39*, 101265.

Damm, C., Münstedt, H., & Rösch, A. (2008). The antimicrobial efficacy of polyamide 6/silver-nano- and microcomposites. *Materials Chemistry and Physics*, *108*(1), 61–66. DOI: 10.1016/J.MATCHEMPHYS.2007.09.002

Danai-Varsou, D., Zhang, P., Afantitis, A., Guo, Z., Lynch, I., & Melagraki, G. (2023). Nanoinformatics and artificial intelligence for nano-enabled sustainable agriculture. In *Nano-Enabled Sustainable and Precision Agriculture* (pp. 503–531). Elsevier., DOI: 10.1016/B978-0-323-91233-4.00015-6

Dang, C., & Morrissey, E. M. (2024). The size and diversity of microbes determine carbon use efficiency in soil. *Environmental Microbiology*, *26*(5). Advance online publication. DOI: 10.1111/1462-2920.16633 PMID: 38733078

Daniel, M. C., & Didier, A. (2004). Gold nanoparticles: Assembly, supramolecular chemistry, quantum-size-related properties, and applications toward biology, catalysis, and nanotechnology. *Chemical Reviews*, *104*, 293–346. PMID: 14719978

Danish, M., Estrella, L., Alemaida, I., Lisin, A., Moiseev, N., Ahmadi, M., Nazari, M., Wali, M., Zaheb, H., & Senjyu, T. (2021). Photocatalytic Applications of Metal Oxides for Sustainable Environmental Remediation. *Metals*. Advance online publication. DOI: 10.3390/met11010080

Darder, M., Lopez-Blanco, M., Aranda, P., Aznar, A. J., Bravo, J., & Ruiz-Hitzky, E. (2006). Microfibrous chitosan–sepiolite nanocomposites. *Chemistry of Materials*, *18*, 1602–1610.

Darie-Ni ă, R. N., & Frąckowiak, S. (2025). An Overview of Potential Applications of Environmentally Friendly Hybrid Polymeric Materials. *Polymers*, *17*(2), 252. DOI: 10.3390/polym17020252 PMID: 39861324

das Neves. P. B. A., Agnelli, J. A. M., Kurachi, C., & de Souza, C. W. O. (2014). Addition of silver nanoparticles to composite resin: effect on physical and bactericidal properties in vitro. *Brazilian Dental Journal, 25*(2), 141–145. https://doi.org/ DOI: 10.1590/0103-6440201302398

Das, N., Panda, S., Das, D. K., Nayak, S. K., Parthasarathy, S., & Patnaik, P. (2025). Polymer Nanocomposites and Nanostructured Polymers. In *Exploring Nanomaterial Synthesis, Characterization, and Applications* (pp. 213-240). IGI Global.

Dasari, A., Yu, Z.-Z., & Mai, Y.-W. (2009). Fundamental aspects and recent progress on wear/scratch damage in polymer nanocomposites. *Materials Science and Engineering R Reports*, *63*(2), 31–80. DOI: 10.1016/j.mser.2008.10.001

Das, R., & Panda, S. N. (2022). Preparation and applications of biochar based nanocomposite: A review. *Journal of Analytical and Applied Pyrolysis*, *167*, 105691.

Das, S. K., Chakraborty, S., Naskar, S., & Rajabalaya, R. (2021). Techniques and methods used for the fabrication of bionanocomposites. In *Bionanocomposites in Tissue Engineering and Regenerative Medicine* (pp. 17–43). Woodhead Publishing.

Davis, A. G., Huggins, D. R., & Reganold, J. P. (2023). Linking soil health and ecological resilience to achieve agricultural sustainability. *Frontiers in Ecology and the Environment*, *21*(3), 131–139. DOI: 10.1002/fee.2594

de Jong, K. P. (2009). *Synthesis of solid catalysts* (Vol. 49). Wiley-VCH.

de Lima, G. G., Zakaluk, I. C. B., Artner, M. A., Pedro, A. C., Gonzalez de Cademartori, P. H., Muniz, G. I. B. D., & Magalhães, W. L. E. (2025). Enhancing Barrier and Antioxidant Properties of Nanocellulose Films for Coatings and Active Packaging: A Review. *ACS Applied Nano Materials*.

De Oliveira, M. L., Rodrigues, L. M., Da Veiga, M. M. S., & Souza, L. R. R. (2024). Advancing nanomaterial synthesis: Harnessing green chemistry for sustainable innovation. *Green Analytical Chemistry*, *11*, 100148. DOI: 10.1016/j.greeac.2024.100148

Dean, K. M., Sangwan, P., Way, C., & Nikolic, M. A. L. (2012). *Biodegradability Characterization of Polymer Nanocomposites. Characterization Techniques for Polymer Nanocomposites*. Portico., DOI: 10.1002/9783527654505.ch13

Defonseka, C. (2014). *Introduction to polymeric composites with rice hulls*. Smithers Rapra.

del Rosario Herrera-Rivera, M., Torres-Arellanes, S. P., Cortés-Martínez, C. I., Navarro-Ibarra, D. C., Hernández-Sánchez, L., Solis-Pomar, F., & Román-Doval, R. (2024). Nanotechnology in food packaging materials: Role and application of nanoparticles. *RSC Advances*, *14*(30), 21832–21858. PMID: 38984259

Deng, J., Zhu, E.-Q., Xu, G.-F., Naik, N., Murugadoss, V., Ma, M.-G., Guo, Z., & Shi, Z.-J. (2022). Overview of renewable polysaccharide-based composites for biodegradable food packaging applications. *Green Chemistry*, *24*(2), 480–492.

Desselberger, U. (2000). Emerging and Re-emerging Infectious Diseases. *The Journal of Infection*, *40*(1), 3–15. DOI: 10.1053/JINF.1999.0624 PMID: 10762105

Devi, H. S., Boda, M. A., Shah, M. A., Parveen, S., & Wani, A. H. (2018). Green synthesis of iron oxide nanoparticles using Platanus orientalis leaf extract for antifungal activity. *Green Processing and Synthesis*, *8*(1), 38–45. DOI: 10.1515/gps-2017-0145

Dey, B., Prabhakar, M. R., Jayaraman, S., Gujjala, L. K. S., Venugopal, A. P., & Balasubramanian, P. (2024). Biopolymer-based solutions for enhanced safety and quality assurance: A review. *Food Research International*, •••, 114723. PMID: 39059918

Dhakal, B., & Kattel, R. R. (2019). Effects of global changes on ecosystems services of multiple natural resources in mountain agricultural landscapes. *The Science of the Total Environment*, *676*, 665–682. DOI: 10.1016/j.scitotenv.2019.04.276 PMID: 31051371

Diaz-Puertas, R., Alvarez-Martinez, F. J., Falco, A., Barrajon-Catalan, E., & Mallavia, R. (2023). Phytochemical-Based Nanomaterials against Antibiotic-Resistant Bacteria: An Updated Review. *Polymers, 15*(6), 1392. DOI: 10.3390/polym15061392 PMID: 36987172

Dijoo, Z. K., & Khurshid, R. (2022). Environmental degradation as a multifaceted consequence of human development. In *Environmental Biotechnology* (pp. 39–56). Apple Academic Press.

Dinesh, A., & Rahul Prasad, B. (2024). Predictive models in machine learning for strength and life cycle assessment of concrete structures. *Automation in Construction, 162*, 105412. DOI: 10.1016/j.autcon.2024.105412

Dizaj, S. M., Lotfipour, F., Barzegar-Jalali, M., Zarrintan, M. H., & Adibkia, K. (2014). Antimicrobial activity of the metals and metal oxide nanoparticles. *Materials Science and Engineering C, 44*, 278–284. DOI: 10.1016/J.MSEC.2014.08.031 PMID: 25280707

Dodero, A., Escher, A., Bertucci, S., Castellano, M., & Lova, P. (2021). Intelligent packaging for real-time monitoring of food-quality: Current and future developments. *Applied Sciences (Basel, Switzerland), 11*(8), 3532.

Dorigato, A., & Fredi, G. (2024). Effect of nanofillers addition on the compatibilization of polymer blends. *Advanced Industrial and Engineering Polymer Research, 7*(4), 405–427.

Drummond, J. L., & Bapna, M. S. (2003). Static and cyclic loading of fiber-reinforced dental resin. *Dental Materials, 19*(3), 226–231. DOI: 10.1016/S0109-5641(02)00034-9 PMID: 12628435

Dufresne, A. (2022). *Nanocellulose: From nature to high performance tailored materials* (3rd ed.). De Gruyter.

Duncan, T. V. (2011). Applications of nanotechnology in food packaging and food safety: Barrier materials, antimicrobials and sensors. *Journal of Colloid and Interface Science, 363*(1), 1–24. DOI: 10.1016/J.JCIS.2011.07.017 PMID: 21824625

Dutta, D., & Sit, N. (2024). *A comprehensive review on types and properties of biopolymers as sustainable bio-based alternatives for packaging*. Food Biomacromolecules.

Du, Y., Shi, B., Luan, X., Wang, Y., & Song, H. (2023). Chitosan/cellulose nanocrystal biocomposite coating for fruit postharvest preservation. *Industrial Crops and Products, 205*, 117543.

Dweib, M. A., Hu, B., Shenton, H. W., & Wool, R. P. (2006). Bio-based composite roof structure: Manufacturing and processing issues. *Composite Structures*, *74*(4), 379–388.

Eduardo, R. H., Katsuhiko, A., & Yuri, M. L. (2008). *Bio-inorganic hybrid nanomaterials: Strategies, synthesis, characterization and applications*. Wiley-VCH.

Eduardo, R. H., Margarita, D., & Pilar, A. (2005). Functional biopolymer nanocomposites based on layered solids. *Journal of Materials Chemistry*, *15*, 3650–3662.

Efremenko, E., Stepanov, N., Aslanli, A., Lyagin, I., Senko, O., & Maslova, O. (2023). Combination of Enzymes with Materials to Give Them Antimicrobial Features: Modern Trends and Perspectives. *Journal of Functional Biomaterials*, *14*(2), 64. DOI: 10.3390/jfb14020064 PMID: 36826863

Eichhorn, S. J., Dufresne, A., Aranguren, M., Marcovich, N. E., Capadona, J. R., Rowan, S. J., Weder, C., Thielemans, W., Roman, M., Renneckar, S., Gindl, W., Veigel, S., Keckes, J., Yano, H., Abe, K., Nogi, M., Nakagaito, A. N., Mangalam, A., Simonsen, J., … Peijs, T. (2010). Review: current international research into cellulose nanofibres and nanocomposites. *Journal of Materials Science 2009 45:1*, *45*(1), 1–33. https://doi.org/DOI: 10.1007/S10853-009-3874-0

El Achaby, M., Essamlali, Y., El Miri, N., Snik, A., Abdelouahdi, K., Fihri, A., Zahouily, M., & Solhy, A. (2014). Graphene oxide reinforced chitosan/polyvinylpyrrolidone polymer bio-nanocomposites. *Journal of Applied Polymer Science*, *131*(22).

El-Ramady, H., Prokisch, J., Mansour, H., Bayoumi, Y. A., Shalaby, T. A., Veres, S., & Brevik, E. C. (2024). Review of Crop Response to Soil Salinity Stress: Possible Approaches from Leaching to Nano-Management. *Soil Systems*, *8*(1), 1. Advance online publication. DOI: 10.3390/soilsystems8010011

Elsabagh, S. S., Elkhatib, E. A., & Rashad, M. (2024). Novel nano-fertilizers derived from drinking water industry waste for sustained release of macronutrients: Performance, kinetics and sorption mechanisms. *Scientific Reports*, *14*(1). Advance online publication. DOI: 10.1038/s41598-024-56274-0 PMID: 38454001

Esmaeilzadeh, P. (2024). Challenges and strategies for wide-scale artificial intelligence (AI) deployment in healthcare practices: A perspective for healthcare organizations. *Artificial Intelligence in Medicine*, *151*, 102861. DOI: 10.1016/j.artmed.2024.102861 PMID: 38555850

Expert Market Research. (2024). Global nanocellulose market report and forecast 2024-2032. Retrieved from https://www.expertmarketresearch.com/reports/nanocellulose-market

Falua, K. J., Pokharel, A., Babaei-Ghazvini, A., Ai, Y., & Acharya, B. (2022). Valorization of Starch to Biobased Materials: A Review. *Polymers*, *14*(11), 2215. PMID: 35683888

Fang, C., Bhattarai, N., Sun, C., & Zhang, M. (2009). Functionalized Nanoparticles with Long-Term Stability in Biological Media. *Small (Weinheim an Der Bergstrasse, Germany)*, *5*(14), 1637. https://doi.org/DOI: 10.1002/SMLL.200801647

Fang, M., Chen, J. H., Xu, X. L., Yang, P. H., & Hildebrand, H. F. (2006). Antibacterial activities of inorganic agents on six bacteria associated with oral infections by two susceptibility tests. *International Journal of Antimicrobial Agents*, *27*(6), 513–517. DOI: 10.1016/J.IJANTIMICAG.2006.01.008 PMID: 16713190

Farooq, K., Kumawat, Y. K., Sharma, K., & Kumar, V. (2024). Bio-nanocomposites: Fundamentals and Recent Advances. Novel Bio-Nanocomposites for Biomedical Applications, 1–35. https://doi.org/DOI: 10.1007/978-3-031-69654-1_1

Farooq, K., Kumawat, Y. K., Sharma, K., & Kumar, V. (2024). Bio-nanocomposites: Fundamentals and Recent Advances. In Sharma, K., Tiwari, S. K., Kumar, V., & Kalia, S. (Eds.), *Novel Bio-nanocomposites for Biomedical Applications* (pp. 1–35). Springer International Publishing.

Farrokhi, H., Koosha, M., Nasirizadeh, N., Salari, M., Abdouss, M., Li, T., & Gong, Y. (2024). The Effect of Nanoclay Type on the Mechanical Properties and Antibacterial Activity of Chitosan/PVA Nanocomposite Films. *Journal of Composites Science*, *8*(7), 255.

Faruk, O., Bledzki, A. K., Fink, H. P., & Sain, M. (2021). Biocomposites reinforced with natural fibers: 2000–2020 developments. *Progress in Polymer Science*, *116*, 1552–1596.

Fathi, M., Karim, M., & Ahmadi, N. (2019). Nanostructures of cellulose for encapsulation of food ingredients. In *Biopolymer Nanostructures for Food Encapsulation Purposes* (pp. 493–519). Elsevier.

Feng, Y., Zhang, X., Shen, Y., Yoshino, K., & Feng, W. (2012). A mechanically strong, flexible and conductive film based on bacterial cellulose/graphene nanocomposite. *Carbohydrate Polymers*, *87*(1), 644–649. PMID: 34663016

Fernandez, E. J., & Monge, M. (2008). In A. Laguna (Ed.), Modern supramolecular gold chemistry: Gold-metal interactions and applications (pp. 131–179). Wiley-VCH.

Ferracane, J. L. (1995). Current trends in dental composites. *Critical Reviews in Oral Biology and Medicine : An Official Publication of the American Association of Oral Biologists*, *6*(4), 302–318. DOI: 10.1177/10454411950060040301 PMID: 8664421

Ferracane, J. L. (2011). Resin composite—State of the art. *Dental Materials, 27*(1), 29–38. DOI: 10.1016/J.DENTAL.2010.10.020 PMID: 21093034

Firouz, M. S., Mohi-Alden, K., & Omid, M. (2021). A critical review on intelligent and active packaging in the food industry: Research and development. *Food Research International, 141*, 110113. PMID: 33641980

Foda, N. H., El-Laithy, H. M., & Tadros, M. I. (2007). Implantable biodegradable sponges: Effect of interpolymer complex formation of chitosan with gelatin on the release behavior of tramadol hydrochloride. *Drug Development and Industrial Pharmacy, 33*(1), 7–17. PMID: 17192246

Fontana, J. D., de Souza, A. M., & Fontana, C. K.. (1990). Acetobacter cellulose pellicle as a temporary skin substitute. *Applied Biochemistry and Biotechnology, 24–25*, 253–264. PMID: 2353811

Forano, C., Hibino, T., Leroux, F., & Taviot-Gueho, C. (2006). In Bergaya, F., Theng, B. K. G., & Lagaly, G. (Eds.), *Handbook of clay science: Developments in clay science* (pp. 1021–1095).

Gámiz, B., Cox, L., Hermosín, M. C., Spokas, K., & Celis, R. (2017). Assessing the effect of organoclays and biochar on the fate of abscisic acid in soil. *Journal of Agricultural and Food Chemistry, 65*(1), 29–38. PMID: 27959547

García-Guzmán, L., Cabrera-Barjas, G., Soria-Hernández, C. G., Castaño, J., Guadarrama-Lezama, A. Y., & Rodríguez Llamazares, S. (2022). Progress in starch-based materials for food packaging applications. *Polysaccharides, 3*, 136–177. DOI: 10.3390/polysaccharides3010007

Gęca, M., Khalil, A. M., Tang, M., Bhakta, A. K., Snoussi, Y., Nowicki, P., & Chehimi, M. M. (2023). Surface treatment of biochar—methods, surface analysis and potential applications: A comprehensive review. *Surfaces, 6*(2), 179–213.

Getahun, M. J., Kassie, B. B., & Alemu, T. S. (2024). Recent advances in biopolymer synthesis, properties, & commercial applications: A review. *Process Biochemistry*.

Ghosh, S., Goudar, V. S., Padmalekha, K. G., Bhat, S. V., Indi, S. S., & Vasan, H. N. (2012). ZnO/Ag nanohybrid: Synthesis, characterization, synergistic antibacterial activity and its mechanism. *RSC Advances, 2*(3), 930–940. DOI: 10.1039/C1RA00815C

Giri, A., Goswami, N., Sarkar, S., & Kumar Pal, S. (2013). Bio-Nanomaterials: Understanding Key Biophysics and Their Applications. *Nanotechnology, 11*, 41–110.

Global, T. W. I. (n.d.). What is a thermoplastic? https://www.twi-global.com/technical-knowledge/faqs/what-is-a-thermoplastic

Gojiya, K. M., Karangiya, B. A., Chavda, S. K., Gaadhe, S. K., Gojiya, D. K., & Rank, H. D. (2024). Soil Moisture Modelling Using Remote Sensing and Artificial Neural Networks: A Study of Devbhumi Dwarka Region, Gujarat, India. *International Journal of Environment and Climate Change*, *14*(1), 1–13. DOI: 10.9734/ijecc/2024/v14i13788

Golomb, D. S., & Fay, J. A. (2004). In Giere, R., & Stille, P. (Eds.), *Energy, waste, and the environment: A geochemical perspective* (Vol. 236, pp. 153–167). Geological Society.

Gomes, M., Azavedo, H., Malafaya, P., Silva, S., Oliveira, J., Silva, G., Sousa, R., Mano, J., & Reis, R. (2013). In van Blitterswijk, C. (Ed.), *Tissue engineering* (pp. 145–192). Academic Press.

Gomiero, T. (2016). Soil Degradation, Land Scarcity and Food Security: Reviewing a Complex Challenge. *Sustainability*, *8*(3), 281. DOI: 10.3390/su8030281

GONG, F. (2004). Thermal properties of poly (vinyl chloride)/montmorillonite nanocomposites. *Polymer Degradation & Stability*, *84*(2), 289–294. DOI: 10.1016/j.polymdegradstab.2003.11.003

Gong, Y.-Z., Niu, Q.-Y., Liu, Y.-G., Dong, J., & Xia, M.-M. (2022). Development of multifarious carrier materials and impact conditions of immobilised microbial technology for environmental remediation: A review. *Environmental Pollution*, *314*, 120232. DOI: 10.1016/j.envpol.2022.120232 PMID: 36155222

Gopi, S., Pius, A., & Thomas, S. (Eds.). (2023). *Handbook of biopolymer-based materials: From blends and composites to gels and complex networks*. Elsevier.

Gorde, P. M., Dash, D. R., Singh, S. K., & Singha, P. (2024). Advancements in sustainable food packaging: A comprehensive review on utilization of nanomaterials, machine learning and deep learning. *Sustainable Chemistry and Pharmacy*, *39*, 101619.

Goswami, L., Kushwaha, A., Kafle, S., & Kim, B. (2022). Surface Modification of Biochar for Dye Removal from Wastewater. *Catalysts*. Advance online publication. DOI: 10.3390/catal12080817

Goyal, R., Devi, M., Gautam, R. K., & Gupta, S. (2024). Insights into the prospects of green chemistry and nanotechnology in environmental sustainability. Role of Green Chemistry in Ecosystem Restoration to Achieve Environmental Sustainability, 3–9. https://doi.org/DOI: 10.1016/b978-0-443-15291-7.00030-4

Grappa, R., Venezia, V., Silvestri, B., Costantini, A., & Luciani, G. (2024). Synthesis of Lignin Nanoparticles: Top-Down and Bottom-Up Approaches. *One Health, 2023*, 3. DOI: 10.3390/msf2024025003

Grassie, N. (1985). *Polymer degradation and stabilization*. Cambridge University Press.

Greenfeld, I., & Wagner, H. D. (2014). Nanocomposite toughness, strength and stiffness: Role of filler geometry. *Nanocomposites*, *1*(1), 3–17. DOI: 10.1179/2055033214y.0000000002

Guirguis, O., & Alharbi, N. D. (2024). Enhancement of Molecular Structural and Linear/Nonlinear Optical Features of Chitosan/Titanium Dioxide Nanocomposite Films for Food Packaging and Optoelectronic Applications. *East European Journal of Physics*, *4*, 469–490.

Gulati, S., Amar, A., Chhabra, L., Katiyar, R., Sahu, T., & Varma, R. S. (2024). Greener nanobiopolymers and nanoencapsulation: Environmental implications and future prospects. *RSC Sustainability*, *2*(10), 2805–2832.

Gunasena, M. D. K. M., Galpaya, G. D. C. P., Abeygunawardena, C. J., Induranga, D. K. A., Priyadarshana, H. V. V., Millavithanachchi, S. S., Bandara, P. K. G. S. S., & Koswattage, K. R. (2025). Advancements in Bio-Nanotechnology: Green Synthesis and Emerging Applications of Bio-Nanoparticles. *Nanomaterials (Basel, Switzerland)*, *15*(7), 7. Advance online publication. DOI: 10.3390/nano15070528 PMID: 40214574

Gupta, R., Swarupa, S., Mayya, C., Bhatia, D., & Thareja, P. (2023). Graphene Oxide–Carbamoylated Chitosan Hydrogels with Tunable Mechanical Properties for Biological Applications. *ACS Applied Bio Materials*, *6*(2), 578–590. PMID: 36655342

Haider, T. P., Völker, C., Kramm, J., Landfester, K., & Wurm, F. R. (2019). Plastics of the Future? The Impact of Biodegradable Polymers on the Environment and on Society. *Angewandte Chemie International Edition*, *58*(1), 50–62. PMID: 29972726

Hamadeen, H. M., & Elkhatib, E. A. (2022). New nanostructured activated biochar for effective removal of antibiotic ciprofloxacin from wastewater: Adsorption dynamics and mechanisms. *Environmental Research*, *210*, 112929. PMID: 35167852

Han, J.-H., & Yun, K.-S. (2024). Real-Time Soil Moisture Monitoring Using a Graphene Oxide Sensor Powered by a Solar Cell-Based Energy Harvester. *Journal of Electrical Engineering & Technology*, *19*(5), 3331–3337. DOI: 10.1007/s42835-024-01790-2

Haroon, S. (2021). Manipulation of Bio-nanotechnology in Agri-Sector for Pest Control with Special Emphasis to Nematode Problems in Egypt. *Annals of Agricultural Sciences, Moshtohor, 59*(2), 33–36. DOI: 10.21608/assjm.2021.183354

Harun-Ur-Rashid, M., Jahan, I., Foyez, T., & Imran, A. Bin. (2023). Bio-Inspired Nanomaterials for Micro/Nanodevices: A New Era in Biomedical Applications. *Micromachines 2023, Vol. 14, Page 1786, 14*(9), 1786. https://doi.org/DOI: 10.3390/MI14091786

Herlekar, M., Barve, S., & Kumar, R. (2014). Plant-Mediated Green Synthesis of Iron Nanoparticles. *Journal of Nanoparticles, 2014*, 1–9. DOI: 10.1155/2014/140614

He, S., Zhong, L., Duan, J., Feng, Y., Yang, B., & Yang, L. (2017). Bioremediation of Wastewater by Iron Oxide-Biochar Nanocomposites Loaded with Photosynthetic Bacteria. *Frontiers in Microbiology, 8*. Advance online publication. DOI: 10.3389/fmicb.2017.00823 PMID: 28588556

Hitzky, E. R., Margarita, D., Ana, C. S., Alcantara, B. W., & Pilar, A. (2013). Functional nanocomposites based on fibrous clays. In Functional polymer composites with nanoclays (Vol. 1, pp. 1–53).

Hlihor, R. M. (2015). *Experimental analysis and mathematical prediction of Cd(II) removal by biosorption using support vector machines and genetic algorithms. 32*(3).

Hoemann, C. D., Lafantaisie-Favreau, C. H., Lascau-Coman, V., Chen, G., & Guzman-Morales, J. (2012). The cartilage-bone interface. *The Journal of Knee Surgery, 25*(2), 85–97. DOI: 10.1055/s-0032-1319782 PMID: 22928426

Holback, H., Yeo, Y., & Park, K. (2011). Hydrogel swelling behavior and its biomedical applications. In *Biomedical hydrogels* (Woodhead Publishing Series in Biomaterials, pp. 3–24). Woodhead Publishing. https://doi.org/DOI: 10.1533/9780857091383.1.3

Hoque, A., & Padhiary, M. (2024). Automation and AI in Precision Agriculture: Innovations for Enhanced Crop Management and Sustainability. *Asian Journal of Research in Computer Science, 17*(10), 95–109. DOI: 10.9734/ajrcos/2024/v17i10512

Hoque, A., Padhiary, M., Kumar, K., & Barbhuiya, J. A. (2025). The Role of Robotics in Sustainable Agriculture and Waste Management. In Rajest, S. S., & Baron, M. (Eds.), *A. J. Obaid, Muthmainnah* (pp. 265–302). Multidisciplinary Advancements in Human-AI Augmentation. IGI Global Scientific Publishing., DOI: 10.4018/979-8-3373-1987-2.ch012

Hoque, A., Padhiary, M., Prasad, G., & Kumar, K. (2025). Real-Time Data Processing in Agricultural Robotics. In Bora, D. J., & Bania, R. K. (Eds.), *Computer Vision Techniques for Agricultural Advancements* (pp. 431–468). IGI Global., DOI: 10.4018/979-8-3693-8019-2.ch014

Hosoya, Y., Shiraishi, T., Odatsu, T., Nagafuji, J., Kotaku, M., Miyazaki, M., & Powers, J. M. (2011). Effects of polishing on surface roughness, gloss, and color of resin composites. *Journal of Oral Science*, *53*(3), 283–291. DOI: 10.2334/JOSNUSD.53.283 PMID: 21959654

Hosseini, R., Rashidi, J., Mokhtariyan, M., & Landarani-Isfahani, A. (2025). Fabrication and Morphology of Biomaterials Based on the Used Synthesis Methods. In *Green Biomaterials in Tissue Engineering* (pp. 89–140). American Chemical Society.

Huang, K., Li, M., Li, R., Rasul, F., Shahzad, S., Wu, C., Shao, J., Huang, G., Li, R., Almari, S., Hashem, M., & Aamer, M. (2023). Soil acidification and salinity: The importance of biochar application to agricultural soils. *Frontiers in Plant Science*, *14*, 1206820. DOI: 10.3389/fpls.2023.1206820 PMID: 37780526

Huang, T., Li, Y., Qian, J., Liu, S., & Wang, J. (2024). Remediation of Cr(VI)-contaminated soil by double-modified nanoscale zero-valent iron: Performance and mechanism. *Journal of Soils and Sediments*, *24*(4), 1724–1738. DOI: 10.1007/s11368-024-03768-z

Hubbe, M. A., Ferrer, A., Tyagi, P., Yin, Y., Salas, C., Pal, L., & Rojas, O. J. (2022). Nanocellulose in thin films, coatings, and plies for packaging applications: A review. *BioResources*, *17*(1), 2143–2233.

Hui, J., Yuanyuan, J., Jiao, W., Yuan, H., Yuan, Z., & Shiru, J. (2009). Potentiality of bacterial cellulose as the scaffold of tissue engineering of cornea. In *Proceedings of the 2nd International Conference on Biomedical Engineering and Informatics (BMEI '09)*, China, October.

Hussain, F., Hojjati, M., Okamoto, M., & Gorga, R. E. (2006). Review article: Polymer-matrix Nanocomposites, Processing, Manufacturing, and Application: An Overview. *Journal of Composite Materials*, *40*(17), 1511–1575. DOI: 10.1177/0021998306067321

Hussain, S., Akhter, R., & Maktedar, S. S. (2024). Advancements in sustainable food packaging: From eco-friendly materials to innovative technologies. *Sustainable Food Technology*, *2*(5), 1297–1364.

Iguchi, M., Yamanaka, S., & Budhiono, A. (2000). Bacterial cellulose—A masterpiece of nature's arts. *Journal of Materials Science*, *35*(2), 261–270.

Ikada, Y., & Tsuji, H. (2000). Biodegradable polyesters for medical and ecological applications. *Macromolecular Rapid Communications, 21*, 117–132.

Ikram, R., Mohamed Jan, B., Abdul Qadir, M., Sidek, A., Stylianakis, M. M., & Kenanakis, G. (2021). Recent advances in chitin and chitosan/graphene-based bio-nanocomposites for energetic applications. *Polymers, 13*(19), 3266. PMID: 34641082

Ilker, S. B., Adam, S., Philip, J. M., Eric, L., & Lance, M. (2009). Superhydrophobic cellulose-based bionanocomposite films from Pickering emulsions. *Applied Physics Letters*, •••, 94.

Inoue, Y. (2020). Satellite- and drone-based remote sensing of crops and soils for smart farming – a review. *Soil Science and Plant Nutrition, 66*(6), 798–810. DOI: 10.1080/00380768.2020.1738899

Irewale, A. T., Dimkpa, C. O., Agunbiade, F. O., Oyetunde, O. A., Elemike, E. E., & Oguzie, E. E. (2024). Unlocking sustainable agricultural development in Africa via bio-nanofertilizer application - challenges, opportunities and prospects. *Scientific African, 25*. Advance online publication. DOI: 10.1016/j.sciaf.2024.e02276

Irfan, S. A., Azeem, B., Irshad, K., & Algarni, S., KuShaari, K., Islam, S., & Abdelmohimen, M. A. H. (2020). Machine Learning Model for Nutrient Release from Biopolymers Coated Controlled-Release Fertilizer. *Agriculture, 10*(11). Advance online publication. DOI: 10.3390/agriculture10110538

Iroegbu, A. O. C., & Ray, S. S. (2022). Recent developments and future perspectives of biorenewable nanocomposites for advanced applications. *Nanotechnology Reviews, 11*(1), 1696–1721.

Israeel, M., Iqbal, J., Abbasi, B. A., Ijaz, S., Ullah, R., Zarshan, F., Yaseen, T., Khan, G., Murtaza, G., Ali, I., Alarjani, K. M., Elshikh, M. S., Rizwan, M., Khan, S., & Iqbal, R. (2024). Potential biological applications of environment friendly synthesized iron oxide nanoparticles using Sageretia thea root extract. *Scientific Reports, 14*(1). Advance online publication. DOI: 10.1038/s41598-024-79953-4 PMID: 39550505

James, J., Choudhary, P., & Singh, S. Archana, & Sharma, N. (2024). Regenerative Agriculture: Potential, Progress, Opportunities, and Challenges. In S. Kumar, R. S. Meena, P. Sheoran, & M. K. Jhariya (Eds.), *Regenerative Agriculture for Sustainable Food Systems* (pp. 49–82). Springer Nature Singapore. https://doi.org/DOI: 10.1007/978-981-97-6691-8_3

James, A., Rahman, M. R., Said, K. A. M., Kanakaraju, D., Sueraya, A. Z., Kuok, K. K., Bakri, M. K. B., & Rahman, M. M. (2024). A review of nanocellulose modification and compatibility barriers for various applications. *Journal of Thermoplastic Composite Materials*, *37*(2), 456–489.

Jamwal, V., & Mittal, A. (2024). Recent Progresses in Nanocomposite Films for Food-Packaging Applications: Synthesis Strategies, Technological Advancements, Potential Risks and Challenges. *Food Reviews International*, •••, 1–32.

Jamzad, M., & Bidkorpeh, M. K. (2020). Green synthesis of iron oxide nanoparticles by the aqueous extract of Laurus nobilis L. leaves and evaluation of the antimicrobial activity. *Journal of Nanostructure in Chemistry*, *10*(3), 193–201. DOI: 10.1007/s40097-020-00341-1

Jawaid, M., Salit, M. S., & Alothman, O. Y. (2017). *Green biocomposites: Design and applications*. Springer.

Jawaid, M., Sapuan, S. M., & Alothman, O. Y. (Eds.). (2023). *Nanocellulose: Preparation, properties and applications*. Springer Nature.

Jayara, A. S., Kumar, R., Pandey, P., Singh, S., Shukla, A., Singh, A. P., Pandey, S., Meena, R. L., & Reddy, K. I. (2023). BOOSTING NUTRIENT USE EFFICIENCY THROUGH FERTILIZER USE MANAGEMENT. *Applied Ecology and Environmental Research*, *21*(4), 2931–2952. DOI: 10.15666/aeer/2104_29312952

Jena, S., Dutta, J., Tulsiyan, K. D., Sahu, A. K., Choudhury, S. S., & Biswal, H. S. (2022). Noncovalent interactions in proteins and nucleic acids: Beyond hydrogen bonding and π-stacking. *Chemical Society Reviews*, *51*(11), 4261–4286. PMID: 35560317

Jenie, S. N., Kristiani, A., Kustomo, K., Simanungkalit, S., & Mansur, D. (2017, November). Preparation of nanobiochar as magnetic solid acid catalyst by pyrolysis-carbonization from oil palm empty fruit bunches. In AIP Conference Proceedings (Vol. 1904, No. 1). AIP Publishing.

Jesus, J. D., Oliveira-Esquerre, K., & Medeiros, D. L. (2022). Environmental model using life cycle assessment and artificial intelligence techniques to predict impacts on industrial water treatment. *IOP Conference Series. Materials Science and Engineering*, *1250*(1), 012002. DOI: 10.1088/1757-899X/1250/1/012002

Jha, P. (2020). Effect of plasticizer and antimicrobial agents on functional properties of bionanocomposite films based on corn starch-chitosan for food packaging applications. *International Journal of Biological Macromolecules*, *160*, 571–582. DOI: 10.1016/J.IJBIOMAC.2020.05.242 PMID: 32485256

Jiang, S., Lin, K., & Cai, M. (2020). ZnO Nanomaterials: Current Advancements in Antibacterial Mechanisms and Applications. *Frontiers in Chemistry, 8*, 552221. DOI: 10.3389/FCHEM.2020.00580/BIBTEX PMID: 32793554

Jia, Y., Yu, X. Y., Luo, T., Zhang, M. Y., Liu, J. H., & Huang, X. J. (2013). PEG aggregation templated porous ZnO nanostructure: Room temperature solution synthesis, pore formation mechanism, and their photoluminescence properties. *CrystEngComm, 15*(18), 3647–3653. DOI: 10.1039/C3CE27091B

Ji, D., Lin, Y., Guo, X., Ramasubramanian, B., Wang, R., Radacsi, N., & Ramakrishna, S. (2024). Electrospinning of nanofibres. *Nature Reviews. Methods Primers, 4*(1), 1.

Jiménez, A. B. P., Aguilar, C. A. H., Ramos, J. M. V., & Thangarasu, P. (2015). Synergistic antibacterial activity of nanohybrid materials ZnO-Ag and ZnO-Au: Synthesis, characterization, and comparative analysis of undoped and doped ZnO nanoparticles. *Australian Journal of Chemistry, 68*(2), 288–297. DOI: 10.1071/CH14123

Jin, J., Wei, G., Hao, A., Zhang, W., Zhang, Y., Wang, Q., & Chen, Q. (2015). ZnO-AlBw-reinforced dental resin composites: The effect of pH level on mechanical properties. *RSC Advances, 5*(33), 26430–26437. DOI: 10.1039/C5RA01413A

Johannson, C. (2011). Bio-nanocomposites for food packaging applications. In Mittal, V. (Ed.), *Nanocomposites with biodegradable polymers: Synthesis, properties, and future perspectives* (pp. 1–xx). Oxford University Press., DOI: 10.1093/acprof:oso/9780199581924.003.0014

John, M. J., & Thomas, S. (2008). Biofibres and biocomposites. *Carbohydrate Polymers, 71*(3), 343–364. DOI: 10.1016/J.CARBPOL.2007.05.040

Jose, J., & Al-Harthi, M. A. (2017). Citric acid crosslinking of poly (vinyl alcohol)/starch/graphene nanocomposites for superior properties. *Iranian Polymer Journal, 26*(8), 579–587.

Joseph, B., Krishnan, S., Sagarika, V. K., Tharayil, A., Kalarikkal, N., & Thomas, S. (2020). Bionanocomposites as industrial materials, current and future perspectives: A review. *Emergent Materials, 3*(5), 711–725.

Joseph, B., Sagarika, V. K., Sabu, C., Kalarikkal, N., & Thomas, S. (2020). Cellulose nanocomposites: Fabrication and biomedical applications. *Journal of Bioresources and Bioproducts, 5*(4), 223–237.

Kadulkar, S., Howard, M. P., Truskett, T. M., & Ganesan, V. (2021). Prediction and Optimization of Ion Transport Characteristics in Nanoparticle-Based Electrolytes Using Convolutional Neural Networks. *The Journal of Physical Chemistry B, 125*(18), 4838–4849. DOI: 10.1021/acs.jpcb.1c02004 PMID: 33914555

Kalantzopoulos, G., Paraskevopoulos, P., Domalis, G., Liopa-Tsakalidi, A., Tsesmelis, D. E., & Barouchas, P. E. (2024). The Western Greece Soil Information System (WESIS)—A Soil Health Design Supported by the Internet of Things, Soil Databases, and Artificial Intelligence Technologies in Western Greece. *Sustainability, 16*(8), 3478. DOI: 10.3390/su16083478

Kalia, S., Dufresne, A., Cherian, B. M., Kaith, B. S., Averous, L., Njuguna, J., & Nassiopoulos, E. (2011). Cellulose-based bio- and nanocomposites: A review. *International Journal of Polymer Science, 837875*. Advance online publication. DOI: 10.1155/2011/837875

Kalia, S., Kaith, B. S., & Kaur, I. (2009). Pretreatments of natural fibres and their application as reinforcing material in polymer composites: A review. *Polymer Engineering and Science, 49*(7), 1253–1272.

Kamali, M., & Ghahremaninezhad, A. (2018). Effect of Biomolecules on the Nanostructure and Nanomechanical Property of Calcium-Silicate-Hydrate. *Scientific Reports, 8*(1). Advance online publication. DOI: 10.1038/S41598-018-27746-X PMID: 29934541

Kandhola, G., Park, S., Lim, J. W., Chivers, C., Song, Y. H., Chung, J. H., & Kim, J. W. (2023). Nanomaterial-based scaffolds for tissue engineering applications: A review on graphene, carbon nanotubes and nanocellulose. *Tissue Engineering and Regenerative Medicine, 20*(3), 411–433. PMID: 37060487

Kang, D. W., Choi, H. R., & Kweon, D. K. (1996). Selective adsorption capacity for metal ions of amidated chitosan bead-g-PAN copolymer. *Polymer (Korea), 20*, 989–995.

Kang, Y., Kim, M., Shim, C., Bae, S., & Jang, S. (2021). Potential of Algae–Bacteria Synergistic Effects on Vegetable Production. *Frontiers in Plant Science, 12*. Advance online publication. DOI: 10.3389/fpls.2021.656662 PMID: 33912211

Kargarzadeh, H., Huang, J., Lin, N., Ahmad, I., Mariano, M., Dufresne, A., Thomas, S., & Gałęski, A. (2018). Recent developments in nanocellulose-based biodegradable polymers, thermoplastic polymers, and porous nanocomposites. *Progress in Polymer Science, 87*, 197–227.

Kargarzadeh, H., Mariano, M., Gopakumar, D. A., Ahmad, I., Thomas, S., Dufresne, A., Huang, J., & Lin, N. (Eds.). (2022). *Handbook of nanocellulose and cellulose nanocomposites* (2nd ed.). Wiley-VCH.

Karki, S., Gohain, M. B., Yadav, D., & Ingole, P. G. (2021). Nanocomposite and bio-nanocomposite polymeric materials/membranes development in energy and medical sector: A review. *International Journal of Biological Macromolecules, 193*, 2121–2139. PMID: 34780890

Karunaratne, T. N., Rodrigo, P. M., Oguntuyi, D. O., Mlsna, T. E., Zhang, J., & Zhang, X. (2023). Unraveling biochar surface area on structure and heavy metal removal performances of carbothermal reduced nanoscale zero-valent iron. *Journal of Bioresources and Bioproducts, 8*(4), 388–398. DOI: 10.1016/j.jobab.2023.06.003

Karwatkar, P. A., Kulkarni, S. J., & Goswami, A. K. (2025). Bionanomaterials in Food Systems: Sources, Synthesis, Properties and Opportunities. *BioNanoScience, 15*(1), 5.

Kasaai, M. R. (2024). Biopolymers as Fillers of Biopolymer Matrices to Fabricate Green Nano-composites for Food Coating and Packaging: A Review. In *Handbook of Nanofillers* (pp. 1–14). Springer.

Kasraei, S., Sami, L., & Hendi, S., AliKhani, M.-Y., Rezaei-Soufi, L., & Khamverdi, Z. (2014). Antibacterial properties of composite resins incorporating silver and zinc oxide nanoparticles on Streptococcus mutans and Lactobacillus. *Restorative Dentistry & Endodontics, 39*(2), 109. DOI: 10.5395/RDE.2014.39.2.109 PMID: 24790923

Kaur, R., & Bhushanwar, K. (2025). Enhancing Sustainability, Climate Resilience, and Resource Efficiency with IoT-Based Precision Agriculture. *International Journal of Research and Review in Applied Science, Humanities, and Technology*, 172–179. https://doi.org/DOI: 10.71143/7db36796

Kausar, A., & Ahmad, I. (2023). Degradable Green Polymers, Green Nano polymers and Green Nanocomposites Derived from Natural Systems: Statistics and Headways. *Nano-Horizons., 2*. Advance online publication. DOI: 10.25159/Nano-Horizons.49f1ef7cdb71

Kawashita, M., Tsuneyama, S., Miyaji, F., Kokubo, T., Kozuka, H., & Yamamoto, K. (2000). Antibacterial silver-containing silica glass prepared by sol–gel method. *Biomaterials, 21*(4), 393–398. DOI: 10.1016/S0142-9612(99)00201-X PMID: 10656321

Kebede, A., Singh, A. K., Rai, P. K., Giri, N. K., Rai, A. K., Watal, G., & Gholap, A. V. (2012). Controlled synthesis, characterization, and application of iron oxide nanoparticles for oral delivery of insulin. *Lasers in Medical Science*, *28*(2), 579–587. DOI: 10.1007/s10103-012-1106-3 PMID: 22581389

Kedang, Y. I., Priyangga, A., Atmaja, L., & Santoso, M. (2022). Characteristics and performance studies of a composite polymer electrolyte membrane based on chitosan/glycerol–sulfosuccinic acid modified montmorillonite clay. *RSC Advances*, *12*(47), 30742–30753. PMID: 36349150

Kelly, K. L., Coronado, E., Zhao, L. L., & Schatz, G. C. (2003). The optical properties of metal nanoparticles: The influence of size, shape, and dielectric environment. *The Journal of Physical Chemistry B*, *107*(3), 668–677. DOI: 10.1021/JP026731Y/ASSET/IMAGES/LARGE/JP026731YF00014.JPEG

Khaki, S., Wang, L., & Archontoulis, S. V. (2020). A CNN-RNN Framework for Crop Yield Prediction. *Frontiers in Plant Science*, *10*, 1750. DOI: 10.3389/fpls.2019.01750 PMID: 32038699

Khalid, M. Y., & Arif, Z. U. (2022). Novel biopolymer-based sustainable composites for food packaging applications: A narrative review. *Food Packaging and Shelf Life*, *33*, 100892.

Khalil, H. P. S. A., Davoudpour, Y., Islam, M. N., Mustapha, A., Sudesh, K., Dungani, R., & Jawaid, M. (2021). Production and modification of nanofibrillated cellulose using various mechanical processes: A review. *Carbohydrate Polymers*, *275*, 649–665. PMID: 24274556

Khane, Y., Albukhaty, S., Sulaiman, G. M., Fennich, F., Bensalah, B., Hafsi, Z., & Al-aqbi, Z. T. (2024). Fabrication, characterization and application of biocompatible nanocomposites: A review. *European Polymer Journal*, *•••*, 113187.

Khan, Y., Sadia, H., Ali Shah, S. Z., Khan, M. N., Shah, A. A., Ullah, N., & Khan, M. I. (2022). Classification, synthetic, and characterization approaches to nanoparticles, and their applications in various fields of nanotechnology: A review. *Catalysts*, *12*(11), 1386.

Khan, Z. I., Mohsin, M. E. A., Habib, U., Mousa, S., Hossain, S. S., Ali, S. S., Mohamad, Z., & Othman, N. (2025). Enhanced Mechanical and Thermal Performance of Sustainable RPET/PA-11/Joncryl® Nanocomposites Reinforced with Halloysite Nanotubes. *Polymers*, *17*(11), 1433. PMID: 40508677

Khor, E. (2001). *Chitin: Fulfilling a biomaterials promise*. Elsevier.

Khor, E., & Lim, L. Y. (2003). Implantable applications of chitin and chitosan. *Biomaterials*, *24*(13), 2339–2349. PMID: 12699672

Khouri, N. G., Bahú, J. O., Blanco-Llamero, C., Severino, P., Concha, V. O., & Souto, E. B. (2024). Polylactic Acid (PLA): Properties, Synthesis, and Biomedical Applications-A Review of the Literature. *Journal of Molecular Structure*, •••, 138243.

Kiatkamjorwong, S., & Meechi, N. (1997). Radiation effects on chitosan. *Radiation Physics and Chemistry*, *49*(6), 689–696.

Kibrete, F., Trzepieciński, T., Gebremedhen, H. S., & Woldemichael, D. E. (2023). Artificial Intelligence in Predicting Mechanical Properties of Composite Materials. *Journal of Composites Science*, *7*(9), 9. Advance online publication. DOI: 10.3390/jcs7090364

Kim, C.-J., Khan, W., Kim, D.-H., Cho, K.-S., & Park, S.-Y. (2011). Graphene oxide/cellulose composite using NMMO monohydrate. *Carbohydrate Polymers*, *86*(2), 903–909.

Klemm, D., Cranston, E. D., Fischer, D., Gama, M., Kedzior, S. A., Kralisch, D., Kramer, F., Kondo, T., Lindström, T., Nietzsche, S., Petzold-Welcke, K., & Rauchfuß, F. (2023). Nanocellulose as a natural source for groundbreaking applications in materials science: Today's state. *Materials Today*, *45*, 720–748.

Klemm, D., Philipp, B., Heinze, T., Heinze, U., & Wagenknecht, W. (1998). *Comprehensive cellulose chemistry: Fundamentals and analytical methods*. Wiley-VCH.

Klemm, D., Schumann, D., Udhardt, U., & Marsch, S. (2001). Bacterial synthesized cellulose—Artificial blood vessels for microsurgery. *Progress in Polymer Science*, *26*(9), 1561–1603.

Köktürk, M., YiĞiT, A., & Sulukan, E. (2023). Green synthesis iron oxide nanoparticles (FE@AV NPs) induce developmental toxicity and Anxiety-Like behavior in zebrafish Embryo-Larvae. *Marine Science and Technology Bulletin*, *12*(1), 39–50. DOI: 10.33714/masteb.1220668

Kolya, H., & Kang, C. (2022). Biogenic Synthesis of Silver-Iron Oxide Nanoparticles Using Kulekhara Leaves Extract for Removing Crystal Violet and Malachite Green Dyes from Water. *Sustainability*, *14*(23), 15800. DOI: 10.3390/su142315800

Konwar, A., Kalita, S., Kotoky, J., & Chowdhury, D. (2016). Chitosan–Iron Oxide Coated Graphene Oxide Nanocomposite Hydrogel: A Robust and Soft Antimicrobial Biofilm. *ACS Applied Materials & Interfaces*, *8*(32), 20625–20634. PMID: 27438339

Korumilli, T., Abdullahi, A., Kumar, T. S., & Rao, K. J. (2024). Nanoclay-reinforced bio-composites and their packaging applications. In *Nanoclay-Based Sustainable Materials* (pp. 467–485). Elsevier.

Kovtun, A., Campodoni, E., Favaretto, L., Zambianchi, M., Salatino, A., Amalfitano, S., Navacchia, M. L., Casentini, B., Palermo, V., Sandri, M., & Melucci, M. (2020). Multifunctional graphene oxide/biopolymer composite aerogels for microcontaminants removal from drinking water. *Chemosphere*, *259*, 127501. DOI: 10.1016/j.chemosphere.2020.127501

Kukowska, S., & Szewczuk-Karpisz, K. (2024). Biochar and Zeolite Uses in Improving Immobilization of Nutrients and Pollutants in Soils. *Separation and Purification Reviews*, •••, 1–24. DOI: 10.1080/15422119.2024.2426157

Kulkarni, D., Sherkar, R., Shirsathe, C., Sonwane, R., Varpe, N., Shelke, S., More, M. P., Pardeshi, S. R., Dhaneshwar, G., Junnuthula, V., & Dyawanapelly, S. (2023). Biofabrication of nanoparticles: Sources, synthesis, and biomedical applications. *Frontiers in Bioengineering and Biotechnology*, *11*, 1159193. DOI: 10.3389/FBIOE.2023.1159193/PDF PMID: 37200842

Kulkarni, S., Mohanty, N., Kadam, N. N., Swain, N., & Thakur, M. (2020). Green synthesis to develop Iron-Nano formulations and its toxicity assays. *Journal of Pharmacopuncture*, *23*(3), 165–172. DOI: 10.3831/kpi.2020.23.3.165 PMID: 33072414

Kumar, A. P. N., Kumar, M., Jose, A., Tomer, V., Oz, E., Proestos, C., Zeng, M., Elobeid, T. K. S., & Oz, F. (2023). Major Phytochemicals: Recent Advances in Health Benefits and Extraction Method. *Molecules (Basel, Switzerland)*, *28*(2), 887. DOI: 10.3390/molecules28020887 PMID: 36677944

Kumar, A., Gautam, Y. K., & Singh, N. (2024). State-of-the-art developments in surface functionalized carbon-based bio/nanocomposites for theranostic antibacterials, advanced bioimaging, and molecular bioelectronics inspired biosensing platforms. *Journal of Industrial and Engineering Chemistry*.

Kumar, A., Mishra, R. K., Verma, K., Aldosari, S. M., Maity, C. K., Verma, S., & Thakur, V. K. (2023). A comprehensive review of various biopolymer composites and their applications: From biocompatibility to self-healing. *Materials Today Sustainability*, *23*, 100431.

Kumar, K., Hoque, A., Padhiary, M., & Saikia, P. (2025). Emerging Trends in Nanotechnology, Autonomous Farms, and Collaborative Agri-Industry Marketing: The Future of AI in Food Production. In Whig, P., & Elngar, A. (Eds.), (pp. 239–278). AI Innovations for Transforming Food Production. IGI Global Scientific Publishing., DOI: 10.4018/979-8-3373-0842-5.ch010

Kumar, P., Sandeep, K., Alavi, S., Truong, V., & Gorga, R. (2010). Preparation and characterization of bio-nanocomposite films based on soy protein isolate and montmorillonite using melt extrusion. *Journal of Food Engineering*, *100*(3), 480–489.

Kumar, R. S., Dhilipkumar, T., Jessie, J. A., Gaayathri, K. K., & Arumugam, S. (2023). Advancements in bio-polymeric composite materials for active and intelligent food packaging: A comprehensive review. *Materials Today: Proceedings*.

Kumar, V., Umrao, D., & Srivastava, A. (2022). Functional Nanomaterials for Food Packaging Applications. In *Food Packaging: The Smarter Way* (pp. 243–269). Springer Nature Singapore.

Kundacina, I., Kundacina, O., Miskovic, D., & Radonic, V. (2025). Advancing microfluidic design with machine learning: A Bayesian optimization approach. *Lab on a Chip*, *25*(4), 657–672. DOI: 10.1039/D4LC00872C PMID: 39887214

Kurita, K. (2006). Chitin and chitosan: Functional biopolymers from marine crustaceans. *Marine Biotechnology (New York, N.Y.)*, *8*, 203–226. PMID: 16532368

Kushwaha, R. K., & Srivastava, A. (2018). Recent developments in bio-nanocomposites: A review. *Research Journal of Nanoscience and Engineering*, *2*(2), 1–4.

Lagarón, J. M., López-Rubio, A., & Fabra, M. J. (2022). Recent advances in sustainable biopolymer-based nanocomposites for smart food packaging applications. *Food Hydrocolloids*, *125*, 106239.

Lages, M., & Nicolas, J. (2023). In situ encapsulation of biologically active ingredients into polymer particles by polymerization in dispersed media. *Progress in Polymer Science*, *137*, 101637.

Lakshminarayanan, S., Shereen, M. F., Niraimathi, K. L., Brindha, P., & Arumugam, A. (2021). One-pot green synthesis of iron oxide nanoparticles from Bauhinia tomentosa: Characterization and application towards synthesis of 1, 3 diolein. *Scientific Reports*, *11*(1). Advance online publication. DOI: 10.1038/s41598-021-87960-y PMID: 33883589

Lallawmkimi, M. C., Patil, S., Upadhyay, D. K., Majumdar, N. B. A., Kumar, G. S., & Panigrahi, C. K. (2025). Application of Nanotechnology in Agriculture: Opportunities and Challenges in the Context of Environmental Sustainability. *Archives of Current Research International*, *25*(1), 37–53. DOI: 10.9734/acri/2025/v25i11035

Lalonde, J. N., Pilania, G., & Marrone, B. L. (2025). Materials designed to degrade: Structure, properties, processing, and performance relationships in polyhydroxyalkanoate biopolymers. *Polymer Chemistry*, *16*(3), 235–265. PMID: 39464417

Lal, R. (2012). Climate Change and Soil Degradation Mitigation by Sustainable Management of Soils and Other Natural Resources. *Agricultural Research, 1*(3), 199–212. DOI: 10.1007/s40003-012-0031-9

Larson, D. B., Harvey, H., Rubin, D. L., Irani, N., Tse, J. R., & Langlotz, C. P. (2021). Regulatory Frameworks for Development and Evaluation of Artificial Intelligence–Based Diagnostic Imaging Algorithms: Summary and Recommendations. *Journal of the American College of Radiology, 18*(3), 413–424. DOI: 10.1016/j.jacr.2020.09.060 PMID: 33096088

Lavado, G. J., Baderna, D., Carnesecchi, E., Toropova, A. P., Toropov, A. A., Dorne, J. L. C. M., & Benfenati, E. (2022). QSAR models for soil ecotoxicity: Development and validation of models to predict reproductive toxicity of organic chemicals in the collembola *Folsomia candida. Journal of Hazardous Materials, 423*, 127236. DOI: 10.1016/j.jhazmat.2021.127236 PMID: 34844354

Lee, D. W., Lim, H., Chong, H. N., & Shim, W. S. (1997). Advances in chitosan material and its hybrid derivatives: A review. *Open Biomaterials Journal, 1*, 10–20.

Lee, S. Y. (1996). Bacterial polyhydroxyalkanoates. *Biotechnology and Bioengineering, 49*, 1–14. PMID: 18623547

Lee, S., & Ryu, C. (2021). Algae as New Kids in the Beneficial Plant Microbiome. *Frontiers in Plant Science, 12*. Advance online publication. DOI: 10.3389/fpls.2021.599742 PMID: 33613596

Lehman, R., Cambardella, C., Stott, D., Acosta-Martinez, V., Manter, D., Buyer, J., Maul, J., Smith, J., Collins, H., Halvorson, J., Kremer, R., Lundgren, J., Ducey, T., Jin, V., & Karlen, D. (2015). Understanding and Enhancing Soil Biological Health: The Solution for Reversing Soil Degradation. *Sustainability, 7*(1), 988–1027. DOI: 10.3390/su7010988

Leite, E. R., Bernardi, M. I. B., Longo, E., Varela, J. A., & Paskocimas, C. A. (2004). Enhanced electrical property of nanostructured Sb-doped SnO2 thin film processed by soft chemical method. *Thin Solid Films, 449*(1–2), 67–72. DOI: 10.1016/J.TSF.2003.10.101

Leuteritz, A., Kretzschmar, B., Pospiech, D., Costa, F. R., Wagenknecht, U., & Heinrich, G. (2007). In Nalwa, H. S. (Ed.), *Polymeric nanostructures and their applications* (pp. 99–151). American Scientific.

Lin, N., & Dufresne, A. (2014). Nanocellulose in biomedicine: Current status and future prospect. *European Polymer Journal, 59*, 302–325.

Li, R., Liu, C., & Ma, J. (2011). Studies on the properties of graphene oxide-reinforced starch biocomposites. *Carbohydrate Polymers, 84*(1), 631–637.

Li, T., Cui, L., Kuhnert, M., McLaren, T. I., Pandey, R., Liu, H., Wang, W., Xu, Z., Xia, A., Dalal, R. C., & Dang, Y. P. (2024). A comprehensive review of soil organic carbon estimates: Integrating remote sensing and machine learning technologies. *Journal of Soils and Sediments, 24*(11), 3556–3571. DOI: 10.1007/s11368-024-03913-8

Liu, C., Shen, J., Yeung, K. W. K., & Tjong, S. C. (2017). Development and antibacterial performance of novel polylactic acid-graphene oxide-silver nanoparticle hybrid nanocomposite mats prepared by electrospinning. *ACS Biomaterials Science & Engineering, 3*(3), 471–486. PMID: 33465942

Liu, H., Li, C., Chen, S., Chen, P., Li, J., Jian, H., Guo, G., Chen, X., Zhu, X., & Wu, J. (2023). Fabrication of 3D Printed Polylactic Acid/Polycaprolactone Nanocomposites with Favorable Thermo-Responsive Cyclic Shape Memory Effects, and Crystallization and Mechanical Properties. *Polymers, 15*(6), 1533. PMID: 36987315

Liu, K., Zhao, L., & Borghei, M. (2024). Wood-based nanocellulose hydrogels for biomedical applications: Recent advances and challenges. *Carbohydrate Polymers, 325*, 120285.

Liu, L., Li, S., Zhu, W., Bao, Q., Liang, Y., Zhao, T., Li, X., & Zhou, J. (2024). Study on the mechanism of ROS-induced oxidative stress injury and the broad-spectrum antimicrobial performance of nickel ion-doped V6O13 powder. *Scientific Reports, 14*(1), 22374. DOI: 10.1038/S41598-024-71959-2 PMID: 39333647

Liu, R., & Lal, R. (2012). Nanoenhanced Materials for Reclamation of Mine Lands and Other Degraded Soils: A Review. *Journal of Nanotechnology, 2012*, 1–18. DOI: 10.1155/2012/461468

Liu, X., Tang, Harper, Steevens, J., Xu, R., & Harper. (2013). Predictive modeling of nanomaterial exposure effects in biological systems. *International Journal of Nanomedicine, 31*. Advance online publication. DOI: 10.2147/IJN.S40742 PMID: 24098077

Liu, X., Wu, Y., Zhao, X., & Wang, Z. (2021). Fabrication and applications of bioactive chitosan-based organic-inorganic hybrid materials: A review. *Carbohydrate Polymers, 267*, 118179. DOI: 10.1016/j.carbpol.2021.118179 PMID: 34119147

Liu, Y., Sun, H., Sun, J., Meng, X., Jiang, Y., & Wang, N. (2021). Transport of micro/nano biochar in quartz sand modified by three different clay minerals. *Environmental Pollutants and Bioavailability, 33*(1), 113–121.

Liz-Marzán, L. M. (2004). Nanometals: Formation and color. *Materials Today*, *7*, 26–31.

Lizundia, E., Luzi, F., & Puglia, D. (2022). Organic waste valorisation towards circular and sustainable biocomposites. *Green Chemistry*, *24*(14), 5429–5459.

Lonkar, S. P., Kumar, A. P., & Singh, R. P. (2011). 16 Commercial aspects associated with bio-nanocomposites. In V. Mittal (Ed.), *Nanocomposites with Biodegradable Polymers: Synthesis, Properties, and Future Perspectives* (p. 0). Oxford University Press.

Lowe, M., Qin, R., & Mao, X. (2022). A Review on Machine Learning, Artificial Intelligence, and Smart Technology in Water Treatment and Monitoring. *Water (Basel)*, *14*(9), 9. Advance online publication. DOI: 10.3390/w14091384

Machado, T. O., Grabow, J., Sayer, C., de Araújo, P. H. H., Ehrenhard, M. L., & Wurm, F. R. (2022). Biopolymer-based nanocarriers for sustained release of agrochemicals: A review on materials and social science perspectives for a sustainable future of agri- and horticulture. *Advances in Colloid and Interface Science*, *303*, 102645. DOI: 10.1016/j.cis.2022.102645 PMID: 35358807

Ma, F., Gao, Y., Xie, W., & Wu, D. (2024). Effect of hydrophobic modification of chitin nanocrystals on role as anti-nucleator in the crystallization of poly (ε-caprolactone)/polylactide blend. *International Journal of Biological Macromolecules*, *269*, 132097. PMID: 38710249

Mahunu, G. K., Osei-Kwarteng, M., Ogwu, M. C., & Afoakwah, N. A. (2024). Safe food handling techniques to prevent microbial contamination. In *Food safety and quality in the global south* (pp. 427–461). Springer Nature Singapore.

Maiti, D., Tong, X., Mou, X., & Yang, K. (2019). Carbon-Based Nanomaterials for Biomedical Applications: A Recent Study. *Frontiers in Pharmacology*, *9*. Advance online publication. DOI: 10.3389/fphar.2018.01401 PMID: 30914959

Manikandan, D., Gandhi, V. C. S., Venkatesan, R., & Nellaiappan, T. A. (2025). Synthesis, Characterization, and Applications of Nanomaterials: Mechanical and Thermal Studies of Polymer Nanocomposites. In *Exploring Nanomaterial Synthesis, Characterization, and Applications* (pp. 197-212). IGI Global.

Mann, S. (1996). Synthesis of inorganic materials with complex form. In Mann, S. (Ed.), *Biomimetic materials chemistry* (pp. 1–40). VCH.

Markus, N., & Nicola, P. (2009). *Metal oxide nanoparticles in organic solvents: Synthesis, formation, assembly and application*. Springer.

Mathew, S., & Radhakrishnan, E. K. (2022). Soil Burial and Biodegradability of Bio nanocomposites. Polymer Based Bio-Nanocomposites, 181–203. https://doi.org/DOI: 10.1007/978-981-16-8578-1_10

Matin, M. M., Amran, M., Islam, M. B., Hasnain, M., Begum, S. H., Rahman, M. R., & Bakri, M. K. B. (2024). Current and future development of nanocarbon and its biocomposites production. In *Advanced Nanocarbon Polymer Biocomposites* (pp. 49–103). Woodhead Publishing.

Mavai, S., Bains, A., Sridhar, K., Rashid, S., Elossaily, G. M., Ali, N., & Sharma, M. (2024). Formulation and application of poly lactic acid, gum, and cellulose-based ternary bioplastic for smart food packaging: A review. *International Journal of Biological Macromolecules*, •••, 131687. PMID: 38642692

Medeiros, E. S., Dufresne, A., & Orts, W. J. (2010). In Bertolini, A. C. (Ed.), *Starches: Characterization, properties, and applications* (pp. 205–251). CRC Press.

Mello, L. R., Feltrin, Y., Selbach, R., Macedo, G.Jr, Spautz, C., & Haas, L. J. (2001). Use of lyophilized cellulose in peripheral nerve lesions with loss of substance. *Arquivos de Neuro-Psiquiatria*, *59*(2), 372–379. PMID: 11460182

Melo, M. A. S., Guedes, S. F. F., Xu, H. H. K., & Rodrigues, L. K. A. (2013). Nanotechnology-based restorative materials for dental caries management. *Trends in Biotechnology*, *31*(8), 459–467. DOI: 10.1016/J.TIBTECH.2013.05.010 PMID: 23810638

Menossi, M., Casalongué, C., & Alvarez, V. A. (2022). Bio-nanocomposites for Modern Agricultural Applications. In Mallakpour, S., & Hussain, C. M. (Eds.), *Handbook of Consumer Nanoproducts* (pp. 1201–1237). Springer Nature., DOI: 10.1007/978-981-16-8698-6_68

Mianehrow, H., Re, G. L., Carosio, F., Fina, A., Larsson, P. T., Chen, P., & Berglund, L. A. (2020). Strong reinforcement effects in 2D cellulose nanofibril–graphene oxide (CNF–GO) nanocomposites due to GO-induced CNF ordering. *Journal of Materials Chemistry. A, Materials for Energy and Sustainability*, *8*(34), 17608–17620. PMID: 33796318

Miklasová, M., Komendová, R., Brtnický, M., Čelan Korošin, N., & Kučerík, J. (2025). Short- and long-term influences of rutile and anatase nanoparticles on soil organic matter structure, ageing and moisture. *Journal of Thermal Analysis and Calorimetry*, *150*(19), 15103–15116. DOI: 10.1007/s10973-025-14741-4

Min, T., Zhou, L., Sun, X., Du, H., Zhu, Z., & Wen, Y. (2022). Electrospun functional polymeric nanofibers for active food packaging: A review. *Food Chemistry, 391*, 133239. PMID: 35609463

Mishra, A. K., Paul, R. A. I., Grover, D., Das, R., Mishra, S., Akshu, A., & Pavithra, P. Roohi, Soni, P. G., & Sharma, S. (2025). Nanotechnology for Soil and Water Management. In K. A. Abd-Elsalam (Ed.), *Agricultural Nanotechnology* (pp. 371–418). Royal Society of Chemistry. https://doi.org/DOI: 10.1039/9781837677313-00371

Mishra, B., Mohanta, Y. K., Reddy, C. N., Reddy, S. D. M., Mandal, S. K., Yadavalli, R., & Sarma, H. (2023). Valorization of agro-industrial biowaste to biomaterials: An innovative circular bioeconomy approach. *Circular Economy, 2*(3), 100050. DOI: 10.1016/j.cec.2023.100050

Mishra, B., Panda, J., Mishra, A. K., Nath, P. C., Nayak, P. K., Mahapatra, U., Sharma, M., Chopra, H., Mohanta, Y. K., & Sridhar, K. (2024). Recent advances in sustainable biopolymer-based nanocomposites for smart food packaging: A review. *International Journal of Biological Macromolecules, 279*(4), 135583. PMID: 39270899

Mishra, P. K., Pavelek, O., Rasticova, M., Mishra, H., & Ekielski, A. (2023). Nanocellulose-based biomedical scaffolds in future bioeconomy: A techno-legal assessment of the state-of-the-art. *Frontiers in Bioengineering and Biotechnology, 11*, 789603. PMID: 35223812

Misra, M., Pandey, J. K., & Mohanty, A. K. (Eds.). (2023). *Biocomposites: Design and mechanical performance*. Woodhead Publishing.

Mistretta, M. C., Botta, L., Arrigo, R., Leto, F., Malucelli, G., & La Mantia, F. P. (2021). Bio nanocomposite Blown Films: Insights on the Rheological and Mechanical Behavior. *Polymers, 13*(7), 1167. DOI: 10.3390/polym13071167 PMID: 33916477

Mittal, V. (2011). *Bio-nanocomposites: Future high-value materials*. Oxford University Press., DOI: 10.1093/acprof:oso/9780199581924.003.0001

Miyashiro, D., Hamano, R., & Umemura, K. (2020). A review of applications using mixed materials of cellulose, nanocellulose and carbon nanotubes. *Nanomaterials (Basel, Switzerland), 10*(2), 186. PMID: 31973149

Mohamed, A., Atta, R. R., Kotp, A. A., El-Ela, F. I. A., El-Raheem, H. A., Farghali, A., Alkhalifah, D. H. M., Hozzein, W. N., & Mahmoud, R. (2023). Green synthesis and characterization of iron oxide nanoparticles for the removal of heavy metals (Cd2+ and Ni2+) from aqueous solutions with Antimicrobial Investigation. *Scientific Reports*, *13*(1). Advance online publication. DOI: 10.1038/s41598-023-31704-7 PMID: 37142660

Mohammadi, B., Pirsa, S., & Mahmudi, M. (2025). Animal and Vegetable Proteins: Applications as Film and Coating in Food Industry. *Journal of Polymers and the Environment*, •••, 1–21.

Mohan, C., Robinson, J., Vodwal, L., & Kumari, N. (2024). Sustainable Development Goals for addressing environmental challenges. Green Chemistry Approaches to Environmental Sustainability, 357–374. https://doi.org/DOI: 10.1016/b978-0-443-18959-3.00007-0

Mohanty, A. K., Misra, M., & Drzal, L. T. (2002). Sustainable bio-composites from renewable resources: Opportunities and challenges in the green materials world. *Journal of Polymers and the Environment*, *10*(1), 19–26. DOI: 10.1023/A:1021013921916

Mohanty, A. K., Misra, M., & Drzal, L. T. (Eds.). (2022). *Natural fibers, biopolymers, and biocomposites* (2nd ed.). CRC Press.

Mohanty, A. K., Misra, M., & Hinrichsen, G. (2000). Biofibres, biodegradable polymers and biocomposites: An overview. *Macromolecular Materials and Engineering*, *276–277*, 1–24.

Mokhena, T. C., Sefadi, J. S., Sadiku, E. R., John, M. J., Mochane, M. J., & Mtibe, A. (2018). Thermoplastic Processing of PLA/Cellulose Nanomaterials Composites. *Polymers*, *10*(12), 1363. PMID: 30961288

Mondal, S. (2025). Sustainable bio-polymeric materials. Advances in Sustainable Materials, 1–22. https://doi.org/DOI: 10.1016/b978-0-443-13849-2.00001-6

Mondal, S. K., Chakraborty, S., Manna, S., & Mandal, S. M. (2024). Antimicrobial nanoparticles: Current landscape and future challenges. *RSC Pharmaceutics*.

Monfared, A. R., Omranpour, H., Tuccitto, A. V., Zaoui, A., Rahman, S. S., Kheradmandkeysomi, M., & Park, C. B. (2024). Sustainable PLA Bio-Nanocomposites: Integration of TPU Nanofibrils and CNC for Enhanced Crystallization, Toughness, Stiffness, Transparency, and Oxygen Barrier Properties. *ACS Sustainable Chemistry & Engineering*, *12*(34), 13017–13029.

Moon, R. J., Martini, A., Nairn, J., Simonsen, J., & Youngblood, J. (2022). Cellulose nanomaterials review: Structure, properties and nanocomposites. *Chemical Society Reviews*, *51*(7), 3941–3994. PMID: 21566801

Moosvi, S. K., Gulzar Naqash, W., & Najar, Mohd. H. (2021). Green Chemistry. https://doi.org/DOI: 10.1515/9783110751895

Motloung, M. P., Ojijo, V., Bandyopadhyay, J., & Ray, S. S. (2019). Cellulose Nanostructure-Based Biodegradable Nanocomposite Foams: A Brief Overview on the Recent Advancements and Perspectives. *Polymers*, *11*(8), 1270. PMID: 31370292

Moulick, D., Mukherjee, A., Mahanta, S., Chowdhara, B., Pattnaik, B. K., Mukhopadhyay, A. K., & Hazra, S. (2025). Nano-composites and nano-biochar: a novel solution to alleviate heavy metal stress in crops. In *Nanomaterials and Nano-Biochar in Reducing Soil Stress* (pp. 183–224). Apple Academic Press.

Mughal, B., Zaidi, S. Z. J., Zhang, X., & Hassan, S. U. (2021). Biogenic Nanoparticles: Synthesis, Characterisation and Applications. *Applied Sciences (Basel, Switzerland)*, *11*(6), 2598. DOI: 10.3390/app11062598

Munir, A., Salah, M. A., Ali, M., Ali, B., Saleem, M. H., Samarasinghe, K. G. B. A., De Silva, S. I. S., Ercisli, S., Iqbal, N., & Anas, M. (2024). Advancing Agriculture: Harnessing Smart Nanoparticles for Precision Fertilization. *BioNanoScience*, *14*(4), 3846–3863. DOI: 10.1007/s12668-024-01597-5

Musa, I. O., Isibor, P. O., Job, O. S., Mustapha, A., Mustapha, A., Akande, S., & Adeniji, H. (2024c). Introduction to Nanotoxicology. In *Environmental Nanotoxicology: Combatting the Minute Contaminants* (pp. 1–22). Springer Nature Switzerland.

Musa, I. O., Job, O. S., Adams, M., Abdulsalam, M., Nathaniel, V., Maude, A. M., Adedayo, O. A., & Tiamiyu, A. G. T. (2024b). Soil Erosion, Mineral Depletion and Regeneration. In *Prospects for Soil Regeneration and Its Impact on Environmental Protection* (pp. 159–172). Springer Nature Switzerland.

Musa, I. O., Mustapha, A., Abdulsalam, M., Livinus, M. U., Job, S. O., Elelu, S. A., Salami, O. L., & Muhammad, A. S. (2024). Soil Microbes and Soil Contamination. In *Soil Microbiome in Green Technology Sustainability* (pp. 3–35). Springer Nature Switzerland.

Mustapha, S., Ndamitso, M. M., Abdulkareem, A. S., Tijani, J. O., Shuaib, D. T., Ajala, A. O., & Mohammed, A. K. (2020). Application of TiO2 and ZnO nanoparticles immobilized on clay in wastewater treatment: A review. *Applied Water Science*, *10*(1), 49. DOI: 10.1007/s13201-019-1138-y

Myung, H. K., Goeun, C., Ahmed, E., Ajayan, V., Young, B. C., & Jin, H. C. (2016). Review of clay–drug hybrid materials for biomedical applications: Administration routes. *Clays and Clay Minerals*, *64*, 115–130. PMID: 32218609

Nafees, A., Saima, S., Sabir, S., Ameer, A., & Mohammad, Z. K. (2018). *Bionanocomposite materials: Concept, applications, and recent advancements*. Jenny Stanford Publishing.

Naghdi, M., Taheran, M., Brar, S. K., Kermanshahi-Pour, A., Verma, M., & Surampalli, R. Y. (2017). Immobilized laccase on oxygen functionalized nanobiochars through mineral acids treatment for removal of carbamazepine. *The Science of the Total Environment*, *584*, 393–401. PMID: 28117156

Naghdi, M., Taheran, M., Brar, S. K., Kermanshahi-Pour, A., Verma, M., & Surampalli, R. Y. (2018). Pinewood nanobiochar: A unique carrier for the immobilization of crude laccase by covalent bonding. *International Journal of Biological Macromolecules*, *115*, 563–571. PMID: 29689286

Naghdi, M., Taheran, M., Pulicharla, R., Rouissi, T., Brar, S. K., Verma, M., & Surampalli, R. Y. (2019). Pine-wood derived nanobiochar for removal of carbamazepine from aqueous media: Adsorption behavior and influential parameters. *Arabian Journal of Chemistry*, *12*(8), 5292–5301.

Nahari, M. H., Ali, A. A., Asiri, A., Mahnashi, M. H., Shaikh, I. A., Shettar, A. K., & Hoskeri, J. (2022). Green Synthesis and Characterization of Iron Nanoparticles Synthesized from Aqueous Leaf Extract of Vitex leucoxylon and Its Biomedical Applications. *Nanomaterials (Basel, Switzerland)*, *12*(14), 2404. DOI: 10.3390/nano12142404 PMID: 35889627

Nah, C., & Kader, M. A. (2010). *Barrier Properties of Rubber Nanocomposites. Rubber Nanocomposites*. Portico., DOI: 10.1002/9780470823477.ch18

Na, J., Kshetrimayum, K. S., Lee, U., & Han, C. (2017). Multi-objective optimization of microchannel reactor for Fischer-Tropsch synthesis using computational fluid dynamics and genetic algorithm. *Chemical Engineering Journal*, *313*, 1521–1534. DOI: 10.1016/j.cej.2016.11.040

Nakhli, S. A. A., Delkash, M., Bakhshayesh, B. E., & Kazemian, H. (2017). Application of Zeolites for Sustainable Agriculture: A Review on Water and Nutrient Retention. *Water, Air, and Soil Pollution*, *228*(12), 464. DOI: 10.1007/s11270-017-3649-1

Namasivayam Sukumaar, V., Arjunan, S., Pandiaraj, L. N., & Narayanan, A. (2025). Experimental investigation of 3D printed polylactic acid and polylactic acid – hydroxy-apatite composite through material extrusion technique for biomedical application. *Journal of Thermoplastic Composite Materials*, *38*(4), 1303–1333.

Nampoothiri, K. M., Nair, N. R., & John, R. P. (2021). An overview of the recent developments in polylactide (PLA) research. *Bioresource Technology*, *320*, 8493–8501. PMID: 20630747

Nandhini, J., Karthikeyan, E., & Rajeshkumar, S. (2024). Eco-friendly bio-nanocomposites: pioneering sustainable biomedical advancements in engineering. *Discover Nano 2024 19:1*, *19*(1), 1–31. https://doi.org/DOI: 10.1186/S11671-024-04007-7

Nandhini, J., Bellarmin, M., Siva Prakash, S., Sowmya Sri, D., & Karthikeyan, E. (2024). Advancements in metal oxide bio-nanocomposites for sustainable food packaging: Fabrication, applications, and future prospectives. *Food Bioengineering*, *3*, 438–463. DOI: 10.1002/fbe2.12106

Nandhini, J., Bellarmin, M., Siva Prakash, S., Sowmya Sri, D., & Karthikeyan, E. (2024). Advancements in metal oxide bio-nanocomposites for sustainable food packaging: Fabrication, applications, and future prospectives. *Food Bioengineering*.

Nandhini, J., Karthikeyan, E., & Rajeshkumar, S. (2024). Eco-friendly bio-nanocomposites: Pioneering sustainable biomedical advancements in engineering. *Discover Nano*, *19*(1), 86. DOI: 10.1186/s11671-024-04007-7 PMID: 38724698

Naseem, T., & Durrani, T. (2021). The role of some important metal oxide nanoparticles for wastewater and antibacterial applications: A review. *Environmental Chemistry and Ecotoxicology*, *3*, 59–75. DOI: 10.1016/j.enceco.2020.12.001

Naseem, T., & Farrukh, M. A. (2015). Antibacterial Activity of Green Synthesis of Iron Nanoparticles Using Lawsonia inermis and Gardenia jasminoides Leaves Extract. *Journal of Chemistry*, *2015*, 1–7. DOI: 10.1155/2015/912342

Naskar, A., Sanyal, I., Nahar, N., Ghosh, D. D., & Chakraborty, S. (2023). Bionano-composites films applied as active and smart food packaging: A review. *Polymer Engineering and Science*, *63*(9), 2675–2699.

Nath, P. C., Sharma, R., Mahapatra, U., Mohanta, Y. K., Rustagi, S., Sharma, M., & Sridhar, K. (2024). Sustainable production of cellulosic biopolymers for enhanced smart food packaging: An up-to-date review. *International Journal of Biological Macromolecules*, *273*, 133090. PMID: 38878920

Nazir, R., & Qayyum, S. A. (2023). Influence of Nanofillers on Biodegradable Composites. Handbook of Nanofillers, 1–23. https://doi.org/DOI: 10.1007/978-981-99-3516-1_57-1

Ncube, L. K., Ude, A. U., Ogunmuyiwa, E. N., Zulkifli, R., & Beas, I. N. (2021). An overview of plastic waste generation and management in food packaging industries. *Recycling*, *6*(1), 12.

Negrao, S. W., Bueno, R. R. L., & Guérios, E. E.. (2006). A eficácia do stent recoberto com celulose biossintética comparado ao stent convencional em angioplastia em coelhos. *Revista Brasileira de Cardiologia Invasiva*, *14*(1), 10–19.

Netravali, A. N. (2005). In Blackburn, R. S. (Ed.), *Biodegradable and sustainable fibres* (pp. 271–309). CRC Press.

Nguong, C. W., Lee, S. N. B., & Sujan, D. (2013). A Review On Natural Fibre Reinforced Polymer Composites. *World Academy of Science, Engineering and Technology, International Journal of Chemical, Molecular, Nuclear, Materials and Metallurgical Engineering, 7*(1), 52–59. https://publications.waset.org/6783/pdf

Nicolopoulou-Stamati, P., Maipas, S., Kotampasi, C., Stamatis, P., & Hens, L. (2016). Chemical pesticides and human health: The urgent need for a new concept in agriculture. *Frontiers in Public Health*, *4*, 148. DOI: 10.3389/fpubh.2016.00148 PMID: 27486573

Nikolaidis, A. K., Achilias, D. S., & Karayannidis, G. P. (2023). Chemical recycling of PLA and PC: A review on methods and products. *Polymers*, *15*(1), 111. PMID: 38201776

Nimbagal, V., Banapurmath, N. R., Sajjan, A. M., Patil, A. Y., & Ganachari, S. V. (2021). Studies on Hybrid Bio-Nanocomposites for Structural Applications. *Journal of Materials Engineering and Performance*, *30*(9), 6461–6480. DOI: 10.1007/s11665-021-05843-9

Ning, W., Xingxiang, Z., Xuechen, W., & Haihui, L. (2009). Ionic liquids modified montmorillonite/thermoplastic starch nanocomposites as ionic conducting biopolymer. *Macromolecular Research*, *17*, 285–288.

Njuguna, J., Ansari, F., Sachse, S., Rodriguez, V. M., Siqqique, S., & Zhu, H. (2021). Nanomaterials, nanofillers, and nanocomposites: types and properties. Health and Environmental Safety of Nanomaterials, 3–37. https://doi.org/DOI: 10.1016/b978-0-12-820505-1.00011-0

Noreen, S., & Abd-Elsalam, K. (2021). Biochar-based nanocomposites: A sustainable tool in wastewater bioremediation., 185-200. .DOI: 10.1016/B978-0-12-821141-0.00023-9

Noreen, S., & Abd-Elsalam, K. A. (2021). Biochar-based nanocomposites: A sustainable tool in wastewater bioremediation. In *Aquananotechnology* (pp. 185–200). Elsevier.

Norizan, M. N., Shazleen, S. S., Alias, A. H., Sabaruddin, F. A., Asyraf, M. R. M., Zainudin, E. S., & Norrahim, M. N. F. (2022). Nanocellulose-based nanocomposites for sustainable applications: A review. *Nanomaterials (Basel, Switzerland)*, *12*(19), 3483. PMID: 36234612

Nurazzi, N. M., Abdullah, N., Norrahim, M. N. F., Kamarudin, S. H., Ahmad, S., Shazleen, S. S., Rayung, M., Asyraf, M. R. M., Ilyas, R. A., & Kuzmin, M. (2022). Thermogravimetric Analysis (TGA) and Differential Scanning Calorimetry (DSC) of PLA/Cellulose Composites. Polylactic Acid-Based Nanocellulose and Cellulose Composites, 145–164. https://doi.org/DOI: 10.1201/9781003160458-7

Okpala, C. C. (2014). The benefits and applications of nanocomposites. *International Journal of Emerging Technology and Advanced Engineering*, *12*, 12–18.

Olayil, R., & Prabu, V. A., DayaPrasad, S., Naresh, K., & Sreekanth, P. R. (2022). A review on the application of bio-nanocomposites for food packaging. *Materials Today: Proceedings*, *56*, 1302–1306.

Olson, D. G., Tripathi, S. A., & Giannone, R. J.. (2010). Deletion of the Cel48S cellulase from *Clostridium thermocellum. Proceedings of the National Academy of Sciences of the United States of America*, *107*(41), 17727–17732. PMID: 20837514

Omigbodun, F. T., Oladapo, B. I., & Osa-uwagboe, N. (2024). Exploring the frontier of Polylactic Acid/Hydroxyapatite composites in bone regeneration and their revolutionary biomedical applications – A review. *Journal of Reinforced Plastics and Composites*, •••, 07316844241278045.

One, K., Saito, Y., & Yura, H.. (2000). Photocrosslinkable chitosan as a biological adhesive. *Journal of Biomedical Materials Research. Part A*, *49*(2), 289–295. PMID: 10571917

Othman, S. H. (2014). Bio-nanocomposite Materials for Food Packaging Applications: Types of Biopolymer and Nano-sized Filler. *Agriculture and Agricultural Science Procedia*, *2*, 296–303. DOI: 10.1016/J.AASPRO.2014.11.042

Othman, S. H., Majid, N. A., Tawakkal, I. S. M. A., Basha, R. K., Nordin, N., & Shapi'i, R. A. (2019). Tapioca starch films reinforced with microcrystalline cellulose for potential food packaging application. *Food Science and Technology (Campinas)*, *39*(3), 605–612. DOI: 10.1590/fst.3607

Othman, S. H., Ronzi, N. D. A., Shapi'i, R. A., Dun, M., Ariffin, S. H., & Mohammed, M. A. P. (2023). Biodegradability of Starch Nanocomposite Films Containing Different Concentrations of Chitosan Nanoparticles in Compost and Planting Soils. *Coatings*, *13*(4), 777. DOI: 10.3390/coatings13040777

Padhiary, M., Kumar, R., & Sethi, L. N. (2024). Navigating the Future of Agriculture: A Comprehensive Review of Automatic All-Terrain Vehicles in Precision Farming. *Journal of The Institution of Engineers (India): Series A*, *105*(3), 767–782. https://doi.org/DOI: 10.1007/s40030-024-00816-2

Padhiary, M. (2025). Benefits, Risks, and Responsible Use of Nanomaterials in Wastewater Remediation. In Agarwal, N., Shah, M. P., Solanki, V. S., & Singh, N. (Eds.), *Nanomaterials in Wastewater Research: Progress and Challenges* (pp. 443–464). Springer Nature., DOI: 10.1007/978-981-96-3908-3_19

Padhiary, M., Hoque, A., Prasad, G., Kumar, K., & Sahu, B. (2025). Precision Agriculture and AI-Driven Resource Optimization for Sustainable Land and Resource Management. In Ruiz-Vanoye, J. A., & Díaz-Parra, O. (Eds.), *Smart Water Technology for Sustainable Management in Modern Cities* (pp. 197–232). IGI Global., DOI: 10.4018/979-8-3693-8074-1.ch009

Padhiary, M., & Kumar, R. (2024). Assessing the Environmental Impacts of Agriculture, Industrial Operations, and Mining on Agro-Ecosystems. In Azrour, M., Mabrouki, J., Alabdulatif, A., Guezzaz, A., & Amounas, F. (Eds.), *Smart Internet of Things for Environment and Healthcare* (Vol. 1165, pp. 107–126). Springer Nature Switzerland., DOI: 10.1007/978-3-031-70102-3_8

Padhiary, M., Powar, S., Asokan, A., Sundar Rajan, M., Al-Said, K., & Al Said, N. (2025). Combining Convolutional Neural Networks with Reinforcement Learning for Autonomous Robotics. *2025 International Conference on Pervasive Computational Technologies (ICPCT)*, 421–425. https://doi.org/DOI: 10.1109/ICPCT64145.2025.10940484

Padhiary, M., Roy, D., & Dey, P. (2024). Mapping the Landscape of Biogenic Nanoparticles in Bioinformatics and Nanobiotechnology: AI-Driven Insights. In Das, S., Khade, S. M., Roy, D. B., & Trivedi, K. (Eds.), *Synthesizing and Characterizing Plant-Mediated Biocompatible Metal Nanoparticles* (pp. 337–376). IGI Global., DOI: 10.4018/979-8-3693-6240-2.ch014

Padhiary, M., Roy, P., Dey, P., & Sahu, B. (2024). Harnessing AI for Automated Decision-Making in Farm Machinery and Operations: Optimizing Agriculture. In Hai-Jew, S. (Ed.), *Enhancing Automated Decision-Making Through AI* (pp. 249–282). IGI Global., DOI: 10.4018/979-8-3693-6230-3.ch008

Pagano, M., Lunetta, E., Belli, F., Mocarli, G., Cocozza, C., & Cacciotti, I. (2025). Advancements in Agricultural Nanotechnology: An Updated Review. *Plants*, *14*(18), 2939. DOI: 10.3390/plants14182939 PMID: 41012090

Pandey, H., & Pandey, R. K. (2025). Ensuring Safe Nanotechnology: A Biosafety and Regulatory Perspective. In Esquivel Escalante, K., Feregrino Pérez, A. A., Rosales Pérez, A., & Prasad, R. (Eds.), *Green Nanomaterials* (pp. 257–272). Springer Nature Switzerland., DOI: 10.1007/978-3-032-04492-1_12

Pandey, M., Nazar, R., Elella, M. H. A., Praharaj, S., Makhado, E., Rout, D., & Gilani, E. H. (2025). A Comprehensive Review of Recent Developments in Biomedical Materials Based on Graphene-Modified Bio-Nanocomposites. *BioNanoScience*, *15*(1), 125.

Pandey, U., Vashisth, A., Mishra, A., & Kumari, M. (2025). *AI and IoT-Driven Soil Health Restoration: A Machine Learning Approach for Sustainable Agriculture*. In Review. https://doi.org/DOI: 10.21203/rs.3.rs-6490610/v1

Pan, Y., Wu, T., Bao, H., & Li, L. (2011). Green fabrication of chitosan films reinforced with parallel aligned graphene oxide. *Carbohydrate Polymers*, *83*(4), 1908–1915.

Parangusan, H., Bhadra, J., Ahmad, Z., Mallick, S., Touati, F., & Al-Thani, N. (2021). Humidity sensor based on poly (lactic acid)/PANI–ZnO composite electrospun fibers. *RSC Advances*, *11*(46), 28735–28743. PMID: 35478584

Park, H. M., Misra, M., Drzal, L. T., & Mohanty, A. K. (2004). Green nanocomposites from cellulose acetate bioplastic and clay: Effect of eco-friendly triethyl citrate plasticizer. *Biomacromolecules*, *5*(8), 2281–2288. DOI: 10.1021/bm049690f PMID: 15530043

Park, J., Lee, Y., Kim, T. Y., Hwang, S., & Seo, J. (2022). Functional Bioelectronic Materials for Long-Term Biocompatibility and Functionality. *ACS Applied Electronic Materials*, *4*(4), 1449–1468. DOI: 10.1021/ACSAELM.1C01212/ASSET/IMAGES/LARGE/EL1C01212_0007.JPEG

Parul, & Das, V. (2024). Environmental Sustainability, Compliance and Life Cycle Assessment of AI-Integrated Systems. In S. Gulati (Ed.), *Application of Artificial Intelligence in Wastewater Treatment* (pp. 225–243). Springer Nature Switzerland. https://doi.org/DOI: 10.1007/978-3-031-69433-2_9

Pascall, M. A., DeAngelo, K., Richards, J., & Arensberg, M. B. (2022). Role and importance of functional food packaging in specialized products for vulnerable populations: Implications for innovation and policy development for sustainability. *Foods*, *11*(19), 3043. PMID: 36230119

Pathak, H. K., Seth, C. S., Chauhan, P. K., Dubey, G., Singh, G., Jain, D., & Khoo, K. S. (2024). Recent advancement of nano-biochar for the remediation of heavy metals and emerging contaminants: Mechanism, adsorption kinetic model, plant growth and development. *Environmental Research*, 119136. PMID: 38740295

Pathan, M., Patel, N., Yagnik, H., & Shah, M. (2020). Artificial cognition for applications in smart agriculture: A comprehensive review. *Artificial Intelligence in Agriculture*, *4*, 81–95. DOI: 10.1016/j.aiia.2020.06.001

Paulo, S. (2005).. . *Brazilian Oral Research*, *19*(3), 1806–8314.

Pawar, Dr. P., Goyal, Dr. J., Gupta, A. K., Y, S. K., Kuttimarks, Dr. M. S., & Desale, Dr. S. V. (2025). Smart Nanomaterials And AI-Enhanced Technologies For Environmental Separation And Sensing. *International Journal of Environmental Sciences*, *11*(8s), 856–877. DOI: 10.64252/hr71wx22

Pechini, M. P. (1967). *Method of preparing lead and alkaline earth titanates and niobates and coating method using the same to form a capacitor.*

Pei, J., Palanisamy, C. P., Srinivasan, G. P., Panagal, M., Kumar, S. S. D., & Mironescu, M. (2024). A comprehensive review on starch-based sustainable edible films loaded with bioactive components for food packaging. *International Journal of Biological Macromolecules*, •••, 133332. PMID: 38914408

Peng, J. min, Lin, J. cheng, Chen, Z. yu, Wei, M. chao, Fu, Y. xiang, Lu, S. shen, Yu, D. sheng, & Zhao, W. (2017). Enhanced antimicrobial activities of silver-nanoparticle-decorated reduced graphene nanocomposites against oral pathogens. *Materials Science and Engineering C*, *71*, 10–16. DOI: 10.1016/J.MSEC.2016.09.070 PMID: 27987652

Peppas, N. A., Bures, P., Leobandung, W., & Ichikawa, H. (2000). Hydrogels in pharmaceutical formulations. *European Journal of Pharmaceutics and Biopharmaceutics*, *50*, 27–46. PMID: 10840191

Perera, K. Y., Jaiswal, S., & Jaiswal, A. K. (2022). A review on nanomaterials and nanohybrids based bio-nanocomposites for food packaging. *Food Chemistry*, *376*, 131912. PMID: 34971895

Pérez-Pacheco, E., Canto-Pinto, J. C., Moo-Huchin, V. M., Estrada-Mota, I. A., Estrada-León, R. J., & Chel-Guerrero, L. (2016). Thermoplastic starch (TPS)-cellulosic fibers composites: Mechanical properties and water vapor barrier: A review. In *Composites from renewable and sustainable materials*. IntechOpen.

Perveen, S., Nadeem, R., Nosheen, F., Asjad, M., Awrejcewicz, J., & Anwar, T. (2022). Biochar-Mediated Zirconium Ferrite Nanocomposites for Tartrazine Dye Removal from Textile Wastewater. *Nanomaterials (Basel, Switzerland)*, *12*. Advance online publication. DOI: 10.3390/nano12162828 PMID: 36014692

Phadkule, S. S., & Sarma, S. (2023). Progress in nanocomposite based flexible temperature sensors: A review. *Measurement. Sensors*, *27*, 100692. DOI: 10.1016/j.measen.2023.100692

Philip, S., Keshavarz, T., & Roy, I. (2007). Polyhydroxyalkanoates: Bioplastics with a green agenda. *Chemical Technology and Biotechnology*, *82*, 233.

Pierre, A. C. (1998). *Introduction to sol–gel processing*. Kluwer.

Pilla, S. (2011). *Handbook of bioplastics and biocomposites engineering applications*. Scrivener.

Pinaeva, L. G., & Noskov, A. S. (2024). Biodegradable biopolymers: Real impact to environment pollution. *The Science of the Total Environment*, •••, 174445. PMID: 38981547

Pires, J. R. A., Rodrigues, C., Coelhoso, I., Fernando, A. L., & Souza, V. G. L. (2023). Current applications of bionanocomposites in food processing and packaging. *Polymers*, *15*(10), 2336. PMID: 37242912

Pöhlein, M., Llamas, A. D. R., Wolf, M., & Van Loosdrecht, M. C. M. (2022). Addressing ecological sustainability in the biotech industry. *Current Opinion in Biotechnology*, *76*, 221–229.

Polaris Market Research. (2024). Nanocellulose market size to hit USD 3162.42 million by 2034. Retrieved from https://www.polarismarketresearch.com/industry-analysis/nanocellulose-market

Popescu, C., Dissanayake, H., Mansi, E., & Stancu, A. (2024). Eco Breakthroughs: Sustainable Materials Transforming the Future of Our Planet. *Sustainability*, *16*(23), 10790.

Prabaharan, M., Reis, R. L., & Mano, J. F. (2005). Chitosan-based particles as controlled drug delivery systems. *Drug Delivery*, *12*(1), 41–57. PMID: 15801720

Precedence Research. (2024). Global nanocellulose market report 2024-2034. Retrieved from https://www.precedenceresearch.com/nanocellulose-market

Priya, A. K., Muruganandam, M., & Suresh, S. (2024). Bio-derived carbon-based materials for sustainable environmental remediation and wastewater treatment. *Chemosphere, 362*, 142731. DOI: 10.1016/j.chemosphere.2024.142731 PMID: 38950744

Priyadarshi, R., Ghosh, T., Roy, S., & Rhim, J. W. (2024). Biopolymer-based antimicrobial nanocomposite materials for food packaging and preservation. In *Food Packaging and Preservation* (pp. 33–52). Academic Press.

Pulido-Reyes, G., Leganes, F., Fernández-Piñas, F., & Rosal, R. (2017). Bio-nano interface and environment: A critical review. *Environmental Toxicology and Chemistry, 36*(12), 3181–3193. DOI: 10.1002/etc.3924 PMID: 28731222

Qian, L., Mei, C., Luo, W., Liu, W., Chen, M., Zhang, X., Li, X., Cheng, B., & Ma, H. (2023). *A Versatile Biochar Fertilizer: Excellent Adsorption Performance for Multi-Heavy Metals and Enhanced Effect on Plant Growth in Metal Contaminated Soil*. Elsevier BV. https://doi.org/DOI: 10.2139/ssrn.4535379

Quan, F., Dan, S., Huaiguo, X., Yuanyuan, H., & Serge, C. (2007). Amperometric phenol biosensor based on laponite clay–chitosan nanocomposite matrix. *Biosensors & Bioelectronics, 22*, 816–821. PMID: 16624546

Quinton, J. N., & Fiener, P. (2024). Soil erosion on arable land: An unresolved global environmental threat. *Progress in Physical Geography, 48*(1), 136–161. DOI: 10.1177/03091333231216595

Rabehi, A., Helal, H., Zappa, D., & Comini, E. (2024). Advancements and Prospects of Electronic Nose in Various Applications: A Comprehensive Review. *Applied Sciences (Basel, Switzerland), 14*(11), 4506. DOI: 10.3390/app14114506

Rabia, Z., Khalid, M. Z., Shazia, T., Farukh, J., Aqdas, N., & Mohammad, Z. (2016). Polysaccharide-based bionanocomposites: Properties and applications—A review. *International Journal of Biological Macromolecules, 92*, 1012–1024. PMID: 27481340

Raczkiewicz, M., Ostolska, I., Mašek, O., & Oleszczuk, P. (2024). Effect of the pyrolysis conditions and type of feedstock on nanobiochars obtained as a result of ball milling. *Journal of Cleaner Production, 458*, 142456.

Ragab, H. M., Diab, N. S., AlElaimi, M., Alghamdi, A. M., Farea, M. O., & Farea, A. (2024). Fabrication and Characterization of Silver Nanoparticle-Doped Chitosan/Carboxymethyl Cellulose Nanocomposites for Optoelectronic and Biological Applications. *ACS Omega, 9*(20), 22112–22122. PMID: 38799360

Raghuvanshi, S., Khan, H., Saroha, V., Sharma, H., Gupta, H. S., Kadam, A., & Dutt, D. (2023). Recent advances in biomacromolecule-based nanocomposite films for intelligent food packaging-A review. *International Journal of Biological Macromolecules*, •••, 127420. PMID: 37852398

Rahim, H. U., Qaswar, M., Uddin, M., Giannini, C., Herrera, M. L., & Rea, G. (2021). Nano-Enable Materials Promoting Sustainability and Resilience in Modern Agriculture. *Nanomaterials (Basel, Switzerland)*, *11*(8), 2068. DOI: 10.3390/nano11082068 PMID: 34443899

Rahman, A., Hasan, K., & Imran, A. B. (2024). Biopolymers for Supercapacitors. In *Bio-Based Polymers: Farm to Industry. Volume 3: Emerging Trends and Applications* (pp. 57-80). American Chemical Society.

Raigond, P., Sood, A., Kalia, A., Joshi, A., Kaundal, B., Raigond, B., Dutt, S., Singh, B., & Chakrabarti, S. K. (2019). Antimicrobial activity of potato starch-based active biodegradable nanocomposite films. *Potato Research*, *62*, 69–83.

Rajeev, A., Yin, L., Kalambate, P. K., Barjini Khabbaz, M., Trinh, B., Kamkar, M., Mekonnen, T. H., & Tang, S. (2024). Nano-enabled smart and functional materials toward human well-being and sustainable developments. *Nanotechnology*, *35*, 352003–352003. DOI: 10.1088/1361-6528/ad4dac PMID: 38768585

Raj, P., & Ghosh, A., Ravi, Jadav, S. K., Patra, H., Shehnaz, Panotra, N., Sharma, V., Sharma, R., Pandey, S. K., & Singh, B. V. (2024). Nanomaterials for Sustainable Soil Remediation and Contaminant Immobilization. *Asian Journal of Soil Science and Plant Nutrition*, *10*(4), 684–697. DOI: 10.9734/ajsspn/2024/v10i4439

Ramadan, M. M., & Abd-Elsalam, K. A. (2020). Micro/nano biochar for sustainable plant health: Present status and future prospects. Carbon nanomaterials for agri-food and environmental applications, 323-357.

Ramanayaka, S., Vithanage, M., Alessi, D. S., Liu, W. J., Jayasundera, A. C., & Ok, Y. S. (2020). Nanobiochar: Production, properties, and multifunctional applications. *Environmental Science. Nano*, *7*(11), 3279–3302.

Ramesan, M. T., Sankar, S., Kalladi, A. J., Labeeba Abdulla, A. C., & Bahuleyan, B. K. (2024). Hydroxyapatite nanoparticles reinforced polyvinyl alcohol/chitosan blend for optical and energy storage applications. *Polymer Engineering and Science*, *64*(3), 1378–1390.

Ramesh, K., Gnanavel, B., & Shkir, M. (2021). Enhanced visible light photocatalytic degradation of bisphenol A (BPA) by reduced graphene oxide (RGO)–metal oxide (TiO2, ZnO and WO3) based nanocomposites. *Diamond and Related Materials, 118*, 108514. DOI: 10.1016/J.DIAMOND.2021.108514

Rana, A., Gautam, D., Kumar, P., & Kumar Das, A. (2025). Architectures, Benefits, Security, and Privacy Issues of Internet of Nano Things: A Comprehensive Survey, Opportunities, and Research Challenges. *IEEE Communications Surveys and Tutorials*, *27*(2), 1152–1190. DOI: 10.1109/COMST.2024.3423477

Ranade, V. V., & Hollinger, M. A. (Eds.). (2003). *Drug delivery systems* (2nd ed., pp. 1–69). CRC Press.

Rando, G., Sfameni, S., Galletta, M., Drommi, D., Cappello, S., & Plutino, M. R. (2022). Functional Nanohybrids and Nanocomposites Development for the Removal of Environmental Pollutants and Bioremediation. *Molecules (Basel, Switzerland)*, *27*(15), 15. Advance online publication. DOI: 10.3390/molecules27154856 PMID: 35956804

Rani, S., & Baloda, S., Dinesh, & Mehta, A. (2024). AI-Driven Insights into Soil Physio-Chemical Properties. *International Journal of Environment and Climate Change*, *14*(12), 834–845. DOI: 10.9734/ijecc/2024/v14i124666

Râpă, M., Mitelu , A. C., Tănase, E. E., Grosu, E., Popescu, P., Popa, M. E., & Darie-Ni ă, R. N. (2016). Influence of chitosan on mechanical, thermal, barrier and antimicrobial properties of PLA-biocomposites for food packaging. *Composites. Part B, Engineering*, *102*, 112–121. DOI: 10.1016/j.compositesb.2016.07.016

Rath, A. P., Krishnan, P. S. G., & Kanny, K. (2024). A comparison of physical, morphological, and mechanical properties of bio-polyester hybrid nanocomposites. Polymers for Advanced Technologies, 35(9). Portico. https://doi.org/DOI: 10.1002/pat.6566

Rawas-Qalaji, M. (2016). Combination of Bottom-Up and Top-Down Techniques for the Fabrication of Epinephrine Nanocrystals. https://doi.org/DOI: 10.26226/morressier.57d6b2b8d462b8028d88cfe0

Rawat, R. S. (2015). Dense plasma Focus - from alternative fusion source to versatile high energy density plasma source for plasma nanotechnology. *Journal of Physics: Conference Series*, *591*, 012021. DOI: 10.1088/1742-6596/591/1/012021

Ray, S. S., & Bousmina, M. (2005). Biodegradable polymers and their layered silicate nanocomposites: Greening the 21st century materials world. *Progress in Materials Science*, *50*, 962–1079.

Ray, S. S., & Bousmina, M. (2005). Biodegradable polymers and their layered silicate nanocomposites: In greening the 21st century materials world. *Progress in Materials Science, 50*(8), 962–1079.

Razavi, R. S., Loghman-Estarki, M. R., & Farhadi-Khouzani, M. (2012). Synthesis and Characterization of ZnO Nanostructures by Polymeric Precursor Route. *Acta Physica Polonica A, 121*(1), 98–100. DOI: 10.12693/APHYSPOLA.121.98

Rech, A., & Daugaard, A. E. (2025). Thermoprocessing Biopolymers and Bio-Waste-Based Materials. *ACS Sustainable Resource Management.*

Redondo, A., Mortensen, N., Djeghdi, K., Jang, D., Ortuso, R. D., Weder, C., Korley, L. T. J., Steiner, U., & Gunkel, I. (2022). Comparing Percolation and Alignment of Cellulose Nanocrystals for the Reinforcement of Polyurethane Nanocomposites. *ACS Applied Materials & Interfaces, 14*(5), 7270–7282. PMID: 35077647

Ren, J. (2011). *Biodegradable poly (lactic acid): Synthesis, modification, processing and applications.* Springer Science & Business Media.

Rhim, J. W., & Ng, P. K. W. (2007). Natural biopolymer-based nanocomposite films for packaging applications. *Critical Reviews in Food Science and Nutrition, 47*, 411–433. PMID: 17457725

Rhim, J. W., Park, H. M., & Ha, C. S. (2013). Bio-nanocomposites for food packaging applications. *Progress in Polymer Science, 38*(10–11), 1629–1652. DOI: 10.1016/J.PROGPOLYMSCI.2013.05.008

Riaz, F., Rasul, I., Azeem, F., Zubair, M., Nadeem, H., Imran, M., Muzammil, A., Afzal, M., & Siddique, M. H. (2024). Environmental impact, health implications, and life cycle assessment of bionanocomposites. Advances in Bionanocomposites, 387–406. https://doi.org/DOI: 10.1016/b978-0-323-91764-3.00009-7

Rivadeneira-Velasco, K. E., Utreras-Silva, C. A., Díaz-Barrios, A., Sommer-Márquez, A. E., Tafur, J. P., & Michell, R. M. (2021). Green nanocomposites based on thermoplastic starch: A review. *Polymers, 13*, 3227. DOI: 10.3390/polym13193227 PMID: 34641042

Rodrigues, C. I. D., Brito, L. M., & Nunes, L. J. R. (2023). Soil Carbon Sequestration in the Context of Climate Change Mitigation: A Review. *Soil Systems, 7*(3), 3. Advance online publication. DOI: 10.3390/soilsystems7030064

Rojas, J., Bedoya, M., & Ciro, Y. (2015). Current trends in the production of cellulose nanoparticles and nanocomposites for biomedical applications. *Cellulose-Fundamental Aspects and Current Trends*, 193–228.

Rong, T., Nathan, P. G., Timothy, M. F., & Jianjun, C. (2012). Polymeric nanomedicines based on poly(lactide) and poly(lactide-co-glycolide). *Current Opinion in Solid State and Materials Science*, *16*, 323–332. PMID: 23914135

Rosato, D. (2003). *Plastics engineered product design*. Elsevier.

Rostamzadeh-Renani, R., Jasim, D. J., Baghoolizadeh, M., Rostamzadeh-Renani, M., Andani, H. T., Salahshour, S., & Baghaei, Sh. (2024). Multi-objective optimization of rheological behavior of nanofluids containing CuO nanoparticles by NSGA II, MOPSO, and MOGWO evolutionary algorithms and group method of data handling artificial neural networks. *Materials Today. Communications*, *38*, 107709. DOI: 10.1016/j.mtcomm.2023.107709

Roy, P., Padhiary, M., Hoque, A., Sahu, B., Roy, D., & Kumar, K. (2025). Machine Learning for Precision Agriculture and Crop Yield Optimization: Techniques and Applications. In Emara, T., Hassan, E., Trinh, T., Li, G., & Saber, A. (Eds.), *The Role of Artificial Intelligence in Advancing Applied Life Sciences* (pp. 189–234). IGI Global., DOI: 10.4018/979-8-3693-9208-9.ch007

Roy, S. B., Shit, S. C., Sengupta, R. A., & Shukla, P. R. (2014). A Review on Bio-Composites: Fabrication, Properties and Applications. *International Journal of Innovative Research in Science, Engineering and Technology*, *03*(10), 16814–16824. DOI: 10.15680/ijirset.2014.0310058

Roy, S., Malik, B., Chawla, R., Bora, S., Ghosh, T., Santhosh, R., & Sarkar, P. (2024). Biocompatible film based on protein/polysaccharides combination for food packaging applications: A comprehensive review. *International Journal of Biological Macromolecules*, •••, 134658. PMID: 39128751

Ruiz Gómez, E. E., Mina Hernández, J. H., & Diosa Astaiza, J. E. (2020). Development of a Chitosan/PVA/TiO2 Nanocomposite for Application as a Solid Polymeric Electrolyte in Fuel Cells. *Polymers*, *12*(8), 1691. PMID: 32751167

Ruiz-Hitzky, E., Darder, M., Fernandes, F. M., Wicklein, B., Alcântara, A. C. S., & Aranda, P. (2013). Fibrous clays based bionanocomposites. *Progress in Polymer Science*, *38*(10–11), 1392–1414. DOI: 10.1016/J.PROGPOLYMSCI.2013.05.004

Sadaf, M., Bragaglia, M., Slemenik Perše, L., & Nanni, F. (2024). Advancements in metal additive manufacturing: A comprehensive review of material extrusion with highly filled polymers. *Journal of Manufacturing and Materials Processing*, *8*(1), 14.

Sahana, M., Rehman, S., Patel, P. P., Dou, J., Hong, H., & Sajjad, H. (2020). Assessing the degree of soil salinity in the Indian Sundarban Biosphere Reserve using measured soil electrical conductivity and remote sensing data–derived salinity indices. *Arabian Journal of Geosciences, 13*(24). Advance online publication. DOI: 10.1007/s12517-020-06310-w

Sahu MK, Yadav R, Tiwari SP (2023) Recent advances in nanotechnology. Int J Nanomater Nanotechnol Nanomed 9(2): 015-023. DOI: https://dx.doi.org/DOI: 10.17352/2455-3492.000053

Saif, S., Tahir, A., & Chen, Y. (2016). Green Synthesis of Iron Nanoparticles and Their Environmental Applications and Implications. *Nanomaterials (Basel, Switzerland), 6*(11), 209. DOI: 10.3390/nano6110209 PMID: 28335338

Saikia, N. (2024). Inorganic-Based Nanoparticles and Biomaterials as Biocompatible Scaffolds for Regenerative Medicine and Tissue Engineering: Current Advances and Trends of Development. *Inorganics 2024, Vol. 12, Page 292, 12*(11), 292. https://doi.org/DOI: 10.3390/INORGANICS12110292

Saini, R., Bajpai, A. K., & Jain, E. (2017). Fundamentals of bionanocomposites. In *Biodegradable and biocompatible polymer composites* (pp. 1–xx). Elsevier., DOI: 10.1016/B978-0-08-100970-3.00012-2

Sain, M., & Oksman, K. (2006). Cellulose nanocomposites. *Journal of the American Chemical Society, 938*, 2–8.

Salazar, R., Domenek, S., & Ducruet, V. (2014). Interactions of flavoured oil in-water emulsions with polylactide. *Food Chemistry, 148*, 138–146. DOI: 10.1016/j.foodchem.2013.10.018 PMID: 24262538

Salem, D. M., Ismail, M. M., & Aly-Eldeen, M. A. (2019). Biogenic synthesis and antimicrobial potency of iron oxide (Fe3O4) nanoparticles using algae harvested from the Mediterranean Sea, Egypt. *Egyptian Journal of Aquatic Research, 45*(3), 197–204. DOI: 10.1016/j.ejar.2019.07.002

Samir, A., Ashour, F. H., Hakim, A. A., & Bassyouni, M. (2022). Recent advances in biodegradable polymers for sustainable applications. *NPJ Materials Degradation, 6*(1), 68.

Samson Prince, H., Macarius, N. M. N., Chandrashekhar Walkikar, S., Das, S., Rajan, A. P., & Deepankumar, K. (2025). Harmonizing Nature and Innovation: Sustainable Environmental Solutions Through Nano-biotech Synergy. In Jeyaseelan, A., & Murugasen, K. (Eds.), *Sustainable Environmental Remediation: Avenues in Nano and Biotechnology* (pp. 35–83). Springer Nature Switzerland., DOI: 10.1007/978-3-031-78483-5_2

Samuel, M. S., Datta, S., Chandrasekar, N., Balaji, R., Selvarajan, E., & Vuppala, S. (2021). Biogenic Synthesis of Iron Oxide Nanoparticles Using Enterococcus faecalis: Adsorption of Hexavalent Chromium from Aqueous Solution and In Vitro Cytotoxicity Analysis. *Nanomaterials, 11*(12), 3290. https://doi.org/ (Creative Commons Attribution 4.0 International (CC BY 4.0)) (https://creativecommons.org/licenses/by/4.0/deed.en)DOI: 10.3390/nano11123290

Sandford, P. A. (1989). In Skjak-Braek, G., Anthonsen, T., & Sandford, P. A. (Eds.), *Chitin and chitosan: Sources, chemistry, biochemistry, physical properties and applications* (pp. 51–69). Elsevier.

Sandhya, J., & Kalaiselvam, S. (2019). Biogenic synthesis of magnetic iron oxide nanoparticles using inedible borassus flabellifer seed coat: characterization, antimicrobial, antioxidant activity and in vitro cytotoxicity analysis. *Materials Research Express, 7*(1), 015045. https://doi.org/ (Creative Commons Attribution 4.0 International (CC BY 4.0)) (https://creativecommons.org/licenses/by/4.0/deed.en)DOI: 10.1088/2053-1591/ab6642

Sandhya, P., Sreekala, M., Padmanabhan, M., Jesitha, K., & Thomas, S. (2019). Effect of starch reduced graphene oxide on thermal and mechanical properties of phenol formaldehyde resin nanocomposites. *Composites. Part B, Engineering*, *167*, 83–92.

Sani, M. A., Priyadarshi, R., Zhang, W., Khezerlou, A., & Rhim, J. W. (2024). Innovative application of laccase enzyme in food packaging. *Trends in Food Science & Technology*, •••, 104623.

Santos, M., Rebola, S., & Evtuguin, D. V. (2025). Soil Remediation: Current Approaches and Emerging Bio-Based Trends. *Soil Systems*, *9*(2), 2. Advance online publication. DOI: 10.3390/soilsystems9020035

Santos, P. A., Biraku, X., Nielsen, E., Ozketen, A. C., Ozketen, A. A., & Hakki, E. E. (2025). Agricultural nanotechnology for a safe and sustainable future: Current status, challenges, and beyond. *Journal of the Science of Food and Agriculture*, *105*(6), 3159–3169. DOI: 10.1002/jsfa.13922 PMID: 39319478

Saod, W. M., Al-Janaby, M. S., Gayadh, E. W., Ramizy, A., & Hamid, L. L. (2024). Biogenic synthesis of iron oxide nanoparticles using Hibiscus sabdariffa extract: Potential for antibiotic development and antibacterial activity against multidrug-resistant bacteria. *Current Research in Green and Sustainable Chemistry*, *8*, 100397. DOI: 10.1016/j.crgsc.2024.100397

Sarangi, P. K., Srivastava, R. K., Sahoo, U. K., Singh, A. K., Parikh, J., Bansod, S., Parsai, G., Luqman, M., Shadangi, K. P., Diwan, D., Lanterbecq, D., & Sharma, M. (2024). Biotechnological innovations in nanocellulose production from waste biomass with a focus on pineapple waste. *Chemosphere*, *349*, 140833. PMID: 38043620

Saraswat, P., Singh, S., Prasad, M., Misra, R., Rajput, V. D., & Ranjan, R. (2023). Applications of bio-based nanomaterials in environment and agriculture: A review on recent progresses. *Hybrid Advances*, *4*, 100097. DOI: 10.1016/j.hybadv.2023.100097

Saravanan, A., Swaminaathan, P., Kumar, P. S., Yaashikaa, P. R., Kamalesh, R., & Rangasamy, G. (2023). A comprehensive review on immobilized microbes - biochar and their environmental remediation: Mechanism, challenges and future perspectives. *Environmental Research*, *236*, 116723. DOI: 10.1016/j.envres.2023.116723 PMID: 37487925

Savci, S. (2012). An agricultural pollutant: Chemical fertilizer. *International Journal of Environmental Sciences and Development*, *3*(1), 73–80. DOI: 10.7763/IJESD.2012.V3.191

Saxena, J., Rana, G., & Pandey, M. (2013). Impact of addition of biochar along with Bacillus sp. on growth and yield of French beans. *Scientia Horticulturae*, *162*, 351–356.

Scarfato, P., Di Maio, L., & Incarnato, L. (2022). Recent advances and migration issues in biodegradable polymers from renewable sources for food packaging. *Journal of Applied Polymer Science*, *139*(48), 42597.

Schoeppler, V., Lemanis, R., Reich, E., Pusztai, T., Gránásy, L., & Zlotnikov, I. (2019). Crystal growth kinetics as an architectural constraint on the evolution of molluscan shells. *Proceedings of the National Academy of Sciences of the United States of America*, *116*(41), 20388–20397. DOI: 10.1073/PNAS.1907229116/SUPPL_FILE/PNAS.1907229116.SM03.MP4 PMID: 31551265

Sen, A., Kurian, M. S., Narayanan, D. P., Abraham, A., Jayalakshmi, P. S., Nair, S. G., & Simon, S. M. (2024). Bio-nanocomposites: Innovative solutions for addressing issues in health, agriculture, energy and environmental domains. *Nano-Structures & Nano-Objects*, *39*, 101270.

Senel, S., & McClure, S. J. (2004). Potential applications of chitosan in veterinary medicine. *Advanced Drug Delivery Reviews*, *56*(10), 1467–1480. PMID: 15191793

Seves, A., Testa, G., Bonfatti, A. M., Paglia, E. D., Selli, E., & Marcandalli, B. (2001). Characterization of native cellulose/poly(ethylene glycol) films. *Macromolecular Materials and Engineering*, *286*(9), 524–528.

Sevinç, B. A., & Hanley, L. (2010). Antibacterial activity of dental composites containing zinc oxide nanoparticles. *Journal of Biomedical Materials Research. Part B, Applied Biomaterials*, *94*(1), 22–31. DOI: 10.1002/JBM.B.31620 PMID: 20225252

Shabbir, M. A., Naveed, M., Rehman, S. U., Ain, N. U., Aziz, T., Alharbi, M., Alsahammari, A., & Alasmari, A. F. (2023). Synthesis of Iron Oxide Nanoparticles from Madhuca indica Plant Extract and Assessment of Their Cytotoxic, Antioxidant, Anti-Inflammatory, and Anti-Diabetic Properties via Different Nanoinformatics Approaches. *ACS Omega*, *8*(37), 33358–33366. DOI: 10.1021/acsomega.3c02744 PMID: 37744851

Shahid, M. A., Rahman, M. M., Hossain, M. T., Hossain, I., Sheikh, M. S., Rahman, M. S., & Hoque, M. I. U. (2025). Advances in Conductive Polymer-Based Flexible Electronics for Multifunctional Applications. *Journal of Composites Science*, *9*(1), 42.

Shah, Y. A., Bhatia, S., Al-Harrasi, A., Tarahi, M., Almasi, H., Chawla, R., & Ali, A. M. M. (2024). Insights into recent innovations in barrier resistance of edible films for food packaging applications. *International Journal of Biological Macromolecules*, •••, 132354. PMID: 38750852

Shar, A. S., Zhang, C., Song, X., Weng, Y., & Du, Q. (2021). Design of Novel PLA/OMMT Films with Improved Gas Barrier and Mechanical Properties by Intercalating OMMT Interlayer with High Gas Barrier Polymers. *Polymers*, *13*(22), 3962. PMID: 34833261

Sharabi, M., Mandelberg, Y., Benayahu, D., Benayahu, Y., Azem, A., & Haj-Ali, R. (2014). A new class of bio-composite materials of unique collagen fibers. *Journal of the Mechanical Behavior of Biomedical Materials*, *36*, 71–81. DOI: 10.1016/J.JMBBM.2014.04.008 PMID: 24801102

Sharma, R. B., Kaur, A., Nautiyal, U., Kabra, A., & Arora, V. (2024). The role of green chemistry and nanotechnology in developing environmental sustainability. Role of Green Chemistry in Ecosystem Restoration to Achieve Environmental Sustainability, 33–40. https://doi.org/DOI: 10.1016/b978-0-443-15291-7.00009-2

Sharma, A., Thakur, M., Bhattacharya, M., Mandal, T., & Goswami, S. (2019). Commercial application of cellulose nano-composites–A review. *Biotechnology Reports (Amsterdam, Netherlands)*, *21*, e00316. PMID: 30847286

Sharma, D., Teli, G., Gupta, K., Bansal, G., Gupta, G. D., & Chawla, P. A. (2022). Nano-biosensors from Agriculture to Nextgen Diagnostic Tools. *Current Nanomaterials*, *7*(2), 110–138. DOI: 10.2174/2405461507666220131104843

Sharma, K., & Shivandu, S. K. (2024). Integrating artificial intelligence and Internet of Things (IoT) for enhanced crop monitoring and management in precision agriculture. *Sensors International*, *5*, 100292. DOI: 10.1016/j.sintl.2024.100292

Sharma, N., & Rana, V. S. (2020). A Review on Polysaccharide Based Nanocomposite Hydrogel Systems Fabrication Using Diverse Reinforcing Materials. *Journal of Polymer & Composites*, *8*(1), 6–17.

Sharma, R., Jafari, S. M., & Sharma, S. (2020). Antimicrobial bio-nanocomposites and their potential applications in food packaging. *Food Control*, *112*, 107086. DOI: 10.1016/j.foodcont.2020.107086

Sharma, R., & Kumar, V. (2024). Nano enabled agriculture for sustainable soil. *Waste Management Bulletin*, *2*(1), 152–161. DOI: 10.1016/j.wmb.2024.01.002

Sharmile, N., Chowdhury, R. R., & Desai, S. (2025). A Comprehensive Review of Quality Control and Reliability Research in Micro–Nano Technology. *Technologies*, *13*(3), 94. DOI: 10.3390/technologies13030094

Shchipunov, Y. (2012). Bio nanocomposites: Green sustainable materials for the near future. *Pure and Applied Chemistry*, *84*(12), 2579–2607. DOI: 10.1351/PAC-CON-12-05-04

Shchipunov, Y. (2012). Bionanocomposites: Green sustainable materials for the near future. *Pure and Applied Chemistry*, *84*, 2579–2607.

Shchipunov, Y., & Irina, P. (2009). One-pot biomimetic synthesis of monolithic titania through mineralization of polysaccharide. *Colloids and Surfaces. B, Biointerfaces*, *74*, 172–177. PMID: 19674874

Sheershwal, A., Singh, A., Sharma, V., & Trivedi, B. (2025). Microbial synthesis of nanoparticles for sustainable agricultural advancements: A comprehensive review. *Nanotechnology for Environmental Engineering*, *10*(1), 16. DOI: 10.1007/s41204-025-00405-z

Sheibani, S., Rozalli, N. H. M., & Nafchi, A. M. (2025). Barrier properties of bionanocomposites. In *Characterization Techniques in Bionanocomposites* (pp. 133–182). Woodhead Publishing.

Sheshmani, S., Ashori, A., & Hasanzadeh, S. (2014). Removal of Acid Orange 7 from aqueous solution using magnetic graphene/chitosan: A promising nano-adsorbent. *International Journal of Biological Macromolecules*, *68*, 218–224. PMID: 24813679

Shibata, M. (2011). Bio-nanocomposites using bio-based epoxy resins. Nanocomposites with Biodegradable Polymers, 329–347. https://doi.org/DOI: 10.1093/acprof:oso/9780199581924.003.0013

Shih, Y. F., & Jeng, R. J. (2011). Biodegradable poly(butylene succinate)/multi-walled carbon nanotube nanocomposites. In Mittal, V. (Ed.), *Nanocomposites with Biodegradable Polymers: Synthesis, Properties, and Future Perspectives*. Oxford University Press.

Shi, Y., Zhang, S., Zhou, H., Dong, Y., Liu, G., Ye, W., He, R., & Zhao, G. (2025). Recent Developments in Heavy Metals Detection: Modified Electrodes, Pretreatment Methods, Prediction Models and Algorithms. *Metals*, *15*(1), 1. Advance online publication. DOI: 10.3390/met15010080

Shi, Y., Zhong, S., Wang, X., & Feng, C. (2022). A review of the removal of heavy metal ions in wastewater by modified montmorillonite. *Water Policy*, *24*(10), 1590–1609. DOI: 10.2166/wp.2022.033

Shou, N. D., Dan, S., Huai, G. X., Dao, B. Z., & Serge, C. (2009). Glucose oxidase immobilized in alginate/layered double hydroxides hybrid membrane and its biosensing application. *Analytical Sciences*, *25*, 1421–1425. PMID: 20009328

Sibalija, T. V., & Majstorovic, V. D. (2012). An integrated approach to optimise parameter design of multi-response processes based on Taguchi method and artificial intelligence. *Journal of Intelligent Manufacturing*, *23*(5), 1511–1528. DOI: 10.1007/s10845-010-0451-y

Siciliano, S., Lopresto, C. G., & Lamonaca, F. (2024). From traditional packaging to smart bio-packaging for food safety: A review. *Euro-Mediterranean Journal for Environmental Integration*, •••, 1–16.

Silva, E. C. (n.d.). *Hidroxiapatita sintética em alvéolo dentário após exodontia em Felis catus: Estudo clínico, radiológico e histomorfométrico* (M.S. Dissertation). Universidade Federal de Viçosa, Brazil.

Singh, S. P., Paneru, P., & Deepak, D. (2025). Nanotechnological approaches for enhancing climate resilience and sustainable adaptation in agriculture. In S. P. Singh (Ed), P. Paneru (Ed), & K. K. Singh (Ed), *Deep Science Publishing*. Deep Science Publishing. https://doi.org/DOI: 10.70593/978-93-49307-12-4_14

Singh, A. V., Ansari, M. H. D., Rosenkranz, D., Maharjan, R. S., Kriegel, F. L., Gandhi, K., Kanase, A., Singh, R., Laux, P., & Luch, A. (2020). Artificial Intelligence and Machine Learning in Computational Nanotoxicology: Unlocking and Empowering Nanomedicine. *Advanced Healthcare Materials*, *9*(17), 1901862. DOI: 10.1002/adhm.201901862 PMID: 32627972

Singh, A. V., Rosenkranz, D., Ansari, M. H. D., Singh, R., Kanase, A., Singh, S. P., Johnston, B., Tentschert, J., Laux, P., & Luch, A. (2020). Artificial Intelligence and Machine Learning Empower Advanced Biomedical Material Design to Toxicity Prediction. *Advanced Intelligent Systems*, *2*(12), 2000084. DOI: 10.1002/aisy.202000084

Singh, D., & Gurjar, B. R. (2022). Nanotechnology for agricultural applications: Facts, issues, knowledge gaps, and challenges in environmental risk assessment. *Journal of Environmental Management*, *322*, 116033. DOI: 10.1016/j.jenvman.2022.116033

Singh, P., Pandey, V. K., Singh, R., Singh, K., Dash, K. K., & Malik, S. (2024). Unveiling the potential of starch-blended biodegradable polymers for substantializing the eco-friendly innovations. *Journal of Agriculture and Food Research*, *15*, 101065.

Singh, S., Patel, M., Schwendemann, D., Zaccone, M., Geng, S., Maspoch, M. L., & Oksman, K. (2020). Effect of chitin nanocrystals on crystallization and properties of poly (lactic acid)-based nanocomposites. *Polymers*, *12*(3), 726. PMID: 32214000

Sing, K. S. W., Everett, D. H., Haul, R. A. W., Moscou, L., Pierotti, R. A., Rouquerol, J., & Siemieniewska, T. (1985). Reporting Physisorption Data for Gas/Solid Systems with Special Reference to the Determination of Surface Area and Porosity. *Pure and Applied Chemistry*, *57*(4), 603–619. DOI: 10.1351/PAC198557040603/MACHINEREADABLECITATION/RIS

Sinthiya, P. A., Johnson, I., Prabu, H. J., Sahayaraj, A. F., & Selvan, M. T. (2024). Extraction of cellulose nanocrystals from Cissus quadrangularis for sustainable biocomposite production. *Biomass Conversion and Biorefinery*, •••, 1–13.

Siracusa, V., & Blanco, I. (2023). Bio-polyethylene (Bio-PE), bio-polypropylene (Bio-PP) and bio-poly(ethylene terephthalate) (Bio-PET): Recent developments in bio-based polymers analogous to petroleum-derived ones for packaging and engineering applications. *Polymers*, *15*(8), 1641. PMID: 32718011

Siró, I., & Plackett, D. (2010). Microfibrillated cellulose and new nanocomposite materials: A review. *Cellulose (London, England), 17*(3), 459–494.

Sivakanthan, S., & Kaumadi, P. H. S. (2022). Environmentally Friendly Bionanocomposites in Food Industry. In *Biodegradable and Environmental Applications of Bionanocomposites* (pp. 237–263). Springer International Publishing.

Solá, F. (2014). *SEM and TEM Characterization of Polymer CNT Nanocomposites. Polymer Nanotube Nanocomposites*. Portico., DOI: 10.1002/9781118945964.ch4

Song, B., Cao, X., Gao, W., Aziz, S., Gao, S., Lam, C. H., & Lin, R. (2022). Preparation of nano-biochar from conventional biorefineries for high-value applications. *Renewable & Sustainable Energy Reviews, 157*, 112057.

Sonia, P., Srinivas, R., Kansal, L., Abdul-Zahra, D. S., Reddy, U., & Kumari, V. (2024). Bioinspired Composites a Review: Lessons from Nature for Materials Design and Performance. *E3S Web of Conferences, 505*, 01024. https://doi.org/DOI: 10.1051/E3SCONF/202450501024

Sorrentino, A., Giuliana, G., & Vittoria, V. (2007). Potential perspectives of bionanocomposites for food packaging applications. *Trends in Food Science & Technology, 18*, 84–95.

Sreenivasan, K. (1988). Synthesis and evaluation of a β-cyclodextrin-based molecularly imprinted copolymer. *Journal of Applied Polymer Science, 69*(6), 1051–1055.

Sreenivasulu, N., Pothula, L., & Reddy, P. D. S. (2016). Growing the importance of bionanocomposites. *International Journal of Advanced Research in Science and Engineering, 5*(1), 374–381.

Srivastava, A. K., Dev, A., & Karmakar, S. (2018). Nanosensors and nanobiosensors in food and agriculture. *Environmental Chemistry Letters, 16*(1), 161–182. DOI: 10.1007/s10311-017-0674-7

Stanley, J., Culliton, D., Jovani-Sancho, A. J., & Neves, A. C. (2025). The Journey of Plastics: Historical Development, Environmental Challenges, and the Emergence of Bioplastics for Single-Use Products. *Eng, 6*(1), 17.

StartUs Insights. (2025). 9 nanotechnology trends in 2025. Retrieved from https://www.startus-insights.com/innovators-guide/nanotechnology-trends/

Sudesh, K., Abe, H., & Doi, Y. (2000). Synthesis, structure and properties of polyhydroxyalkanoates: Biological polyesters. *Progress in Polymer Science, 25*, 1503–1555.

Sudesh, K., & Iwata, T. (2008). Sustainability of biobased and biodegradable plastics. *Clean (Weinheim), 36*, 433–442.

Sumper, M., & Brunner, E. (2006). Learning from diatoms: Nature's tools for the production of nanostructured silica. *Advanced Functional Materials*, *16*, 17.

Sumper, M., & Brunner, E. (2006). Silicon uptake and metabolism of the marine diatom Thalassiosira pseudonana: Solid-state 29Si NMR and fluorescence microscopic studies. *Advanced Functional Materials*, *16*, 17.

Sun, J., Forster, A. M., Johnson, P. M., Eidelman, N., Quinn, G., Schumacher, G., Zhang, X., & Wu, W. L. (2011). Improving performance of dental resins by adding titanium dioxide nanoparticles. *Dental Materials*, *27*(10), 972–982. DOI: 10.1016/J.DENTAL.2011.06.003 PMID: 21774976

Sun, T., Xu, P., Liu, Q., Xue, J., & Xie, W. (2003). Graft copolymerization of methacrylic acid onto carboxymethyl chitosan. *European Polymer Journal*, *39*(1), 189–192.

Surya Rao, G., Debnath, K., & Mahapatra, R. (2021). Development and Characterization of PLA-Based Green Composites: Experimental and Simulation Studies. In *Green Composites* (pp. 209–223). Springer.

Suryanegara, L., Nakagaito, A. N., & Yano, H. (2009). The effect of crystallization of PLA on the thermal and mechanical properties of microfibrillated cellulose-reinforced PLA composites. *Composites Science and Technology*, *69*(7–8), 1187–1192.

Susan, Md. A. B. H., Harun-Ur-Rashid, M., & Imran, A. B. (2023). Green Polymer Nanocomposites in Automotive and Packaging Industries. *Current Pharmaceutical Biotechnology*, *24*(1), 145–163. DOI: 10.2174/1389201023666220506111027 PMID: 35524658

Swain, S. K., Sethy, P. K., & Pattanayak, A. J. (2017a). Synthesis of Soy Protein Based Biocomposites for Packaging Applications. In Jawaid, M., Salit, M. S., & Alothman, O. Y. (Eds.), *Green Biocomposites: Design and Applications* (pp. 143–166). Springer International Publishing.

Swain, S. K., Sethy, P. K., & Pattanayak, A. J. (2017b). Synthesis of soy protein based biocomposites for packaging applications. In *Green Biocomposites* (pp. 143–166). Springer.

Tabassum, Z., Girdhar, M., Kumar, A., Malik, T., & Mohan, A. (2023). ZnO Nanoparticles-Reinforced Chitosan–Xanthan Gum Blend Novel Film with Enhanced Properties and Degradability for Application in Food Packaging. *ACS Omega*, *8*(34), 31318–31332. PMID: 37663466

Taha, M. F., Mao, H., Zhang, Z., Elmasry, G., Awad, M. A., Abdalla, A., Mousa, S., Elwakeel, A. E., & Elsherbiny, O. (2025). Emerging Technologies for Precision Crop Management Towards Agriculture 5.0: A Comprehensive Overview. *Agriculture*, *15*(6), 582. DOI: 10.3390/agriculture15060582

Takahashi, C., Matsubara, N., Akachi, Y., Ogawa, N., Kalita, G., Asaka, T., Tanemura, M., Kawashima, Y., & Yamamoto, H. (2017). Visualization of silver-decorated poly (DL-lactide-co-glycolide) nanoparticles and their efficacy against Staphylococcus epidermidis. *Materials Science and Engineering C*, *72*, 143–149. DOI: 10.1016/J.MSEC.2016.11.051 PMID: 28024570

Tan, H. L., Lim, Y. C., Ng, L. Y., & Lim, Y. P. (2023). Plant-mediated synthesis of iron nanoparticles for environmental application: Mini review. *Materials Today: Proceedings*, *87*, 64–69. DOI: 10.1016/j.matpr.2023.02.101

TavallaieNejad, A., Vila, M. C., Paneiro, G., & Baptista, J. S. (2025). A Systematic Review of Machine Learning Algorithms for Soil Pollutant Detection Using Satellite Imagery. *Remote Sensing*, *17*(7), 1207. DOI: 10.3390/rs17071207

Tavassoli Hojati, S., Alaghemand, H., Hamze, F., Ahmadian Babaki, F., Rajab-Nia, R., Rezvani, M. B., Kaviani, M., & Atai, M. (2013). Antibacterial, physical and mechanical properties of flowable resin composites containing zinc oxide nanoparticles. *Dental Materials*, *29*(5), 495–505. DOI: 10.1016/J.DENTAL.2013.03.011 PMID: 23562673

Teixeira, W. G., Víctor Hugo Alvarez, V., & Neves, J. C. L. (2020). New methods for estimating lime requirement to attain desirable pH values in Brazilian soils. *Revista Brasileira de Ciência do Solo*, *44*. Advance online publication. DOI: 10.36783/18069657rbcs20200008

Teo, W. E., & Ramakrishna, S. (2006). A review on electrospinning design and nanofibre assemblies. *Nanotechnology*, *17*, R89–R106. PMID: 19661572

Thakur, A., & Kumar, A. (2022). Recent advances on rapid detection and remediation of environmental pollutants utilizing nanomaterials-based (bio)sensors. *The Science of the Total Environment*, *834*, 155219. DOI: 10.1016/j.scitotenv.2022.155219 PMID: 35421493

Thakur, V. K. (Ed.). (2023). *Bio-based polymers and nanocomposites: Preparation, properties and applications*. Springer Nature.

Thakur, V. K., Sharma, N., Vetcher, A., & Mishra, P. K. (2023). Recent developments in nanocellulose-based biodegradable polymers, thermoplastic polymers, and porous nanocomposites. *Frontiers in Nanotechnology*, *5*, 1186613.

Thapliyal, D., Karale, M., Diwan, V., Kumra, S., Arya, R. K., & Verros, G. D. (2024). Current status of sustainable food packaging regulations: Global perspective. *Sustainability, 16*(13), 5554.

Thapliyal, D., Verros, G. D., & Arya, R. K. (2025). Nanoparticle-Doped Antibacterial and Antifungal Coatings. *Polymers, 17*(2), 247. PMID: 39861318

Tharanathan, R. N., & Kittur, F. S. (2003). Chitin—The undisputed biomolecule of great potential. *Critical Reviews in Food Science and Nutrition, 43*, 61–87. PMID: 12587986

Thirugnanasambandan, T. (2022). Bio-nanocomposites in Biomedical Application. Polymer Based Bio-Nanocomposites, 275–291. https://doi.org/Bottom of FormDOI: 10.1007/978-981-16-8578-1_15

Thirumalaisamy, R., Nimithap, S., Harini, S., Rajiniganth, R., Haritha, S., Ameer, K., Vadivel, P., & Selvankumar, T. (2025). Potential Application of Nanoemulsion in Agriculture: A Current Perspective. *International Journal of Advanced Science and Engineering, 11*(3), 4328–4350. DOI: 10.29294/IJASE.11.3.2025.4328-4350

Thivya, P., Gururaj, P. N., Reddy, N. B. P., & Rajam, R. (2024). Recent advances in protein-polysaccharide based biocomposites and their potential applications in food packaging: A review. *International Journal of Biological Macromolecules*, •••, 131757. PMID: 38657934

Thomas, D., & Thomas, S. (2013). Chemical modification of chitosan and its biomedical application. *Biopolymer Nanocomposites: Processing, Properties, and Applications*, 33–51.

Thomas, L. M., & Sreekanth, M. S. (2025). Enhanced water-induced shape memory and swelling characteristics of graphene oxide-cellulose nanofiber reinforced polyvinyl alcohol nanocomposites. *Journal of Vinyl and Additive Technology, 31*(3), 728–755.

Thomas, S., Mishra, R. K., & Asiri, A. M. (Eds.). (2022). *Sustainable polymer composites and nanocomposites*. Springer.

Tiwari, K. N. (2022). Nano technology based P fertilizers for higher efficiency and agriculture sustainability. *ANNALS OF PLANT AND SOIL RESEARCH, 24*(2), 198–207. DOI: 10.47815/apsr.2022.10149

Tojima, T., Katsura, H., Han, S. M., Tanida, F., Nishi, N., & Tokura, S. (1998). Preparation of a β-cyclodextrin-linked chitosan derivative via reductive amination. *Journal of Polymer Science. Part A, Polymer Chemistry, 36*(11), 1965–1968.

Tripathy, S. (2023). Top-down and Bottom-up Approaches for Synthesis of Nanoparticles. Nano biomaterials, 92–130. https://doi.org/DOI: 10.21741/9781644902370-4

Trotta, F., Da Silva, S., Massironi, A., Mirpoor, S. F., Lignou, S., Ghawi, S. K., & Charalampopoulos, D. (2023). Silver bionanocomposites as active food packaging: Recent advances & future trends tackling the food waste crisis. *Polymers, 15*(21), 4243. PMID: 37959923

Tsakiridis, N. L., Samarinas, N., Kalopesa, E., & Zalidis, G. C. (2023). Cognitive Soil Digital Twin for Monitoring the Soil Ecosystem: A Conceptual Framework. *Soil Systems*, *7*(4), 88. DOI: 10.3390/soilsystems7040088

Tyagi, P. K., Gupta, S., Tyagi, S., Kumar, M., Pandiselvam, R., Daştan, S. D., Sharifi-Rad, J., Gola, D., & Arya, A. (2021). Green Synthesis of Iron Nanoparticles from Spinach Leaf and Banana Peel Aqueous Extracts and Evaluation of Antibacterial Potential. *Journal of Nanomaterials*, *2021*, 1–11. DOI: 10.1155/2021/4871453

Uday, V., Harikrishnan, P. S., Deoli, K., Zitouni, F., Mahlknecht, J., & Kumar, M. (2022). Current trends in production, morphology, and real-world environmental applications of biochar for the promotion of sustainability. *Bioresource Technology*, *359*, 127467. PMID: 35710048

Uddin, Md. H., Mulla, M. H., Abedin, T., Manap, A., Yap, B. K., Rajamony, R. K., Shahapurkar, K., Khan, T. M. Y., Soudagar, M. E. M., & Nur-E-Alam, M. (2025). Advances in natural fiber polymer and PLA composites through artificial intelligence and machine learning integration. *Journal of Polymer Research*, *32*(3), 76. DOI: 10.1007/s10965-025-04282-7

Udomkun, P., Chandi, K., Boonupara, T., & Kaewlom, P. (2024). Innovative approaches: Exploring nano-biochar technology's impact on soil properties, alachlor retention, and microbial populations. *Environmental Technology & Innovation*, *35*, 103659.

Ujcic, A., Krejcikova, S., Nevoralova, M., Zhigunov, A., Dybal, J., Krulis, Z., Fulin, P., Nyc, O., & Slouf, M. (2020). Thermoplastic starch composites with titanium dioxide and vancomycin antibiotic: Preparation, morphology, thermomechanical properties, and antimicrobial susceptibility testing. *Frontiers in Materials*, *7*, 9. DOI: 10.3389/fmats.2020.00009

UPM Biomedicals. (2024). FibGel natural injectable hydrogel for soft tissue repair. Company press release, October 2024.

Uskoković, V., & Bertassoni, L. E. (2010). Nanotechnology in dental sciences: Moving towards a finer way of doing dentistry. *Materials (Basel), 3*(3), 1674–1691. DOI: 10.3390/MA3031674 PMID: 27103959

Usman, A., Hussain, Z., Riaz, A., & Khan, A. N. (2016). Enhanced mechanical, thermal and antimicrobial properties of poly (vinyl alcohol)/graphene oxide/starch/silver nanocomposites films. *Carbohydrate Polymers, 153*, 592–599. PMID: 27561532

Utracki, L. A., Sepehr, M., & Boccaleri, E. (2007). Synthetic layered nanoparticles for polymeric nanocomposites (PNCs). *Polymers for Advanced Technologies, 18*, 1–37.

Vatansever, E., Arslan, D., & Nofar, M. (2023). Recent advances in biodegradable nanocomposites: A comprehensive review. *International Journal of Biological Macromolecules, 252*, 126427.

Venkateswarlu, S., Yoon, M., & Kim, M. (2021). An environmentally benign synthesis of Fe3O4 nanoparticles to Fe3O4 nanoclusters: Rapid separation and removal of Hg(II) from an aqueous medium. *Chemosphere, 286*(Pt 2), 131673. DOI: 10.1016/j.chemosphere.2021.131673 PMID: 34358889

Verified Market Reports. (2024). Plant-based nanocellulose market growth & trends 2024. Retrieved from https://www.verifiedmarketreports.com/product/plant-based-nanocellulose-market/

Verlinden, R. A. J., Hill, D. J., Kenward, M. A., Williams, C. D., & Radecka, I. (2007). Bacterial synthesis of biodegradable polyhydroxyalkanoates. *Journal of Applied Microbiology, 102*, 1437–1449. PMID: 17578408

Vidal, C. L. A., & Govan, J. E. (2024). Machine Learning Techniques for Improving Nanosensors in Agroenvironmental Applications. *Agronomy (Basel), 14*(2), 2. Advance online publication. DOI: 10.3390/agronomy14020341

Vilaplana, F., Strömberg, E., & Karlsson, S. (2010). Environmental and resource aspects of sustainable biocomposites. *Polymer Degradation & Stability, 95*(11), 2147–2161. DOI: 10.1016/j.polymdegradstab.2010.07.016

Vishnuvarthanan, M. (2024). Bio-nanocomposite and biofilm: An overview. In Elsevier Reference Module in Materials Science and Materials Engineering. https://doi.org/DOI: 10.1016/B978-0-443-15738-7.00051-9

Volk, A. A., Epps, R. W., Yonemoto, D. T., Masters, B. S., Castellano, F. N., Reyes, K. G., & Abolhasani, M. (2023). AlphaFlow: Autonomous discovery and optimization of multi-step chemistry using a self-driven fluidic lab guided by reinforcement learning. *Nature Communications, 14*(1), 1403. DOI: 10.1038/s41467-023-37139-y PMID: 36918561

Wahab, A., Muhammad, M., Ullah, S., Abdi, G., Shah, G. M., Zaman, W., & Ayaz, A. (2024). Agriculture and environmental management through nanotechnology: Eco-friendly nanomaterial synthesis for soil-plant systems, food safety, and sustainability. *The Science of the Total Environment*, *926*, 171862. DOI: 10.1016/j.scitotenv.2024.171862 PMID: 38527538

Wai, K. P., Chia, M. Y., Koo, C. H., Huang, Y. F., & Chong, W. C. (2022). Applications of deep learning in water quality management: A state-of-the-art review. *Journal of Hydrology (Amsterdam)*, *613*, 128332. DOI: 10.1016/j.jhydrol.2022.128332

Wang, B., Torres-Rendon, J. G., Yu, J., Zhang, Y., & Walther, A. (2015). Aligned Bioinspired Cellulose Nanocrystal-Based Nanocomposites with Synergetic Mechanical Properties and Improved Hygromechanical Performance. *ACS Applied Materials & Interfaces*, *7*(8), 4595–4607. PMID: 25646801

Wang, C., Ji, Y., Cao, X., Yue, L., Chen, F., Li, J., Yang, H., Wang, Z., & Xing, B. (2022). Carbon Dots Improve Nitrogen Bioavailability to Promote the Growth and Nutritional Quality of Soybeans under Drought Stress. *ACS Nano*, *16*(8), 12415–12424. DOI: 10.1021/acsnano.2c03591 PMID: 35946591

Wang, G., Li, L., Medina, L., Harish, S., Liu, J., Xu, B., Guo, R., Shao, C., Kodama, T., Berglund, L. A., & Shiomi, J. (2025). Large Anisotropy of Thermal Conductivity in Oriented Cellulose–Clay Composites. *ACS Omega*, *10*(25), 26560–26566. PMID: 40620995

Wang, H., Gao, B., Wang, S., Fang, J., Xue, Y., & Yang, K. (2015). Removal of Pb (II), Cu (II), and Cd (II) from aqueous solutions by biochar derived from KMnO4 treated hickory wood. *Bioresource Technology*, *197*, 356–362. PMID: 26344243

Wang, H., Li, D., & Zhang, R. (2023). Sustainable nanocellulose-based composites for advanced applications: Recent progress and future perspectives. *Composites. Part A, Applied Science and Manufacturing*, *168*, 107459.

Wang, H., Zhang, L., & Zhao, J. (2023). Application of a Fusion Attention Mechanism-Based Model Combining Bidirectional Gated Recurrent Units and Recurrent Neural Networks in Soil Nutrient Content Estimation. *Agronomy (Basel)*, *13*(11), 2724. DOI: 10.3390/agronomy13112724

Wang, L., Hu, C., & Shao, L. (2017). The antimicrobial activity of nanoparticles: Present situation and prospects for the future. *International Journal of Nanomedicine*, *12*, 1227–1249. DOI: 10.2147/IJN.S121956 PMID: 28243086

Wang, L., Kelly, P. V., Ozveren, N., Zhang, X., Korey, M., Chen, C., & Gardner, D. J. (2023). Multifunctional polymer composite coatings and adhesives by incorporating cellulose nanomaterials. *Matter*, *6*(2), 344–372.

Wang, N., Ding, E., & Cheng, R. (2007). Surface modification of cellulose nanocrystals. *Frontiers of Chemical Engineering in China*, *1*(3), 228–232.

Wang, R., Feng, Y., Li, D., Li, K., & Yan, Y. (2024). Towards the sustainable production of biomass-derived materials with smart functionality: A tutorial review. *Green Chemistry*.

Wang, W., Ma, X., Jiang, P., Hu, L., Zhi, Z., Chen, J., Ding, T., Ye, X., & Liu, D. (2016). Characterization of pectin from grapefruit peel: A comparison of ultrasound-assisted and conventional heating extractions. *Food Hydrocolloids*, *61*, 730–739. DOI: 10.1016/j.foodhyd.2016.06.019

Wang, X., Bai, H., Yao, Z., Liu, A., & Shi, G. (2010). Electrically conductive and mechanically strong biomimetic chitosan/reduced graphene oxide composite films. *Journal of Materials Chemistry*, *20*(41), 9032–9036.

Wang, X., Tarahomi, M., Sheibani, R., Xia, C., & Wang, W. (2023). Progresses in lignin, cellulose, starch, chitosan, chitin, alginate, and gum/carbon nanotube (nano) composites for environmental applications: A review. *International Journal of Biological Macromolecules*, *241*, 124472. DOI: 10.1016/j.ijbiomac.2023.124472 PMID: 37076069

Wang, Y. Y., Liu, Y. X., Lu, H. H., Yang, R. Q., & Yang, S. M. (2018). Competitive adsorption of Pb (II), Cu (II), and Zn (II) ions onto hydroxyapatite-biochar nanocomposite in aqueous solutions. *Journal of Solid State Chemistry*, *261*, 53–61.

Wani, N. R., Dar, A. H., Dash, K. K., Pandey, V. K., Srivastava, S., Jan, S. Y., & Sabahi, N. (2023). Recent advances in the production of bionanomaterials for development of sustainable food packaging: A comprehensive review. *Environmental Research*, •••, 116948. PMID: 37611789

Watanabe, Y., Mukai, B., & Kawamura, K. I.. (2002). Preparation and evaluation of press-coated aminophylline tablet using crystalline cellulose and polyethylene glycol in the outer shell for timed-release dosage forms. *Yakugaku Zasshi*, *122*(2), 157–162. PMID: 11857956

Wei, Y. J., Silikas, N., Zhang, Z. T., & Watts, D. C. (2011). Hygroscopic dimensional changes of self-adhering and new resin-matrix composites during water sorption/desorption cycles. *Dental Materials*, *27*(3), 259–266. DOI: 10.1016/J.DENTAL.2010.10.015 PMID: 21111476

Williams, D. F. (2008). On the mechanisms of biocompatibility. *Biomaterials, 29*(20), 2941–2953. DOI: 10.1016/j.biomaterials.2008.04.023 PMID: 18440630

Winters, D., Boakye, K., & Simske, S. (2022). Toward carbon-neutral concrete through biochar–cement–calcium carbonate composites: A critical review. *Sustainability, 14*(8), 4633.

Wong, M. S. (2004). In Lu, G. Q., & Zhao, X. S. (Eds.), *Nanoporous materials: Science and technology* (pp. 125–164). Imperial College Press.

Wronka, A., & Kowaluk, G. (2024). Multiphase Biopolymers Enriched with Suberin Extraction Waste: Impact on Properties and Sustainable Development. *Materials (Basel), 17*(22), 5472. PMID: 39597296

Wu, B., Wu, Y., Zhang, M., Guo, H., Liu, T., Lin, G., & Kuang, T. (2025). Enhanced Mechanical and Multifunctional Properties of GNPs/CNTs Hybridized PLA Nanocomposites by Implementing Dual-Processing of Pickering Emulsion-Melt Blending Methods. *Macromolecular Materials and Engineering, 310*(3), 2400306.

Xia, C., Liang, Y., Li, X., Al Garalleh, H., Garaleh, M., Hill, J. M., & Pugazhendhi, A. (2023). Remediation competence of nanoparticles amalgamated biochar (nanobiochar/nanocomposite) on pollutants: A review. *Environmental Research, 218*, 114947. PMID: 36462692

Xiaobo, C., & Samuel, S. M. (2007). Titanium dioxide nanomaterials: Synthesis, properties, modifications, and applications. *Chemical Reviews, 107*, 2891–2959. PMID: 17590053

Xinming, L., Qingshen, S., Huajie, W., Lei, Z., & Jin, Y. W. (2005). Microspheres of corn protein, zein, for an ivermectin drug delivery system. *Biomaterials, 26*, 109–115. PMID: 15193886

Xiuquan, Z., Wang, G., Li, Z., Zheng, D., & Song, H. (2025). *Soil Organic Matter Estimation Using Fusion Raman Spectroscopy Based on Multimodal Deep Learning* (SSRN Scholarly Paper No. 5244519). Social Science Research Network. https://doi.org/DOI: 10.2139/ssrn.5244519

Xu, D., Wang, S., Berglund, L. A., & Zhou, Q. (2021). Surface Charges Control the Structure and Properties of Layered Nanocomposite of Cellulose Nanofibrils and Clay Platelets. *ACS Applied Materials & Interfaces, 13*(3), 4463–4472. PMID: 33428385

Xueping, X. (2015). *Organic fertilizer capable of improving fertilizer efficiency of soil and preparation method thereof.*

Xue, Y., Mou, Z., & Xiao, H. (2023). Nanocellulose as a sustainable biomass material: Structure, properties, present status and future prospects in biomedical applications. *Nanomaterials (Basel, Switzerland), 13*(18), 2611. PMID: 37764641

Xu, J., Xie, Y., Yao, Q., Lv, L., & Chu, H. (2024). Advances in sustainable nano-biochar: Precursors, synthesis methods and applications. *Nanoscale, 16*(32), 15009–15032. PMID: 39041285

Yang, E., Miao, S., Zhong, J., Zhang, Z., Mills, D. K., & Zhang, L. G. (2018). Bio-based polymers for 3D printing of bioscaffolds. *Polymer Reviews (Philadelphia, Pa.), 58*(4), 668–687. DOI: 10.1080/15583724.2018.1484761 PMID: 30911289

Yang, J., Hou, B., Wang, J., Tian, B., Bi, J., Wang, N., Li, X., & Huang, X. (2019). Nanomaterials for the Removal of Heavy Metals from Wastewater. *Nanomaterials (Basel, Switzerland), 9*(3), 3. Advance online publication. DOI: 10.3390/nano9030424 PMID: 30871096

Yang, J., Li, C., Lo, L. S. H., Zhang, X., Chen, Z., Gao, J. U. C., Dai, Z., Nakaoka, M., Yang, H., & Cheng, J. (2024). Artificial Intelligence-Assisted Environmental DNA Metabarcoding and High-Resolution Underwater Optical Imaging for Noninvasive and Innovative Marine Environmental Monitoring. *Journal of Marine Science and Engineering, 12*(10), 1729. DOI: 10.3390/jmse12101729

Yang, W., Qi, G., Kenny, J. M., Puglia, D., & Ma, P. (2023). Bio-based polymers for 3D printing of bioresorbable vascular scaffolds. *Polymers, 15*(16), 3336. PMID: 37631394

Yang, X., Zhou, Y., Hu, J., Zheng, Q., Zhao, Y., Lv, G., & Liao, L. (2024). Clay minerals and clay-based materials for heavy metals pollution control. *The Science of the Total Environment, 954*, 176193. DOI: 10.1016/j.scitotenv.2024.176193 PMID: 39278488

Yano, H., Sugiyama, J., & Nakagaito, A. N.. (2005). Optically transparent composites reinforced with networks of bacterial nanofibers. *Advanced Materials, 17*(2), 153–155.

Yasmin, K., & Hossain, Md. S., & Li, W. C. (2024). Simultaneous immobilization strategy of anionic metalloids and cationic metals in agricultural systems: A review. *Chemosphere, 364*, 143106. DOI: 10.1016/j.chemosphere.2024.143106 PMID: 39153530

Yeo, J. C. C., Muiruri, J. K., Fei, X., Wang, T., Zhang, X., Xiao, Y., & Li, Z. (2024). Innovative biomaterials for food packaging: Unlocking the potential of polyhydroxyalkanoate (PHA) biopolymers. *Biomaterials Advances*, •••, 213929. PMID: 39024863

Yin, H., Cao, Y., Marelli, B., Zeng, X., Mason, A. J., & Cao, C. (2021). Soil Sensors and Plant Wearables for Smart and Precision Agriculture. *Advanced Materials, 33*(20), 2007764. DOI: 10.1002/adma.202007764 PMID: 33829545

Yosri, N., Khalifa, S. A., Attia, N. F., Du, M., Yin, L., Abolibda, T. Z., & El-Seedi, H. R. (2024). Advancing sustainability in the green engineering of nanocomposites based on marine-derived polymers and their applications: A comprehensive review. *International Journal of Biological Macromolecules*, •••, 133249. PMID: 38906361

Youden, B., Jiang, R., Carrier, A. J., Servos, M. R., & Zhang, X. (2022). A Nanomedicine Structure–Activity Framework for Research, Development, and Regulation of Future Cancer Therapies. *ACS Nano, 16*(11), 17497–17551. DOI: 10.1021/acsnano.2c06337 PMID: 36322785

Young, R. J., Kinloch, I. A., Gong, L., & Novoselov, K. S. (2012). The mechanics of graphene nanocomposites: A review. *Composites Science and Technology, 72*(12), 1459–1476.

Youssef, A. M., & El-Sayed, S. M. (2022). Bionanocomposites materials for food packaging applications: Concepts and future outlook. *Carbohydrate Polymers, 295*, 19–27. PMID: 29773372

Yu, F., Tian, F., Zou, H., Ye, Z., Peng, C., Huang, J., Zheng, Y., Zhang, Y., Yang, Y., Wei, X., & Gao, B. (2021). ZnO/biochar nanocomposites via solvent free ball milling for enhanced adsorption and photocatalytic degradation of methylene blue. *Journal of Hazardous Materials, 415*, 125511. DOI: 10.1016/j.jhazmat.2021.125511 PMID: 33740715

Yu, J., Wang, D., Geetha, N., Khawar, K. M., Jogaiah, S., & Mujtaba, M. (2021). Current trends and challenges in the synthesis and applications of chitosan-based nanocomposites for plants: A review. *Carbohydrate Polymers, 261*, 117904. PMID: 33766382

Yu, X., Keitel, C., Zhang, Y., Wangeci, A. N., & Dijkstra, F. A. (2022). Global meta-analysis of nitrogen fertilizer use efficiency in rice, wheat and maize. *Agriculture, Ecosystems & Environment, 338*, 108089. DOI: 10.1016/j.agee.2022.108089

Zaaba, N. F., & Jaafar, M. (2020). *A review on degradation mechanisms of polylactic acid: Hydrolytic, photodegradative, microbial, and enzymatic degradation*. https://doi.org/DOI: 10.1002/pen.25511

Zafar, R., Zia, K. M., Tabasum, S., Jabeen, F., Noreen, A., & Zuber, M. (2016). Polysaccharide based bionanocomposites, properties and applications: A review. *International Journal of Biological Macromolecules, 92*, 1012–1024. DOI: 10.1016/J.IJBIOMAC.2016.07.102 PMID: 27481340

Zhai, J., Duan, S., Luo, B., Jin, X., Dong, H., & Wang, X. (2024). Classification techniques of ion selective electrode arrays in agriculture: A review. *Analytical Methods, 16*(47), 8068–8079. DOI: 10.1039/D4AY01346H PMID: 39543972

Zhai, Y., Feng, Y., & Zhou, L. (Eds.). (2023). *Sustainable nanocellulose and nanohydrogels from natural sources: Chemistry, characterization and applications*. Elsevier.

Zhang, F., Si, Y., Yu, J., & Ding, B. (2023). Electrospun porous engineered nanofiber materials: A versatile medium for energy and environmental applications. *Chemical Engineering Journal, 456*, 140989.

Zhang, K., Feng, Q., Fang, Z., Gu, L., & Bian, L. (2022). Structurally dynamic hydrogels for biomedical applications: Pursuing a fine balance between macroscopic stability and microscopic dynamics. *Chemical Reviews, 122*(18), 11149–11193. PMID: 34189903

Zhang, P., Guo, Z., Ullah, S., Melagraki, G., Afantitis, A., & Lynch, I. (2021). Nanotechnology and artificial intelligence to enable sustainable and precision agriculture. *Nature Plants, 7*(7), 864–876. DOI: 10.1038/s41477-021-00946-6 PMID: 34168318

Zhang, X., Wells, M., Niazi, N. K., Bolan, N., Shaheen, S., Hou, D., & Wang, Z. (2022). Nanobiochar-rhizosphere interactions: Implications for the remediation of heavy-metal contaminated soils. *Environmental Pollution, 299*, 118810. PMID: 35007673

Zhang, Y., Gao, X., Zhi, L., Liu, X., Jiang, W., Sun, Y., & Yang, J. (2014). The synergetic antibacterial activity of Ag islands on ZnO (Ag/ZnO) heterostructure nanoparticles and its mode of action. *Journal of Inorganic Biochemistry, 130*(1), 74–83. DOI: 10.1016/J.JINORGBIO.2013.10.004 PMID: 24176922

Zhang, Y., & Jiang, W. (2023). Effective strategies to enhance ultraviolet barrier ability in biodegradable polymer-based films/coatings for fruit and vegetable packaging. *Trends in Food Science & Technology*, •••, 104139.

Zhang, Y., Wang, S., & Ji, G. (2015). A Comprehensive Survey on Particle Swarm Optimization Algorithm and Its Applications. *Mathematical Problems in Engineering, 2015*, 1–38. DOI: 10.1155/2015/931256

Zhao, J., & Wan, S. (2023). Artificial Intelligence and Hyperspectral Modeling for Soil Management. In Clarke, N., Peng, D., & Clarke, J. L. (Eds.), *Innovation for Environmentally-friendly Food Production and Food Safety in China* (pp. 67–91). Springer Nature Singapore., DOI: 10.1007/978-981-99-2828-6_4

Zhao, R., Torley, P., & Halley, P. (2008). Emerging biodegradable materials: Starch- and protein-based bio-nanocomposites. *Journal of Materials Science*, *43*, 3058–3071.

Zhao, R., Torley, P., & Halley, P. J. (2008). Emerging biodegradable materials: Starch-and protein-based bio-nanocomposites. *Journal of Materials Science*, *43*(9), 3058–3071.

Zhao, Y., Qamar, S., Qamar, M., Bilal, M., & Iqbal, H. (2021). Sustainable remediation of hazardous environmental pollutants using biochar-based nanohybrid materials. *Journal of Environmental Management*, *300*, 113762. DOI: 10.1016/j.jenvman.2021.113762 PMID: 34543967

Zhou, J., Du, B., Liu, H., Cui, H., Zhang, W., Fan, X., Cui, J., & Zhou, J. (2020). The bioavailability and contribution of the newly deposited heavy metals (copper and lead) from atmosphere to rice (Oryza sativa L.). *Journal of Hazardous Materials*, *384*, 121285. DOI: 10.1016/j.jhazmat.2019.121285 PMID: 31577969

Zhou, T., Deng, J., Zeng, Y., Liu, X., Song, B., Ye, S., Li, M., Yang, Y., Wang, Z., & Zhou, C. (2024). Biochar Meets Single-Atom: A Catalyst for Efficient Utilization in Environmental Protection Applications and Energy Conversion. *Small*, *2404254*. Advance online publication. DOI: 10.1002/smll.202404254 PMID: 38984755

Zhou, Y., Tang, S., Zhang, Q., Jiang, Y., Tan, L., Xia, Z., Wang, Z., Liu, L., Zhang, Y., & Wang, G. (2024). Preparation and Characterization of Cellulose Nanocrystal-Doped Cellulose Triacetate Optical Films by Sol-Gel Process. *ChemistrySelect*, *9*(27), e202400262.

Zhuang, W., Liu, J., Zhang, J. H., Hu, B. X., & Shen, J. (2009). Preparation, characterization, and properties of TiO2/PLA nanocomposites by in situ polymerization. *Polymer Composites*, *30*(8), 1074–1080.

Ziabari, M., Mottaghitalab, V., & Haghi, A. K. (2009). Application of direct tracking method for measuring electrospun nanofiber diameter. *Brazilian Journal of Chemical Engineering*, *26*, 53–62.

About the Contributors

Kaizar Hossain, an ardent and dynamic personality, completed his Bachelor of Science (B.Sc) in Environment and Water Management from Burdwan University, West Bengal and Master of Science (M.Sc) in Environmental Studies from Andhra University, Andhra Pradesh. He is a doctorate with his Ph.D in Environmental Sciences from GITAM University collaboration with ILU-6, RTDD/BARC/DAE, Mumbai, and has awarded as Post-Doctoral Fellow by USM- TWAS Italy. Dr. Kaizar's expertise in pollution mitigation, waste management, and advanced bioremediation technologies has positioned him as a pioneer in crafting innovative solutions to the world's most critical environmental challenges. He has numerous publications in globally renowned Q1 journals. One of his core research focuses is bioremediation technology for wastewater treatment—a method that uses biological agents such as bacteria, algae, fungi, and plants to remove contaminants from wastewater. Dr. Kaizar is also at the forefront of exploring nanotechnology applications for pollution mitigation, pioneering sustainable technologies that are both effective and accessible for diverse global contexts. Dr. Kaizar's academic achievements and impactful contributions earned him a prestigious appointment as an Editorial Board Member of Scientific Reports, Nature Portfolio in 2024. In this capacity, he plays a vital role in guiding and promoting cutting-edge research that aligns with the United Nations' Sustainable Development Goals, particularly SDG 6: Clean Water and Sanitation. Beyond his research, Dr. Kaizar is a strong advocate for global collaboration in addressing environmental issues. His participation as a Guest Editor for Scientific Reports' special issue on "Bioremediation and Biotreatment of wastewater" underscores his dedication to fostering innovation and practical solutions for wastewater management, a critical area in the fight against water pollution and water scarcity. Dr. Kaizar's contributions reflect his unwavering commitment to environmental sustainability and highlight the importance of integrating advanced science with practical applications to create a cleaner, healthier world for future generations. He was the Member of International Steering Committee of 2nd 3rd &

4th International Conference on Sustainable Development, ICSD-2014, 2015, 2016, 2017, 2018, 2019 Rome, Italy. He is National Panel of Expert of Engineers Without Borders (EWB) – India and Life member of National Association for Application of Radio-isotopes and Radiation in Industry (NAARRI) Mumbai.

Akil Ahmad currently working at Prince Sattam bin Abdulaziz University, Saudi Arabia as Associate Professor in Chemistry and having the experience of seven years as Research Fellow, Teaching Fellow, Postdoc and Visiting Researcher from Universiti Teknologi Malaysia, Universiti Sains Malaysia, University of KwaZulu-Natal, South Africa and Universiti Kebangsaan Malaysia, Malaysia. He has completed Ph.D. in Analytical Chemistry (2011) with the topic "Modification of resin for their use in the separation, preconcentration and determination of metal ions" from Aligarh Muslim University (AMU), India. His research interest in the areas of environmental pollutants and their safe removal, synthesis of nanoparticles and Nano-sorbents (GO, CNT), photo-degradation and antimicrobial effects, water and wastewater treatment and adsorption and ion-exchange. He has published more than 200 research articles and chapters in the journals and publishers of international repute such as Scientific reports, Talanta, Chemical Engineering Journal, Journal of Industrial and Engineering Chemistry, Journal of Molecular Liquids etc. He has also edited more than ten books of Springer and Elsevier. H-index and citation in Scopus are 45 and 8892 and in Google scholar, H-index and citation are 41 and 6853. He is guest editor of many reputed journal namely Adsorption Science and Technology-Hindawi, Polymers MDPI, Frontiers in Environmental Chemistry and Journal of Chemistry, Hindawi.

Mohammad Jawaid is currently working as Professor at Chemical and Petroleum Department, College of Engineering, United Arab Emirates University, Al Ain, UAE, and also has been Fellow researcher, Visiting Professor at Universiti Putra Malaysia and Department of Chemical Engineering, College of Engineering, King Saud University, Riyadh, Saudi Arabia since June 2013. He has more than 20 years of experience in teaching, research, and industries. His area of research interests includes hybrid composites, lignocellulosic reinforced/filled polymer composites, advance materials: graphene/nanoclay/fire retardant, modification and treatment of lignocellulosic fibres and solid wood, biopolymers and biopolymers for packaging applications, nanocomposites and nanocellulose fibres, and polymer blends. So far, he has published 40 books, 65 book chapters, more than 500 peer-reviewed international journal papers, and several published review papers under top 25 hot articles in science direct during 2013–2019. He also obtained 3 Patents and 6 Copyrights. H-index and citation in Scopus are 123 and 60423 and in Google scholar, H-index and citation are 104 and 44917. Recently he joined as Chief Executive

Editor of Pertanika UPM Journals. He is founding Series Editor of Composite Science and Technology Book Series from Springer-Nature, and also Series Editor of Springer Proceedings in Materials, Springer-Nature and also International Advisory board member of Springer Series on Polymer and Composite Materials. He worked as guest editor of special issues of SN Applied Science, Frontiers in Sustainable Food Systems, Current Organic Synthesis and Current Analytical Chemistry, International Journal of Polymer Science, IOP Conference Proceeding. He also in Editorial Board Member of Journal of Polymers and The Environment, Journal of Natural fibres, Journal of Plastics Technology, Applied Science and Engineering Progress Journal, Journal of Asian Science, Technology and Innovation and the Recent Innovations in Chemical Engineering. Besides that, he is also reviewer of several high-impact international peer-reviewed journals of Elsevier, Springer, Wiley, Saga, ACS, RSC, Frontiers, etc. Presently, he is supervising 12 Ph.D. students (5 Ph.D. as Chairman, and 7 Ph.D. as Member) and 6 Master's students (1 Master as Chairman, and 5 Master as Member) in the fields of hybrid composites, green composites, nanocomposites, natural fiber-reinforced composites, nanocellulose, etc. 26 Ph.D. and 13 Master's students graduated under his supervision in 2014–2020. He has several research grants at university, national, and international levels on polymer composites of around 3 million Malaysian ringgits (USD 700,000). He also delivered plenary and invited talks in international conferences related to composites in India, Turkey, Malaysia, Thailand, the United Kingdom, France, Saudi Arabia, Egypt, and China. Besides that, he is also a member of technical committees of several national and international conferences on composites and material science. Recently Dr. Mohammad Jawaid received Excellent Academic Award in Category of International Grant-Universiti Putra Malaysia-2018 and also Excellent Academic Staff Award in industry High Impact Network (ICAN 2019) Award. Beside that Gold Medal-Community and Industry Network (JINM Showcase) at Universiti Putra Malaysia. He also Received Publons Peer Review Awards 2017, and 2018 (Materials Science), Certified Sentinel of science Award Receipient-2016 (Materials Science) and 2019 (Materials Science and Cross field). He is also Winner of Newton-Ungku Omar Coordination Fund: UK-Malaysia Research and Innovation Bridges Competition 2015. Recently he recognized with Fellow and Charted Scientist Award from Institute of Materials, Minerals and Mining (IOM), UK. He is also life member of Asian Polymer Association, and Malaysian Society for Engineering and Technology. He has professional membership of American Chemical Society (ACS), and Society for polymers Engineers (SPE), USA.

***

Abiodun Nathaniel is a committed academic, researcher and a publisher in microbiology. He earned his Bachelor of Science (BSc) in Microbiology from the University of Ilorin, where he developed a keen interest in scientific research and innovation. Currently, Nathaniel is a MTECH student at the Department of Microbiology, Federal University of Technology, Akure, where he continues to deepen his research in Environmental Microbiology, contributing to advancements in the field and expanding his expertise in Environmental Microbiology. His academic journey is driven by a passion for knowledge, a commitment to excellence, and a determination to solve real-world problems through scientific inquiry.

Bala Jeremiah David is a Professor of Microbiology at the Federal University of Technology (FUT), Minna, Niger State, Nigeria. He also serves as the Deputy Dean of the School of Life Sciences at FUT, Minna. Dr. Bala Jeremiah David obtained his PhD in Environmental Technology from Universiti Sains Malaysia (USM). He also holds a B.Tech (Hons) in Microbiology from FUT, Yola, Nigeria, and an MSc in Industrial Microbiology from ATBU, Bauchi. Area of Specialization: Environmental and industrial Microbiology, Microbial Water Quality Monitoring, Industrial Wastewater and Solid Waste Biotreatment Technology and Management, Bioremediation and Industrial Bioprocesses. Collaboration/Award: Lead-Researcher: Award of TET Fund Institution-Based Research Intervention (IBRI) Fund, N1,937,200.00, 2024. Co-Researcher: Award of TET Fund Institution-Based Research Intervention (IBRI) Fund, N1,908,500.00, 2024. Co-Researcher: Award of World Bank Grant for African Center of Excellence for Mycotoxin and Food Safety, $6 Million USD, 2018. Co-Researcher: Award of TET Fund Institution-Based Research Intervention (IBRI) Fund, N1 107,700.00, 2017. Co-Researcher: FUT Board of Research, (FUTBR Grant), N600 000.00, 2012 Membership of Professional Associations/Bodies: Nigerian Society for Microbiology(NSM) Biotechnology Society of Nigeria (BSN) American Society for Microbiology(ASM) Microbiology Society United Kingdom (MSUK)

Hari Shankar Biswas is Assistant Professor in Chemistry Surendranath College, Kolkata. Dr. Biswas is an esteemed Assistant Professor in Chemistry at Surendranath College, Kolkata, with a multifaceted academic background and a strong dedication to research and education. He embarked on his academic journey by obtaining his B.Sc. degree in Chemistry Honours from R.K.M.V.C. College in 2005, followed by his M.Sc. degree from Presidency College, Kolkata, in 2008. Dr. Biswas further pursued his scholarly pursuits by earning a Ph.D. in Material Science from the Saha Institute of Nuclear Physics, University of Calcutta, in 2015. His professional experience spans various roles, including a stint as a School teacher at Keshub Academy, Kolkata, from 2008 to 2010, before transitioning into academia as an Assistant Professor at Surendranath College. Dr. Biswas specializes in the research domain of Carbon thin Film, Graphene, and Graphene Oxide-based nano Composites, focusing on their synthesis, characterization, and application across Environmental, Biological, and Technological fields. Dr. Biswas's scholarly contributions are significant, with authorship of 16 book chapters and over 30 papers published in esteemed International Journals. He is actively involved in academic societies, holding life membership in the Indian Science Congress Association and the Indian Chemical Society, along with an affiliated membership with the Royal Society of Chemistry. In addition to his research and academic endeavors, Dr. Biswas plays a vital role in the dissemination of scientific knowledge as a series book Editor and reviewer for reputable international publishers and journals. He serves as an Editorial board member for the American Journal of Chemical Engineering, Reseapro Journal, Lincoln University College, etc., and is affiliated with Rotary International. Dr. Biswas's editorial contributions extend to editing various books, including titles such as "Emerging Concepts in Chemical and Biological Science" and "Modern Approaches in Chemical and Biological Sciences," among others. Through his comprehensive involvement in research, education, and editorial responsibilities, Dr. Hari Shankar Biswas continues to make valuable contributions to the field of Chemistry, fostering advancements and innovations for the betterment of society.

Susanta Das is a seasoned educator with a proven record of teaching and mentoring diverse students, working at multiple institutions, and providing service to the various initiatives of organizations. He received his Ph.D. and M.A. degrees from Western Michigan University (WMU), USA, and M.Sc. degree from Banaras Hindu University (BHU), India all in Physics. He continued his research as a Marie-Curie post-doctoral fellow at Stockholm University, Stockholm, Sweden on beam diagnostics for the DESIREE (Double ElectroStatic Ion Ring ExpEriment) in the project DITANET (Diagnostic Techniques for particle Accelerators – a Marie-Curie initial training NETwork), at the Indian Institute of Science Education and Research-Kolkata, India on high-pressure physics as a Project Scientist-B, and at the University of Electro-Communications, Tokyo, Japan on ion-surface interactions as a Post-doctoral fellow. Throughout his career, Dr. Das worked and collaborated with many researchers, Ph.D., master, and visiting students from different countries. He further visited many countries to discuss research and an international conference participant (UK, Italy, Germany, Greece, Belgium, Bulgaria, Romania, Brazil etc.). He co-authored several research articles in WoS/Scopus indexed international journals, conference proceedings, and scholarly book chapters published by renowned international publishing houses. Beyond his research endeavors, Dr. Das has a storied history of service to academic and administrative committees, exhibiting his commitment to the growth and development of educational institutions. His experience includes tenure at Central University South Bihar, Sri Sri University, and P.K. University before assuming his current role at Ajeenkya DY Patil University. He received the Marie-Curie post-doctoral fellowship in Sweden, Science Academies' Summer Research Fellowship in India, Gwen Frostic Doctoral Fellowship, Department Graduate Research and Creative Scholar Award by WMU, and Leo R. Parpart Doctoral Fellowship by Dept. of Physics, WMU, among many others, throughout his academic journey. At present, Dr. Das continues to delve into cutting-edge fields, with a keen interest in nanotechnology, UN SDGs, and data science. His multifaceted contributions, spanning teaching, research, and administrative leadership, showcase a dedicated professional who is instrumental in advancing the vision and mission of the institutions he serves.

Nathaniel N. Friday is a Microbiologist passionate about Food and industrial processes especially as it pertains to indigenous sources. Alongside his colleagues, several publications have been made on the topic. Currently, Nathaniel Friday is carrying out more research on ways to improve the production of indigenous fermented foods meeting global best practices especially microbiologically. He resides in Abuja.

D. Gnanasangeetha, currently serve as Assistant Professor in the Department of Chemistry at PSNA College of Engineering and Technology, Dindigul. She possesses 15 years of teaching experience and earned her Ph.D. from Bharathiar University, Coimbatore. Her research portfolio, has been authored with 32 research papers among that 6 indexed in SCI, 18 in Scopus and 6 in WoS. Acting as Reviewer for few SCI and Scopus journal. Have received one Copyright and published one patent through PSNA IPR cell.

Job Oloruntoba Samuel is a dedicated microbiologist with a strong academic background in Environmental Microbiology. He holds a Bachelor of Science degree in Microbiology from Prince Abubakar Audu University (formerly Kogi State University), a Postgraduate Diploma in Education from the University of Maiduguri, and a Master of Technology (M. Tech.) in Environmental Microbiology from the Federal University of Technology, Minna, where he graduated with distinction. His research interests span across biosciences, environmental microbiology, biostatistics, computational biology, biotechnology, and bioinformatics. Notably, her master's research focused on "Faecal Contamination and Identification of Pathogenic Escherichia coli from Drinking Water Sources in Maikunkele, Niger State, Nigeria."

Sunil Sankathala is pursuing B.Tech. degree in computer science in the School of Engineering of Ajeenkya DY Patil University, Pune, Maharashtra, India. He actively participates in various extracurricular activities, such as drama, technical, and cultural programs of the university. He participated in the Global Hyperloop Competition (GHC) at IIT Madras and significantly contributed to the overall performance of the team Vac-Vectoor. The competition was held from 21st through 24th February. The team achieved 2nd position in business and socio-economic impact category (Innoquest) of hyperloop and Special Mention Certificate in Traction sub-system (DesignX). His research interests are Artificial Intelligence, Data Science, and UNSDG.

Amit Kumar Kundu is Assistant Professor & HOD Department of Chemistry Sripat Singh College, Jiaganj, Murshidabad. Dr. Amit Kumar Kundu is an Assistant professor and HOD at the Department of Chemistry, Sripat Singh College, Jiaganj, Murshidabad. Dr. Kundu did B.Sc. from Krishnagar Government College, University of Kalyani in 2007 and completed MSc. Degree from Department of Chemistry, University of Kalyani in 2009. After that he received his PhD in Analytical Chemistry in 2017 from University of Kalyani. His research mainly focuses on Environmental Analytical Chemistry, carbon nano materials, Graphene, Graphene Oxide-basednano composites Synthesis, Characterization and its application. He has huge experience in Groundwater and surface water onsite sampling, sediment sampling from vadose zone and aquifer. He is also proficient in operating and managing various advanced analytical instruments. To date, he has published near about eighteen peer-reviewed journal papers and he is the author of five book chapters. He received World Environment and Livelihood Award form 'International Foundation for Environment and Ecology'.

Dilip Kumar Maiti is an eminent chemist with a distinguished academic and professional career. He completed his B.Sc. in Chemistry in 1991 from Ramakrishna Mission Vidyamandir, University of Calcutta, followed by an M.Sc. in 1993 from the University College of Science, Calcutta University. In 1998, he earned his Ph.D. from the Indian Institute of Chemical Biology, Jadavpur University. Dr. Maiti pursued postdoctoral research at Jadavpur University (2000–2002) and later at Wayne State University, Detroit, USA (2002–2004). He began his career as a Scientist in the R&D departments of RPG Life Sciences and Ciba India. Transitioning to academia, he served as a Senior Lecturer at Dumkal College, Kalyani University, before joining the University of Calcutta as a Professor in 2005. Currently, he is the Vice-Chancellor of Biswa Bangla Biswabidyalay, Bolpur. Dr. Maiti's research interests include Organic Synthesis and Material Chemistry. He has been honored with several prestigious awards, including the FRSC (London), UGC Mid-Career Award (2017), and Basudev Banerjee Memorial Award. He is actively involved with numerous scientific societies and editorial boards, contributing significantly to the global scientific community.

Murali Monangi Ph.D (Civil and Environmental Engineering) from JNTU, Hyderabad. M.E..(Environmental Engineering) at Andhra University College of Engineering, Visakhapatnam. B.Tech. (Civil Engineering) at JNTU College of Engineering, Kakinada. Teaching Engineering students since 1988. Served as Principal of Engineering College for 6 years. Completed successfully one UGC major research project one DST-NRDMS research project as Principal Investigator. More than 20 SCOPUS publications.

Innocent Ojeba Musa is a notable Ph.D holder in Environmental Microbiology in the Department of Microbiology, Federal University of Technology, Minna, Nigeria. He obtained his first degree (B.Tech) and Master Degree (M.Tech) from the same Department in 2010 and 2015 respectively. He has demonstrated his research expert in the use of Phytoremediation technique for Hydrocarbon pollution clean up. Currently, he is the Head of Department of Biological Science, Skyline University Nigeria. his expertise spans environmental health, phytoremediation, vermicomposting, marine resources and bioremediation of pollutants soil including nano- plastics. Innocent's research is deeply rooted in innovative solutions for soil and water remediation, with significant contributions to microbial biotechnology and waste management. He is an award-winning researcher and has over 70 publications including book chapters, research and review articles of good international repute with high impact factor. Dr. Innocent's academic journey is complemented by a robust teaching career and active involvement in community services, where he leverages his knowledge to foster sustainability. He has Supervised Several Undergraduate and Post graduate Students.His contributions continue to impact both academia and the broader environmental sciences community. Innocent is a member of Nigerian Society for Microbiology and American Society for Microbiology among others.

Mustapha Adamu is a committed academic, researcher and a publisher in microbiology. He earned his Bachelor of Science (BSc) in Microbiology from the Usmanu Danfodiyo University, Sokoto, where he developed a keen interest in scientific research and innovation. He further pursued a Master of Technology (MTech) at the Federal University of Technology, Minna, expanding his expertise in Environmental Microbiology. Mustapha worked on the research "Bioremediation of Spent Engine Oil (SEO) by bacterial isolates" during his MTech research work. Currently, Mustapha is a PhD student at the Department of Microbiology, Federal University of Technology, Minna, where he continues to deepen his research in Environmental Microbiology, contributing to advancements in the field. His academic journey is driven by a passion for knowledge, a commitment to excellence, and a determination to solve real-world problems through scientific inquiry. Beyond his studies, Mustapha is a serious-minded and disciplined individual who values hard work, critical thinking, and continuous self-improvement. He believes in using education as a tool for societal development and is dedicated to mentoring and inspiring the next generation of scientists.

N.S.C. Chaitanya is an accomplished academician and mechanical engineering expert serving as an Associate Professor in the Department of Mechanical Engineering at Raghu Engineering College. His expertise spans across robotics, alternative fuels, energy systems, and non-conventional energy sources, making him a versatile contributor to the field of mechanical engineering. His research work has been particularly impactful in the field of alternative fuels and engine performance. He has published several significant papers in Scopus-indexed journals, with groundbreaking research on binary diesel/JME blends and their emissions characteristics. His innovative work on nano-metallic oxide additives and heterogeneous base catalysts has advanced the understanding of alternative fuel technology. As an Institution Innovation Ambassador, he has played a crucial role in fostering an environment of innovation and entrepreneurship. Dr. Chaitanya continues to contribute to the academic community through his research publications, conference presentations, and organization of technical events, making him a respected figure in mechanical engineering education.

Mrutyunjay Padhiary is an Assistant Professor in the Department of Agricultural Engineering at Assam University, Silchar, India, with a deep academic background. His previous positions included significant teaching and research duties at the Faculty of Technology, UBKV in West Bengal, and Tribhuvan University in Nepal, as well as research experience at IIT Kharagpur. He has been awarded multiple awards, including an International NUS Research Fellowship and an excellent score on the 2001 GATE Exam, demonstrating his proficiency in precision agriculture. He has made substantial contributions to biofuel production, artificial intelligence vision, aquaculture, precision agriculture, agricultural automation, and artificial intelligence. He is currently working on several research projects, which range from autonomous surface vessels to precision agriculture all-terrain vehicles.

Aishwarya R. is presently employed as an Analytical Technology Specialist in cell culture for a pharmaceutical company located in Ireland. With extensive experience in research projects and an academic publication on the bioactivities of phytochemicals, she has made significant contributions to this chapter. She eagerly anticipates the remarkable applications of bionanocomposites in the near future.

R. Venkatesh has established a distinguished academic career, culminating in significant achievements. He holds an M.Phil. in Physics, an M.Sc. in Applied Physics, and a B.Sc. in Applied Physics from esteemed institutions including Nehru Memorial College (Autonomous), Puthanampatti, Tiruchirappalli, and Adaikala Matha College, Vallam, Thanjavur, all affiliated with Bharathidasan University, Tiruchirappalli, Tamil Nadu, earned in 2006, 2005, and 2002 respectively. Currently, he is actively pursuing a part-time Ph.D. program in Physics at Nehru Memorial College (Autonomous), Puthanampatti, Tiruchirappalli, under Bharathidasan University, demonstrating his ongoing dedication to scholarly pursuits. As an Assistant Professor of Physics at PSNA College of Engineering and Technology (Autonomous) in Dindigul, Tamil Nadu, India, R. Venkatesh leverages over 19 years of teaching experience to enrich the academic environment. His commitment to the field is evident through numerous publications in respected journals and conferences, as well as authored chapters for reputable publishers. He is widely recognized in the academic community for his profound dedication and expertise.

N. Sampathkumar earned his graduation, post-graduation and Doctoral Degrees from Bharathidasan University Trichy in the field of Chemistry. He is currently serving as Professor in Department of Chemistry at SSM College of Arts and Science, Dindigul since 2022. He is a Life member of ISTE. He has published more than 20 research papers in reputed international journals including ACS and Luminescence. His fields of research include Phytochemistry, Crystallography, Nanomaterial, Material Science and Biological Science.

Aditya Shrivastav is currently studying B.Tech. in Computer Engineering at the School of Engineering, Ajeenkya DY Patil University. His research interests are centered around nanoparticle biosynthesis, exploring ionic liquids' use in various applications, and applying Artificial Intelligence to different fields. Aditya is particularly focused on understanding how these technologies can be utilized together to solve practical problems. He aims to advance research in these areas and contribute to innovative solutions that address real-world challenges. He recently participated at the Global Hyperloop Competition (GHC) at IIT Madras as a member of the team Vac-Vectoor. The event took place from 21st February through 24th February 2025. He contributed significantly to the performance of his team. The team Vac-Vectoor achieved 2nd position in Business and Socio-Economic Impact of Hyperloop in the Innoquest category and Special Mention Certificate in Traction sub-system under Design X category.

|T Srinivas - Contributing Author|is a distinguished academic in the Department of Civil Engineering at GITAM (Deemed to be University), Visakhapatnam. With

a career spanning over two decades, he has significantly contributed to the field of environmental engineering through both teaching and research. His scholarly work encompasses areas such as water and wastewater treatment, environmental pollution control, and sustainable construction practices. As a dedicated educator, Prof. Srinivas has mentored numerous undergraduate and postgraduate students, fostering a culture of academic excellence. His commitment to research is evident from his publications in reputed journals and his active participation in national and international conferences. Beyond academia, he has been instrumental in initiating community outreach programs aimed at promoting environmental awareness and sustainable development.

Kalyani Teku, serves as an Associate Professor in the Department of Mechanical Engineering at Raghu Engineering College (A) in Visakhapatnam, India. She completed her academic journey at Andhra University, Visakhapatnam, India, where she earned Bachelor of Engineering (BE), Master of Engineering (ME) and Doctor of Philosophy (PhD) in Mechanical Engineering. Her principal area of research interests includes Thermal, Bio fuels, Algae Biodiesel, CI Engines, Performance, Emissions and Optimization techniques. She published E-SCI in Elsevier, Scopus journals and also published some papers in national, international journals, conferences. She received one patent grant through Raghu Engineering College(A). She has actively participated in FDP sessions and published book chapter in Taylor and Francis. She actively contributes to the scholarly community as a reviewer for peer-reviewed journals, providing valuable feedback and evaluation of research papers. She has membership from International Association of Engineers (IAENG) and Institute for Engineering Research and Publication (IFERP). She can be contacted at email: kalyani.teku@raghuenggcollege.in kalyanithermal@gmail.com

V Surya Subrahmanyam is having 26 years of experience, present he is working as Assistant professor in department of Mechanical engineering. He is pursuing his PhD in KIT University, Bhubaneswar. He research area is on Nano composites. He participated in several workshops and FDP

Index

A

Agricultural Applications 181, 182, 183, 184, 185, 188, 192, 193, 195, 196, 197, 280, 303, 315, 321
Antimicrobial activity 14, 65, 77, 107, 108, 121, 122, 128, 133, 182, 190, 192, 240, 241, 246, 247, 249, 250
Artificial intelligence 144, 275, 277, 283, 285, 289, 290, 291, 292, 295, 296, 300, 303, 304, 305, 307, 308, 310, 311, 313, 314, 315, 318, 319, 320, 321, 322, 324, 325

B

Bio-based Matrices 31, 135, 139, 140, 142, 167, 170, 203, 205, 207, 230, 279, 284
Biochar 191, 194, 277, 281, 282, 283, 289, 291, 292, 299, 301, 308, 309, 312, 313, 318, 320, 327, 328, 329, 330, 331, 332, 333, 334, 335, 336, 337, 338, 339, 340, 341, 342, 343, 344, 345, 346, 347, 348, 349, 350, 351, 352
Biodegradability 3, 22, 31, 32, 36, 38, 47, 59, 61, 63, 68, 70, 124, 138, 140, 155, 158, 159, 180, 182, 183, 187, 189, 191, 201, 207, 210, 211, 212, 215, 216, 221, 224, 225, 226, 228, 229, 230, 231, 233, 234, 240, 243, 244, 246, 254, 255, 256, 281, 282, 316
Biodegradable 2, 3, 4, 9, 15, 16, 17, 18, 19, 20, 21, 22, 23, 24, 32, 39, 40, 41, 49, 51, 52, 54, 55, 56, 57, 61, 63, 65, 66, 68, 71, 72, 74, 75, 77, 124, 137, 139, 154, 155, 158, 172, 174, 175, 181, 184, 186, 195, 201, 204, 205, 206, 207, 210, 215, 216, 223, 225, 226, 228, 230, 231, 234, 236, 239, 240, 241, 243, 246, 249, 250, 255, 257, 258, 261, 263, 270, 271, 272, 273, 279, 280, 283, 284, 285, 292
Biodegradable polymers 4, 17, 18, 52, 54, 55, 56, 61, 65, 66, 75, 174, 234, 236, 272, 273, 280, 283, 292
Biomedical Applications 14, 17, 29, 43, 50, 51, 52, 54, 55, 63, 66, 67, 68, 72, 74, 75, 84, 92, 110, 124, 129, 130, 159, 172, 173, 175, 211, 233, 263, 266, 267, 349
Biomineralization 113, 118
Biomolecules 5, 22, 44, 112, 114, 115, 116, 117, 125, 126, 130, 182, 279, 285
Bio-nano composites 275, 276, 278, 279, 280, 281, 283, 284, 285, 286, 287, 289, 290, 291, 292, 293, 294, 295, 296, 299, 300, 302, 303, 304, 305, 306
Bionanocomposites 1, 2, 5, 9, 10, 13, 18, 52, 60, 61, 63, 65, 72, 75, 77, 78, 132, 133, 175, 235, 264, 265, 269, 270, 272, 320
Bio-Nanocomposites 3, 4, 5, 6, 10, 11, 13, 14, 19, 20, 22, 23, 24, 25, 26, 27, 28, 29, 30, 31, 32, 35, 36, 37, 38, 39, 40, 42, 43, 44, 45, 46, 47, 48, 49, 50, 51, 54, 57, 59, 65, 66, 75, 76, 78, 124, 131, 132, 160, 166, 175, 176, 200, 203, 204, 205, 206, 207, 208, 209, 210, 211, 212, 213, 214, 215, 216, 217, 218, 219, 220, 221, 222, 223, 224, 225, 226, 227, 228, 229, 230, 231, 232, 233, 234, 236, 239, 240, 241, 242, 244, 245, 246, 247, 248, 249, 250, 251, 252, 253, 254, 255, 256, 257, 258, 259, 260, 261, 262, 264, 266, 269, 270, 271, 280, 281, 285, 315, 316, 327
Biopolymers 1, 3, 4, 6, 9, 13, 17, 24, 26, 30, 31, 32, 36, 47, 50, 53, 59, 60, 65, 66, 72, 124, 164, 173, 180, 181, 185, 186, 187, 188, 191, 196, 197, 201, 206, 213, 214, 215, 216, 221, 222, 223, 225, 226, 227, 228, 240, 241, 242, 243, 244, 245, 246, 247, 248, 249, 252, 253, 254, 257, 259, 260, 261, 263, 266, 268, 269, 270, 271, 273, 279, 284, 285, 290, 295, 312

C

Cellulose 4, 6, 7, 9, 10, 13, 16, 17, 18, 21, 22, 27, 28, 32, 33, 34, 35, 36, 38, 39, 47, 49, 52, 53, 54, 55, 56, 57, 58, 59, 61, 66, 68, 69, 70, 71, 72, 73, 74, 75, 76, 77, 78, 79, 117, 118, 128, 136, 139, 140, 141, 146, 149, 154, 158, 159, 172, 173, 176, 180, 181, 182, 185, 189, 191, 194, 199, 203, 204, 205, 208, 209, 210, 211, 212, 213, 216, 221, 222, 223, 225, 226, 227, 229, 235, 240, 242, 244, 245, 248, 252, 254, 255, 257, 259, 261, 263, 266, 269, 272, 273, 279, 284, 290, 323, 331

Cellulose Nanocrystals 32, 33, 36, 38, 39, 47, 55, 57, 136, 141, 154, 176, 211, 213, 221, 225, 229, 240, 244, 245, 248, 252, 254, 255, 257, 259, 261, 272

Characterization Techniques 87, 99, 136, 183, 204, 216, 233, 264, 272, 333

Chemical Recycling 151, 152, 174, 187, 229

Circular Economy 23, 47, 136, 139, 145, 157, 167, 176, 201, 203, 208, 212, 216, 229, 230, 231, 241, 255, 259, 261, 316

Controlled Release Fertilizers 179, 180, 181, 182, 185, 188, 189, 196, 201

D

Degradation Mechanisms 135, 137, 139, 148, 151, 168, 169, 176, 324

Doped Nanoparticles 119, 123

E

Environment 2, 3, 14, 16, 52, 64, 82, 83, 84, 85, 103, 104, 105, 108, 118, 130, 180, 183, 184, 194, 195, 196, 197, 198, 199, 206, 207, 213, 215, 216, 224, 225, 239, 255, 264, 269, 270, 275, 278, 279, 280, 290, 310, 311, 317, 318, 319, 320, 322, 323, 324, 327, 328, 331, 339, 340, 341, 343, 345, 350

Environmental Applications 90, 110, 272, 273, 281, 305, 323, 329, 336, 339, 351

Environmental Sustainability 206, 233, 234, 235, 245, 257, 258, 261, 275, 277, 278, 289, 292, 314, 318, 328

F

Food Packaging 19, 23, 39, 51, 52, 56, 59, 60, 65, 71, 72, 74, 75, 76, 77, 78, 128, 129, 131, 132, 153, 158, 163, 169, 173, 174, 175, 199, 200, 206, 210, 211, 226, 233, 234, 239, 240, 241, 242, 243, 244, 245, 246, 247, 249, 250, 253, 254, 255, 258, 259, 260, 261, 263, 264, 265, 266, 267, 268, 269, 270, 271, 272, 273

Food Security 179, 192, 196, 197, 198, 264, 276, 311

Functional Materials 19, 22, 203, 235, 279

G

Graphene 5, 36, 38, 43, 44, 45, 46, 47, 49, 50, 51, 52, 53, 54, 55, 56, 57, 131, 141, 142, 181, 182, 185, 189, 199, 204, 205, 211, 214, 215, 220, 221, 222, 223, 225, 226, 227, 228, 246, 248, 249, 250, 253, 254, 266, 267, 270, 279, 282, 295, 298, 301, 309, 311, 313, 330, 331, 335, 338, 351

Green Chemistry 47, 51, 82, 91, 145, 146, 177, 203, 204, 205, 207, 208, 211, 212, 213, 223, 231, 233, 234, 235, 247, 250, 261, 268, 273, 279

Green Synthesis 1, 81, 82, 83, 84, 85, 87, 88, 90, 92, 94, 97, 100, 102, 107, 108, 109, 110, 111, 182, 196, 201, 228, 279, 284, 285, 311

I

Intelligent packaging 240, 244, 246, 250, 255, 256, 261, 262, 265

Interface Phenomena 142, 176, 177

Internet of things 164, 277, 290, 313, 317, 320

Iron Nanoparticles 81, 82, 85, 87, 89, 90, 91, 92, 93, 94, 95, 98, 100, 103, 104,

105, 107, 108, 110, 111

L

Lifecycle Assessment 136, 154, 155, 168, 176, 177

M

Mechanical Properties 2, 5, 14, 22, 24, 31, 35, 38, 39, 40, 42, 43, 44, 45, 50, 51, 52, 53, 55, 56, 57, 61, 64, 65, 70, 71, 77, 119, 130, 133, 138, 139, 140, 141, 148, 153, 158, 160, 176, 183, 191, 206, 211, 213, 214, 215, 216, 218, 219, 221, 222, 223, 225, 226, 227, 229, 235, 237, 243, 245, 246, 248, 313

N

Nanobiochar 309, 327, 328, 329, 330, 331, 332, 333, 334, 335, 336, 337, 338, 339, 340, 341, 342, 343, 344, 345, 346, 348, 349, 350, 351, 352
Nanocomposites 2, 3, 4, 5, 6, 10, 11, 13, 14, 15, 16, 17, 18, 19, 20, 22, 23, 24, 25, 26, 27, 28, 29, 30, 31, 32, 33, 35, 36, 37, 38, 39, 40, 42, 43, 44, 45, 46, 47, 48, 49, 50, 51, 52, 53, 54, 55, 56, 57, 58, 59, 60, 61, 63, 65, 66, 70, 71, 72, 73, 74, 75, 76, 77, 78, 119, 121, 124, 125, 128, 129, 131, 132, 135, 136, 137, 138, 139, 142, 143, 144, 145, 146, 147, 148, 149, 150, 151, 152, 153, 154, 155, 156, 158, 159, 160, 161, 163, 165, 166, 167, 168, 169, 170, 171, 172, 173, 174, 175, 176, 179, 180, 181, 182, 183, 184, 185, 186, 187, 188, 189, 190, 191, 192, 193, 194, 195, 196, 197, 198, 199, 200, 201, 203, 204, 205, 206, 207, 208, 209, 210, 211, 212, 213, 214, 215, 216, 217, 218, 219, 220, 221, 222, 223, 224, 225, 226, 227, 228, 229, 230, 231, 232, 233, 234, 235, 236, 239, 240, 241, 242, 244, 245, 246, 247, 248, 249, 250, 251, 252, 253, 254, 255, 256, 257, 258, 259, 260, 261, 262, 264, 265, 266, 267, 268, 269, 270, 271, 273, 279, 280, 281, 285, 291, 295, 299, 303, 307, 309, 314, 315, 316, 319, 327, 328, 329, 330, 331, 339, 340, 341, 342, 343, 344, 345, 346, 347, 348, 349, 350, 351, 352
Nanofillers 14, 22, 26, 28, 30, 31, 32, 33, 36, 37, 38, 39, 40, 42, 43, 46, 47, 48, 53, 181, 183, 184, 186, 187, 189, 191, 192, 193, 194, 197, 204, 205, 209, 210, 211, 213, 214, 215, 216, 217, 218, 219, 220, 221, 222, 223, 224, 225, 226, 227, 228, 229, 230, 231, 234, 263, 265
Nanomaterials 1, 4, 15, 16, 19, 22, 44, 49, 54, 108, 110, 111, 113, 114, 115, 119, 120, 121, 123, 124, 129, 136, 141, 142, 143, 161, 167, 173, 175, 176, 177, 181, 184, 185, 186, 191, 194, 195, 196, 197, 201, 204, 205, 207, 208, 209, 211, 212, 215, 217, 218, 220, 223, 226, 228, 230, 234, 240, 241, 244, 245, 246, 247, 248, 249, 250, 251, 252, 253, 254, 255, 256, 257, 258, 259, 260, 261, 262, 263, 266, 268, 269, 270, 273, 278, 279, 281, 282, 284, 292, 295, 303, 305, 309, 311, 316, 317, 318, 319, 320, 322, 323, 329, 330, 331, 332, 335, 336, 337, 338, 339, 342, 347, 349, 350, 351
Nano Reinforcements 141, 142, 143, 151, 152, 154, 177
Nanotechnology 16, 52, 65, 78, 81, 82, 105, 106, 108, 111, 114, 125, 127, 128, 129, 131, 133, 138, 160, 174, 188, 197, 203, 204, 205, 206, 207, 208, 231, 232, 233, 235, 239, 241, 265, 277, 278, 296, 304, 305, 308, 309, 311, 314, 315, 317, 320, 321, 323, 324, 329, 336, 341, 342, 346, 347, 349

P

Phytochemicals 81, 83, 87, 90, 95, 99, 109
Practical Utility 59

Predictive modeling 165, 284, 285, 287, 299, 301, 315
Properties 1, 2, 3, 4, 5, 6, 9, 11, 13, 14, 15, 16, 17, 18, 19, 21, 22, 23, 24, 26, 31, 32, 34, 35, 36, 37, 38, 39, 40, 41, 42, 43, 44, 45, 46, 49, 50, 51, 52, 53, 54, 55, 56, 57, 58, 59, 60, 61, 63, 64, 65, 66, 67, 69, 70, 71, 72, 73, 74, 75, 77, 78, 81, 82, 83, 84, 87, 90, 93, 95, 97, 98, 101, 102, 103, 104, 105, 111, 113, 114, 115, 116, 119, 120, 121, 124, 125, 127, 128, 129, 130, 132, 133, 138, 139, 140, 141, 142, 143, 144, 145, 147, 148, 149, 151, 152, 153, 154, 156, 158, 160, 161, 162, 164, 165, 166, 168, 172, 173, 174, 175, 176, 177, 180, 181, 182, 183, 184, 186, 188, 189, 190, 191, 194, 195, 196, 197, 201, 203, 204, 205, 206, 207, 208, 209, 210, 211, 213, 214, 215, 216, 217, 218, 219, 220, 221, 222, 223, 225, 226, 227, 228, 229, 230, 231, 232, 233, 234, 235, 236, 237, 240, 241, 242, 243, 244, 245, 246, 247, 248, 249, 250, 251, 253, 254, 255, 257, 259, 260, 261, 264, 265, 266, 267, 268, 269, 272, 273, 278, 282, 285, 286, 291, 295, 304, 313, 319, 327, 328, 329, 330, 331, 332, 333, 334, 335, 336, 337, 338, 339, 341, 343, 344, 345, 347, 351

R

Regenerable Materials 135, 142, 143, 144, 145, 146, 147, 149, 155, 157, 159, 162, 163, 164, 166, 167, 168, 169, 170, 176

S

Soil Conditioning 179, 180, 181, 182, 188, 189, 191, 196, 282
Sustainable agriculture 179, 181, 188, 193, 196, 197, 201, 291, 299, 309, 310, 312, 316, 318, 345, 348
Sustainable food packaging 234, 239, 240, 241, 260, 261, 266, 269, 272, 273
Sustainable Materials 18, 48, 60, 77, 78, 136, 138, 154, 166, 167, 205, 206, 207, 224, 234, 235, 243, 264, 267, 270, 320
Synthesis 1, 9, 11, 15, 16, 17, 18, 19, 21, 22, 23, 24, 32, 33, 35, 37, 38, 39, 41, 42, 44, 45, 51, 54, 55, 56, 57, 73, 75, 78, 81, 82, 83, 84, 85, 87, 88, 89, 90, 91, 92, 93, 94, 96, 97, 98, 99, 100, 101, 102, 105, 106, 107, 108, 109, 110, 111, 113, 115, 116, 117, 118, 119, 126, 129, 130, 131, 135, 137, 145, 146, 166, 172, 179, 181, 182, 183, 185, 194, 196, 197, 201, 204, 205, 206, 207, 208, 209, 210, 211, 212, 213, 220, 223, 228, 230, 233, 236, 240, 241, 250, 251, 253, 254, 259, 260, 263, 265, 266, 267, 268, 275, 278, 279, 280, 284, 285, 286, 288, 295, 307, 311, 316, 321, 323, 327, 331, 332, 333, 335, 347, 351, 352
Synthesis Methods 24, 81, 100, 106, 119, 137, 145, 196, 208, 228, 240, 241, 251, 266, 279, 284, 352

T

Thermoplastic-Based bio Nanocomposites 59

www.ingramcontent.com/pod-product-compliance
Lightning Source LLC
LaVergne TN
LVHW062328300926
849161LV00007B/1611